Algebras, Rings and Modules

Non-commutative Algebras and Rings

Algebras, Rings and Modules

Non-commutative Algebras and Rings

Michiel Hazewinkel
Dept. of Pure and Applied Mathematics
Centrum Wiskunde & Informatica
Amsterdam, Netherlands

and

Nadiya Gubareni
Institute of Mathematics
Częstochowa University of Technology
Częstochowa, Poland

CRC Press
Taylor & Francis Group
Boca Raton London New York

CRC Press is an imprint of the
Taylor & Francis Group, an **informa** business
A SCIENCE PUBLISHERS BOOK

Reprint 2025

CRC Press
Taylor & Francis Group
6000 Broken Sound Parkway NW, Suite 300
Boca Raton, FL 33487-2742

ISBN: 9780367240066 (hbk)

Library of Congress Cataloging-in-Publication Data

Hazewinkel, Michiel.
Algebras, rings, and modules : non-commutative algebras and rings / Michiel Hazewinkel, Nadiya Gubareni.
pages cm
"A CRC title."
Includes bibliographical references and index.
ISBN 978-1-4822-4503-5 (hardcover : alk. paper) 1. Noncommutative algebras. 2. Rings (Algebra) I. Gubareni, Nadezhda Mikhailovna. II. Title.

QA251.4.H39 2016
512'.46--dc23 2015028914

Visit the Taylor & Francis Web site at
http://www.taylorandfrancis.com

and the CRC Press Web site at
http://www.crcpress.com

Printed and bound in India

PREFACE

The theory of rings and modules is one of the most fundamental domains of modern algebra.

This volume is a continuation of the first two volumes of *Algebra, Rings and Modules* by M. Hazewinkel, N. Gubareni and V.V. Kirichenko. Volume 3, published by the American Mathematical Society, is about Lie algebras and Hopf algebras and largely independent of the other volumes.

It systematizes and presents the main results of the structure theory of some special classes of non-commutative rings. The book presents both the basic classical theory and more recent results related to current research such as the structure theory of some special classes of rings, which arise in many applications. Some of the topics covered include quivers, partially ordered sets and their representations, as well as such special rings as hereditary and semihereditary rings, serial rings, semidistributive rings and modules over them.

Some results of this book are new and have, so far, been published in journals only. All results are given with complete proofs which are based on the material contained in the book.

We assume that the reader is familiar with the basic concepts of theory of rings and modules. Nevertheless for the reader's convenience Chapter 1 summarizes the basic ring-theoretic notions and results considered in our previous books. Proofs of all results presented can be found in these books via corresponding citations which have been inserted in parentheses.

In mathematics, specifically in the area of abstract algebra, it is often interesting to construct new objects using objects already known. In group theory, ring theory, Lie algebra theory there are a variety of different constructions such as crossed, skew, smash products which are very important as sources of various counter examples.

Some of the main ring constructions, such as a finite direct product of rings, group rings, matrix rings, path algebras and graded algebras were considered in our previous books. Chapter 2 represents the definitions and some properties of some of these basic general constructions of rings and modules, such as direct product, semidirect product, direct sum, crossed product of rings, polynomial and skew polynomial rings, power and skew power series rings, Laurent polynomial rings and Laurent power series rings, generalized matrix rings, formal triangular matrix rings, and G-graded rings.

The theory of valuation rings was first related only with commutative fields. Discrete valuation domains are, excepting only fields, the simplest class of rings. Nevertheless they play an important role in algebra and algebraic geometry.

But there is also a noncommutative side of this theory. In the noncommutative case there are different generalizations of valuation rings. The first generalization, valuation rings of division rings, was obtained by Schilling in [283], who introduced the class of invariant valuation rings and systematically studied them in [284].

Another significant contribution in non-commutative valuation rings was made by N.I. Dubrovin who introduced a more general concept of a valuation ring for simple Artinian rings and proved a number of significant nontrivial properties about them [73], [71], [72]. These rings are named Dubrovin valuation rings after him. Dubrovin valuation rings found a large number of applications.

In Chapter 3 most of the basic results for valuation rings and discrete valuation rings are described and discussed briefly.

Section 3.1 is devoted to valuation rings of fields. In this section we give the main properties of these rings and describe different definitions of them. Section 3.2 presents the basic results about discrete valuation domains. We give the structure of these rings and a number of equivalent definitions. In Section 3.3 we describe noncommutative invariant valuation rings of division rings, and the main properties of these rings and their equivalent definitions are described. Some examples of noncommutative non-discretely-valued valuation rings are given in Section 3.4. The main properties and structure of noncommutative discrete valuation rings are presented in Section 3.5. In Section 3.6 we briefly discuss total valuation rings which are more general than invariant valuation rings. Section 3.7 is devoted to other types of valuation rings with zero divisors. We consider some valuation rings of commutative rings with zero divisors and the Dubrovin valuation rings. Finally, in Section 3.8 we consider the approximation theorems for special kinds of noncommutative valuation rings, in particular for locally invariant rings and Dubrovin valuation rings. We also give some corollaries from these theorems for noncommutative discrete valuation rings.

Chapter 3 may be considered as a short introduction to the theory of valuation rings. More information about valuation rings of division rings can be found in [284], and more about Dubrovin valuation rings, semihereditary and Prüfer orders in simple Artinian rings can be found in the book [233].

The concepts of homological dimensions of rings and modules were discussed in [146, Chapter 6] and [147, Chapter 4]. Chapter 4 considers some other questions connected with these notions.

For the reader's convenience Section 4.1 summarizes the basic concepts and results on projective and injective dimensions of modules, and global dimensions of rings. The concepts and results on flat dimensions of modules and weak dimensions of rings are presented in Section 4.2. In Section 4.3 we present various examples of rings with different global dimensions.

The homological characterization of some classes of rings, such as semisimple, right Noetherian, right hereditary, right semihereditary, semiperfect, right perfect and quasi-Frobenius rings are considered in Section 4.4.

Duality over Noetherian rings, which is given by the covariant functor $* = \mathrm{Hom}_A(-, A)$ was considered in [147, Section 4.10]. For an arbitrary ring A this functor induces a duality between the full subcategories of finitely generated projective right A-modules and left A-modules. The main properties of this functor and torsionless modules, and the relationship between them are studied in Section 4.5.

The basic properties of flat modules were considered in [146, Section 5.4], and in [147, Section 5.1]. Further properties of flat modules are studied in Section 4.6. In particular, the main theorem of this section, which was proved by S. Chase [41], gives equivalent conditions for a direct product of any family of flat modules to be flat. As the corollary of this theorem we obtain homological characterization of semihereditary rings, proved by S. Chase in [41]. Section 4.7 gives necessary and sufficient conditions under which a formal triangular matrix ring is right (left) hereditary.

Chapter 5 contains a short introduction to the theory of uniform, Goldie and Krull dimensions of rings and modules. Uniform modules and their main properties are considered in Section 5.1, where the uniform dimension of modules is also introduced and studied. Modules of finite Goldie dimension, a notion due to A. Goldie, are considered in Section 5.2.

The notions of singular and nonsingular modules are introduced in Section 5.3, where there are also studied the main properties of such modules and some properties of nonsingular rings. In Section 5.4 the results of this theory are applied to prove a theorem which gives equivalent conditions for a ring being a Goldie ring. This theorem includes the famous Goldie theorem which was proved in [100], [101], [102] (see also [146, Section 9.3]). This well-known theorem gives necessary and sufficient conditions for a ring to have a semisimple classical quotient ring. In 1966 L. Small generalized this theorem and described Noetherian rings which have Artinian classical rings of quotients [292]. A variant of Small's theorem without the Noetherian hypothesis was obtained by R.B. Warfield, Jr. [321]. In Section 5.5 the proofs of these results are given.

The notion of the classical Krull dimension as a powerful tool for arbitrary commutative Noetherian rings was considered by W. Krull. There are a few different generalizations of this concept for the case of noncommutative rings. One of them, the classical Krull dimension introduced by G. Krause in [197], is considered in Section 5.6. The more important generalization, the concept of Krull dimension is considered in Section 5.7. The module-theoretic form of this notion in the general case for noncommutative rings was introduced by R. Rentschler and P. Gabriel in 1967, [270]. Note that not all modules have Krull dimension, but each Noetherian module has Krull dimension. The Krull dimension of any Artinian module is equal to 0. So in some sense the Krull dimension of a module can be considered as a measure which shows of how far the module is from being Artinian. The basic properties of Krull dimension are studied in this section. In Section 5.8 we consider some relationships between the concepts of classical Krull dimension and Krull dimension.

An important role in the theory of rings and modules is played by various finiteness conditions. Many types of finiteness conditions on rings can be formulated in terms of d.c.c. (descending chain condition) or a.c.c. (ascending chain condition) on suitable classes of one-sided ideals. The d.c.c. (minimal condition) on right (resp., left) ideals defines right Artinian (resp., left Artinian) rings. Analogously, right (resp., left) Noetherian rings are defined as rings which satisfy the maximal condition, or the a.c.c. on right (resp. left) ideals. These rings were considered in [146, Chapter 3]. Section 6.1 gives some examples of Noetherian rings connected with the basic constructions of rings considered in Chapter 2.

Section 6.2 considers various finiteness conditions for rings and modules, and relations between them. Dedekind-finite rings, orthogonally finite rings, stably finite rings, unit-regular rings and IBN rings are examples of rings which are considered in this section.

FDI-rings, i.e., rings with a finite decomposition of the identity into a sum of pairwise orthogonal primitive idempotents, form the next class of rings with finiteness conditions. These rings are considered in Section 6.3. In Section 6.4 we prove the main theorem which gives a criterium for a semiprime FDI-ring to be decomposable into a direct product of prime rings.

Chapter 7 is devoted to the important problems connected with uniqueness of decompositions of modules into direct sums of indecomposable modules. The famous Krull-Remak-Schmidt theorem was already considered in [146], where in Section 10.4 this theorem was proved for the case of finite direct sums of modules with local endomorphism rings. Actually, G. Azumaya proved this theorem in [12] for infinite direct summands in the general case for Abelian categories with some additional condition. In this chapter the proof of this theorem is given for the case of infinite direct sums of modules with local endomorphism rings following to Peter Crawley and Bjarni Jónsson. They proved this theorem using the exchange property, which was introduced in 1964 for general algebras, [61]. From that time this notion has become an important theoretical tool for studying rings and modules.

Some properties of modules having the exchange property are studied in Section 7.1. It is proved that the 2-exchange property is equivalent to the finite exchange property for arbitrary modules, and that the 2-exchange property is equivalent to the exchange property for indecomposable modules. These results were obtained by P. Crawley and B. Jónsson for general algebras in [61] and R.B. Warfied, Jr. for Abelian categories in [321]. In this section we also prove the important result obtained by R.B. Warfied, Jr. in [317], which states that an indecomposable module has the exchange property if and only if its endomorphism ring is local.

The proof of the Azumaya theorem for infinite direct sums of modules is given in Section 7.2. The cancellation property notion and some properties of modules having the cancellation property are considered in Section 7.3.

At the end of this chapter we consider different classes of rings connected with exchange rings. Section 7.4 is devoted to the study of properties and some structural theorems for exchange rings.

Generally speaking, the class of semisimple rings has been studied most extensively. Their structure is completely described by the Wedderburn-Artin theorem. Semisimple rings are also very simple from the point of view of homological properties of modules over them. These are rings whose global homological dimension is equal to zero. Hereditary rings immediately follow semisimple ones in terms of a homological classification. According to theorem 4.3.8, r.gl.dim $A \leqslant 1$ if and only if A is a right hereditary ring. The structure of hereditary rings is not so well studied as in the case of semisimple rings. Chapter 8 is devoted to the study of the structure and main properties of hereditary rings. In addition, semihereditary rings, which are close to hereditary rings, are considered in this chapter.

In Section 4.5 it was shown that for a large class of rings (coherent, semiperfect, right serial, right Noetherian) being right semihereditary implies being left semihereditary. In Section 8.1 this result is proved for orthogonally finite rings.

The main results of Chapters 4 and 5 are applied to study the properties of right hereditary and right semihereditary rings in Section 8.2. In particular the Goldie theorem it is proved there. This theorem gives equivalent conditions for a domain A

to be right Ore. There is also the important Small theorem which states that a right Noetherian right hereditary ring is a right order in a right Artinian ring. The structure and properties of some classes of right hereditary (semihereditary) prime rings are considered in Section 8.3. In particular we prove an important theorem which states the relationship between Dubrovin valuation rings, semihereditary orders and Bézout orders in simple Artinian rings. We also consider right Noetherian hereditary (semihereditary) semiperfect prime rings.

The next generalization, following hereditary and semihereditary rings, are piecewise domains. These rings were first introduced and studied by R. Gordon and L.W. Small in 1972. Section 8.4 considers properties of piecewise domains and their relationships with hereditary and semihereditary rings. It is proved that a piecewise domain is a nonsingular ring. This section also gives a proof of the theorem which states that a right perfect piecewise domain is semiprimary. This theorem first was proved by M. Teply in 1991.

The notion of a triangular ring was first introduced by S.U. Chase in 1961 for semiprimary rings. In 1966 L. Small extended this notion to Noetherian rings and proved that a right Noetherian right hereditary ring is triangular. M. Harada in 1964 introduced the notion of generalized triangular rings and proved that any hereditary semiprimary ring is isomorphic to a generalized triangular ring with simple Artinian blocks along the main diagonal. In 1980 Yu.A. Drozd extended the notion of a triangular ring to FDI-rings and described the structure of right hereditary (semihereditary) FDI-rings.

Section 8.5 introduces the notions of triangular and primely triangular rings which includes all notions of a triangular ring mentioned above. The main result of this section gives the structure of piecewise domains in terms of primely triangular rings. This theorem was proved by R. Gordon and L.W. Small in 1972 and it states that any piecewise domain is a primely triangular ring. From this statement there easily follows the theorem obtained by L.W. Small in 1966 about the structure of right Noetherian right hereditary rings.

Section 8.6 gives the criterion for a triangular FDI-ring to be right hereditary or right semihereditary, which was obtained by Yu.A. Drozd in 1980.

In Section 8.7 the results of Section 8.6 are applied to different concrete classes of rings. In particular, we give the criterion for a right Noetherian primely triangular ring to be right hereditary. From this result there follows the famous decomposition theorem of Chatters which states that a Noetherian hereditary ring is a direct sum of rings each of which is either an Artinian hereditary ring or a prime Noetherian hereditary ring. Section 8.8 is devoted to the study of hereditary species and tensor algebras, which were introduced by Yu.A. Drozd.

Chapter 9 is devoted to the further study of serial rings which were considered in [146, Sections 12, 13]. In this chapter we present the structure theorems for various different classes of serial nonsingular rings.

In Section 9.1 we consider serial right Noetherian piecewise domains. We prove that for a serial right Noetherian ring being a piecewise domain is equivalent to being a right hereditary ring. Section 9.2 is devoted to the study of the structure of serial nonsingular rings. In particular, it is given the main result of R.B. Warfield, Jr. who proved that for a right serial ring being right semihereditary is equivalent to be right nonsingular. We also give another equivalent conditions for right serial right semihereditary rings. In this section we study the structure of serial nonsingular rings and show that any serial nonsingular ring has a classical ring of quotients which is an Artinian ring. Section 9.3 is devoted to serial rings with Noetherian diagonal. We consider the prime quiver of such rings and describe the structure of serial nonsingular rings with Noetherian diagonal.

The Krull intersection theorem is very important and well known for Noetherian commutative rings. In Section 9.4 we consider some versions of this theorem for noncommutative rings.

Section 9.5 is devoted to the problems connected with the Jacobson conjecture which states that for any Noetherian ring with Jacobson radical R the intersection $\bigcap_{n\geq 0} R^n = 0$. This conjecture is true for any commutative Noetherian ring, but it is still open for noncommutative Noetherian rings in general. There are various classes of noncommutative Noetherian rings for which the Jacobson conjecture holds. In particular, the Jacobson conjecture holds for discrete valuation rings which is shown in Section 3.5. A.V. Jategaonkar [167] and G. Cauchon [39] have shown that Jacobson's conjecture hold for fully bounded Noetherian rings. T.H. Lenagan [216] have proved that this conjecture also holds for Noetherian rings with right Krull dimension one. In this section we prove that the Jacobson conjecture holds for Noetherian SPSD-rings and Noetherian serial rings.

The book is written on a level accessible to advanced students who have some experience with modern algebra. It will be useful for those new to the subject as well for researchers and serves as a reference volume.

CONTENTS

CHAPTER 1

Preliminaries

We assume that the readers are familiar with the basic concepts of the theory of rings and modules. For the readers' convenience this chapter summarizes the basic ring-theoretic notions and results considered in our previous books [146], [147]. The proofs of all results presented can be found in these books by the corresponding citations which are inserted in square brackets.

All rings in this book are assumed to be associative (but not necessarily commutative) with $1 \neq 0$. A subring of a ring A with 1 is required to have the same identity 1. All modules are assumed to be unitary.

1.1 Basic Concepts Concerning Rings and Modules

Recall that a **ring** is a nonempty set A together with two binary algebraic operations, that are denoted by $+$ and $\cdot$ and called addition and multiplication, respectively, such that, for all $a, b, c \in A$ the following axioms are satisfied:

1. $a + (b + c) = (a + b) + c$
2. $a + b = b + a$
3. There exists an element $0 \in A$, such that $a + 0 = 0 + a = a$
4. There exists an element $x \in A$, such that $a + x = 0$
5. $(a + b) \cdot c = a \cdot c + b \cdot c$
6. $a \cdot (b + c) = a \cdot b + a \cdot c$

We will usually write simply ab rather than $a \cdot b$ for $a, b \in A$.

A ring A is **associative** if the multiplication satisfies the associative law, that is, $(a_1a_2)a_3 = a_1(a_2a_3)$ for all $a_1, a_2, a_3 \in A$. A ring A is **commutative** if the multiplication satisfies the commutative law, that is, $a_1a_2 = a_2a_1$ for all $a_1, a_2 \in A$.

An element e of a ring A is called a **left identity** if $ea = a$ for every $a \in A$. Similarly $f \in A$ is a **right identity** if $af = f$ for every $a \in A$. If an associative ring possesses a left identity e and a right identity f, then necessarily $e = f$. If e is both a right and left identity of a ring A, we will say that e is a **two-sided identity**. If a two-sided identity exists in an associative ring A then it is unique. In this case a two-sided identity is called an **identity** which we will denote as 1, and a ring A is usually called a **ring with identity** or, shortly, a **ring with** 1. A ring having only one element 1=0 is called **trivial**, otherwise it is called non-trivial (or **non-zero**).

In this book all rings are always assumed to be associative with $1 \neq 0$, but we will not generally make the assumption about commutativity of rings.

An element u of a ring A with 1 is called **right invertible** if there exists an element $v \in A$ such that $uv = 1$. In this case the element v is called **right inverse** for u. Similarly, $u \in A$ is **left invertible** if $wu = 1$ for some $w \in A$, and the element w is **left inverse** for u. If an element u of a ring A is both right invertible and left invertible we will say that it is **two-sided invertible** (or it is a **unit**). If an element u of an associative ring is invertible with a right inverse v and a left inverse w, then necessarily $v = w$.

The following example shows that there exist rings with elements which are right invertible but not left invertible.

Example 1.1.1. (See [14, Example 1].)

Let V be an infinite dimension vector space over a field K which elements are countably infinite sequences with elements from a field K of the form: $(a_1, a_2, \dots, a_n, \dots)$ where $a_i \in K$. Assume that $A = \mathrm{End}_K(V)$ is a set of all linear transformations of V over K. Then A is an associative ring with $1 \neq 0$. Consider two elements $\varphi, \psi \in A$ of the form:

$$\varphi(a) = (0, a_1, a_2, \dots, a_n, \dots), \quad \psi(a) = (a_2, \dots, a_n, \dots)$$

for each element $a = (a_1, a_2, \dots, a_n, \dots) \in V$. Then $\psi\varphi(a) = a$, i.e., $\psi\varphi = 1$, but $\varphi\psi(a) = (0, a_2, \dots, a_n, \dots) \neq a$, i.e., $\varphi\psi \neq 1$.

The set of all units in a ring A will be denoted by $U(A)$, or A^*. It is easy to show that if A is an associative ring with $1 \neq 0$ then $U(A)$ is a group under the multiplication of A.

A non-zero ring A is said to be a **domain**[1] if it has neither right nor left zero divisors, i.e., $ab \neq 0$ for any non-zero elements $a, b \in A$. A commutative domain A is called an **integral domain.**[2] A non-zero ring A with identity is called a **division ring** (or a **skew field**) if every non-zero element of A is invertible, i.e., $U(A) = A \backslash \{0\}$. A commutative division ring is called a **field**.

---

[1]Note that a domain may be a non-commutative ring. J.H.M. Wedderburn in [324] considered such rings and he called them **domains of integrity**.

[2]Some authors, e.g. [58], use this term also for non-commutative rings, but we will use this term only for the commutative case.

A right ideal I is called **right principal** if it is of the form xA for some $x \in A$. Analogously an ideal $I = Ax$ is called **left principal**. A ring A whose all right and left ideals are principal is called a **principal ideal ring**. If A is a domain which is a principal ideal ring then it is called a **principal ideal domain** (abbreviated PID).

An element e of a ring A is said to be an **idempotent** if $e^2 = e$. Two idempotents e and f are called **orthogonal** if $ef = fe = 0$. An idempotent $e \in A$ is said to be **primitive** if e has no decomposition $e = e_1 + e_2$ into a sum of non-zero orthogonal idempotents $e_1, e_2 \in A$. An idempotent e of a ring A is called **central** if $ea = ae$ for any element $a \in A$. A central idempotent $e \in A$ is called **centrally primitive** if it cannot be written as a sum of two non-zero orthogonal central idempotents.

An element x of a ring A is said to be **nilpotent** if $x^n = 0$ for some natural number $n \in \mathbf{N}$. A ring A is called **reduced** if it has no non-zero nilpotent elements. A one-sided (or two-sided) ideal of a ring A for which all elements are nilpotent is called a **nil-ideal.** A one-sided (or two-sided) ideal I of A is called **nilpotent** if $I^n = 0$ for some natural number $n \in \mathbf{N}$. Note that $I^n = 0$ means that $a_1 a_2 \cdots a_n = 0$ for any set of elements $a_1, a_2, \ldots, a_n \in I$.

An equality $1 = e_1 + e_2 + ... + e_n$, where e_1, e_2,...,e_n are pairwise orthogonal idempotents of a ring A, will be called a **decomposition of the identity** of A. Such a decomposition defines a decomposition of the ring A into a direct sum of Abelian groups $e_i A e_j = A_{ij}$ $(i, j = 1, 2, ..., n)$:

$$A = \bigoplus_{i,j=1}^{n} e_i A e_j,$$

which can be conveniently represented as a matrix ring

$$A = \begin{pmatrix} A_{11} & A_{12} & \ldots & A_{1n} \\ A_{21} & A_{22} & \ldots & A_{2n} \\ \vdots & \vdots & \ddots & \vdots \\ A_{n1} & A_{n2} & \ldots & A_{nn} \end{pmatrix}$$

with the usual operations of addition and multiplication. Note that it is idempotency and orthogonality that make this work. Such a decomposition is called a **two-sided Peirce decomposition**, or simply a **Peirce decomposition** of the ring A. Since there is an isomorphism θ between the additive groups $\mathrm{Hom}_A(e_j A, e_i A)$ and $e_i A e_j$, the elements of $e_i A e_j$ are naturally identified with homomorphisms from $e_j A$ to $e_i A$ by the following way. If $\psi \in \mathrm{Hom}(e_j A, e_i A)$ then $\theta(\psi) = e_i a e_j \in e_i A e_j$, where a is an element in A such that $\psi(e_j) = e_i a$.

A **right module** over a ring A (or **right A-module**) is an additive Abelian group M together with a map $M \times A \to M$ so that to every pair (m, a), where $m \in M$, $a \in A$, there corresponds a uniquely determined element $ma \in M$ and such that the following conditions are satisfied:

1. $m(a_1 + a_2) = ma_1 + ma_2$
2. $(m_1 + m_2)a = m_1a + m_2a$
3. $m(a_1a_2) = (ma_1)a_2$
4. $m \cdot 1 = m$

for any $m, m_1, m_2 \in M$ and any $a, a_1, a_2 \in A$.

In a similar way one can define the notion of a **left A-module**.

In what follows, an A-module will always mean a right A-module.

Let A and B be two rings. An Abelian group X is called an **(A,B)-bimodule**, which is denoted by ${}_AX_B$, if X is both a left A-module and a right B-module such that $(ax)b = a(xb)$ for all $a \in A$, $x \in X$, and $b \in B$.

Note also the following important fact about submodules.

Theorem 1.1.2 (Modular Law). (See [146, theorem 1.3.6].) *Let U, V and W be submodules of X with $V \subseteq U$. Then:*

$$U \cap (V + W) = (U \cap V) + (U \cap W) = V + (U \cap W). \tag{1.1.3}$$

In particular, if $V \cap W = 0$, then this law has the following form:

$$U \cap (V \oplus W) = (U \cap V) \oplus (U \cap W) = V \oplus (U \cap W). \tag{1.1.4}$$

Remark 1.1.5. Note that without the condition $V \subseteq U$ the modular law does not necessarily hold. Take for example $X = K^3$, where K is a field of characteristic $\neq 2$, and submodules U, V, W of X of the form: $U = (1,0,0)K + (0,1,0)K$, $V = (1,0,1)K$, $W = (1,0,-1)K$. Then $U \cap V = 0 = U \cap W$, but $U \cap (V + W) = (1,0,0)K$.

Let $M = A$ and the map $\varphi : M \times A \to M$ the usual multiplication, i.e., $\varphi(m,a) = ma \in M$. Then one obtains a right module A_A which is called the **right regular module**. Analogously, one can construct the left regular module ${}_AA$.

A (right) A-module M is **free** if it is isomorphic to a direct sum of (right) regular modules, i.e., $M \simeq \bigoplus_{i \in I} M_i$. where $M_i \simeq A_A$ for all $i \in I$.

A non-zero module M is **simple** (or **irreducible**) if it has no non-zero proper submodules, i.e., its only submodules are the module M and the zero module.

A module M is **semisimple** (or **completely reducible**) if it can be decomposed into a direct sum of simple modules. A ring A is a **right (left) semisimple** if it is semisimple as a right (left) module over itself. Since A has an identity and any right submodule of A is just a right ideal, A is right semisimple if A is a direct sum of a finite number of simple right ideals.

A ring A is said to be **indecomposable** if $A \neq 0$ and A cannot be decomposed into a direct product of two non-zero rings.

Theorem 1.1.6 (Wedderburn-Artin's Theorem). (See [146, theorem 2.2.2].) *A ring A is right (left) semisimple if and only if A is isomorphic to a direct sum of a* *finite number of full matrix rings over division rings:*

$$A \cong M_{n_1}(D_1) \times M_{n_2}(D_2) \times \cdots \times M_{n_k}(D_k),$$

where the D_i *are division rings for* $i = 1, 2, \ldots, k$.

In view of this theorem the concepts of right and left semisimple rings are the same. Therefore one can simply talk about **semisimple rings**.

Theorem 1.1.7. (See [146, theorem 2.2.5].) *The following conditions are equivalent for a ring A:*

a. *A is right semisimple*
b. *A is left semisimple*
c. *Any right A-module M is semisimple*
d. *Any left A-module M is semisimple.*

A module M is **Artinian** if it satisfies the following equivalent conditions:

1. **Descending chain condition** (or **d.c.c.**): there does not exist an infinite strictly descending chain

$$M_1 \supset M_2 \supset M_3 \supset \cdots$$

   of submodules of M;
2. **Minimal condition**: any non-empty family of submodules of M has a minimal element with respect to inclusion.

Analogously, a module M is **Noetherian** if it satisfies the following equivalent conditions:

1. **Ascending chain condition** (or **a.c.c.**): there does not exist an infinite strictly descending chain

$$M_1 \subset M_2 \subset M_3 \subset \cdots$$

   of submodules of M;
2. **Maximal condition**: any non-empty family of submodules of M has a maximal element with respect to inclusion.

A ring A is a **right (left) Noetherian** if the right regular module A_A (left regular module ${}_AA$) is Noetherian. A ring A is **Noetherian,** if it is both right and left Noetherian[3].

In a similar way one can define Artinian rings. A ring A is a **right (left) Artinian** if the right regular module A_A (left regular module ${}_AA$) is Artinian. A ring A is **Artinian,** if it is both right and left Artinian.

---

[3]Note that one sometimes can found in the published literature the notion of a "two-sided Noetherian ring" which means that a.c.c. holds for two-sided ideals of this ring. In this book we use the notion of a (two-sided) Noetherian ring as was given above.

Note that a semisimple ring is both Artinian and Noetherian.

A ring is **simple** if it has no two-sided ideals different from zero and the ring itself. The following proposition which is follows from Wedderburn-Artin's theorem and [146, Proposition 3.5.7] states the relationship between simple, semisimple and Artinian rings.

Proposition 1.1.8. *The following statements are equivalent for a ring* A:

1. *A is simple and Artinian.*
2. *A is simple and semisimple.*
3. *A is isomorphic to the full matrix ring $M_n(D)$, where D is a division ring.*

A module M which is isomorphic to a direct sum $M_1 \oplus M_2$ of non-zero modules M_1 and M_2 is called **decomposable**, otherwise it is **indecomposable**.

Proposition 1.1.9. (See [146, corollary 3.1.9].) *If a module M is indecomposable and both Artinian and Noetherian, then any endomorphism of M is either an automorphism or nilpotent.*

An A-module M is called **finitely generated** if it has a finite set of generators, i.e., there exists a set of elements $X = \{m_1, m_2, \ldots, m_n\} \subset M$ such that every element $m \in M$ can be written as $m = \sum\limits_{i=1}^{n} m_i a_i$ for some $a_i \in A$.

It is easy to show that an A-module M is finitely generated if and only if whenever there is an epimorphism $\bigoplus\limits_{i \in I} M_i \to M \to 0$ for a family M_i, $i \in I$, of submodules of M, then there is an epimorphism $\bigoplus\limits_{i \in J} M_j \to M \to 0$ for some finite subset J of I.

An A-module M is said to be **cyclic** if it is generated by one element, i.e., there is an element $m_0 \in M$ such that every element $m \in M$ has the form $m = m_0 a$ for some $a \in A$.

There exists a close connection between right (left) Noetherian rings and finitely generated modules over them.

Proposition 1.1.10. (See [146, proposition 3.1.12].) *A module is Noetherian if and only if each its submodule is finitely generated. In particular, a ring is right (left) Noetherian if and only if each its right (left) ideal is finitely generated.*

Proposition 1.1.11. (See [146, proposition 3.1.12].) *Let A be a right (left) Noetherian ring. Then any finitely generated right (left) A-module M is Noetherian.*

Proposition 1.1.12. (See [146, corollary 3.1.13].) *If A is a right (left) Noetherian ring, then any submodule of a finitely generated right (left) A-module M is finitely generated.*

The ring of integers $\mathbf{Z}$ is an important example of a Noetherian ring. Other examples of Noetherian rings are given by the famous Hilbert theorem:

Theorem 1.1.13 (Hilbert Basis Theorem). (See [146, theorem 3.3.1].) *Let A be a right (left) Noetherian ring. Then the ring $A[x]$ is right (left) Noetherian as well.*

The famous theorem proved by Levitzki and Hopkins states the relationship between Artinian and Noetherian ring:

Theorem 1.1.14 (Hopkins-Levitzki Theorem). (See [146, theorem 3.5.6].) *A right (left) Artinian ring A is right (left) Noetherian.*

The following proposition is similar as proposition 1.1.11 for Noetherian rings:

Proposition 1.1.15. (See [146, proposition 3.1.12].) *Let A be a right (left) Artinian ring. Then any finitely generated right (left) A-module M is Artinian.*

A dual notion to a finitely generated module is the notion of a finitely cogenerated module.

An A-module M is called **finitely cogenerated** if whenever $\bigcap_{i \in I} M_i = 0$ for any family M_i, $i \in I$, of submodules of M then $\bigcap_{i \in J} M_i = 0$ for some finite subset J of I.

One can show that an A-module M is finitely cogenerated if and only if whenever there is a monomorphism $M \to \prod_{i \in I} A$ then there is a monomorphism $M \to \prod_{i \in J} A$ for some finite subset J of I.

The following proposition is dual to proposition 1.1.10:

Proposition 1.1.16. (See [146, Proposition 3.1.6].) *A module M is Artinian if and only if each its factor module M/N is finitely cogenerated. In particular, a ring A is right Artinian if and only if each finitely generated (cyclic) right A-module is finitely cogenerated.*

The next equivalent definition of Artinian rings is connected with annihilators. Let A be a ring and $x \in A$. Then

$$\text{r.ann}_A(x) = \{a \in A \ : xa = 0\}$$

is the **right annihilator** of x and

$$\text{l.ann}_A(x) = \{a \in A \ : ax = 0\}$$

is the **left annihilator** of x.

Proposition 1.1.17. *Let A be a right Artinian ring. Then $x \in A$ is invertible in A if and only if* $\text{r.ann}_A(x) = 0$.

Proof. It is obvious that if $x \in A$ is invertible in A then $\text{r.ann}_A(x) = 0$. Suppose $\text{r.ann}_A(x) = 0$. Consider a map $\varphi : A \to xA$ given by $\varphi(a) = xa$. Then φ is an A-module isomorphism, since $\text{r.ann}_A(x) = 0$. In particular, A and xA has the same composition length. Therefore A/xA has length 0, i.e., $A = xA$. So there exists

$y \in A$ such that $xy = 1$, which implies that $\mathrm{r.ann}_A(y) = 0$. Analogously we can show that there exists $z \in A$ such that $yz = 1$, which implies that $x = z$, i.e., x is invertible. □

An important role in the theory of rings is played by the notion of a radical. There are different kinds of radicals. The best known among them is the Jacobson radical. The intersection of all maximal right ideals in a ring A is called the **Jacobson radical** of A and denoted by $\mathrm{rad}A$ or $J(A)$. The Jacobson radical of a ring A is a two-sided ideal which is described by the following:

Proposition 1.1.18. (See [146, proposition 3.4.6].) *The Jacobson radical of a ring is the largest (with respect to inclusion) two-sided ideal among all two-sided ideals I such that $1 - x$ is two-sided invertible for all $x \in I$.*

An important fact, which in many cases helps to calculate the Jacobson radical of a ring, is given by the following proposition:

Proposition 1.1.19. (See [146, proposition 3.4.8].) *Let A be a ring with Jacobson radical R and $e^2 = e \in A$. Then* $\mathrm{rad}(eAe) = eRe$.

The following Nakayama lemma plays an important role in many circumstances.

Lemma 1.1.20 (Nakayama's Lemma). (See [146, lemmas 3.4.11, 3.4.12].) *Let R be the Jacobson radical of a ring A.*

1. *If M is a finitely generated A-module and $MR = M$ then $M = 0$.*
2. *If N is a submodule of a finitely generated A-module M and $N + MR = M$ then $N = M$.*

One of the main properties of the Jacobson radical of a right Artinian ring is described by a well-known theorem of Hopkins.

Theorem 1.1.21 (Hopkins Theorem). (See [146, proposition 3.5.1].) *The Jacobson radical of a right (left) Artinian ring is nilpotent.*

The next important class of rings are semiprimary rings. A ring A with Jacobson radical R is called **semiprimary** if A/R is semisimple and R is nilpotent. The following theorem of Hopkins and Levitzki gives a criterion for a ring to be Artinian:

Theorem 1.1.22 (Hopkins-Levitzki Theorem). (See [146, corollary 3.7.2].) *A ring A is right (left) Artinian if and only if A is right (left) Noetherian and semiprimary.*

The following useful criterion helps to decide whether a ring is Artinian (or Noetherian).

Theorem 1.1.23. (See [146, theorem 3.6.1].) *Let A be an arbitrary ring with an idempotent $e^2 = e \in A$. Set $f = 1 - e$, $eAf = X$, $fAe = Y$, and let*

$$A = \begin{pmatrix} eAe & X \\ Y & fAf \end{pmatrix}$$

be the corresponding two-sided Peirce decomposition of the ring A. Then the ring A is right Noetherian (Artinian) if and only if the rings eAe and fAf are right Noetherian (Artinian), X is a finitely generated fAf-module and Y is a finitely generated eAe-module.

A ring A is called **semiprimitive** if the Jacobson radical of A is equal to zero.

Theorem 1.1.24. (See [146, theorem 3.5.5].) *The FSAE for a ring A:*

1. *A is semisimple;*
2. *A is right (left) Artinian and semiprimitive*[4].

Besides the Jacobson radical of a ring there exist many other different radicals. One other important example of a radical is the prime radical of a ring.

The **prime radical** of a ring A is the intersection of all prime ideals in A. It will be denoted by $Pr(A)$. The prime radical of A contains all nilpotent one-sided ideals of A. It is easy to show that in the general case

$$Pr(A) \subseteq \mathrm{rad}(A).$$

For an Artinian ring a strong equality holds:

Proposition 1.1.25. (See [146, proposition 11.2.3].) *The Jacobson radical* $\mathrm{rad}(A)$ *of a right Artinian ring A is equal to the prime radical $Pr(A)$, i.e.,* $\mathrm{rad}(A) = Pr(A)$.

A ring A is **prime** if the product of any two non-zero two-sided ideals of A is not equal to zero. A ring A is **semiprime** if A does not contain non-zero nilpotent ideals.

---

[4]Note that some authors define semisimple rings as rings which Jacobson radical is equal to zero, i.e., as semiprimitive rings. Throughout this book the concept of semisimple ring A assumes that the ring A is Artinian and that the Jacobson radical of A is equal to 0.

Proposition 1.1.26. (See [146, proposition 11.2.4].) *For any ring A the following statements are equivalent:*

1. *A is a semiprime ring.*
2. *The prime radical of A is equal to zero.*
3. *A has no non-zero nilpotent ideals.*
4. *A has no non-zero right nilpotent ideals.*
5. *A has no non-zero left nilpotent ideals.*

Proposition 1.1.27. (See [146, corollary 11.2.7].) *The prime radical of a ring A is a nil-ideal.*

Taking into account the proof of [146, proposition 11.2.11] one can write this theorem in the following form.

Theorem 1.1.28. *The prime radical of a right Noetherian ring is the largest one-sided nilpotent ideal of A.*

Recall the well known theorem of J. Levitzki.

Theorem 1.1.29 (Levitzki's Theorem). (See [147, Theorem 4.12.7]). *If A is a right Noetherian ring then each of its nil one-sided ideal is nilpotent.*

1.2 Categories and Functors

C is said to be a **category** if there are defined:

1. A class ObC, whose elements are called the **objects** of the category C;
2. A class Mor$C$, whose elements are called the **morphisms** of the category C;
3. For any morphism $f \in \mathrm{Mor}C$ there is an ordered pair (X,Y) of objects X,Y of C (f is said to be a morphism from an object X to an object Y and this is written by $f : X \to Y$). The set of all morphisms from X to Y is denoted by $\mathrm{Hom}_C(X,Y)$, or shortly $\mathrm{Hom}(X,Y)$;
4. For any ordered triple $X,Y,Z \in \mathrm{Ob}C$ and any pair of morphisms $f : X \to Y$ and $g : Y \to Z$ there is a uniquely defined morphism $gf : X \to Z$, which is called the **composition** or **product** of morphisms f and g.
   Objects, morphisms and compositions of morphisms are required to satisfy the following conditions:
5. Composition of morphisms is associative, i.e., for any triple of morphisms f,g,h one has $h(gf) = (hg)f$ whenever these products are defined;
6. If $X \neq X'$ or $Y \neq Y'$, then $\mathrm{Hom}(X,Y)$ and $\mathrm{Hom}(X',Y')$ are disjoint sets;
7. For any object $X \in \mathrm{Ob}C$ there exists a morphism $1_X \in \mathrm{Hom}(X,X)$ such that $f \cdot 1_X = f$ and $1_X \cdot g = g$ for any morphisms $f : X \to Y$ and $g : Z \to X$.

It is easy to see that a morphism 1_X with the above properties is unique. It is called the **identity morphism** of the object X.

If in a category C the class ObC is actually a set, then the category C is called **small**. In this case Mor$C$ is also a set.

The main examples of categories, which will be considered in this book are the categories of groups, Abelian groups and modules over a ring. **Ab** is the category Abelian groups, where Ob**Ab** is a class of all Abelian groups and Hom(A, B) is a set of all Abelian group homomorphisms from A to B.

If ObC is the class of all right (left) A-modules and Hom(X,Y) is the set of all module homomorphisms from X to Y then there results the category C of all right (left) A-modules which is denoted by $\mathbf{Mod}_A$ ($_A\mathbf{Mod}$) or $\mathbf{M}_A$ ($_A\mathbf{M}$). If ObC is the class of all right (left) finitely generated A-modules and Hom(X,Y) is the set of all module homomorphisms from X to Y then there results the category of all right (left) finitely generated A-modules which is denoted by $\mathbf{mod}_A$ ($_A\mathbf{mod}$).

One of the most important concepts in the theory of categories is the concept of a functor.

A **covariant functor** F from a category C to a category D is a pair of maps F_{ob} : Ob$C \to$ ObD and F_{mor} : Mor$C \to$ Mor$D$ satisfying the following conditions:

1. If $X,Y \in$ ObC, then to each morphism $f : X \to Y$ in Mor$C$ there corresponds a morphism $F_{mor}(f) : F_{ob}(X) \to F_{ob}(Y)$ in Mor$D$;
2. $F_{mor}(1_X) = 1_{F_{ob}(X)}$ for all $X \in$ ObC;
3. If the product of morphisms gf is defined in C, then

$$F_{mor}(gf) = F_{mor}(g)F_{mor}(f).$$

Usually, instead of $F_{mor}(f)$ and $F_{ob}(X)$ one simply writes $F(f)$ and $F(X)$.

Given a category C, one can form the opposite category C^{op} in which ObC^{op} = ObC, while $\mathrm{Hom}_{C^{op}}(X,Y) = \mathrm{Hom}_C(Y,X)$. Composition of morphisms in C^{op} is defined as the reversed version of composition in C, i.e., if $*$ denotes composition in C^{op} then $f * g = g \cdot f$.

A contravariant functor from a category C to a category D is literally a covariant functor from C to D^{op}. That is, there results the following definition:
A **contravariant functor** F from a category C to a category D is a pair of maps F_{ob} : Ob$C \to$ ObD and F_{mor} : Mor$C \to$ Mor$D$ satisfying the following conditions:

1. If $X,Y \in$ ObC, then to each morphism $f : X \to Y$ in Mor$C$ there corresponds a morphism $F_{mor}(f) : F_{ob}(Y) \to F_{ob}(X)$ in Mor$D$;
2. $F_{mor}(1_X) = 1_{F_{ob}(X)}$ for all $X \in$ ObC;
3. If the product of morphisms gf is defined in C, then

$$F_{mor}(gf) = F_{mor}(f)F_{mor}(g).$$

A **functor** is defined as either a covariant functor or a contravariant functor.

Consider the category $\mathbf{Mod}_A$ of right A-modules over a given ring A. This setting holds for all the remainder of this section 1.2.

A functor $F : \mathbf{Mod}_A \longrightarrow \mathbf{Ab}$ of either variance is called an **additive functor** if for any pair of A-morphisms $f_1 : X \longrightarrow Y$ and $f_2 : X \longrightarrow Y$ the identity $F(f_1 + f_2) = F(f_1) + F(f_2)$ holds.

A sequence of A-modules and homomorphisms

$$\cdots \longrightarrow M_{i-1} \xrightarrow{f_i} M_i \xrightarrow{f_{i+1}} M_{i+1} \longrightarrow \cdots \tag{1.2.1}$$

is said to be **exact at** M_i if $\mathrm{Im} f_i = \mathrm{Ker} f_{i+1}$. If the sequence (1.2.1) is exact at every M_i, then it is called **exact**.

An exact sequence of the form

$$0 \longrightarrow N \xrightarrow{f} M \xrightarrow{g} L \longrightarrow 0 \tag{1.2.2}$$

is called a **short exact sequence**.

A monomorphism $f : N \longrightarrow M$ is called **split** if there exists a homomorphism $\overline{f} : M \longrightarrow N$ such that $\overline{f} f = 1_N$. An epimorphism $g : M \longrightarrow L$ is called **split** if there exists a homomorphism $\overline{g} : L \longrightarrow M$ such that $g\overline{g} = 1_L$.

A short exact sequence (1.2.2) is said to be **split** if f is a split monomorphism and g is a split epimorphism.

The following proposition gives a useful criterion to check when an exact sequence is split.

Proposition 1.2.3. (See [146, proposition 4.2.1].) *The following statements for an exact sequence*

$$0 \longrightarrow X \xrightarrow{f} M \xrightarrow{g} Y \longrightarrow 0 \tag{1.2.4}$$

are equivalent:

1. *The sequence is split;*
2. *There exists a homomorphism* $\overline{g} : Y \longrightarrow M$ *such that* $g\overline{g} = 1_Y$;
3. *There exists a homomorphism* $\overline{f} : M \longrightarrow X$ *such that* $\overline{f} f = 1_X$;
4. $M \simeq X \oplus Y$.

The following lemma which is often called as "Five Lemma" is often useful:

Lemma 1.2.5 (Five Lemma). (See [146, lemma 4.2.4].) *Let*

$$\begin{array}{ccccccccc} M_1 & \xrightarrow{f_1} & M_2 & \xrightarrow{f_2} & M_3 & \xrightarrow{f_3} & M_4 & \xrightarrow{f_4} & M_5 \\ \downarrow{\scriptstyle \varphi_1} & & \downarrow{\scriptstyle \varphi_2} & & \downarrow{\scriptstyle \varphi_3} & & \downarrow{\scriptstyle \varphi_4} & & \downarrow{\scriptstyle \varphi_5} \\ N_1 & \xrightarrow{g_1} & N_2 & \xrightarrow{g_2} & N_3 & \xrightarrow{g_3} & N_4 & \xrightarrow{g_4} & N_5 \end{array} \tag{1.2.6}$$

be a commutative diagram with exact rows and isomorphisms φ_i, $i = 1,2,4,5$. *Then* φ_3 *is also an isomorphism.*

As immediately corollaries of this lemma there result the following statements:

Corollary 1.2.7. (See [146, corollary 4.2.5].) *Let*

$$\begin{array}{ccccccccc} 0 & \longrightarrow & M_1 & \longrightarrow & M_2 & \longrightarrow & M_3 & \longrightarrow & 0 \\ & & \downarrow{\scriptstyle \varphi_1} & & \downarrow{\scriptstyle \varphi_2} & & \downarrow{\scriptstyle \varphi_3} & & \\ 0 & \longrightarrow & N_1 & \longrightarrow & N_2 & \longrightarrow & N_3 & \longrightarrow & 0 \end{array} \qquad (1.2.8)$$

be a commutative diagram with exact rows and isomorphisms φ_1 and φ_3. Then φ_2 is also an isomorphism.

Corollary 1.2.9. (See [146, corollary 4.2.6].) *Let*

$$\begin{array}{ccccccc} M_1 & \longrightarrow & M_2 & \longrightarrow & M_3 & \longrightarrow & 0 \\ \downarrow{\scriptstyle \varphi_1} & & \downarrow{\scriptstyle \varphi_2} & & \downarrow{\scriptstyle \varphi_3} & & \\ N_1 & \longrightarrow & N_2 & \longrightarrow & N_3 & \longrightarrow & 0 \end{array} \qquad (1.2.10)$$

be a commutative diagram with exact rows and isomorphisms φ_1 and φ_2. Then φ_3 is also an isomorphism.

Let A and A' be rings and let $F : \mathbf{M}_A \longrightarrow \mathbf{M}_{A'}$ be a covariant functor. Suppose $0 \longrightarrow B_1 \xrightarrow{\varphi} B \xrightarrow{\psi} B_2$ is an arbitrary exact sequence in $\mathbf{M}_A$, then the functor F is said to be **left exact** if for any such sequence the sequence

$$0 \longrightarrow F(B_1) \xrightarrow{F\varphi} F(B) \xrightarrow{F\psi} F(B_2)$$

is exact in $\mathbf{M}_{A'}$. Analogously, a functor F is said to be **right exact** if from the exactness of an arbitrary sequence of right A-modules $B_1 \xrightarrow{\varphi} B \xrightarrow{\psi} B_2 \longrightarrow 0$ there follows the exactness of the sequence

$$F(B_1) \xrightarrow{F\varphi} F(B) \xrightarrow{F\psi} F(B_2) \longrightarrow 0$$

in $\mathbf{M}_{A'}$. If F is both left and right exact, i.e., if exactness of a sequence

$$0 \longrightarrow B_1 \xrightarrow{\varphi} B \xrightarrow{\psi} B_2 \longrightarrow 0$$

always implies exactness of a sequence

$$0 \longrightarrow F(B_1) \xrightarrow{F\varphi} F(B) \xrightarrow{F\psi} F(B_2) \longrightarrow 0$$

then F is said to be an **exact functor**.

The most important examples of functors considered in this book are the functors Hom, Ext and Tor.

Let M and N be right A-modules, then the set $\mathrm{Hom}_A(M,N)$ of all A-homomorphisms from M to N forms an additive Abelian group. $\mathrm{Hom}_A(M,*)$ is a covariant functor from the category $\mathbf{M}_A$ of right A-modules to the category **Ab** of Abelian groups and $\mathrm{Hom}_A(*,M)$ is a contravariant functor from $\mathbf{M}_A$ to **Ab** for each fixed right A-module M.

The following theorem gives some of the basic properties of the functor Hom (See [146, propositions 4.3.1, 4.3.2, 4.3.3]).

Theorem 1.2.11. *The* Hom *functor is left exact in each variable. This is, a sequence of right A-modules B_1, B, B_2*

$$0 \longrightarrow B_1 \xrightarrow{\varphi} B \xrightarrow{\psi} B_2 \tag{1.2.12}$$

is exact if and only if for any right A-module M the sequence

$$0 \longrightarrow \mathrm{Hom}_A(M,B_1) \xrightarrow{\overline{\varphi}} \mathrm{Hom}_A(M,B) \xrightarrow{\overline{\psi}} \mathrm{Hom}_A(M,B_2) \tag{1.2.13}$$

is exact. (Here $\overline{\varphi} = \mathrm{Hom}_A(M,\varphi)$ and $\overline{\psi} = \mathrm{Hom}_A(M,\psi)$.)

A sequence of right A-modules B_1, B, B_2

$$B_1 \xrightarrow{\varphi} B \xrightarrow{\psi} B_2 \longrightarrow 0 \tag{1.2.14}$$

is exact if and only if for any right A-module M the sequence

$$0 \longrightarrow \mathrm{Hom}_A(B_2,M) \xrightarrow{\overline{\psi}} \mathrm{Hom}_A(B,M) \xrightarrow{\overline{\varphi}} \mathrm{Hom}_A(B_1,M) \tag{1.2.15}$$

is exact. (Here $\overline{\varphi} = \mathrm{Hom}_A(\varphi,M)$ and $\overline{\psi} = \mathrm{Hom}_A(\psi,M)$.)

The following proposition shows the behavior of the Hom functor with respect to direct sums and direct products.

Proposition 1.2.16. (See [146, propositions 4.3.4, 4.3.5].)

1. *Let A be a ring and let Y, X_i, $(i \in I)$ be A-modules. Then there exists a natural isomorphism*

$$\mathrm{Hom}_A(\bigoplus_{i\in I} X_i, Y) \simeq \prod_{i\in I} \mathrm{Hom}_A(X_i,Y). \tag{1.2.17}$$

2. *Let A be a ring and let X, Y_i, $(i \in I)$ be A-modules. Then there exists a natural isomorphism*

$$\mathrm{Hom}_A(X, \prod_{i\in I} Y_i) \simeq \prod_{i\in I} \mathrm{Hom}_A(X,Y_i). \tag{1.2.18}$$

1.3 Tensor Product of Modules

Let A be a ring and let G be an additive Abelian group. Suppose $X \in \mathbf{M}_A$ and $Y \in {}_A\mathbf{M}$. An A-**balanced map** from $X \times Y$ to G is a map $\varphi : X \times Y \to G$ satisfying the following identities:

1. $\varphi(x, y + y') = \varphi(x, y) + \varphi(x, y')$,
2. $\varphi(x + x', y) = \varphi(x, y) + \varphi(x', y)$,
3. $\varphi(xa, y) = \varphi(x, ay)$
   for all $x, x' \in X$, $y, y' \in Y$, and $a \in A$.[5])

Consider the free Abelian group F, whose free generators are the elements of $X \times Y$. Let H be the subgroup of F generated by all elements of the form:

$$(x + x', y) - (x, y) - (x', y)$$

$$(x, y + y') - (x, y) - (x, y')$$

$$(xa, y) - (x, ay)$$

Consider the canonical projection $\pi : F \to F/H$. Write $\varphi = \pi i$, where $i : X \times Y \to F$ is the natural monomorphism. We write $\varphi(x, y) = x \otimes y$ and write $X \otimes Y$ (or $X \otimes_A Y$ if A is to be emphasized) for the Abelian quotient group F/H. With these notations we have that $X \otimes Y$ is an Abelian group, whose generators $x \otimes y$ satisfy the following identities:

$$(x_1 + x_2) \otimes y = (x_1 \otimes y) + (x_2 \otimes y)$$

$$x \otimes (y_1 + y_2) = (x \otimes y_1) + (x \otimes y_2)$$

$$xa \otimes y = x \otimes ay$$

for all $x, x' \in X$, $y, y' \in Y$, and $a \in A$.

The Abelian group $X \otimes_A Y$ is a **tensor product** of modules X and Y over A. It has the following universal property. For any Abelian group G and any A-balanced maps $f : X \times Y \to G$ there exists a unique morphism of Abelian groups $g : X \otimes_A Y \to G$ such that the diagram

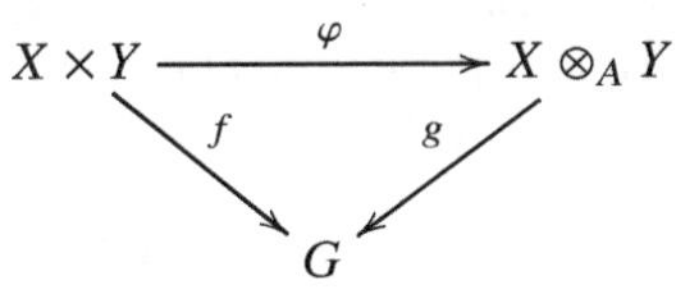

commutes.

[5]Conditions 1 and 2 state that φ is a bilinear map of Abelian groups.

Fix a left A-module Y and construct the functor $* \otimes_A Y : \mathbf{M}_A \to \mathbf{Ab}$ as follows. Assign to every right A-module X the Abelian group $X \otimes Y$ and to every homomorphism $f : X \longrightarrow X'$ the homomorphism

$$f \otimes_A 1 : X \otimes_A Y \longrightarrow X' \otimes_A Y.$$

Then $f \otimes_A 1$ is a group homomorphism and one obtains a functor

$$* \otimes_A Y : X \longrightarrow X \otimes_A Y$$

from the category of right A-modules to the category of Abelian groups. If in addition Y is an (A, B)-bimodule for some ring B, then $f \otimes_A 1$ is a homomorphism of right B-modules and there results a functor from the category of right A-modules to the category of right B-modules. Similarly, for a fixed right A-module X one can construct the functor

$$X \otimes_A * : Y \longrightarrow X \otimes_A Y$$

from the category of left A-modules to the category of Abelian groups (respectively, to the category of left B-modules when X is a (B, A)-bimodule for some ring B). Thus one obtains a functor of two variables which is called the **tensor product functor** and denoted by $* \otimes *$. This functor is covariant in both variables. Indeed,

$$f'f \otimes 1 = (f' \otimes 1)(f \otimes 1)$$

and

$$1 \otimes g'g = (1 \otimes g')(1 \otimes g)$$

The following proposition shows the important connection between the functor Hom and the tensor product functor.

Proposition 1.3.1 (Adjoint isomorphism). (See [146, proposition 4.6.3].)

1. *In a situation X_A, ${}_AY_B$ and Z_B, there exists a canonical isomorphism:*

$$\mathrm{Hom}_B(X \otimes_A Y, Z) \simeq \mathrm{Hom}_A(X, \mathrm{Hom}_B(Y, Z)), \qquad (1.3.2)$$

assigning to a homomorphism $f : X \otimes_A Y \longrightarrow Z$ the homomorphism $\bar{f} : X \longrightarrow Hom_B(Y, Z)$ such that $\bar{f}(x)(y) = f(x \otimes_A y)$.

2. *In a situation ${}_AX$, ${}_BY_A$ and ${}_BZ$, there exists a canonical isomorphism:*

$$\mathrm{Hom}_B(Y \otimes_A X, Z) \simeq \mathrm{Hom}_A(X, \mathrm{Hom}_B(Y, Z)), \qquad (1.3.3)$$

assigning to a homomorphism $g : Y \otimes_A X \longrightarrow Z$ the homomorphism $\bar{g} : X \longrightarrow Hom_B(Y, Z)$ such that $\bar{g}(x)(y) = g(y \otimes_A x)$.

Proposition 1.3.4 (See [146, proposition 4.6.4].) *The tensor product functor is right exact in both variables.*

1.4 Direct and Inverse Limits

A partially ordered set S is called (upwards) **directed** if for any pair $a, b \in S$ there exists an element $c \in S$ such that $a \leq c$ and $b \leq c$.

Let I be a directed partially ordered set, and let $\{M_i \ : \ i \in I\}$ be a set of A-modules. Suppose that a homomorphism $\varphi_{ij} \ : \ M_i \to M_j$ is given for any pair of indexes $i, j \in I$, where $i \leq j$, such that for all $i \leq j \leq k$ and $n \in I$ the following conditions hold:

1. $\varphi_{nn} : M_n \to M_n$ is the identity on M_n;
2. $\varphi_{ik} = \varphi_{jk}\varphi_{ij}$, i.e., the diagram

$$\begin{array}{ccc} M_i & \xrightarrow{\varphi_{ij}} & M_j \\ & \searrow^{\varphi_{ik}} & \downarrow{\scriptstyle \varphi_{jk}} \\ & & M_k \end{array} \tag{1.4.1}$$

commutes.

In this case the triple

$$\mathbf{M} = \{\{I, \leq\}; \{M_i \ : \ i \in I\}; \{\varphi_{ij} \mid i \leq j \in I\}\} \tag{1.4.2}$$

is called a **direct system** of A-modules.

Let $\mathbf{M}$ be a direct system and $M = \bigoplus_{i \in I} M_i$, where $M_i \in \mathbf{M}$. Consider the submodule $N \subset M$, which is generated by all elements $m_j - \varphi_{ij} m_i$ for $i \leq j$. The quotient module M/N is called the **direct limit** (also called **injective limit** or **inductive limit**) of the directed system $\mathbf{M}$ and denoted by $\varinjlim M_i$.

Theorem 1.4.3. (See [146, theorem 4.7.1].) *The direct limit $\varinjlim M_i$ of a directed system* $\mathbf{M}$ *has the following universal property:*

For any module X and a set of homomorphisms $f_i \ : \ M_i \longrightarrow X$ such that all diagrams

$$\begin{array}{ccc} M_i & \xrightarrow{\varphi_{ij}} & M_j \\ {\scriptstyle f_i}\downarrow & \swarrow_{f_j} & \\ X & & \end{array} \tag{1.4.4}$$

commute for $i \leq j$, there exists a unique homomorphism $\sigma : \varinjlim M_i \longrightarrow X$ such that all diagrams

$$\begin{array}{ccc} M_i & & \\ {\scriptstyle \pi_i}\downarrow & \searrow^{f_i} & \\ \varinjlim M_i & \xrightarrow[\sigma]{} & X \end{array} \tag{1.4.5}$$

commute. Here π_i is the composite of the injection $M_i \longrightarrow \bigoplus_{i\in I} M_i$ with the quotient morphism $M \longrightarrow M/N$. The module $\varinjlim M_i$ together with the homomorphisms π_i is determined uniquely up to isomorphism.

In a dual way one can define the notion of the inverse limit.

Let I be a partially ordered set, and $\{M_i \mid i \in I\}$ a set of modules. Suppose that a homomorphism $\varphi_{ji} : M_j \to M_i$ is given for any pair of indexes $i, j \in I$, where $i \leq j$, such that for all $i \leq j \leq k$ and $n \in I$ the following conditions hold:

1. $\varphi_{nn} : M_n \to M_n$ is the identity on M_n;
2. $\varphi_{ki} = \varphi_{ji}\varphi_{kj}$, i.e., the diagram

$$\begin{array}{ccc} M_k & \xrightarrow{\varphi_{kj}} & M_j \\ & \searrow^{\varphi_{ki}} & \downarrow{\scriptstyle \varphi_{ji}} \\ & & M_i \end{array} \tag{1.4.6}$$

commutes.

In this case the triple

$$\mathbf{M} = \{(I, \leq); \{M_i \mid i \in I\}; \{\varphi_{ij} \mid i \leq j \in I\}\} \tag{1.4.7}$$

is called an **inverse system** of A-modules.

Suppose there is an inverse system (1.4.7) of A-modules. Set $M = \prod\limits_i M_i$, where $M_i \in \mathbf{M}$. Let π_i be the system of canonical projections. Consider the submodule N of M which is generated by all elements $m \in M$ such that

$$\pi_i(m) = \varphi_{ji}(\pi_j(m)) \tag{1.4.8}$$

whenever $i \leq j$. The submodule N is called the **inverse limit** (or **projective limit**) of the inverse system $\mathbf{M}$ and denoted by $\varprojlim M_i$. When the elements of M are written in the form $m = (m_i)_{i\in I}$, condition (1.4.8) takes the form $m_i = \varphi_{ji}(m_j)$.

Exactly like the direct limit, the inverse limit has a universal property, which determines it uniquely up to isomorphism.

Theorem 1.4.9 (See [146, theorem 4.7.4].) *The inverse limit $\varprojlim M_i$ of an inverse system $\mathbf{M}$ has the following property:*

For any module X and any family of homomorphisms $f_i : X \to M_i$ such that all diagrams

$$\begin{array}{ccc} X & \xrightarrow{f_i} & M_i \\ {\scriptstyle f_j}\downarrow & \swarrow_{\varphi_{ij}} & \\ M_j & & \end{array}$$

commute for $i \leq j$, there exists a unique homomorphism $\tau : X \longrightarrow \varprojlim M_i$ such that all diagrams

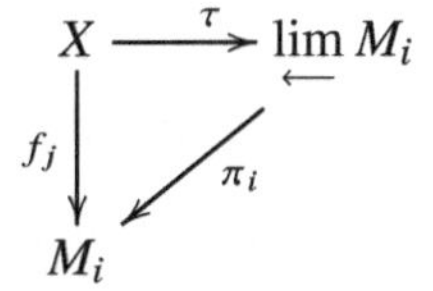

commute (where π_i is also used for the restriction of $\pi_i : \prod_{i \in I} M_i \longrightarrow M_i$ to $\varprojlim M_i$). The module $\varprojlim M_i$ together with the homomorphisms π_i is determined uniquely up to isomorphism.

The following proposition illustrates some connections of inverse and direct limits with exact functors.

Proposition 1.4.10. (See [146, propositions 4.7.5, 4.7.7, corollaries 4.7.6, 4.7.8].)

1. *If F is a left exact functor that preserves direct products, then F preserves inverse limits. In particular, for any A-module X the functor* $\mathrm{Hom}_A(X, *)$ *preserves inverse limits.*
2. *If F is a right exact functor that preserves direct sums, then F preserves direct limits. In particular, for any A-module X the functor $X \otimes_A *$ preserves direct limits.*

1.5 Projective, Injective, and Flat Modules

A module P is called **projective** if for any epimorphism $\varphi : M \longrightarrow N$ and for any homomorphism $\psi : P \longrightarrow N$ there is a homomorphism $h : P \rightarrow M$ such that $\psi = \varphi h$. This means that any diagram of the form

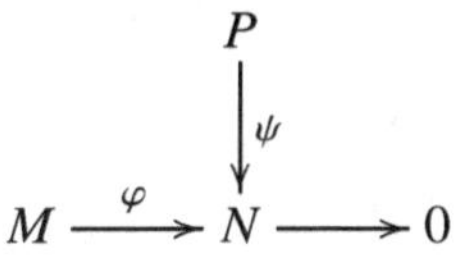

with the bottom row exact can be completed to a commutative diagram

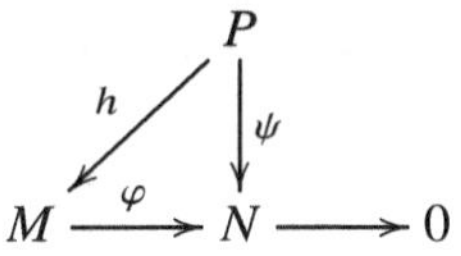

Here are some basic results for projective modules.

Proposition 1.5.1. (See [146, propositions 5.1.1, 5.1.2; corollary 5.1.3].)

1. *An A-module P is projective if and only if* $\mathrm{Hom}_A(P, *)$ *is an exact functor.*
2. *Any free module is projective.*
3. *Every module is isomorphic to a factor module of a projective module.*

The connection of projective modules with direct sums is given by the following:

Proposition 1.5.2. (See [146, proposition 5.1.4; corollary 5.1.5].)
A direct sum $P = \bigoplus_{\alpha \in I} P_\alpha$ *of modules* P_α *is a projective module if and only if each* P_α *is projective.*

The following proposition gives another equivalent definitions of a projective module.

Proposition 1.5.3. (See [146, proposition 5.1.6].) *Let A be a ring. For an A-module P the following statements are equivalent*:

1. *P is projective;*
2. *Every short exact sequence* $0 \longrightarrow N \longrightarrow M \longrightarrow P \longrightarrow 0$ *splits*;
3. *P is a direct summand of a free A-module F.*

From this proposition one immediately obtains the following corollary.

Corollary 1.5.4. (See [146, corollary 5.1.7].) *Let A be a ring. For any idempotent* $e = e^2 \in A$, eA *is a projective A-module.*

There is also another useful test for a module to be projective which was proved by I. Kaplansky.

Theorem 1.5.5 (Kaplansky's Theorem). (See [146, proposition 8.3.1].) *An A-module P is projective if and only if there is a system of elements* $\{p_\alpha\}$ *of P and a system of homomorphisms* $\{\varphi_\alpha\}$, $\varphi_\alpha : P \longrightarrow A$ *such that any element* $p \in P$ *may be written in the form:*

$$p = \sum_\alpha p_\alpha(\varphi_\alpha(p)), \tag{1.5.6}$$

where only a finite number of elements $\varphi_\alpha(p) \in A$ *are not equal to zero.*

"Dual" to the notion of projectivity is that of injectivity. Under "duality" (in this setting) one means "inverting all arrows" (maps) and interchanging "epimorphism" with "monomorphism".

A module Q is called **injective** if for any monomorphism $\varphi : M \longrightarrow N$ and any homomorphism $\psi : M \longrightarrow Q$ there exists a homomorphism $h : N \longrightarrow Q$ such that

$\psi = h\varphi$. This means that any diagram of the form

$$\begin{array}{ccccc} 0 & \longrightarrow & M & \xrightarrow{\varphi} & N \\ & & {\scriptstyle\psi}\downarrow & & \\ & & Q & & \end{array}$$

with the top row exact can be completed to a commutative diagram

$$\begin{array}{ccccc} 0 & \longrightarrow & M & \xrightarrow{\varphi} & N \\ & & {\scriptstyle\psi}\downarrow & \swarrow{\scriptstyle h} & \\ & & Q & & \end{array}$$

Here are some basic results for injective modules.

Proposition 1.5.7. (See [146, propositions 5.2.1, 5.2.2; corollaries 5.2.9, 5.2.10].)

1. *An A-module Q is injective if and only if* $\mathrm{Hom}_A(*, Q)$ *is an exact functor.*
2. *An A-module Q is injective if and only if every exact sequence of A-modules*

   $$0 \longrightarrow Q \longrightarrow M \longrightarrow N \longrightarrow 0$$

   splits.
3. *A direct product $Q = \prod_{\alpha \in I} Q_\alpha$ of injective modules Q_α is injective if and only if each Q_α is injective.*
4. *A module Q is injective if and only if it is a direct summand of every module that contains it.*

The following theorem gives a useful criterion for a module to be injective.

Theorem 1.5.8 (Baer's Criterion). (See [146, proposition 5.2.4].) *Let Q be a right module over a ring A. Then the following statements are equivalent:*

1. *Q is injective;*
2. *For any right ideal $\mathcal{I} \subset A$ and each $f \in \mathrm{Hom}_A(\mathcal{I}, Q)$ there exists an extension $\varphi \in \mathrm{Hom}_A(A, Q)$ of f, i.e., $\varphi i = f$ where i is the natural embedding from $\mathcal{I}$ to A;*
3. *For any right ideal $\mathcal{I} \subset A$ and each $f \in \mathrm{Hom}_A(\mathcal{I}, Q)$ there exists an element $q \in Q$ such that $f(a) = qa$ for all $a \in \mathcal{I}$.*

The following theorem gives a characterization of semisimple rings in terms of projective and injective modules.

Theorem 1.5.9. (See [146, theorem 5.2.13].) *For a ring A the following statements are equivalent:*

1. *A is a semisimple ring.*
2. *Any A-module M is projective.*
3. *Any A-module M is injective.*

If N is a submodule of a module M, then M is said to be an **extension** of N. A submodule N of M is called **essential** (or **large**) in M if it has non-zero intersection with every non-zero submodule of M. In this case M is also said to be an **essential extension** of N.

The next simple lemma gives a very useful test for essential extensions.

Lemma 1.5.10. (See [146, lemma 5.3.1].) *An A-module M is an essential extension of an A-module N if and only if for any* $0 \neq x \in M$ *there exists an* $a \in A$ *such that* $0 \neq xa \in N$.

There is another criterion for a module to be injective which is connected with essential extensions.

Theorem 1.5.11. (B. Eckmann-A. Schopf.) (See [146, theorem 5.3.3].) *A module Q is injective if and only if it has no proper essential extensions.*

A module Q is called an **injective hull** (or **injective envelope**) of a module M if it is both an essential extension of M and an injective module.

Theorem 1.5.12. (See [146, theorem 5.3.4].) *Every module M has an injective hull, which is unique up to isomorphism extending the identity of M.*

Let M be a right A-module. The **socle** of M, denoted by soc(M), is the sum of all simple right submodules of M. If there are no such submodules, then soc(M) = 0.

The notion which is dual to the injective hull is that of a projective cover.

A submodule N of a module M is **small** (or **superfluous**) if the equality $N + X = M$ implies $X = M$ for any submodule X of the module M.

A projective module P is a **projective cover** of a module M and denoted by $P(M)$ if there is an epimorphism $\varphi : P \longrightarrow M$ such that $\mathrm{Ker}\varphi$ is a small submodule in P.

Proposition 1.5.13. (See [146, corollary 10.4.6; proposition 5.3.8].) *If a module M has a projective cover P(M), then the cover is unique up to isomorphism. In other words, if* $P \xrightarrow{f} M$, $P' \xrightarrow{f'} M$ *are two projective covers of M then there is an isomorphism* $\varphi : P \longrightarrow P'$ *such that* $f'\varphi = f$.

An A-module X is **flat** if $X \otimes_A *$ is an exact functor.

Proposition 1.5.14. (See [146, corollaries 5.4.4, 5.4.5].)

1. *Every free module is flat.*
2. *Every projective module is flat.*

The following theorem gives useful tests for a module to be flat.

Theorem 1.5.15. (Flatness tests). (See [146, theorem 5.4.10; propositions 5.4.11, 5.4.12].) *Let B be a right A-module. Then the following statements are equivalent:*

1. *B is flat.*
2. *The character module $B^* = \mathrm{Hom}_{\mathbf{Z}}(B, \mathbf{Q}/\mathbf{Z})$ is injective as a left A-module.*
3. *For each left ideal $\mathcal{I} \subseteq A$ the natural map $B \otimes_A \mathcal{I} \longrightarrow B\mathcal{I}$ is an isomorphism of Abelian groups.*
4. *For each finitely generated left ideal $\mathcal{I} \subseteq A$ the natural map $B \otimes_A \mathcal{I} \longrightarrow B\mathcal{I}$ is an isomorphism of Abelian groups.*
5. *Let a sequence of right A-modules $0 \longrightarrow M \longrightarrow F \longrightarrow B \longrightarrow 0$ be exact, where F is flat. Then $M \cap F\mathcal{I} = M\mathcal{I}$ for every finitely generated left ideal $\mathcal{I} \subseteq A$.*

1.6 The Functor Tor

Let M be an A-module. A **projective resolution** of M is an exact sequence of A-modules

$$\cdots \longrightarrow P_2 \xrightarrow{d_2} P_1 \xrightarrow{d_1} P_0 \xrightarrow{\pi} M \longrightarrow 0 \tag{1.6.1}$$

for which the P_n are projective for all $n \geq 0$.

It is easy to prove that every module has a projective resolution.

In a dual way one can define an **injective resolution** of an A-module M as an exact sequence of A-modules

$$0 \longrightarrow M \xrightarrow{i} Q_0 \xrightarrow{d_0} Q_1 \xrightarrow{d_1} Q_2 \longrightarrow \cdots \tag{1.6.2}$$

for which the Q_n are injective for all $n \geq 0$.

Applying the construction of derived functors (considered, e.g., in [146, section 6.2]) to the functors $* \otimes_A Y$ and $X \otimes_A *$ one can construct the functors Tor.

Let X, Y be A-modules and $F = * \otimes_A Y$, then by definition

$$\mathrm{Tor}_n^A(X,Y) = \mathrm{Ker}(d_n \otimes 1)/\mathrm{Im}(d_{n+1} \otimes 1), \tag{1.6.3}$$

where

$$\cdots \longrightarrow P_2 \xrightarrow{d_2} P_1 \xrightarrow{d_1} P_0 \to X \to 0$$

is a projective resolution of the A-module X.

One can show (see [146, proposition 6.2.3]) that this definition of $\operatorname{Tor}_n^A(X,Y)$ is independent of the choice of a projective resolution of X. It is easy to see that $\operatorname{Tor}_n^A(*,Y)$ is an additive covariant functor.

Analogously one can introduce the functors $\operatorname{Tor}_n^A(X,*)$ as the left derived functors of the functor $F = * \otimes_A Y$, i.e.,

$$\operatorname{Tor}_n^A(X,Y) = \operatorname{Ker}(1 \otimes d_n)/\operatorname{Im}(1 \otimes d_{n+1}), \tag{1.6.4}$$

where

$$\cdots \longrightarrow P_2 \xrightarrow{d_2} P_1 \xrightarrow{d_1} P_0 \to Y \to 0$$

is a projective resolution of the A-module Y.

$\operatorname{Tor}_n^A(X,*)$ is also an additive covariant functor.

One can show (see, e.g. [146, proposition 6.3.1]) that these constructions give the same results and the common value of these functors is denoted by $\operatorname{Tor}_n^A(X,Y)$. This functor is right exact in both variables.

Here are some other main properties of the functor Tor.

Proposition 1.6.5. (See [146, propositions 6.3.2, 6.3.3, 6.3.5].)

1. $\operatorname{Tor}_0^A(X,Y)$ *is naturally equivalent to* $X \otimes_A Y$. *If* $n < 0$ *then* $\operatorname{Tor}_n^A(X,Y) = 0$ *for all* X, Y.
2. *If* P *is projective then* $\operatorname{Tor}_n^A(P,Y) = 0$ *for all* Y *and all* $n > 0$. *Similarly,* $\operatorname{Tor}_n^A(X,P) = 0$ *for all* X *and all* $n > 0$.

The following theorem states a connection of flat modules with the functors Tor.

Theorem 1.6.6. (See [146, propositions 6.3.6, 6.3.7, 6.3.8, 6.3.9].)

1. *If* X *is a flat A-module, then* $\operatorname{Tor}_n^A(X,Y) = 0$ *for all* Y *and all* $n > 0$.
2. *If* $\operatorname{Tor}_1^A(X,Y) = 0$ *for all* Y, *then* X *is flat.*
3. *If* $0 \longrightarrow X' \longrightarrow X \longrightarrow X'' \longrightarrow 0$ *is exact with* X *flat, then* $\operatorname{Tor}_n^A(X',Y) \simeq \operatorname{Tor}_{n+1}^A(X'',Y)$ *for all* Y *and all* $n > 0$.
4. *Suppose* Y *is a left A-module and* $\operatorname{Tor}_1^A(A/\mathcal{I},Y) = 0$ *for every finitely generated right ideal* $\mathcal{I}$. *Then* Y *is flat.*

1.7 The Functor Ext

This section recalls the construction of the functors $\operatorname{Ext}_A(*,Y)$ and $\operatorname{Ext}_A(X,*)$ and considers the main properties of these functors.

Let X, Y be A-modules and $F = \operatorname{Hom}_A(*,Y)$, then

$$\operatorname{Ext}_A^n(X,Y) = \operatorname{Ker}\operatorname{Hom}(d_{n+1},Y)/\operatorname{Im}\operatorname{Hom}(d_n,Y),$$

where

$$\cdots \longrightarrow P_2 \xrightarrow{d_2} P_1 \xrightarrow{d_1} P_0 \longrightarrow X \longrightarrow 0$$

is a projective resolution of the A-module X.

One can show that the definition of $\mathrm{Ext}^n_A(X,Y)$ is independent of the choice of a projective resolution of X. It is easy to see that $\mathrm{Ext}^n_A(*,Y)$ is an additive contravariant functor.

Using injective resolutions one can also make the following definition.

Let X, Y be A-modules and $F = \mathrm{Hom}_A(X,*)$, then

$$\mathrm{Ext}^n_A(X,Y) = \mathrm{Ker\,Hom}(X,d_n)/\mathrm{Im\,Hom}(X,d_{n-1}),$$

where

$$0 \longrightarrow Y \longrightarrow Q_0 \xrightarrow{d_0} Q_1 \xrightarrow{d_1} Q_2 \longrightarrow \cdots$$

is an injective resolution of the A-module Y.

$\mathrm{Ext}^n_A(X,*)$ is also an additive covariant functor and it is independent of the choice of an injective resolution of Y.

As in the case of the functor Tor one can show that these two constructions for the functor Ext are (basically) the same and the common value of these functors is denoted by $\mathrm{Ext}^A_n(X,Y)$.

From the construction of the functors Ext it immediately follows that $\mathrm{Ext}^0_A(X,Y)$ is naturally equivalent to $\mathrm{Hom}_A(X,Y)$ and $\mathrm{Ext}^n_A(X,Y) = 0$ for all X, Y and all $n < 0$.

Theorem 1.7.1. (See [146, theorem 6.4.4].)

If $0 \longrightarrow M \xrightarrow{f} N \xrightarrow{g} L \longrightarrow 0$ *is an exact sequence of modules, then there exists a long exact sequence*

$$0 \longrightarrow \mathrm{Hom}_A(X,M) \xrightarrow{h(f)} \mathrm{Hom}_A(X,N) \xrightarrow{h(g)} \mathrm{Hom}_A(X,L) \xrightarrow{\partial} \mathrm{Ext}^1_A(X,M) \longrightarrow \cdots$$

$$\cdots \longrightarrow \mathrm{Ext}^n_A(X,M) \longrightarrow \mathrm{Ext}^n_A(X,N) \longrightarrow \mathrm{Ext}^n_A(X,L) \longrightarrow \mathrm{Ext}^{n+1}_A(X,M) \longrightarrow \cdots$$

where ∂ is a connecting homomorphism.[6]

Theorem 1.7.2. (See [146, theorem 6.4.5].)

If $0 \longrightarrow M \xrightarrow{f} N \xrightarrow{g} L \longrightarrow 0$ *is an exact sequence of modules, then there exists a long exact sequence*

$$0 \longrightarrow \mathrm{Hom}_A(L,Y) \xrightarrow{h(g)} \mathrm{Hom}_A(N,Y) \xrightarrow{h(f)} \mathrm{Hom}_A(M,Y) \xrightarrow{\partial} \mathrm{Ext}^1_A(L,Y) \longrightarrow \cdots$$

$$\cdots \longrightarrow \mathrm{Ext}^n_A(L,Y) \longrightarrow \mathrm{Ext}^n_A(M,Y) \longrightarrow \mathrm{Ext}^n_A(N,Y) \longrightarrow \mathrm{Ext}^{n+1}_A(L,Y) \longrightarrow \cdots$$

where ∂ is a connecting homomorphism.

---

[6]Here $h(f)$ is short for $\mathrm{Hom}_A(X, f)$.

The following proposition shows a basic connection of projective modules with the functors Ext.

Proposition 1.7.3. (See [146, theorem 6.4.9].) *Suppose X,Y are A-modules. The following conditions are equivalent:*

1. *X is projective.*
2. $\mathrm{Ext}^n_A(X,Y) = 0$ *for all Y and all $n > 0$.*
3. $\mathrm{Ext}^1_A(X,Y) = 0$ *for all Y.*

Dually there is the following proposition:

Proposition 1.7.4 (See [146, theorem 6.4.10].) *Suppose X,Y are A-modules. The following conditions are equivalent:*

1. *Y is injective.*
2. $\mathrm{Ext}^n_A(X,Y) = 0$ *for all X and all $n > 0$.*
3. $\mathrm{Ext}^1_A(X,Y) = 0$ *for all X.*

1.8 Hereditary and Semihereditary Rings

A ring A is said to be **right (left) hereditary** if each right (left) ideal of A is a projective A-module. If a ring A is both right and left hereditary, it is called **hereditary**.

There are many other equivalent definitions of a right (left) hereditary ring. The following theorem gives some of these equivalent conditions.

Theorem 1.8.1 (See [146, theorem 5.5.6, proposition 6.5.4].) *The following conditions are equivalent for a ring A:*

a. *A is a right hereditary ring.*
b. *Any submodule of a right projective A-module is projective.*
c. *Any quotient of a right injective A-module is injective.*
d. $\mathrm{Ext}^k_A(X,Y) = 0$ *for all $k > 1$ and all right A-modules X and Y.*

Examples 1.8.2.

a. Any semisimple ring is hereditary.
b. Any principal ideal domain is hereditary.

Theorem 1.8.3. (Kaplansky's Theorem). (See, e.g. [146, theorem 5.5.1].) *If a ring is a right hereditary then any submodule of a free A-module is isomorphic to a direct sum of right ideals of A.*

A ring A is said to be **right (left) semihereditary** if each right (left) finitely generated ideal of A is a projective A-module. If a ring A is both right and left semihereditary, it is called **semihereditary**.

The following theorem gives some of other equivalent condition for a ring to be right (left) semihereditary.

Theorem 1.8.4. (See [146, corollary 5.5.10].) *A ring A is right (left) semihereditary if and only if every finitely generated submodule of a right (left) projective A-module is projective.*

Now consider the case of commutative rings.

Let A be an integral domain with a quotient field K. A **fractional ideal** of a ring A in the field K is an A-module $M \subset K$ such that $xM \subset A$ for some element $0 \neq x \in A$. In particular, any ordinary ideal $I \subset A$ is a fractional ideal.

A fractional ideal M of an integral domain A in the field K is called **invertible** if there exists a fractional ideal M^{-1} such that $MM^{-1} = A$.

Proposition 1.8.5. (See [146, proposition 8.3.2].) *A non-zero ideal I of an integral domain A invertible if and only if it is projective.*

Proposition 1.8.6. (See [146, proposition 8.3.3].) *Any invertible ideal I of an integral domain A is finitely generated.*

A commutative hereditary integral domain is called a **Dedekind domain**.

The following theorem gives other equivalent definition of a Dedekind domain.

Theorem 1.8.7. (See [146, theorem 8.3.4].) *Let A be an integral domain. Then the FSAE:*

1. *A is a Dedekind domain.*
2. *A is a Noetherian integrally closed domain in which any non-zero prime ideal is maximal.*

An A-module over a ring A is **torsion-free** if $mx = 0$ for $m \in M$, $x \in A$ implies $m = 0$. If A is an integral domain then a free A-module is torsion-free.

Proposition 1.8.8. (See [146, corollary 8.5.4].) *A finitely generated torsion-free module M over a Dedekind domain A is projective.*

A commutative semihereditary integral domain is called a **Prüfer domain.** Taking into account [146, theorem 8.6.1], proposition 1.8.5 and proposition 1.8.6 we obtain the following theorem which gives other equivalent definition of a Prüfer domain.

Theorem 1.8.9. (See [146, theorem 8.6.1].) *Let A be an integral domain. Then the FSAE:*

1. *A is a Prüfer domain.*
2. *Every finitely generated non-zero ideal of A is invertible.*
3. *Every finitely generated non-zero ideal of A is projective.*

4. *Every finitely generated torsion-free A-module is projective.*

Note that the statement (2) of this theorem was an original definition of a Prüfer domain given by Prüfer [267] in 1932. He also showed that to verify this condition it suffices to check that it holds for all two-generated ideals of a ring. Prüfer domains play a central role in the theory of non-Noetherian commutative algebra. The theorem which collects the most full equivalent definitions counting 22 equivalent conditions to the Prüfer domain notion is given in [19, theorem 1.1].

1.9 Semiperfect and Perfect Rings

A ring A is called **local** if it has a unique maximal right ideal. The following theorem gives various equivalent definitions for a ring to be local.

Proposition 1.9.1 (See [146, proposition 10.1.1].) *The following conditions are equivalent for a ring A with Jacobson radical R:*

1. *A is local;*
2. *R is the unique maximal right ideal in A;*
3. *All non-invertible elements of A form a proper ideal;*
4. *R is the set of all non-invertible elements of A;*
5. *A/R is a division ring.*

Note also the following important result about modules over local rings.

Proposition 1.9.2. (Kaplansky's Theorem). (See [146, theorem 10.1.8].) *Every finitely generated projective module over a local ring is free.*

A ring A with Jacobson radical R is called **semilocal** if $\bar{A} = A/R$ is a right Artinian ring.

One says that **idempotents may be lifted modulo an ideal** $\mathcal{I}$ of a ring A (or an ideal $\mathcal{I}$ admits **idempotent-lifting**) if from the fact that $g^2 - g \in \mathcal{I}$, where $g \in A$, it follows that there exists an idempotent $e^2 = e \in A$ such that $e - g \in \mathcal{I}$.

A semilocal ring A is called **semiperfect** if the Jacobson radical R of the ring A admits idempotent-lifting.

An idempotent $e \in A$ is called **local** if eAe is a local ring.

There are many different criteria for a ring to be semiperfect, some of them are given by the following theorem.

Theorem 1.9.3. (See, e.g. [146, theorems 10.3.7, 10.3.8, 10.4.8].) *The following conditions are equivalent for a ring A:*

1. *A is semiperfect.*

2. *The ring A can be decomposed into a direct sum of right ideals each of which has exactly one maximal submodule.*
3. *$1 \in A$ can be decomposed into a sum of a finite number of pairwise orthogonal local idempotents.*
4. *Any finitely generated right A-module has a projective cover.*
5. *Any cyclic right A-module has a projective cover.*

Projective modules over a semiperfect ring are described by the following important theorem.

Theorem 1.9.4. (See [146, theorem 10.4.10].) *Any indecomposable projective module over a semiperfect ring A is finitely generated, it is a projective cover of a simple A-module, and it has exactly one maximal submodule. There is a one-to-one correspondence between mutually nonisomorphic indecomposable projective A-modules $P_1, ..., P_s$ and mutually nonisomorphic simple A-modules which is given by the following correspondences: $P_i \mapsto P_i/P_iR = U_i$ and $U_i \mapsto P(U_i)$.*

An indecomposable projective right module over a semiperfect ring A is called a **principal right module**. A **principal left module** can be defined analogously.

Any principal right (left) A-module has the form eA (Ae), where e is a local idempotent.

The famous Krull-Remak-Schmidt theorem for semiperfect rings can be written in the following two forms:

Theorem 1.9.5. (**Krull-Remak-Schmidt Theorem**). (See [146, theorems 10.4.11, 10.4.13].)

1. *Let an A-module M have two different decompositions as a direct sum of submodules $M = \bigoplus_{i=1}^{n} M_i = \bigoplus_{i=1}^{m} N_i$ whose endomorphism rings are local. Then $m = n$ and there is a permutation τ of the numbers $i = 1, 2, ..., n$ such that $M_i \simeq N_{\tau(i)}$ $(i = 1, ..., n)$.*
2. *If the endomorphism ring $End_A(M)$ of an A-module M is semiperfect, then the module M has a unique decomposition up to isomorphism into a direct sum of indecomposable modules.*

As an immediate corollary we obtain the following result.

Proposition 1.9.6. (See [146, corollary 10.4.14].) *Any finitely generated projective right module over a semiperfect ring can be uniquely decomposed up to isomorphism into a direct sum of principal right modules.*

An ideal (right, left, two-sided) $\mathcal{J}$ is called **right** (resp. **left**) **T-nilpotent** if for any sequence $a_1, a_2, \ldots, a_n \ldots$ of elements $a_i \in \mathcal{J}$ there exists a positive integer k

such that $a_k a_{k-1}...a_1 = 0$ (resp. $a_1 a_2 \dots a_k = 0$). An ideal $\mathcal{J}$ is called **T-nilpotent** if it is right and left T-nilpotent.

The following theorem may be considered as some kind of generalization of Nakayama's lemma for arbitrary right modules.

Theorem 1.9.7. (See [146, corollary 10.5.1].) *For any right ideal $\mathcal{I}$ in a ring A the following conditions are equivalent:*

1. *$\mathcal{I}$ is right T-nilpotent;*
2. *A right A-module M satisfying the equality $M\mathcal{I} = M$ is equal to zero;*
3. *$M\mathcal{I}$ is a small submodule in M for any non-zero right A-module M;*
4. *$A^{\mathcal{N}}\mathcal{I}$ is a small submodule in $A^{\mathcal{N}}$, where $A^{\mathcal{N}}$ is a free module of countable rank.*

A ring A with Jacobson radical R is called **right (left) perfect** if A/R is semisimple and R is right (left) T-nilpotent. If A is both right and left perfect, R is called a **perfect ring**.

Theorem 1.9.8. (H. Bass [15]). (See [146, corollaries 10.5.3, 10.5.5].) *Let A be a ring with Jacobson radical R. Then the following are equivalent:*

1. *A is right perfect.*
2. *Every right A-module has a projective cover.*
3. *Every flat right A-module is projective.*
4. *A satisfies the descending chain condition on principal left ideals.*

The Jacobson radical of a semiperfect ring has a simple form of a two-sided Peirce decomposition which is described by the following proposition.

Proposition 1.9.9. (See [146, proposition 11.1.1].) *Let $A = P_1^{n_1} \oplus \cdots \oplus P_s^{n_s}$ be the decomposition of a semiperfect ring A into a direct sum of principal right A-modules and $1 = f_1 + \cdots + f_s$ the corresponding decomposition of the identity of A into a sum of pairwise orthogonal idempotents, i.e., $f_i A = P_i^{n_i}$. Then the Jacobson radical of the ring A has a two-sided Peirce decomposition of the following form:*

$$R = \begin{pmatrix} R_{11} & A_{12} & \dots & A_{1n} \\ A_{21} & R_{22} & \dots & A_{2n} \\ \vdots & \vdots & \ddots & \vdots \\ A_{n1} & A_{n2} & \dots & R_{nn} \end{pmatrix}, \qquad (1.9.10)$$

where $R_{ii} = \mathrm{rad}(A_{ii})$, $A_{ij} = f_i A f_j$ for $i, j = 1, ..., n$.

Let A be a semiperfect ring with Jacobson radical R. An idempotent $f \in A$ is said to be **canonical** if $\bar{f}\bar{A} = \bar{A}\bar{f} = M_n(\mathcal{D})$, where $\mathcal{D}$ is a division ring, and $\bar{f} = f + R$. Equivalently, f is a minimal central idempotent modulo R. A decomposition

$1 = f_1 + \cdots + f_s$ into a sum of pairwise orthogonal canonical idempotents is said to be a **canonical decomposition** of the identity of a ring A.

It is clear that the decomposition of the identity used in proposition 1.9.9 is a canonical decomposition of the identity of a ring A.

Lemma 1.9.11. (Annihilation lemma). (See [146, lemma 11.1.2].) *Let A be a semiperfect ring and $1 = f_1+\cdots+f_s$ a canonical decomposition of $1 \in A$. For every simple right A-module U_i and each f_j one has $U_i f_j = \delta_{ij} U_i$ for $i,j = 1,\dots,s$. Similarly, $f_j V_i = \delta_{ij} V_i$ for every simple left A-module V_i, each f_j, and $i,j = 1,\dots,s$.*

A semiperfect ring A is called **basic** if the quotient ring A/R, where R is the Jacobson radical of A, is a direct sum of division rings.

Lemma 1.9.12. (Q-Lemma). (See [146, lemma 11.1.3].) *Let A be a basic semiperfect ring, and let $1 = e_1 + \cdots + e_s$ be a decomposition of $1 \in A$ into a sum of mutually orthogonal local idempotents. The simple module $U_k = e_k A/e_k R$ (resp. $V_k = Ae_k/Re_k$) appears in the direct sum decomposition of the module $e_i R/e_i R^2$ (resp. $Re_i/R^2 e_i$) if and only if $e_i R^2 e_k$ (resp. $e_k R^2 e_i$) is strictly contained in $e_i R e_k$ (resp. $e_k R e_i$).*

1.10 Serial and Semidistributive Rings

A module is **uniserial** if the lattice of its submodules is a chain, i.e., the set of all its submodules is linearly ordered by inclusion. A module is **serial** if it is decomposed into a direct sum of uniserial submodules.

A ring is said to be **right (left) uniserial** if it is a right (left) uniserial module over itself, i.e., the lattice of right (left) ideals is linearly ordered. A ring is **right (left) serial** if it is a right (left) serial module over itself. A ring which is both a right and left serial ring is said to be a **serial ring.**

Recall that a module M is **finitely presented** if M is a quotient of a finitely generated free module with finitely generated kernel.

The following well known theorem characterizes serial rings in terms of finitely presented modules.

Theorem 1.10.1. (Drozd-Warfield Theorem, [67], [320]). (See, e.g. [146, theorem 13.2.1].) *For a ring A the following conditions are equivalent:*

1. *A is serial;*
2. *Any finitely presented right A-module is serial;*
3. *Any finitely presented left A-module is serial.*

Recall that $\mathcal{O}$ is a **discrete valuation ring**[7] if it can be embedded into a division ring D with discrete valuation ν such that

$$\mathcal{O} = \{x \in D^* : \nu(x) \geq 0\} \cup \{0\}.$$

The following proposition gives a description of right Noetherian uniserial rings.

Proposition 1.10.2. (See [146, proposition 13.3.1].) *A local right Noetherian ring $\mathcal{O}$ is serial if and only if it is either a discrete valuation ring or an Artinian uniserial ring.*

Let $\mathcal{O}$ be a discrete valuation ring with unique maximal ideal $\mathcal{M}$ and a division ring of fractions D. Denote by $H(\mathcal{O},m,n)$ the formal triangular matrix ring [8] of the form

$$H(\mathcal{O},m,n) = \begin{pmatrix} H_m(\mathcal{O}) & X \\ 0 & T_n(D) \end{pmatrix}, \tag{1.10.3}$$

where

$$H_m(\mathcal{O}) = \begin{pmatrix} \mathcal{O} & \mathcal{O} & \dots & \mathcal{O} \\ \mathcal{M} & \mathcal{O} & \dots & \mathcal{O} \\ \vdots & \vdots & \ddots & \vdots \\ \mathcal{M} & \mathcal{M} & \dots & \mathcal{O} \end{pmatrix}, \tag{1.10.4}$$

$T_n(D)$ is the ring of upper triangular $n \times n$ matrices over D, and $X = M_{m \times n}(D)$ is a set of all rectangular $m \times n$ matrices over D.

This ring, as was shown in [146], is a right Noetherian right hereditary serial ring.

A full description of serial right Noetherian rings is given by the following theorem.

Theorem 1.10.5. (See [146, theorem 13.4.3].) *Any serial right Noetherian ring is Morita equivalent*[9] *to a direct product of a finite number of rings of the following types:*

1. *Artinian serial rings;*
2. *Rings isomorphic to rings of the form $H_m(\mathcal{O})$;*
3. *Rings isomorphic to quotient rings of a ring $H(\mathcal{O},m,n)$.*
   Conversely, all rings of these forms are serial and right Noetherian.

A full description of serial right hereditary rings is given by the following theorem.

Theorem 1.10.6. (See [146, theorem 13.5.2].) *Any serial right hereditary ring A is Morita equivalent to a direct product of rings isomorphic to rings of upper*

---

[7] Other equivalent definitions of a discrete valuation ring is given in Chapter 3.

[8] For more about formal triangular matrix rings see section H2.6.

[9] Recall that two rings are said to be **Morita equivalent** if their categories of modules are equivalent. (See, e.g. [146, Section 10.2].)

triangular matrices over divisions rings, rings of the form $H_m(O)$ and rings of the form $H(O,m,n)$, where O is a discrete valuation ring.

The following theorem gives the description of serial semiprime and right Noetherian rings.

Theorem 1.10.7. (See [146, theorem 13.5.3].) *A serial semiprime and right Noetherian ring can be decomposed into a direct product of prime rings. A serial prime and right Noetherian ring is also left Noetherian and hereditary. In the Artinian case such a ring is Morita equivalent to a division ring and in the non-Artinian case it is Morita equivalent to a ring isomorphic to $H_m(O)$, where O is a discrete valuation ring. Conversely, all such rings are prime hereditary and Noetherian serial rings.*

Recall that a ring A is called **prime** if a product of any two non-zero ideals of A is not equal to zero. The structure of two-sided Noetherian serial rings yields the following theorem.

Theorem 1.10.8. (See [146, theorem 12.3.8].) *A serial Noetherian ring can be decomposed into a finite direct product of an Artinian serial ring and a finite number of semiperfect Noetherian prime hereditary rings. Conversely, all such rings are serial and Noetherian.*

Recall that a module M is **distributive** if

$$K \cap (L + N) = K \cap L + K \cap N$$

for all submodules K, L, N of M. A module is **semidistributive** if it is a direct sum of distributive modules. A ring A is said to be **right (left) semidistributive** if the right (left) regular module A_A $({}_AA)$ is semidistributive. A right and left semidistributive ring is called **semidistributive**. A semiperfect semidistributive ring will be written as an SPSD-ring.

In the study of the structure of SPSD-rings an important role is played by the following theorems, which give criteria for a ring being right (left) semidistributive.

Theorem 1.10.9. (A.Tuganbaev). (See [146, theorem 14.2.1].) *A semiperfect ring A is right (left) semidistributive if and only if for any local idempotents $e, f \in A$ the set eAf is a uniserial right fAf-module (left eAe-module).*

Theorem 1.10.10. (Reduction theorem for $SPSD$-rings). (See, e.g. [146, corollary 14.2.3].) *Let A be a semiperfect ring, and let $1 = e_1 + \cdots + e_n$ be a decomposition of $1 \in A$ into a sum of mutually orthogonal local idempotents. The ring A is right (left) semidistributive if and only if for any idempotents e_i and e_j of the above decomposition the ring $(e_i + e_j)A(e_i + e_j)$ is right (left) semidistributive for $i \neq j$.*

Proposition 1.10.11. (See [146, corollary 14.2.4].) *Let A be a Noetherian $SPSD$-ring, and let $1 = e_1 + \cdots + e_n$ be a decomposition of the identity $1 \in A$*

into a sum of mutually orthogonal local idempotents. If $A_{ij} = e_i A e_j$ and R_i is the Jacobson radical of the ring A_{ii}, then $R_i A_{ij} = A_{ij} R_j$ for $i, j = 1, \ldots, n$.

Proposition 1.10.12. (See [146, proposition 14.4.7].) *Let A be an $SPSD$-ring. Then for any non-zero idempotent $e = e^2 \in A$ the ring eAe is also SPSD.*

The next theorem states a connection between semidistributive and Artinian rings.

Theorem 1.10.13. (See [146, theorem 14.1.6].) *A semiprimary right (left) semidistributive ring is a right (left) Artinian.*

1.11 Classical Rings of Fractions

An element y of a ring A is called **regular** if $ay \neq 0$ and $ya \neq 0$ for any non-zero element $a \in A$. We denote by $C_A(0)$ the set of all regular elements of a ring A.

Let A be a subring of a ring Q. The ring Q is said to be a **classical right ring of fractions** (or **classical right ring of quotients**) of the ring A if and only if the following conditions hold:

1. All regular elements of the ring A are invertible in the ring Q;
2. Each element of the ring Q has the form ab^{-1}, where $a, b \in A$ and $b \in C_A(0)$.

In this case the ring A is said to be a **right order** in Q.

Analogously one can define a **classical left ring of fractions** and a **left order** in Q. In the case when a right classical ring of fractions and a left classical ring of fractions are the same, this common ring is called a **classical ring of fractions** (or a **classical quotient ring**). If A is both a right and left order in Q then A is called an **order** in Q.

The (**right**) **Ore condition:** *Let A be a ring with the nonempty set $C_A(0)$ of all regular elements in A. For any element $a \in A$ and any element $r \in C_A(0)$ there exists a regular element $y \in C_A(0)$ and an element $b \in A$ such that $ay = rb$.*

Analogously one can define the left Ore condition.

Definition 1.11.1. A ring A satisfying the right (left) Ore condition is called a **right (left) Ore ring**. A ring which is both a right and left Ore ring is called an **Ore ring**. If, in addition, the ring is a domain, then it is called an **Ore domain**.

For any subset S of a ring A we denote with

$$\text{r.ann}_A(S) = \{x \in A \ : \ Sx = 0\},$$

the **right annihilator** of S, and with

$$\text{l.ann}_A(S) = \{x \in A \ : \ xS = 0\},$$

the **left annihilator** of S. Note, that a right (left) annihilator is always a right (left) ideal of A.

We say that a ring A satisfies the **a.c.c. (d.c.c.) on right annihilators** if A contains no infinite properly ascending (descending) chain of right ideals of the form $\mathrm{r.ann}_A(S)$, where S is a subset of A.

Theorem 1.11.2. (See [146, theorem 9.1.1].) *A ring A has a classical right ring of fractions if and only if it satisfies the right Ore condition, i.e., A is a right Ore ring.*

A ring A is said to be a **right Goldie ring** if

1. A satisfies the ascending chain condition on right annihilators;
2. A contains no infinite direct sum of non-zero right ideals.

Analogously one can define a **left Goldie ring.** A ring A, which is both a right and left Goldie ring, is called a **Goldie ring.**

Theorem 1.11.3. (Goldie's Theorem). (See [146, theorem 9.3.9].) *A ring A has a classical right ring of fractions which is a semisimple ring if and only if A is a semiprime right Goldie ring.*

Theorem 1.11.4 (A.W. Goldie, L.Lesieur-R.Croisot). (See [146, theorem 9.3.11].) *A ring A is a right order in a simple ring Q if and only if A is a prime right Goldie ring.*

Theorem 1.11.5. (See [146, theorem 13.2.2].) *Every serial ring satisfies the Ore condition, i.e., every serial ring has a classical ring of fractions which is also a serial ring.*

1.12 Quivers of Algebras and Rings

Let A be a semiperfect right Noetherian ring, and let $P_1, \ldots, P_s$ be all pairwise nonisomorphic principal right A-modules. Consider the projective cover of $R_i = P_i R$ $(i = 1, \ldots, s)$, which will be denoted by $P(R_i)$. Let $P(R_i) = \bigoplus_{j=1}^{s} P_j^{t_{ij}}$. Assign to the principal modules $P_1, \ldots, P_s$ points $1, .., s$ in the plane and join the point i with the point j by t_{ij} arrows. The thus constructed graph is called the **right quiver** (or simply the **quiver**) of the semiperfect right Noetherian ring A and denoted by $Q(A)$. Note that $Q(A) = Q(A/R^2)$.

Analogously, one can define the left quiver $Q'(A)$ of a left Noetherian semiperfect ring. If A is a Noetherian semiperfect ring then it is possible to construct both its right quiver and its left quiver.

The quiver $Q(A)$ of a ring A is called **connected** if it cannot be represented in the form of a union of two non-empty disjoint subsets of vertices Q_1 and Q_2 which are not connected by any arrows.

Theorem 1.12.1. (See [146, theorem 11.1.9].) *The following conditions are equivalent for a semiperfect Noetherian ring A:*

a. *A is an indecomposable ring;*
b. *A/R^2 is an indecomposable ring;*
c. *The quiver of A is connected.*

Theorem 1.12.2. (See [146, theorem 12.1.2] and [146, corollary 12.1.3].)

1. *The quiver of a serial ring A is a disconnected union of cycles and chains.*
2. *The quiver of a serial Noetherian indecomposable ring is either a cycle or a chain.*

A quiver without multiple arrows and multiple loops is called a **simply laced quiver**.

Theorem 1.12.3. (See [146, theorem 14.3.1].) *The quiver $Q(A)$ of an SPSD-ring A is simply laced. Conversely, for any simply laced quiver Q there exists an SPSD-ring A such that $Q(A) = Q$.*

For a semiperfect ring A with prime radical $Pr(A)$ one can define the prime quiver $PQ(A)$ in the following way.

Write $\mathcal{J} = Pr(A)$. Consider the quotient ring $\bar{A} = A/\mathcal{J} = \bar{A}_1 \times \cdots \times \bar{A}_t$, where all the rings $\bar{A}_1, \ldots, \bar{A}_t$ are indecomposable and $\bar{1} = \bar{f}_1 + \cdots + \bar{f}_t \in \bar{A}$ is the corresponding decomposition of $\bar{1} \in \bar{A}$ into a sum of pairwise orthogonal central idempotents. Put $W = \mathcal{J}/\mathcal{J}^2$ and represent the idempotents $\bar{f}_1, \ldots, \bar{f}_t$ by the corresponding points $1, \ldots, t$. Join the point i with the point j by an arrow if and only if $\bar{f}_i W \bar{f}_j \neq 0$. The thus obtained finite directed graph $Q(A, \mathcal{J})$ is called the **prime quiver** of A and denoted by $PQ(A)$. The set of points $\{1, 2, ..., t\}$ is the set of vertices and the set of arrows between these points is the set of arrows of the quiver $PQ(A)$.

CHAPTER 2

Basic General Constructions of Groups and Rings

In mathematics, specifically in the area of abstract algebra, it is often interesting to construct new objects using the objects already known. In group theory, ring theory, Lie algebra theory there exist a variety of different constructions such as crossed and skew products which are very important, for instance as sources of various examples and counterexamples.

Some of the main ring constructions, such as a finite direct product of rings, group rings, matrix rings, path algebras and graded algebras were considered in [146], [147]. This chapter presents the definitions and properties of some more basic general constructions of groups and rings, such as direct product, semidirect product, direct sum, crossed product of groups and rings, polynomial and skew polynomial rings, formal power and skew formal power series rings, Laurent polynomial and Laurent formal power series rings, generalized matrix rings, formal triangular matrix rings, and G-graded rings.

2.1 Direct and Semidirect Products

One of the basic general constructions in mathematics is the direct product. The idea behind it is due to R. Descartes; therefore the direct product is also called the **Cartesian product**. The Cartesian product of sets is defined as follows.

Let I be an index set, and $\{X_i\}_{i\in I}$, an arbitrary family of sets indexed by I. The **direct product** (or **Cartesian product**) X of $\{X_i\}_{i\in I}$ is the set of functions

$$f : I \to \bigcup_{i\in I} X_i$$

where $\bigcup$ denotes disjoint union, such that $f(i) \in X_i$ for each $i \in I$. The direct product X is denoted by $\prod_{i\in I} X_i$. If $I = \{1,2,\ldots,n\}$ is a finite index set then the direct product is denoted by $\prod_{i=1}^{n} X_i$ or $X_1 \times X_2 \times \cdots \times X_n$.

For any arbitrary factor X_i of the direct product $X = \prod_{i \in I} X_i$ there exists a natural projection $\pi_i : X \to X_i$ defined by $\pi_i(f) = f(i)$. The main property of the direct product is as follows:

(i) If $f, g \in X = \prod_{i \in I} X_i$ then $f = g$ if and only if $\pi_i(f) = \pi_i(g)$ for each $i \in I$.

The set X and the family of projections $\{\pi_i\}_{i \in I}$ have the following universal property:

Proposition 2.1.1 (Universal property of the direct product). *Let $X = \prod_{i \in I} X_i$ be the direct product of sets X_i, $i \in I$, with projections $\pi_i : X \to X_i$. For every family of mappings $g_i : Y \to X_i$ there exists a unique mapping $\varphi : Y \to X$ such that $g_i = \pi_i(\varphi)$, i.e., such that the following diagram*

$$\begin{array}{ccc} Y & \xrightarrow{\varphi} & X \\ & {\scriptstyle g_i}\searrow & \downarrow{\scriptstyle \pi_i} \\ & & X_i \end{array}$$

commutes for every $i \in I$.

Proof. Let $y \in Y$ and $g_i(y) = x_i \in X_i$. Define a function $f : I \to \bigcup_{i \in I} X_i$ by $f(i) = x_i = g_i(y)$. Then by the properties of the direct product the natural projection π_i satisfies $\pi_i(f) = f(i) = g_i(y)$ for all $i \in I$. Set $\varphi(y) = f \in X$. Then $\pi_i\varphi(y) = \pi_i(f) = f(i) = g_i(y)$ for all $y \in Y$ and $i \in I$. So $g_i = \pi_i\varphi$ for all $i \in I$. If there exists another mapping $\psi : Y \to X$ such that $g_i = \pi_i\psi$ for all $i \in I$, then $\pi_i\varphi(y) = \pi_i\psi(y) = g_i(y)$ for all $y \in Y$ and all $i \in I$. Therefore all components of the elements $\varphi(y), \psi(y) \in X$ are equal and so $\varphi = \psi$ for all $y \in Y$, i.e., $\varphi = \psi$ as required. □

This universal property allows the possibility to define the Cartesian product in categorial terms in the following way:

Let I be an index set and $\{X_i\}_{i \in I}$ an arbitrary family of objects in a category $\mathfrak{K}$. An object $X \in \mathfrak{K}$ together with morphisms $\pi_i : X \longrightarrow X_i$, $i \in I$, is called the **product** of the family of objects $\{X_i\}_{i \in I}$, if for every family of morphisms $g_i : Y \longrightarrow X_i$ there exists a unique morphism $\varphi : Y \longrightarrow X$ such that $g_i = \pi_i\varphi$ for every $i \in I$. The product X is denoted by $\prod_{i \in I} X_i$, and the morphisms π_i are called **product projections** (or **canonical projections**).

In categories with zero morphisms, for any product $X = \prod_{i \in I} X_i$ there exist uniquely defined morphisms $\sigma_i : X_i \to X$, $i \in I$, such that $\sigma_i\pi_i = 1_{X_i}$, $\sigma_i\pi_j = 0$ for $i \neq j$. σ_i is defined by the family of morphisms $id : X_i \to X_i$, $o : X_j \to X_i$ if $j \neq i$. If the set I is finite, $|I| = n < \infty$ and the category $\mathfrak{K}$ is additive, then $\pi_1\sigma_1 + \cdots + \pi_n\sigma_n = 1_X$.

The concept of the product of a family of objects is dual to that of a coproduct of a family of objects.

An object $X \in \mathfrak{K}$ together with morphisms $\sigma_i : X_i \to X$, $i \in I$, is called the **coproduct** of the family of objects $\{X_i\}_{i\in I}$, if for every family of morphisms $g_i : X_i \to Y$ there exists a unique morphism $\varphi : X \to Y$ such that $g_i = \varphi\sigma_i$, i.e., the following diagram

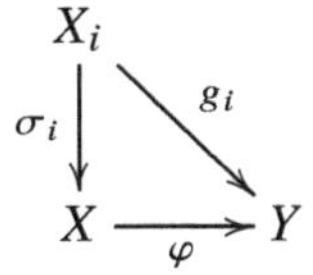

commutes for every $i \in I$. The coproduct X is denoted by $\coprod\limits_{i\in I} X_i$, and the morphisms σ_i are called **embeddings of the coproduct**.

In an Abelian category[1] the coproduct is often called the **direct sum** of the family $\{X_i\}_{i\in I}$, and denoted by $\bigoplus\limits_{i\in I} X_i$.

Since a category of modules is an Abelian category, the coproduct of the family of modules is the direct sum of modules.

In categories with zero morphisms, for any coproduct $X = \coprod\limits_{i\in I} X_i$ there exist uniquely defined morphisms $\pi_i : X \to X_i$, $i \in I$, such that $\sigma_i\pi_i = 1_X$, $\sigma_i\pi_j = 0$ for $i \neq j$. If the set I is finite, $|I| = n < \infty$, and the category $\mathfrak{K}$ is additive, then $\pi_1\sigma_1 + \cdots + \pi_n\sigma_n = 1_X$ and the product $\prod\limits_{i=1}^{n} X_i$ is equal to their coproduct $\coprod\limits_{i=1}^{n} X_i$. In this case this common object is called the **direct sum** and denoted by $\bigoplus\limits_{i=1}^{n} X_i$. Point out that this is not the case in a category of rings.

The direct product of a finite number of rings were considered in [146]. Now consider this construction for an infinite number of rings.

Definition 2.1.2. Let $\{A_i\}_{i\in I}$, be a family of rings. The **direct product** of the rings $\{A_i\}_{i\in I}$ is the Cartesian product $A = \prod\limits_{i\in I} A_i$ with operations of addition and multiplication defined componentwise:

$$(f + g)(i) = f(i) + g(i)$$

$$(fg)(i) = f(i)g(i)$$

for any $f, g \in A$ and for each $i \in I$.

---

[1]Recall that a category C is **Abelian** if its morphisms and objects can be added, there exist kernels and cokernels, and moreover, every monomorphism is a kernel of some morphism and every epimorphism is a cokernel of some morphism.

We can also consider any element $f \in A$ as a vector $(a_i) = (\ldots, a_i, \ldots)$, where $a_i = f(i) \in A_i$. Writing $f(i) = a_i \in A_i$ and $g(i) = b_i$ one can rewrite the ring operations in the usual way:

$$(a_i + b_i) = (a_i) + (b_i)$$

$$(a_i b_i) = (a_i)(b_i)$$

If each A_i is a ring with identity 1_i then A is also a ring with identity 1 such that $1(i) = 1_i$.

The natural projections $\pi_i : A \to A_i$ with $\pi_i f = f(i) = a_i$ are ring morphisms, while the injections $\sigma_i : A_i \to A$ with $\sigma_i f = (\ldots, 0, a_i, 0, \ldots)$ preserve addition and multiplication, but not the identity, so they are not ring morphisms. For the same reasons the images of the σ_i are not subrings, they are ideals in A.

The direct product of rings $A = \prod_{i \in I} A_i$ with projections $\pi_i : A \to A_i$ has the following universal property:

Proposition 2.1.3 (Universal property). *Let $A = \prod_{i \in I} A_i$ be the direct product of the rings A_i, $i \in I$, with projections $\pi_i : A \to A_i$. If B is any ring and $g_i : B \to A_i$ is a ring morphism for all $i \in I$, then there exists a unique ring morphism $\varphi : B \to A$ such that $\pi_i \circ \varphi = g_i$, i.e., the following diagram commutes*

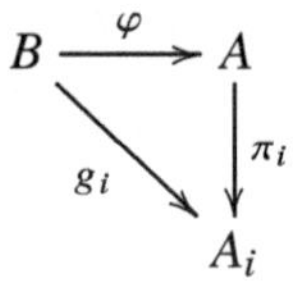

for each $i \in I$.

Proof. By proposition 2.1.1 there exists a unique mapping $\varphi : B \to A$ such that $\pi_i \varphi = g_i$. So it only remains to show that φ is a ring morphism which follows from the fact that g_i and π_i are ring morphisms for each i. Indeed, for any $b_1, b_2 \in B$ one has

$$\pi_i \varphi(b_1 + b_2) = g_i(b_1 + b_2) = g_i(b_1) + g_i(b_2) = \pi_i \varphi(b_1) + \pi_i \varphi(b_2) = \pi_i(\varphi(b_1) + \varphi(b_2))$$

whence $\varphi(b_1 + b_2) = \varphi(b_1) + \varphi(b_2)$. Analogously one obtains that $\varphi(b_1 b_2) = \varphi(b_1)\varphi(b_2)$; also $\varphi(1) = 1$, i.e., φ is a ring morphism. □

Remark 2.1.4. The direct product is sometimes called the **complete direct product** to distinguish it from the discrete direct product (or direct sum).

All groups above we will assumed to be multiplicative[2].

---

[2] 'Multiplicative' as used here has no technical meaning. It simply emphasizes that the group multiplication is written multiplicatively.

There are two notions for the direct product of groups, the inner (or internal) direct product and the outer (or external) direct product.

Definition 2.1.5. A group $G = N \times H$ is the **outer direct product** of two groups of N and H, if G is a set of ordered pairs (n, h) such that $n \in N$ and $h \in H$ with operation of multiplication defined by

$$(n,h)(n_1,h_1) = (nn_1, hh_1).$$

In this case there are two natural embeddings $\iota_1 : N \longrightarrow G$ and $\iota_2 : H \longrightarrow G$. The subgroups $\iota_1(N) = N \times 1 \simeq N$ and $\iota_2(H) = 1 \times H \simeq H$ are normal subgroups in G. Moreover, these subgroups, $\iota_1(N)$ and $\iota_2(H)$, commute, i.e., if $x = (n, 1)$, $y = (1, h)$ then $xy = yx$.

Definition 2.1.6. Let N, H be normal subgroups of a group G. Then G is the **internal direct product** of N and H if

1. $G = NH$;
2. $N \cap H = \{1\}$.

Note that here $NH = \{nh \ : \ n \in N, h \in H\}$. In general NH is not subgroup of G. However if either N or H is a normal subgroup in G, then NH is a subgroup in G.

Two definitions 2.1.5 and 2.1.6 are closely connected with each other. Namely, it is easy to show that the groups determined by these definitions are the same up to isomorphism.

Proposition 2.1.7. *Let G be the internal direct product of two normal subgroups N and H. Then subgroups N and H commute with each other and there is a group isomorphism $\varphi : G \longrightarrow N \times H$ given by $\varphi(nh) = (n, h)$.*

The construction considered below is a generalization of the direct product of two groups. We consider the case when N is a normal subgroup of G but a subgroup H is not necessarily normal in G.

Definition 2.1.8. Let H be a subgroup of a group G, and N a normal subgroup of G. If $G = NH$ and $N \cap H = \{1\}$ then the group G is called the **internal semidirect product** of the subgroup H by N and it is denoted by $G = N \rtimes H$.

Remark 2.1.9. In the definition of the semidirect group the subgroups N and H are not entered symmetrically, so the notation $G = N \rtimes H$ is not symmetrical. If we want to change the place of the groups we use the other notation: $G = H \ltimes N$. If H is also a normal subgroup of G then the semidirect product $G = N \rtimes H$ becomes the direct product $G = N \times H$ of subgroups. Since any subgroup of an Abelian group is normal, for Abelian groups the semidirect product is always the direct product.

Proposition 2.1.10. *If $G = N \rtimes H$ is the semidirect product of subgroups N and H then $NH = HN$ and every element of G can be written as a unique product nh where $n \in N$ and $h \in H$.*

Proof. If $nh = n_1h_1$ then $n_1^{-1}n = h_1h^{-1} \in N \cap H = \{1\}$, which implies that $n = n_1$ and $h = h_1$. Since N is a normal subgroup of G then NH and HN are subgroups in G, so $HN \subseteq G = NH$. On the other hand, each element $nh = nhn^{-1}n = h_1n \in HN$, since N is normal in G. So $NH \subseteq HN$, which implies $NH = HN$. □

Let $G = N \rtimes H$, and consider two elements in G, $g_1 = n_1h_1$, $g_2 = n_2h_2$. Since N is a normal subgroup in G, one may multiply these elements by the rule:

$$g_1g_2 = (n_1h_1)(n_2h_2) = (n_1h_1n_2)(h_1^{-1}h_1)h_2 = n_1(h_1n_2h_1^{-1})(h_1h_2) =$$

$$= (n_1\varphi(h_1)(n_2))(h_1h_2) = n_3h_3 \in NH = G,$$

where $\varphi(h) \in \mathrm{Aut}(N)$ given by the conjugation $\varphi(h)(n) = hnh^{-1}$. So the multiplication in G is not given by 'componentwise', it is 'twisted' by the conjugation in H.

Consider now some generalization of this construction.

Definition 2.1.11. Let N and H be two groups with identity elements e_N and e_H together with a group homomorphism $\varphi : H \to \mathrm{Aut}(N)$. The set of elements of the Cartesian product $N \times H$ together with multiplication defined by the formula

$$(n_1, h_1) * (n_2, h_2) = (n_1\varphi(h_1)(n_2), h_1h_2) \tag{2.1.12}$$

for all $h_1, h_2 \in H$ and $n_1, n_2 \in N$ forms a multiplicative group which is denoted by $N \rtimes_\varphi H$ and called the **external semidirect product** of N by H with respect to φ. The identity element of $N \rtimes_\varphi H$ is (e_N, e_H) and the inverse element of (n, h) is $(\varphi(h^{-1})(n^{-1}), h^{-1})$.

It is easy to show that the subsets $N' = \{(n, e_H) : n \in N\}$ and $H' = \{(e_N, h) : h \in H\}$ are subgroups in G that are isomorphic to N and H. They satisfy conditions $G = N'H'$ and $N' \cap H' = \{1\}$ and N' is normal in G. So the group G we have constructed is the internal semidirect product $N' \rtimes H'$ of N' by H'.

Remark 2.1.13. If φ is the trivial homomorphism for which $\varphi(h)(n) = id_N(n) = n$ for all $h \in H$ and $n \in N$, then the semidirect product becomes the direct product of groups, i.e., $N \rtimes_\varphi H = N \times H$.

Suppose that a group G is the internal semidirect product of a normal subgroup N and a subgroup H, i.e., every element of G can be written uniquely in the form $g = nh$, where $n \in N$ and $h \in H$. Let $\varphi : H \to \mathrm{Aut}(N)$ be the group homomorphism given by

$$\varphi(h)(n) = hnh^{-1}$$

for all $h \in H$ and all $n \in N$. Then $G \cong N \rtimes_{\varphi} H$. The isomorphism $\alpha : G \to N \rtimes_{\varphi} H$ is given by $\alpha(g) = \alpha(nh) = (n, h)$. The multiplication in G is given by

$$(n_1h_1)(n_2h_2) = (n_1\varphi(h_1)(n_2))(h_1h_2) = (n_1(h_1n_2h_1^{-1}))(h_1h_2). \tag{2.1.14}$$

Example 2.1.15. Let

$$D_{2n} = \{a, b \ : a^2 = e, \ b^n = e, \ aba^{-1} = b^{-1}\}.$$

This is the dihedral group which is the group of symmetries of a regular n-gon in the plane. If C_n is the cyclic group of order n, then $D_{2n} \cong C_n \rtimes_{\varphi} C_2$, where $\varphi(a)(b) = aba^{-1} = b^{-1}$.

Use now the language of short exact sequences of groups for the equivalent definition of the semidirect product of groups.

Definition 2.1.16. A sequence of groups

$$1 \longrightarrow N \xrightarrow{f} G \xrightarrow{\alpha} H \longrightarrow 1 \tag{2.1.17}$$

is called a **short exact sequence** if the group homomorphism f is injective, the group homomorphism g is surjective, and $\mathrm{Im}(f) = \mathrm{Ker}(\alpha)$.

A short exact sequence (2.1.17) is called **split** if there exists a homomorphism $\beta : H \longrightarrow G$ such that $\alpha\beta = id_H$.

If N is a normal subgroup in a group G, then there is an exact sequence of groups

$$1 \longrightarrow N \longrightarrow G \longrightarrow H = G/N \longrightarrow 1 \tag{2.1.18}$$

This sequence is split if G is isomorphic to the direct product of groups $N \times H$. The inverse statement is not true in general, i.e., if the sequence (2.1.17) is exact and split, then the group G is not necessary isomorphic to a direct product of groups N and H, since in general H is not isomorphic to a normal subgroup in G. In this case G is a semidirect product, which shows the next statement considered as an equivalent definition for the semidirect product of groups.

Proposition 2.1.19. *A group G is isomorphic to the semidirect product of two groups N and H if and only if there exists an exact split sequence of groups* (2.1.17).

Proof. Suppose G is the semidirect product of subgroups N and H, then by definition N is a normal subgroup in G, $G = NH = HN$ and $N \cap H = \{1\}$. Consider the canonical projection $\pi : G \longrightarrow G/N$, and the canonical injection $\iota : H \longrightarrow G$. Then the group homomorphism $\psi = \iota\pi$ is injective since $N \cap H = \{1\}$. On the other hand, ψ is a surjection. Indeed, since π is a surjection, for each $gN \in G/N$ there exists $x \in G$ such that $\pi(x) = gN$. If $x = hn$ with $h \in H$ and $n \in N$, then $\psi(h) = \pi\iota(h) = \pi(h) = hN = hnN = gN$, so $gN \in \mathrm{Im}(\psi)$, i.e., ψ is a group isomorphism. Therefore one obtains the exact sequence

$$1 \longrightarrow N \xrightarrow{f} G \xrightarrow{\pi} G/N \longrightarrow 1 \tag{2.1.20}$$

with $G/N \simeq H$. Setting $\beta = \iota\psi^{-1}$ we obtain that $\pi\beta = \pi\iota\psi^{-1} = \psi\psi^{-1} = id_{G/N}$, i.e., the sequence (2.1.20) is exact and split.

Inversely, suppose we have an exact split sequence (2.1.17) and there is a group homomorphism $\beta : H \longrightarrow G$ such that $\alpha\beta = id_H$. Let $x \in G$. Consider elements $z = \beta\alpha(x) \in G$ and $y = xz^{-1} = x(\beta\alpha(x))^{-1} \in G$. Then $\alpha(z) = \alpha(\beta\alpha(x)) = \alpha(x)$, $\alpha(y) = \alpha(xz^{-1}) = \alpha(x)(\alpha(z))^{-1} = 1$, which implies $y \in \mathrm{Ker}(\alpha) = \mathrm{Im}(f)$. By construction $z \in \mathrm{Im}(\beta)$. So any element $x \in G$ can be written in the form $x = yz$, where $y \in \mathrm{Im}(f)$, $z \in \mathrm{Im}(\beta)$. Let $g \in \mathrm{Im}(f) \cap \mathrm{Im}(\beta)$, then $g = f(n) = \beta(h)$, and $\alpha(g) = \alpha f(g) = 1$. On the other hand $\alpha(g) = \alpha\beta(h) = h$, which implies $h = 1$ and so $g = 1$. Therefore $G = (\mathrm{Im}(f))(\mathrm{Im}(\beta))$ and $\mathrm{Im}(f) \cap \mathrm{Im}(\beta) = \{1\}$. Since $\mathrm{Im}(f) = \mathrm{Ker}(\alpha)$ is a normal subgroup in G, we obtain that $G = \mathrm{Im}(f) \rtimes \mathrm{Im}(\beta)$. □

Remark 2.1.21. In the case of modules we obtain some more from the existence of exact split sequence. Namely, if there is an exact split sequence of modules over a ring A:

$$0 \longrightarrow N \longrightarrow M \longrightarrow G/N \longrightarrow 0, \tag{2.1.22}$$

then $M \simeq N \oplus G/N$. It is obvious because for Abelian groups there are no distinction between the semidirect product and the direct product.

2.2. Group Rings, Skew Group Rings and Twisted Group Rings

Let A be an associative ring, G a monoid under multiplication. The **monoid ring** of the monoid G over the ring A is the set $A[G]$ of all formal finite sums of the form

$$\sum_{g \in G} a_g g, \tag{2.2.1}$$

where $a_g \in A$, with operations being defined by the formulas:

$$\sum_{g \in G} a_g g + \sum_{g \in G} b_g g = \sum_{g \in G} (a_g + b_g) g, \tag{2.2.2}$$

$$(\sum_{g \in G} a_g g)(\sum_{g \in G} b_g g) = \sum_{h \in G} (\sum_{\substack{xy=h \\ x,y \in G}} a_x b_y) h. \tag{2.2.3}$$

(All sums in these formulas are finite.) Any element of $A[G]$ is uniquely written in the form (2.2.1), i.e., $\sum\limits_{g \in G} a_g g = \sum\limits_{g \in G} b_g g$ iff $a_g = b_g$ for all $g \in G$.

It is easy to verify that $A[G]$ is an associative ring with the identity $1_A e_G$, where 1_A is the identity of A and e_G is the identity of G.

If G is a monoid then there are two natural canonical embeddings: a ring A into the ring $A[G]$ via the ring homomorphism $i : A \longrightarrow A[G]$ defined by $a \mapsto a e_G$, and a monoid G into the monoid of the group ring $A[G]$ via the monoid homomorphism $\alpha : G \longrightarrow A[G]$ such that $\alpha(g) = 1_A g$. In this case from the rule of multiplication (2.2.3) it follows that $i(a)\alpha(g) = \alpha(g)i(a)$ for all $a \in A$ and $g \in G$. After identification $i(a)$

with a and $\alpha(g)$ with g one can assume that $A \hookrightarrow A[G]$ and $G \hookrightarrow A[G]$. Moreover, we will assume that $ag = ga$ for all $a \in A$ and $g \in G$.

There is a natural way to turn a monoid ring $A[G]$ into a left A-module setting $b\alpha = \sum_{g\in G}(ba_g)g$ for any $\alpha = \sum_{g\in G} a_g g \in A[G]$ and $b \in A$. In this case the set of elements $\{1\cdot g\}_{g\in G}$ form a bases of $A[G]$ over A, and $A[G]$ becomes a free A-module. Since $ag = ga$, in a similar way we can also turn $A[G]$ into a right A-module. So $A[G]$ is a $(A$-$A)$-bimodule.

The ring $A[G]$ is commutative if and only if both A and G are commutative. If G is a semigroup, the same construction yields a **semigroup ring** $A[G]$ of the semigroup G over a ring A. If G is a group, then $A[G]$ is called the **group ring** of G over A. If $A = K$ is a field and G is a group, then the group ring $K[G]$ is called the **group algebra** of G over K.

The monoid group $A[G]$ is defined uniquely by a ring A and a monoid G by the following universal property.

Proposition 2.2.4 (Universal property). *Given rings A, S and a monoid G with natural embeddings $i : A \longrightarrow A[G]$ and $\alpha : G \longrightarrow A[G]$. If $\varphi : A \longrightarrow S$ is a ring homomorphism, and $\beta : G \longrightarrow S$ is a monoid homomorphism such that $\varphi(a)\beta(g) = \beta(g)\varphi(a)$ for all $a \in A$ and $g \in G$, then there exists a unique ring homomorphism $\psi : A[G] \longrightarrow S$ such that $\psi i = \varphi$ and $\psi\alpha = \beta$, i.e., the diagram*

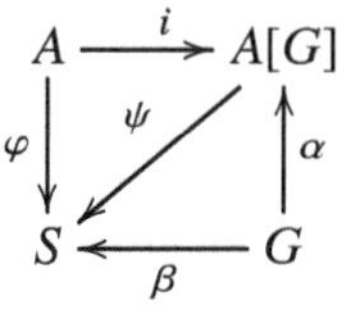

is commutative.

Proof. Given $\varphi : A \longrightarrow S$ and $\beta : G \longrightarrow S$, consider $\psi : A[G] \longrightarrow S$ by

$$\psi(u) = \sum_{g\in G} \varphi(a_g)\beta(g) \quad \text{where} \quad u = \sum_{g\in G} a_g g \in A[G]. \tag{2.2.5}$$

If $v = \sum_{g\in G} b_g g \in A[G]$, then $uv = \sum_{h\in G}(\sum_{\substack{xy=h\\x,y\in G}} a_x b_y)h$.

Now taking into account that φ is a ring homomorphism, β is a monoid homomorphism, and $\varphi(a)\beta(g) = \beta(g)\varphi(a)$ for all $a \in A$ and all $g \in G$, we obtain:

$$\psi(uv) = \sum_{h\in G} \varphi(\sum_{\substack{xy=h\\x,y\in G}} a_x b_y)\beta(h) = \sum_{h\in G}(\sum_{\substack{xy=h\\x,y\in G}} \varphi(a_x)\varphi(b_y))\beta(h) =$$

$$= \sum_{x\in G} \varphi(a_x)\beta(x) \cdot \sum_{y\in G} \varphi(b_y)\beta(y) = \psi(u)\psi(v).$$

The other conditions of the statement are verified similarly. □

Corollary 2.2.6. *Given a ring A and monoids G, G_1 with natural embeddings $\iota : A \longrightarrow A[G]$, $\iota_1 : A \longrightarrow A[G_1]$, $\alpha : G \longrightarrow A[G]$, $\alpha_1 : G_1 \longrightarrow A[G_1]$. If $f : G \longrightarrow G_1$ is a monoid homomorphism such that $(\alpha_1 f)(g)\iota_1(a) = \iota_1(a)(\alpha_1 f)(g)$ for all $g \in G$ and $a \in A$, then there exists a unique ring homomorphism $f^* : A[G] \longrightarrow A[G_1]$ such that $f^*\alpha = \alpha_1 f$, i.e., the diagram*

$$\begin{array}{ccc} G & \xrightarrow{f} & G_1 \\ {\scriptstyle\alpha}\downarrow & & \downarrow{\scriptstyle\alpha_1} \\ A[G] & \xrightarrow{f^*} & A[G_1] \end{array}$$

is commutative. Moreover, if f is an isomorphism then f^* is an isomorphism, as well.

Remark 2.2.7. The statement in this corollary is not invertible in the general case, i.e., given a ring A and monoids G, H, an isomorphism $f^* : A[G] \longrightarrow A[H]$ of monoid rings does not imply an isomorphism $f : G \longrightarrow H$ of monoids.

Now consider more general constructions for obtaining associative rings.

Let A be an associative ring, G a monoid and let $\sigma : G \to \mathrm{Aut}(A)$ be a monoid homomorphism. Write $a^{\sigma(g)}$ for $\sigma(g)(a)$.

Definition 2.2.8. The **skew monoid ring** of a monoid G and an associative ring A with respect to a monoid homomorphism $\sigma : G \to \mathrm{Aut}(A)$ is the free left A-module, denoted by $A *_\sigma G$, or $A[G, \sigma]$, with elements of G as a basis over A assuming that

$$ga = a^{\sigma(g)}g \tag{2.2.9}$$

for all $a \in A$ and $g \in G$. (This rule is often called a **skewing** or an **action**). Any element of $A *_\sigma G$ has a unique representation in the form

$$\sum_{g \in G} a_g g \tag{2.2.10}$$

(where $a_g = 0$ for all but finitely many $g \in G$) with the binary operation of addition defined componentwise, i.e.,

$$\sum_{g \in G} a_g g + \sum_{g \in G} b_g g = \sum_{g \in G} (a_g + b_g)g \tag{2.2.11}$$

while a multiplication is defined distributively by the formula:

$$\Big(\sum_{g \in G} a_g g\Big)\Big(\sum_{g \in G} b_g g\Big) = \sum_{h \in G} \Big(\sum_{\substack{xy=h \\ x,y \in G}} a_x b_y^{\sigma(x)} \Big) h. \tag{2.2.12}$$

Remark 2.2.13. If σ maps G onto the identity automorphism of A, i.e., $\sigma = id_A$, then the skew monoid ring $A *_{id_A} G = A[G, id_A]$ so obtained is the ordinary monoid ring $A[G]$. If G is a group, then $A *_\sigma G = A[G,\sigma]$ is called the **skew group ring**.

Definition 2.2.14. Let G be a multiplicative monoid and A an associative ring with identity. Consider a mapping

$$\rho : G \times G \longrightarrow U(A) \tag{2.2.15}$$

satisfying the associativity conditions:

$$\rho_{g,h}\rho_{gh,f} = \rho_{h,f}\rho_{g,hf} \tag{2.2.16}$$

for all $g, h, f \in G$. Such a mapping ρ is often called a **twisting**.

The **twisted monoid ring** $A[G,\rho]$ of a monoid G and an associative ring A with respect to a twisting ρ, defined by (2.2.15) and (2.2.16), is the set of all formal finite sums of the form

$$\sum_{g \in G} a_g t_g, \tag{2.2.17}$$

where $a_g \in A$, and the t_g are symbols. The elements t_g form a basis for $A[G,\rho]$ over A. The multiplication of elements t_g and $a \in A$ is basically defined by

$$t_g t_h = \rho_{g,h} t_{gh} \tag{2.2.18}$$

and

$$t_g a = a t_g \tag{2.2.19}$$

for all $g, h \in G$.

There are two binary operations in $A[G,\rho]$: addition, defined componentwise:

$$\sum_{g \in G} a_g t_g + \sum_{g \in G} b_g t_g = \sum_{g \in G} (a_g + b_g) t_g$$

and multiplication, defined by

$$(\sum_{g \in G} a_g t_g)(\sum_{g \in G} b_g t_g) = \sum_{h \in G} (\sum_{\substack{xy=h,\\ x,y \in G}} a_x b_y \rho_{x,y}) t_h. \tag{2.2.20}$$

Taking into account the condition (2.2.16) for the mapping ρ it is easy to show the associative condition for elements of $A[G,\rho]$. Indeed, it suffices to show for elements $t_g, t_h, t_f \in A[G,\rho]$. We have

$$(t_g t_h) t_f = \rho_{g,h} t_{gh} t_f = \rho_{g,h}\rho_{gh,f} t_{ghf}$$

and

$$t_g(t_h t_f) = t_g \rho_{h,f} t_{hf} = \rho_{h,f} \rho_{g,hf} t_{ghf},$$

which implies that $(t_g t_h)t_f = t_g(t_h t_f)$.

Remark 2.2.21. If $\rho = 1$, i.e., $\rho_{g,h} = 1$ for all $g, h \in G$, then $A[G, 1]$ becomes the monoid ring $A[G]$. If G is a group, then $A[G, \rho]$ is called the **twisted group ring**. Twisted monoid (group) rings are most used in the case where A is a field.

The next construction is a further generalization of both skew monoid rings and twisted monoid rings.

Definition 2.2.22. Let G be a multiplicative monoid and A an associative ring with identity. Suppose there is a monoid homomorphism:

$$\sigma : G \to \mathrm{Aut}(A).$$

Write $\sigma(g)(a) = a^{\sigma(g)}$ for any $g \in G$ and $a \in A$.

The **crossed product** $A[G, \rho, \sigma]$ of a monoid G and an associative ring A with respect to a twisting ρ, defined by (2.2.15)–(2.2.16), and a monoid homomorphism $\sigma : G \longrightarrow \mathrm{Aut}(A)$, is the set of all formal finite sums of the form

$$\sum_{g \in G} a_g t_g, \tag{2.2.23}$$

where $a_g \in A$, and the t_g are symbols. The elements t_g form a basis for $A[G, \rho, \sigma]$ over A. The multiplication of elements t_g and $a \in A$ is basically defined by

$$t_g a = a^{\sigma(g)} t_g \tag{2.2.24}$$

and

$$t_g t_h = \rho_{g,h} t_{gh} \tag{2.2.25}$$

for all $g, h \in G$, and $a \in A$. There are two binary operations in $A[G, \rho, \sigma]$: addition, defined componentwise:

$$\sum_{g \in G} a_g t_g + \sum_{g \in G} b_g t_g = \sum_{g \in G} (a_g + b_g) t_g \tag{2.2.26}$$

and multiplication, defined by

$$(\sum_{g \in G} a_g t_g)(\sum_{g \in G} b_g t_g) = \sum_{h \in G} (\sum_{\substack{xy=h, \\ x,y \in G}} a_x b_y^{\sigma(x)} \rho_{x,y}) t_h. \tag{2.2.27}$$

Remark 2.2.28. If σ is trivial, i.e., $\sigma(g)(a) = a$ for all $a \in A$ and all $g \in G$, then $A[G, \rho, 1] = A[G, \rho]$ becomes the twisted monoid ring. If $\rho = 1$, then $A[G, 1, \sigma]$ becomes the skew monoid ring $A *_\sigma G$. If σ is trivial and $\rho = 1$, then $A[G, 1, 1]$

becomes the monoid ring $A[G]$ of G over A. If G is a group, then $A[G,\rho,\sigma]$ is called the **crossed product** of the group G and the ring A with respect to ρ and σ.

Example 2.2.29. Let A be a ring, and let N be a normal subgroup of a group G. We can define the group rings $B = A[G]$ and $C = A[N]$. Let $H = G/N$, then $A[G] = A[N] * G/N = C[H,\sigma,\rho]$ is the crossed product of the group $H = G/N$ and the ring $C = A[N]$ with respect to an action $\sigma : H \longrightarrow \mathrm{Aut}(C)$ and the twisting $\rho : H\times H \longrightarrow U(B)$. Indeed, we have a natural projection $\pi : G \longrightarrow G/N = H$. Let $h \in H$, and let $g \in G$ be such that $\pi(g) = h \in H$. Then $A[G]$ can be presented in the form

$$A[G] = \bigoplus_{\substack{h\in H\\ \pi(g)=h}} gB$$

where $X = \{g \in G : \pi(g) = h \in H$ is a basis of $A[G]$ over A. Since $N\lhd G$, $g^{-1}Ng = N$ for all $g \in G$, therefore $g^{-1}Cg = C$, i.e., g induces conjugate automorphism $\sigma(h) \in \mathrm{Aut}(C)$. If $h \in H$, $\pi(g) = h$ and $b \in B$, then

$$gb = gb(g^{-1}g) = (gbg^{-1})g = \sigma(g)(b)g.$$

On the other hand, the multiplication is given by $(g_1N)(g_2N) = g_3N$, where $\pi(g_3) = \pi(g_1g_2) = \pi(g_1)\pi(g_2)$. So, $g_1g_2 = \rho(h_1,h_2)g_3$, where $\rho(h_1,h_2) \in N \subseteq U(B)$. So we have a twisting $\rho : H \times H \longrightarrow U(B)$.

2.3. Polynomial and Skew Polynomial Rings

The polynomial ring (in one or more variables) over a field (or the integers) and its properties are standard topics of a first course in abstract algebra. In this section we consider the construction of polynomial and skew polynomial rings over an arbitrary associative ring A.

Let A be an associative ring with $1 \neq 0$, and let x be indeterminate. Define the polynomial ring $A[x]$ in one variable x over an associative ring A as a monoid ring $A[G]$, where $G = \{x\} = \{1, x, x^2, \ldots, x^n, \ldots\}$ is a cyclic monoid generating by one generator x, i.e., $A[x]$ is a set of formal expressions (called **polynomials in one variable** x) of the form $\sum\limits_{i=0}^{\infty} a_i x^i$, where $a_i \in A$ and there is $n \in \mathbf{N}$ such that $a_i = 0$ for all $i > n$. Moreover it is assumed that $ax = xa$ for all $a \in A$.

The operations of addition and multiplications of polynomials are defined by the following way:

$$\sum_{i=0}^{\infty} a_i x^i + \sum_{i=0}^{\infty} b_i x^i = \sum_{i=0}^{\infty} (a_i + b_i)x^i \tag{2.3.1}$$

and

$$\left(\sum_{i=0}^{\infty} a_i x^i\right)\left(\sum_{i=0}^{\infty} b_i x^i\right) = \sum_{i=0}^{\infty} c_i x^i \tag{2.3.2}$$

where

$$c_i = \sum_{j=0}^{i} a_j b_{n-j}, \quad i = 0,1,2,... \tag{2.3.3}$$

Then $A[x]$ is an associative ring with identity $1 \neq 0$. The ring A can be considered as a subring of $A[x]$ if we identify A with $\iota(A) \subset A[x]$, where $\iota : A \longrightarrow A[x]$ is a ring homomorphism defined by $\iota(a) = \sum_{i=0}^{\infty} a_i x^i$ with $a_0 = a$ and $a_i = 0$ for $i > 0$. If the ring A is commutative, then so is $A[x]$.

If $A = K$ is a field then $K[x]$ is a principal ideal ring. Note that this is not true in the general case. For example, let $A = \mathbf{Z}$ be the ring of integers and $\mathcal{I}$ be the set of all polynomials with even constant terms. It is easy to see that $\mathcal{I}$ is an ideal in $\mathbf{Z}[x]$ but it is not a principal ideal.

By definition, each polynomial $f \in A[x]$ is uniquely written in the form

$$f = a_0 + a_1 x + \cdots + a_n x^n \tag{2.3.4}$$

with $a_i \in A$ and $a_n \neq 0$. We call this n the **degree** of f and write $\deg(f) = n$ (by assumption $\deg(0) = -\infty$), a_i are called the **coefficients** of f and a_n is called the **leading coefficient** of f.

Analogously for any associative ring A one can define the polynomial ring in several variables. Let

$$G = \{x_1^{\alpha_1} x_2^{\alpha_2} \cdots x_n^{\alpha_n} \ : \ \alpha_i \geq 0, \ i = 1,2,...,n\} \tag{2.3.5}$$

be a commutative (multiplicative) monoid generated by distinct elements $x_1, x_2, \ldots, x_n$ that commute with each other. Let $\alpha = (\alpha_1, \alpha_2, \ldots, \alpha_n)$, where $\alpha_i \in \mathbf{N} \cup \{0\}$ for $i = 1,2,\ldots,n$. The elements of G of the form

$$x^\alpha = x_1^{\alpha_1} x_2^{\alpha_2} \cdots x_n^{\alpha_n} \tag{2.3.6}$$

are called **monomials**. Such monomials form the basis of the monoid ring $A[G]$. By definition any element of $A[G]$ can be uniquely written in the following form:

$$f = \sum_{\alpha} a_\alpha x^\alpha, \tag{2.3.7}$$

where $a_\alpha \in A$. The sum above is supposed to be finite and $ax_i = x_i a$ for all $a \in A$ and all $i, j = 1,2,\ldots,n$. The ring $A[G]$ is also denoted by $A[x_1, x_2, \ldots, x_n]$ and called the **ring of polynomials in** n **(commuting) variables** $x_1, x_2, \ldots, x_n$ over the ring A. The elements of this ring are called **polynomials in** n **variables** $x_1, x_2, \ldots, x_n$ over a ring A.

The constants a_α are called the **coefficients** of the polynomial f. The **degree of the monomial** (2.3.6) is defined as

$$|\alpha| = \alpha_1 + \alpha_2 + \cdots + \alpha_n \geq 0. \tag{2.3.8}$$

The largest degree of all monomials occurring with non-zero coefficients in a non-zero polynomial f is called the **degree of the polynomial** and denoted by $\deg(f)$. The zero polynomial, by convention, has degree equal to $-\infty$. If a polynomial f is a constant then $\deg(f) = 0$. The polynomial of the form

$$f_k = \sum_{|\alpha|=k} a_\alpha x^\alpha \tag{2.3.9}$$

is called a **homogeneous polynomial**. Using this notion any polynomial $f \in A[x_1, x_2, \ldots, x_n]$ can be uniquely written in the form:

$$f = \sum_{k\geq 0} f_k \tag{2.3.10}$$

and the operations of addition and multiplication in $A[x_1, x_2, \ldots, x_n]$ are defined by:

$$f + g = \sum_{k\geq 0} f_k + \sum_{k\geq 0} g_k = \sum_{k\geq 0} (f_k + g_k) \tag{2.3.11}$$

$$fg = (\sum_{k\geq 0} f_k)(\sum_{k\geq 0} g_k) = \sum_{k\geq 0} h_k, \tag{2.3.12}$$

where

$$h_k = \sum_{m+n=k} f_m g_n \tag{2.3.13}$$

which are similar to that (2.3.1)–(2.3.3).

One can also define recursively the ring

$$B[x_n] = A[x_1, x_2, \ldots, x_{n-1}][x_n].$$

In other words, elements of $B[x_n]$ are considered as polynomials in x_n whose coefficients are polynomials in the first $n-1$ variables. In this case one also assumes that $x_i x_j = x_j x_i$ and $a x_i = x_i a$ for all $a \in A$ and all $i, j = 1, 2, \ldots, n$.

Proposition 2.3.14. *Let A be a ring, and let $x_1, x_2, \ldots, x_{n-1}, x_n$ be distinct indeterminates. Then*

$$A[x_1, x_2, \ldots, x_{n-1}, x_n] \simeq A[x_1, x_2, \ldots, x_{n-1}][x_n], \tag{2.3.15}$$

where $x_i x_j = x_j x_i$ and $a x_i = x_i a$ for all $a \in A$ and $i, j = 1, 2, \ldots, n$.

Proof. There is a natural embedding $\psi : B \longrightarrow B[x_n]$, given by

$$\psi(g) = \sum_{k\geq 0} a_k x_n^k$$

with $a_0 = g$, and $a_k = 0$ for $k > 0$, which is a ring monomorphism. Any element $f \in A[x_1, x_2, \ldots, x_{n-1}, x_n]$ is uniquely written in the form

$$f = \sum_\alpha a_\alpha x^\alpha = \sum_{\alpha_1, \alpha_2, \ldots, \alpha_n} a_\alpha x_1^{\alpha_1} x_2^{\alpha_2} \cdots x_n^{\alpha_n},$$

where $a_\alpha \in A$.

Consider a mapping

$$\varphi : A[x_1, x_2, \dots, x_{n-1}, x_n] \longrightarrow A[x_1, x_2, \dots, x_{n-1}][x_n],$$

given by:

$$\varphi(\sum_{\alpha} a_\alpha x^\alpha) = \sum_{\alpha_n} \psi(\sum_{\alpha_1,\alpha_2,\dots,\alpha_{n-1}} a_\alpha x_1^{\alpha_1} x_2^{\alpha_2} \cdots x_{n-1}^{\alpha_{n-1}}) x_n^{\alpha_n} \tag{2.3.16}$$

Taking into account that $x_i x_j = x_j x_i$ and $a x_i = x_i a$ for all $a \in A$ and all $i, j = 1, 2, \dots, n$ one can easy verify that φ is a ring isomorphism. □

This proposition gives possibility to prove properties of polynomials in several variables $x_1, x_2, \dots, x_n$ by induction on the number n of variables. In particular, using finite induction to theorem 1.1.13 one can immediately obtain the following famous result.

Theorem 2.3.17 (Hilbert Basis Theorem). *If A is a right Noetherian ring then the polynomial ring $A[x_1, x_2, \dots, x_n]$ is also right Noetherian.*

Remark 2.3.18. Note that in the ring $A[x_1, x_2, \dots, x_n]$ one assumes that $x_i x_j = x_j x_i$ and $a x_i = x_i a$ for all $a \in A$ and all $i, j \in \{1, 2, \dots, n\}$. It is possible to assume that $x_i x_j \neq x_j x_i$ for all $i, j \in \{1, 2, \dots, n\}$ and $i \neq j$. In this case G is a free monoid generated by elements $1, x_1, x_2, \dots, x_n$, and the ring obtained $A[G]$ is called a **polynomial ring in n noncommutating variables** $x_1, x_2, \dots, x_n$, and it is denoted by $A\langle x_1, x_2, \dots, x_n\rangle$. Note that here we still assume that $a x_i = x_i a$ for all $a \in A$ and $i \in \{1, 2, \dots, n\}$.

Now consider a class of rings which are similar to the ring of polynomials over a ring A in one variable x considered above, but in these rings the variable x is not assumed to commute with all elements of A.

Let σ be a ring endomorphism of A, and $\delta : A \to A$ a (left) σ-**derivation** on A, that is, an additive map of A such that

$$\delta(ab) = \delta(a)b + \sigma(a)\delta(b) \tag{2.3.19}$$

for all $a, b \in A$. Such a pair (σ, δ) is called a (left) **skew derivation** on the ring A. Note that $\delta(1) = 0$ since $\delta(1) = \delta(1 \cdot 1) = \delta(1) \cdot 1 + \sigma(1)\delta(1) = 2\delta(1)$, whence $\delta(1) = 0$.

In the case that σ is the identity map on A, a σ-derivation of A is called a **derivation**. Thus a derivation δ of A is any additive map $\delta : A \to A$ such that

$$\delta(ab) = \delta(a)b + a\delta(b) \tag{2.3.20}$$

for all $a, b \in A$.

Example 2.3.21.

If $A = K[x]$ is a polynomial ring over a field K then the usual formal derivation $\frac{d}{dx}$ given by

$$\frac{d}{dx}(a_0 + a_1x + a_2x^2 + \cdots + a_nx^n) = a_1 + 2a_2x + \cdots + na_nx^{n-1}, \qquad (2.3.22)$$

where $a_i \in K$, is a derivation of A.

Let A be a ring with a (left) skew derivation (σ,δ). Consider a set of all polynomials in one variable x over A of the form

$$p(x) = \sum_{i=0}^{n} a_i x^i, \qquad (2.3.23)$$

where $a_i \in A$ and $a_n \neq 0$, and with multiplication defined by the distributive law and the rule

$$xa = \sigma(a)x + \delta(a). \qquad (2.3.24)$$

Then this set forms a ring which is called a (left) **skew polynomial ring** (or **Ore extension**) and denoted by $A[x;\sigma,\delta]$. The integer n is called the **degree** of $p(x)$ and written $\deg(p)$, the element a_n is called the **leading coefficient** of p.

If $\delta = 0$ the ring $A[x;\sigma,0] = A[x;\sigma]$ is also called the **twisted polynomial ring**. If $\sigma = 1$ is the identity map on A, $A[x;1,\delta] = A[x;\delta]$ is often called the **formal differential operator ring**. If $\delta = 0$ and $\sigma = 1$, then $A[x;1,0] = A[x]$ becomes the usual polynomial ring over A. If σ is not the identity map on A, then $A[x;\sigma,\delta]$ is a not necessarily commutative ring.

If $\delta = 0$ the equality (2.3.24) has the following form

$$xa = \sigma(a)x,$$

and is often called as "Hilbert's twist". The rule of multiplication in $A[x;\sigma]$ is given by

$$(ax^i)(bx^j) = a\sigma^i(b)x^{i+j}$$

for all $a,b \in A$ and integers $i,j \geq 0$. Any element of $A[x;\sigma]$ is written in the form (2.3.23) and it is called the **right polynomial**. A left polynomial of the special form

$$p(x) = \sum_{i=0}^{n} \sigma^i(b_i)x^i$$

can be written as a right polynomial:

$$p(x) = \sum_{i=0}^{n} b_i x^i.$$

However, not every left polynomial can be written as a right polynomial. It is the case only if σ is a surjective ring endomorphism.

The skew polynomial ring $A[x;\sigma,\delta]$ is defined uniquely by the element x and one relation (2.3.24), that exhibits the following universal property:

Proposition 2.3.25. (Universal property). *Let S be a ring and let $A[x;\sigma,\delta]$ be the skew polynomial ring with (left) skew derivation (σ,δ). So (2.3.19) holds. If $\varphi : A \to S$ is a ring homomorphism, and $i : A \to A[x;\sigma,\delta]$ denotes the obvious (canonical) ring embedding, and there exists an element $y \in S$ which has the property:*

$$y\varphi(a) = \varphi(\sigma(a))y + \varphi(\delta(a)), \tag{2.3.26}$$

for all $a \in A$, then there exists a unique ring homomorphism $\psi : A[x;\sigma,\delta] \to S$ such that $\psi(x) = y$ and the diagram

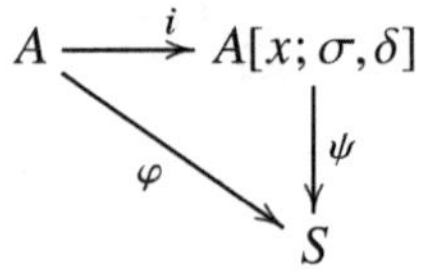

is commutative.

Proof. Set $\psi(ax) = \varphi(a)y$ and $\psi(xa) = \varphi(\sigma(a))y + \varphi(\delta(a))$. In particular, $\psi(x) = y$. To prove that ψ is a ring homomorphism it is sufficient to show that $\psi(ax \cdot bx) = \psi(ax)\psi(bx)$ for all $a,b \in A$.

$$\psi(ax \cdot bx) = \psi(a(xb)x) = \psi(a(\sigma(b)x + \delta(b))x) = \psi(a\sigma(b)x^2 + a\delta(b)x) =$$

$$= \varphi[a\sigma(b)]y^2 + \varphi[a\delta(b)]y = \varphi(a)\varphi(\sigma(b))y^2 + \varphi(a)\varphi(\delta(b))y.$$

On the other hand,

$$\psi(ax)\psi(bx) = \varphi(a)y \cdot \varphi(b)y = \varphi(a)[\varphi(\sigma(b))y + \varphi(\delta(b))]y =$$

$$= \varphi(a)\varphi(\sigma(b))y^2 + \varphi(a)\varphi(\delta(b))y.$$

So $\psi(ax \cdot bx) = \psi(ax)\psi(bx)$. □

Lemma 2.3.27. *If A is a domain and σ is an injective ring endomorphism of A then the skew polynomial ring $A[x;\sigma,\delta]$ is also a domain.*

Proof. It is easy to show that $x^n a = \sigma^n(a)x^n + \cdots$ for any $a \in A$. If σ is injective and $a \neq 0$ then $\sigma^n(a) \neq 0$. Therefore if $f(x) = a_n x^n + \cdots$ with $a_n \neq 0$, and $g(x) = b_m x^m + \cdots$ with $b_m \neq 0$, then $f(x)g(x) = a_n\sigma^n(b_m)x^{n+m} + \cdots$ with $a_n\sigma^n(b_m) \neq 0$, since A is a domain. Therefore $A[x;\sigma,\delta]$ is also a domain and $\deg(fg) = \deg(f) + \deg(g)$ for any $f,g \in A[x;\sigma,\delta]$. □

In the general case, if σ is not injective, the skew polynomial ring $A[x;\sigma,\delta]$ is not necessarily a domain, as is shown by the following example.

Example 2.3.28. ([237, 2.11].)

Let K be a field and $A = K[y]$. Define an endomorphism $\sigma : A \longrightarrow A$ by $\sigma(f(y)) = f(0)$. Then the ring $T = A[x;\sigma]$ is not an integral domain since $xy = \sigma(y)x = 0$.

Since $f(y)x \neq 0$ for any $0 \neq f(y) \in A$, we obtain the example of a ring with element $x \in T$ which is a right zero divisor but not left zero divisor.

Proposition 2.3.29. *If A is a domain and σ is an injective ring endomorphism of A then the group of units $U(T)$ of $T = A[x;\sigma]$ is equal to the group of units $U(A)$ of A. If A is a division ring then the Jacobson radicals of the skew polynomial rings $A[x;\sigma]$ and $A[x;\delta]$ are equal to 0.*

Proof. Note that $\sigma^i(a) \in U(A)$ for all $a \in U(A)$ and $i \geq 0$. Suppose that $f = \sum\limits_{i=0}^{n} a_i x^i \in U(T)$. Then there exists $g = \sum\limits_{j=0}^{m} b_j x^j \in U(T)$ such that $fg = \sum\limits_{i+j=0}^{n+m} a_i \sigma^i(b_j) x^{i+j} = 1$. Therefore $a_0 b_0 = 1$ and $a_i = b_i = 0$ for all $i > 0$, since T is a domain, by lemma 2.3.10. Thus, $U(T) = U(A)$.

Let $T = A[x;\sigma]$. Suppose that A is a division ring, then $\text{rad(A)} = 0$. By the previous statement $U(T) = U(A)$. Therefore $A = U(T) \cup \{0\}$. It is obvious that A is an ideal in T. Let $b \in \text{rad(T)}$, then $1 - b \in U(T)$. Therefore $b \in T \cap \text{rad(A)} = 0$, i.e., $\text{rad(T)} = 0$. The proof for the case when $T = A[x;\delta]$ is similar. □

For any associative ring A and any ring automorphism $\sigma \in \text{Aut}(A)$ one can consider the ring $A[x_1, x_2, \ldots, x_n; \sigma]$ of all skew polynomials in n commuting variables $x_1, x_2, \ldots, x_n$ with coefficients in A which is defined inductively by

$$A[x_1, x_2, \ldots, x_n; \sigma] = A[x_1, x_2, \ldots, x_{n-1}; \sigma][x_n; \sigma] =$$

$$= ((\ldots((A[x_1;\sigma])[x_2;\sigma])\ldots)[x_{n-1};\sigma])[x_n;\sigma], \qquad (2.3.30)$$

where the automorphism σ is naturally extends to an automorphism of each ring $A[x_1, x_2, \ldots, x_k; \sigma]$ for $k \in \{1, 2, \ldots, n-1\}$. In this case one also assumes that $x_i x_j = x_j x_i$ for all $i, j \in \{1, 2, \ldots, n\}$.

2.4 Formal Power and Skew Formal Power Series Rings

Let A be an associative ring and let x be an indeterminate. Denote by $A[[x]]$ the set of all formal infinite sums of the form

$$\sum_{n=0}^{\infty} a_n x^n, \quad a_n \in K \text{ for } n = 0, 1, 2, \ldots \qquad (2.4.1)$$

Basically one assumes that $xa = ax$ for any $a \in A$. Then addition and multiplication in $A[[x]]$ are defined as follows:

$$\sum_{n=0}^{\infty} a_n x^n + \sum_{n=0}^{\infty} b_n x^n = \sum_{n=0}^{\infty} (a_n + b_n) x^n, \tag{2.4.2}$$

and

$$\left(\sum_{n=0}^{\infty} a_n x^n\right)\left(\sum_{n=0}^{\infty} b_n x^n\right) = \sum_{n=0}^{\infty} d_n x^n, \tag{2.4.3}$$

where

$$d_n = \sum_{k=0}^{n} a_k b_{n-k}, \quad n = 0, 1, 2, \ldots \tag{2.4.4}$$

By definition, $\sum_{n=0}^{\infty} a_n x^n = \sum_{n=0}^{\infty} b_n x^n$ if and only if $a_n = b_n$ for all $n \geq 0$. It is easy to verify that the set $A[[x]]$ forms an associative ring under the operations of addition and multiplication as specified above. The set $A[[x]]$ with operations of addition and multiplication as defined by formulas (2.4.2)–(2.4.4) is called the **formal power series ring** in one variable x over the ring A. The elements of A and $A[x]$ themselves can be considered as elements of $A[[x]]$. So, the ring A and the polynomial ring $A[x]$ may naturally be considered as subrings of $A[[x]]$ by natural canonical embeddings $\iota : A \longrightarrow A[x]$ and $\nu : A[x] \longrightarrow A[[x]]$. In particular, the identity of A is the identity of $A[[x]]$. If A is a commutative ring then $A[[x]]$ is also commutative.

A ring $A[[x]]$ can be turned into a left A-module setting $b(\sum_{n=0}^{\infty} a_n x^n) = \sum_{n=0}^{\infty} (ba_n) x^n$ for all $b \in A$ and $\sum_{n=0}^{\infty} a_n x^n \in A[[x]]$. This module is free with basis $1, x, x^2, \ldots$ over A. Since $ax = xa$ for all $a \in A$, in the similar way $A[[x]]$ can be turned into a right A-module. Therefore $A[[x]]$ is an $(A$-$A)$-bimodule, and it is sometimes written as $A\langle x\rangle$. If A is a commutative ring, $A\langle x\rangle$ is called a **free A-algebra**.

Let $f(x) = \sum_{n=0}^{\infty} a_n x^n$ be a non-zero power series in $A[[x]]$. The least power d of x for which $a_d \neq 0$ is called the order of $f(x)$ and denoted by $o(f)$ and the coefficient a_d is called the **leading coefficient of the power series** $f(x)$.

If $A = K$ is a field, $K[[x]]$ has the following simple properties (See, [146]):

1. An element $f \in K[[x]]$ is invertible in $K[[x]]$ if and only if $a_0 \neq 0$;
2. Any ideal I in $K[[x]]$ is principal;
3. The ideals in $K[[x]]$ form a descending chain of ideals:

$$K[[x]] \supset (x) \supset (x^2) \supset (x^3) \supset \ldots \supset (x^n) \supset \ldots.$$

4. $K[[x]]$ is a local ring with unique maximal ideal $M = (x)$, $M^n = (x^n)$, and $\bigcap_{n=0}^{\infty} M^n = 0$.

   If A is not a field we have the following results.

Proposition 2.4.5. *Let A be an associative ring. A formal power series $\sum_{n=0}^{\infty} a_n x^n \in A[[x]]$ is invertible if and only if $a_0 \in U(A)$.*

Proof. If $f(x) = \sum_{n=0}^{\infty} a_n x^n$ is right invertible in $A[[x]]$ then there exists $g(x) = \sum_{n=0}^{\infty} b_n x^n \in A[[x]]$ such that $f(x)g(x) = 1$. In particular, $a_0 b_0 = 1$. Analogously, if $f(x)$ is left invertible in $A[[x]]$ then there exist $c_0 \in A$ such that $c_0 a_0 = 1$. Therefore $a_0 \in U(A)$.

Conversely, assume that $a_0 \in U(A)$. Then there is $b_0 \in U(A)$ such that $a_0 b_0 = 1$. In order to prove that $f(x) = \sum_{n=0}^{\infty} a_n x^n$ is right invertible in $A[[x]]$ one must find $g(x) = \sum_{n=0}^{\infty} b_n x^n \in A[[x]]$ such that $f(x)g(x) = 1$. This gives the following infinite system of equations:

$$a_0 b_0 = 1$$

$$\sum_{i=0}^{n} a_i b_{n-i} = 0$$

for $n \geq 1$, which has a solution obtained inductively:

$$b_0 = a_0^{-1}$$

$$b_n = -a_0^{-1} \sum_{i=1}^{n} a_i b_{n-i}$$

for $n \geq 1$. This shows that $f(x)$ is right invertible. Similar we can show that $f(x)$ is left invertible. So, $f(x)$ is invertible in $A[[x]]$. □

Proposition 2.4.6. *For any ring A the Jacobson radical of $T = A[[x]]$ is equal to the set of expressions of the form $a + f(x)x$, where $a \in \text{rad}(A)$ and $f(x) \in T$. If A is a local ring, then $T = A[[x]]$ is also local.*

Proof. By the previous proposition 2.4.5,

$$U(T) = \{\sum_{i=0}^{\infty} a_i x^i \ : \ a_0 \in U(A)\}.$$

Let $b \in \text{rad(T)}$. Then $1 - b \in U(T)$. So $b = 1 + y$, where $y = a_0 + a_1 x + \cdots = a_0 + f(x)x$ and $a_0 \in U(A)$. Since $a = 1 + a_0 \in \text{rad(A)}$, $b = a + f(x)x$ with $a \in \text{rad(A)}$ and $f(x) \in T$.

Suppose that A is a local ring. Since $T/\mathrm{rad(T)} \cong \mathrm{A/rad(A)}$ and $A/\mathrm{rad}(A)$ is a division ring, T is a local ring as well. □

As in the case of polynomial rings rejecting the assumption of commutativity of the variable x with coefficients of the ring A one is lead (among others) to the following definition.

Definition 2.4.7. Let A be a ring, and σ a ring endomorphism of A. The set of all formal series $\sum_{i=0}^{\infty} a_i x^i$, where $a_i \in A$, with operations of addition

$$\sum_{i=0}^{\infty} a_i x^i + \sum_{i=0}^{\infty} b_i x^i = \sum_{i=0}^{\infty} (a_i + b_i) x^i$$

and multiplication defined by the distributive law and the rule

$$xa = \sigma(a)x$$

for any $a \in A$, is called the (left) **skew** (or **twisted**) **power series ring** and denoted by $A[[x,\sigma]]$.

Let $f(x) = \sum_{n=0}^{\infty} a_n x^n$ be a non-zero skew power series of $A[[x,\sigma]]$. The least power d of x for which $a_d \neq 0$ is called the **order** of $f(x)$ and denoted by $o(f)$ and the coefficient a_d is called the **leading coefficient of a skew power series** $f(x)$.

Proposition 2.4.8. *The group of units of $T = A[[x,\sigma]]$ is equal to the set of expressions of the form $\sum_{i=0}^{\infty} a_i x^i$, where $a_i \in A$ and $a_0 \in U(A)$. Let R be the Jacobson radical of A. Denote by $J = \mathrm{rad}(A[[x,\sigma]])$ the Jacobson radical of $A[[x,\sigma]]$. Then $J = R + A[[x,\sigma]]x$ and $A[[x,\sigma]]/A[[x,\sigma]]x \cong A$.*

Proof.
The first statement follows from the observation that for any $a_0 \in U(A)$ one has $\sigma^i(a_0) \in U(A)$ for any $i \geq 0$. So $U(T) = \{\sum_{i=0}^{\infty} a_i x^i : a_0 \in U(A)\}$. Let $b \in J$, where $J = \mathrm{rad}(T)$. Then $1 - b \in U(T)$. So $b = 1 + y$, where $y = \sum_{i=0}^{\infty} a_i x^i = a_0 + f(x)x$ and $f(x) \in T$. Since $c = 1 + a_0 \in \mathrm{rad}(A)$, $b = c + f(x)x$ with $a \in \mathrm{rad}(A)$ and $f(x) \in T$.
Therefore $T/Tx \cong A$. □

Proposition 2.4.9. *The ring $A[[x,\sigma]]$ is semilocal (resp., local) if and only if A is a semilocal (resp., local) ring. The ring $A[[x,\sigma]]$ is semiperfect if and only if A is a semiperfect ring.*

Proof. This follows from the fact that $T/\mathrm{rad(T)} \cong \mathrm{A/rad(A)}$. □

Proposition 2.4.10. *Let A be a ring and σ an injective ring endomorphism of A. Then the ring $T = A[[x,\sigma]]$ is a domain if and only if A is a domain.*

Proof. This follows from the fact that if $y = \sum_{i=n}^{\infty} a_i x^i \in T$ with leading coefficient $a_n \neq 0$ and $z = \sum_{i=m}^{\infty} b_i x^i$ with leading coefficient $b_m \neq 0$, then the leading coefficient of $yz \in T$ is $\sigma^m(a_n)b_m \neq 0$, since $\sigma^m(a_n) \neq 0$ and A is a domain. □

Similar as for polynomial rings one can introduce the formal power series rings and skew formal power series rings in several variables.

Let A be a ring, $I = \{1,2,\dots,n\}$ and $\{x_i\}_{i\in I}$ be independent variables. We will assume that $x_i x_j = x_j x_i$ for all $i,j \in I$ and $ax_i = x_i a$ for all $a \in A$ and $i \in I$.

Consider the monomials of the form (2.3.6):

$$x^\alpha = x_1^{\alpha_1} x_2^{\alpha_2} \cdots x_n^{\alpha_n},$$

where $\alpha = (\alpha_1,\alpha_2,\dots,\alpha_n)$ with $\alpha_i \in \mathbf{N} \cup \{0\}$ for $i = 1,2,\dots,n$. The degree of this monomial is defined as (2.3.8): $|\alpha| = \alpha_1 + \alpha_2 + \cdots + \alpha_n$.

Now we can consider the set of elements of the form

$$f = \sum_{i\geq 0} f_i \tag{2.4.11}$$

where

$$f_k = \sum_{|\alpha|=k} a_\alpha x^\alpha, \tag{2.3.12}$$

is a homogeneous polynomial of degree k in n variables $x_1,x_2,\dots,x_n$.

Addition and multiplication of these formal power series is defined as follows:

$$f + g = \sum_{i\geq 0} f_i + \sum_{i\geq 0} g_i = \sum_{i\geq 0}(f_i + g_i) \tag{2.4.13}$$

$$fg = (\sum_{i\geq 0} f_i)(\sum_{j\geq 0} g_j) = \sum_{k\geq 0} h_k, \tag{2.4.14}$$

where

$$h_k = \sum_{i+j=k} f_i g_j.$$

It is easy to verify that these elements under operations of addition and multiplication defined above form a ring which is called the **ring of formal power series** in n variables and denoted by $A[[x_1,x_2,\dots,x_n]]$ or $A[[x_i]]_{i\in I}$, where $I = \{1,2,\dots,n\}$, and elements of this ring are called **formal power series** in n variables.

Analogously as the proof of proposition 2.4.5 we can prove the following result.

Proposition 2.4.15. *Let A be a ring. A formal power series $f \in A[[x_1,x_2,\dots,x_n]]$ of the form* (2.4.11) *is invertible if and only if $f_0 \in U(A)$.*

Since $A[[x_1,x_2,\dots,x_{n-1}]]$ is an associative ring with identity one can construct the ring $B = A[[x_1,x_2,\dots,x_{n-1}]][[x_n]]$, which elements are formal power series of the form

$$f = \sum_{i\geq 0} a_i x_n^i \tag{2.4.16}$$

with $a_i \in A[[x_1,x_2,\dots,x_{n-1}]]$. An in this case, similar to proposition 2.3.14, one can prove the following result.

Proposition 2.4.17. *Let A be a ring, and let $x_1,x_2,\dots,x_{n-1},x_n$ be the indeterminates. Then*

$$A[[x_1,x_2,\dots,x_{n-1},x_n]] \simeq A[[x_1,x_2,\dots,x_{n-1}]][[x_n]], \tag{2.4.18}$$

where $x_i x_j = x_j x_i$ and $a x_i = x_i a$ for all $a \in A$ and all $i,j = 1,2,\dots,n$.

Remark 2.4.19. Algebraically if J is a non-empty subset of $I = \{1,2,\dots,n\}$ which is not equal to I and $K = I \setminus J$ then writing $B = A[[x_j]]_{j\in J}$ one has

$$B[[x_k]]_{k\in K} \simeq A[[x_i]]_{i\in I} \tag{2.4.20}$$

(but there are problems with the topologies in these two power series rings; on the left side in (2.4.20) B has the discrete topology and that does not fit with the induced topology in B as a subring of $A[[x_i]]_{i\in I}$).

Remark 2.4.21. It can be shown that $A[x] \otimes_A A[y] \simeq A[x,y]$ for polynomials rings over a commutative ring A. However this fails for formal power series.

There is still natural morphism:

$$A[[x]] \otimes_A A[[y]] \longrightarrow A[[x,y]] \tag{2.4.22}$$

which is injective but not surjective. The element $\sum\limits_{i=0}^{\infty} x^i y^i$, e.g., is not in the image of (2.4.22).

The construction above for the ring $A[[x_i]]_{i\in I}$ for a finite set I can be generalized for the case of an infinite set of indices I.

Let I be an index set, finite or infinite. Consider a special function

$$\mathbf{n} : I \longrightarrow \mathbf{N} \cup \{0\}$$

such that $\mathbf{n}(i) \neq 0$ for only finitely many $i \in I$. This function will be called **multi-index**.

Given multi-indices $\mathbf{n}$, $\mathbf{k}$ we define

$$|\mathbf{n}| = \sum_{i\in I} \mathbf{n}(i) \tag{2.4.23}$$

$$\mathbf{n} \leq \mathbf{k} \iff \mathbf{n}(i) \leq \mathbf{k}(i) \text{ for all } i \in I \tag{2.4.24}$$

$$\mathbf{n} + \mathbf{k} = \mathbf{m} \iff \mathbf{m}(i) = \mathbf{n}(i) + \mathbf{k}(i) \text{ for all } i \in I \tag{2.4.25}$$

$$\mathbf{n} < \mathbf{k} \quad \Longleftrightarrow \quad \mathbf{n} \leq \mathbf{k} \text{ and } |\mathbf{n}| < |\mathbf{k}| \tag{2.4.26}$$

We write $\mathbf{0}$ for the multi-index $\mathbf{0}(i) = 0$ for all $i \in I$. Then $\mathbf{n} > \mathbf{k}$ iff $\mathbf{n} = \mathbf{k} + \mathbf{m}$ for some multi-index $\mathbf{m} \neq \mathbf{0}$.

We write Ω_I for the set of all multi-indices indexed by I. If $I = \{1, 2, \ldots, m\}$ we simply write Ω_m or even Ω for the set of multi-indices indexed by $\{1, 2, \ldots, m\}$, i.e., $\Omega_m = \{(n_1, n_2, \ldots, n_m) : n_i \in \mathbf{N} \cup \{0\}\}$. Consider a set of different elements $\{x_i\}_{i \in I}$. For each $\mathbf{n} \in \Omega_I$, we define $x^{\mathbf{n}}$ as follows:

$$x^{\mathbf{n}} = \prod_{i \in I, \mathbf{n}(i) \neq 0} x_i^{\mathbf{n}(i)} \tag{2.4.27}$$

Let A be an associative ring with $1 \neq 0$. Then consider a set of formal sums of the form:

$$f(x) = \sum_{\mathbf{n} \in \Omega_I} a_{\mathbf{n}} x^{\mathbf{n}}, \quad a_{\mathbf{n}} \in A \tag{2.4.28}$$

with addition and multiplication defined by:

$$\sum a_{\mathbf{n}} x^{\mathbf{n}} + \sum b_{\mathbf{n}} x^{\mathbf{n}} = \sum (a_{\mathbf{n}} + b_{\mathbf{n}}) x^{\mathbf{n}} \tag{2.4.29}$$

$$\left(\sum a_{\mathbf{n}} x^{\mathbf{n}}\right)\left(\sum b_{\mathbf{m}} x^{\mathbf{m}}\right) = \sum c_{\mathbf{k}} x^{\mathbf{k}}, \text{ where } c_{\mathbf{k}} = \sum_{\mathbf{n}+\mathbf{m}=\mathbf{k}} a_{\mathbf{n}} b_{\mathbf{m}} \tag{2.4.30}$$

We also assume that $x_i x_j = x_j x_i$ for all $i, j \in I$. Such set of elements (2.4.28) with addition (2.4.29) and multiplication (2.4.30) forms an associative ring which is called the **formal power series ring**, and denoted by $A[[x_i]]_{i \in I}$. Elements of this ring are called **formal power series**. If A is a commutative ring, then so is $A[[x_i]]_{i \in I}$.

There are two natural ring homomorphisms

$$\iota : A \longrightarrow A[[x_i]]_{i \in I} \quad \text{and} \quad \varepsilon : A[[x_i]]_{i \in I} \longrightarrow A \tag{2.4.31}$$

defined by $\iota(a) = \sum a_{\mathbf{n}} x^{\mathbf{n}}$ with $a_{\mathbf{n}} = 0$ if $|\mathbf{n}| \geq 1$, $a_{\mathbf{0}} = a$, and $\varepsilon(\sum a_{\mathbf{n}} x^{\mathbf{n}}) = a_{\mathbf{0}}$. Of course, $\varepsilon \circ \iota = id_A$. We will identify the elements of A with elements $\iota(A) \subset A[[x_i]]_{i \in I}$. Then from (2.4.30) it follows that $a x_i = x_i a$ for all $a \in A$ and $i \in I$.

Lemma 2.4.32. *An element $f(x) = \sum_{\mathbf{n} \in \Omega_I} a_{\mathbf{n}} x^{\mathbf{n}} \in A[[x_i]]_{i \in I}$ is invertible if and only if $\varepsilon(f(x)) = a_{\mathbf{0}} \in U(A)$.*

Proof. Since ε is a ring homomorphism, the "only if" part is trivial. So suppose that $\varepsilon(f(x)) = a_{\mathbf{0}} \in U(A)$, where $f(x) = \sum_{\mathbf{n} \in \Omega_I} a_{\mathbf{n}} x^{\mathbf{n}}$. Then there is $b_{\mathbf{0}} \in U(A)$ such that $a_{\mathbf{0}} b_{\mathbf{0}} = 1$. In order to prove that $f(x)$ is right invertible in $A[[x_i]]_{i \in I}$ one must find $g(x) = \sum_{\mathbf{m} \in \Omega_I} b_{\mathbf{m}} x^{\mathbf{m}} \in A[[x_i]]_{i \in I}$ such that $f(x) g(x) = 1$. This gives the following infinite system of equations:

$$a_{\mathbf{0}} b_{\mathbf{0}} = 1 \tag{2.4.33}$$

$$0 = \sum_{\mathbf{n}+\mathbf{m}=\mathbf{k}} a_{\mathbf{n}} b_{\mathbf{m}} = a_{\mathbf{0}} b_{\mathbf{k}} + \sum_{\substack{\mathbf{n}+\mathbf{m}=\mathbf{k} \\ \mathbf{m}<\mathbf{k}}} a_{\mathbf{n}} b_{\mathbf{m}}, \tag{2.4.34}$$

which has a solution obtained inductively:

$$b_{\mathbf{0}} = a_{\mathbf{0}}^{-1},$$

and given $b_{\mathbf{m}}$ with $\mathbf{m} < \mathbf{k}$ we can solve (2.4.34) with $b_{\mathbf{k}}$ again. This shows that $f(x)$ is right invertible. Similar we can show that $f(x)$ is left invertible. So $f(x)$ is invertible, i.e., it is a unit.

Remark 2.4.35. If $I = \{1,2,\dots,n\}$ we obtain the ring of formal power series $A[[x_i]]_{i\in I} = A[[x_1,x_2,\dots,x_n]]$ in n variables defined by (2.4.11) above.

Remark 2.4.36. If in the definition of a set of formal sums of the form (2.4.28) we will require that $x_i x_j \neq x_j x_i$ for all $i,j \in I$, $i \neq j$, we obtain a ring which is a free A-module generated by the set of variables $\{x_i\}_{i\in I}$. This ring is denoted by $A\langle\langle x_i\rangle\rangle_{i\in I}$, and it is called the **free associative power series ring**.

For any associative ring A and any ring automorphism $\sigma \in \mathrm{Aut}(A)$ one can consider the ring $A[x_1,x_2,\dots,x_n;\sigma]$ of all skew power series in n commuting variables $x_1,x_2,\dots,x_n$ with coefficients in a ring A which is defined inductively by

$$A[[x_1,x_2,\dots,x_n;\sigma]] = A[[x_1,x_2,\dots,x_{n-1};\sigma]][[x_n;\sigma]] =$$
$$= A[[x_1;\sigma]][[x_2;\sigma]]\dots[[x_n;\sigma]], \tag{2.4.37}$$

where the automorphism σ is naturally extends to an automorphism of each ring $A[[x_1,x_2,\dots,x_k;\sigma]]$ for $k = 1,2,\dots,n-1$. In this case one also assumes that $x_i x_j = x_j x_i$ for all $i,j = 1,2,\dots,n$.

2.5 Laurent Polynomial Rings and Laurent Power Series Rings

Let A be a ring. The set of elements which are finite sums of the form

$$f = \sum_{i=m}^{n} a_i x^i, \tag{2.5.1}$$

where $m = m(f) \in \mathbf{Z}$, $n \geq m$, $n \in \mathbf{Z}$, with the operation of addition componentwise and multiplication defined by the distributive law and the rule $ax = xa$ for all $a \in A$, forms a ring which is called the **Laurent polynomial ring**, and denoted by $A[x,x^{-1}]$.

Let A be a ring. Consider the set of formal expressions of the form

$$f = \sum_{i=m}^{\infty} a_i x^i, \tag{2.5.2}$$

where $m = m(f) \in \mathbf{Z}$, $a_i \in A$. The element a_m is called a_m the **lowest coefficient** of f. We will also assume that $ax = xa$ for all $a \in A$. The expressions are added componentwise and multiplied as follows:

$$\left(\sum_{i=n}^{\infty} a_i x^i\right)\left(\sum_{j=m}^{\infty} b_j x^j\right) = \sum_{\substack{k=n+m \\ n,m\in\mathbf{Z}}}^{\infty} c_k x^k,$$

where

$$c_k = \sum_{\substack{k=i+j \\ i,j\in\mathbf{Z}}} a_i b_j$$

(note that this sum is finite). Thus this set forms a ring which is called the **formal Laurent series ring** and denoted by $A[[x,x^{-1}]]$, or $A((x))$ and its elements (2.5.2) are called **formal Laurent series.**

Remark 2.5.3. The Laurent polynomial ring $A[x,x^{-1}]$ is obviously a subring in $A[[x,x^{-1}]]$.

Proposition 2.5.4. *If $A = D$ is a division ring, then $D[[x,x^{-1}]]$ is also a division ring. In particular, if $A = K$ is a field then $K[[x,x^{-1}]]$ is also a field and it is the quotient field of the ring of formal power series $K[[x]]$.*

Proof. Let $A = D$ be a division ring and $f = \sum_{i=m}^{\infty} a_i x^i \in D[[x,x^{-1}]]$ with $a_m \neq 0$. Then $fx^{-m} = \sum_{i=m}^{\infty} a_i x^{i-m} \in D[[x]]$. Since $a_m \neq 0$, $fx^{-m} \in U(D[[x]])$ by proposition 2.4.5. Then $f \in U(D[[x]])$ as well, since $x^m \in U(D[[x,x^{-1}]])$. □

As in the previous cases removing the assumption of commutativity of the variable x with coefficients of the ring A leads, e.g., to the construction of the following ring.

Let A be a ring with a (left) skew derivation (σ,δ), so that (2.3.19) holds. Consider the set of finite sums

$$f = \sum_{i=m}^{n} a_i x^i \tag{2.5.5}$$

where $m = m(f) \in \mathbf{Z}$, $n \geq m$, $n \in \mathbf{Z}$, with addition defined componentwise and multiplication defined by the distributive law and the rule

$$xa = \sigma(a)x + \delta(a) \tag{2.5.6}$$

for all $a \in A$. Then this set forms a ring which is called a (left) **skew** (or **twisted**) **Laurent polynomial ring**, and denoted by $A[x,x^{-1};\sigma,\delta]$, and elements of this ring are called **skew Laurent polynomials**.

Note, that if $\sigma = 1$ the ring $A[x,x^{-1};\delta] = A[x,x^{-1};1,\delta]$ is called a **skew Laurent polynomial ring of derivative type**. If $\sigma = 1$ and $\delta = 0$ one obtains the Laurent polynomial ring $A[x,x^{-1};1,0] = A[x,x^{-1}]$.

Example 2.5.7. If $G = \{x^n : n \in \mathbf{Z}\} \cong \mathbf{Z}$ and $\varphi(x) = \sigma$ then $A[x,x^{-1};\sigma] \simeq A *_\varphi G$.

Proposition 2.5.8. *Let A be a ring and σ an automorphism of A. Then a ring $T = A[x,x^{-1};\sigma]$ is a domain if and only if A is a domain.*

Proof. If $T = A[x,x^{-1};\sigma]$ is a domain then A is obviously a domain as well. Suppose that A is a domain. Let $u,v \in T$, then $u = \sum\limits_{i=n}^{k} a_i x^i$ and $v = \sum\limits_{j=m}^{k} b_j x^j$ with lowest coefficients $a_n \neq 0$ and $b_m \neq 0$ respectively. Then the lowest coefficient of uv is equal to $c = a_n\sigma^n(b_m)$ and $c \neq 0$, since σ is a automorphism of A. So A is a domain. □

Definition 2.5.9. Let A be a ring, and let σ be an automorphism of A. Then the set of formal expressions of the form $f = \sum\limits_{i=m}^{\infty} a_i x^i$, where $m = m(f) \in \mathbf{Z}$ and $xa = \sigma(a)x$ for all $a \in A$, is called a (left) **skew** (or **twisted**) **Laurent power series ring** over A and denoted by $A[[x,x^{-1};\sigma]]$, or $A((x,\sigma))$.

Note, that if $\sigma = 1$ one obtains the Laurent power series ring $A[[x,x^{-1}]] = A[[x,x^{-1};1]]$.

Proposition 2.5.10. *Let A be a ring and σ an automorphism of A. Then:*

1. *The ring $T = A[[x,x^{-1};\sigma]]$ is a division ring if and only if A is a division ring.*
2. *The ring $T = A[[x,x^{-1};\sigma]]$ is a domain if and only if A is a domain.*

Proof. 1. It is obvious that if $A[[x,x^{-1};\sigma]]$ is a division ring then A is also a division ring. Conversely, suppose that A is a division ring. Take a non-zero element $u = \sum\limits_{i=n}^{\infty} a_i x^i \in T$ with lowest coefficient $a_n \neq 0$. We must show that there is a non-zero element $v = \sum\limits_{i=m}^{\infty} b_i x^i \in T$ such that $uv = 1$. Since $a_n \neq 0$, and A is a division ring, there exists an element $b \in A$ such that $a_n b = 1$. Set

$$w = u \cdot (x^{-n}b) = \sum_{i=n}^{\infty} a_i x^i \cdot (x^{-n}b) = \sum_{i=n}^{\infty} a_i \sigma^{i-n}(b) x^{i-n}.$$

Then $w = \sum\limits_{j=0}^{\infty} c_j x^j \in A[[x;\sigma]]$, where $c_j = a_{n+j}\sigma^j(b) \in A$. Since $c_0 = a_n b = 1$, w is invertible in $A[[x;\sigma]]$, by proposition 2.4.8. Therefore there is an element $w^{-1} \in A[[x;\sigma]]$, such that $w \cdot w^{-1} = 1$. Set

$$v = (x^{-n}b)w^{-1} \in A[[x,x^{-1};\sigma]].$$

Then $uv = u \cdot (x^{-n}b)w^{-1} = w \cdot w^{-1} = 1$, which means that u is invertible in $A[[x, x^{-1}; \sigma]]$.

2. If $T = A[[x, x^{-1}; \sigma]]$ is a domain then A is obviously a domain as well. Suppose that A is a domain. Let $u, v \in T$, then $u = \sum_{i=n}^{\infty} a_i x^i$ and $v = \sum_{i=m}^{\infty} b_i x^i$ with lowest coefficients $a_n \neq 0$ and $b_m \neq 0$ respectively. Then the lowest coefficient of uv is equal to $c = a_n \sigma^n(b_m)$ and $c \neq 0$, since σ is a automorphism of A and A is a domain. □

2.6 Generalized Matrix Rings. Generalized Triangular Matrix Rings

This section contains the important construction of generalized matrix rings. Generalized matrix rings form one of the largest classes of matrix rings and is studied extensively in various contexts.

Let $A_i = X_{ii}$ be rings, and let X_{ij} be an A_i-A_j-bimodule. Assume that there are bimodule homomorphisms

$$f_{ijk} : X_{ij} \otimes X_{jk} \to X_{ik}. \tag{2.6.1}$$

Put

$$x_{ij} \cdot x_{jk} = f_{ijk}(x_{ij} \otimes x_{jk}) \tag{2.6.2}$$

and assume also that bimodule morphisms (2.6.1) satisfy the associativity conditions:

$$(x_{ij} \cdot x_{jk}) \cdot x_{ks} = x_{ij} \cdot (x_{jk} \cdot x_{ks}) \tag{2.6.3}$$

for all $x_{rt} \in X_{rt}$.

Consider the set

$$A = \begin{pmatrix} A_1 & X_{12} & \cdots & X_{1n} \\ X_{21} & A_2 & \cdots & X_{2n} \\ \vdots & \vdots & \ddots & \vdots \\ X_{n1} & X_{n2} & \cdots & A_n \end{pmatrix} \tag{2.6.4}$$

of all matrices of the form

$$\begin{pmatrix} a_{11} & a_{12} & \cdots & a_{1n} \\ a_{21} & a_{22} & \cdots & a_{2n} \\ \vdots & \vdots & \ddots & \vdots \\ a_{n1} & a_{n2} & \cdots & a_{nn} \end{pmatrix}$$

where $a_{ij} \in X_{ij}$. Addition in this set is defined componentwise and multiplication is defined as in the usual matrix ring taking into account (2.6.1) and (2.6.2).

With respect to these operations A becomes a ring which is called a **generalized matrix ring** (of order n).

A ring A of the form (2.6.4) is said to be a **generalized upper (lower) triangular matrix ring** (or **formal triangular matrix ring**) if $X_{ij} = 0$ for all $1 \le j < i \le n$ (resp. $1 \le i < j \le n$).

Examples 2.6.5.

1. Any ring A can be written trivially as a formal triangular matrix ring, since
$$A \cong \begin{pmatrix} A & 0 \\ 0 & 0 \end{pmatrix} \cong \begin{pmatrix} 0 & 0 \\ 0 & A \end{pmatrix}.$$
2. If a ring A has a non-trivial idempotent e then there are two subrings eAe and $(1-e)A(1-e)$ in A. The two-sided Peirce decomposition of A gives a generalized matrix ring of the form
$$\begin{pmatrix} eAe & eA(1-e) \\ (1-e)Ae & (1-e)A(1-e) \end{pmatrix}.$$
If e is a central idempotent of a ring A then one has a direct product decomposition which is a formal triangular matrix ring of the form
$$\begin{pmatrix} eAe & 0 \\ 0 & (1-e)A(1-e) \end{pmatrix} \cong eAe \times (1-e)A(1-e).$$
3. If a ring A has a finite set of pairwise orthogonal idempotents $e_1, e_2, \dots, e_n$ such that $1 = e_1 + e_2 + \cdots + e_n$, then its two-sided Peirce decomposition[3] in the usual way can be considered as a generalized matrix ring.
4. If $A_1 = X_{12} = A_2 = A$ is a ring, $X_{21} = 0$, then the formal triangular matrix ring $\begin{pmatrix} A & A \\ 0 & A \end{pmatrix}$ is exactly the ring of all upper triangular 2×2 matrices over A.
5. If $A_1 = \mathbf{Z}$, $A_2 = \mathbf{Q}$, $X_{21} = 0$ and $X_{12} = \mathbf{Q}$, considered as a $\mathbf{Z}$-$\mathbf{Q}$-bimodule, then we obtain the formal triangular matrix ring $\begin{pmatrix} \mathbf{Z} & \mathbf{Q} \\ 0 & \mathbf{Q} \end{pmatrix}$. This ring was used in many papers as an example of a ring which is right Noetherian but not left Noetherian, neither right nor left Artinian (See e.g. [146, section 3.6]).

Consider the structure of all ideals of the formal triangular matrix ring $A = \begin{pmatrix} S & M \\ 0 & T \end{pmatrix}$. Let $1 = e_1 + e_2$ be a decomposition of 1 into a sum of orthogonal idempotents, where $e_1 = \begin{pmatrix} 1 & 0 \\ 0 & 0 \end{pmatrix}$ and $e_2 = \begin{pmatrix} 0 & 0 \\ 0 & 1 \end{pmatrix}$. Then $e_1 A = \begin{pmatrix} S & M \\ 0 & 0 \end{pmatrix}$, $e_2 A = \begin{pmatrix} 0 & 0 \\ 0 & T \end{pmatrix}$, $Ae_1 = \begin{pmatrix} S & 0 \\ 0 & 0 \end{pmatrix}$, and $Ae_2 = \begin{pmatrix} 0 & M \\ 0 & T \end{pmatrix}$ are obviously two-sided ideals of A.

[3]See Chapter 1.

Let $\mathcal{I} = \begin{pmatrix} I_1 & N \\ 0 & I_2 \end{pmatrix}$ be a right ideal of A. From the definition of addition in A, one immediately obtains that I_1, I_2 are additive subgroups in S, T, respectively, and N is a sub-bimodule in M. From the definition of multiplication in A it follows that

$$\mathcal{I}A = \begin{pmatrix} I_1 & N \\ 0 & I_2 \end{pmatrix}\begin{pmatrix} S & M \\ 0 & T \end{pmatrix} = \begin{pmatrix} I_1S & I_1M + NT \\ 0 & I_2T \end{pmatrix}$$

which means that the I_1, I_2 are right ideals in S, T respectively, and N is a sub-bimodule in M such that $I_1M \subseteq N$.

Conversely, any set $\mathcal{I} = \begin{pmatrix} I_1 & N \\ 0 & I_2 \end{pmatrix}$, where I_1, I_2 are right ideals in S, T respectively, and N is a sub-bimodule in M such that $I_1M \subseteq N$, is a right ideal of A.

Analogously one can check that if $\mathcal{J} = \begin{pmatrix} J_1 & K \\ 0 & J_2 \end{pmatrix}$ is a left ideal of A then J_1, J_2 must be left ideals in S, T respectively, and K a sub-bimodule in M such that $MJ_2 \subseteq K$.

From the considerations above one can conclude that a two-sided ideal of A has the form $\begin{pmatrix} X_1 & Y \\ 0 & X_2 \end{pmatrix}$ where X_1, X_2 are two-sided ideals in S, T respectively, and Y is a sub-bimodule in M such that $X_1M + MX_2 \subseteq Y$.

Therefore we obtain the following result.

Proposition 2.6.6 (K.R. Goodearl [109]). *Let* $A = \begin{pmatrix} S & M \\ 0 & T \end{pmatrix}$ *be a formal triangular matrix ring. Then*

1. *If* I_2 *is a right ideal in* T, $I_1 \oplus N$ *is a right* T*-submodule in* $S \oplus M$, *and* $I_1M \subseteq N$, *then* $\begin{pmatrix} I_1 & N \\ 0 & I_2 \end{pmatrix}$ *is a right ideal of A. Conversely, any right ideal of A has this form.*
2. *If* J_1 *is a left ideal in* S, $K \oplus J_2$ *is a left* S*-submodule in* $M \oplus T$, *and* $MJ_2 \subseteq K$, *then* $\begin{pmatrix} J_1 & K \\ 0 & J_2 \end{pmatrix}$ *is a left ideal of A. Conversely, any left ideal of A has this form.*
3. *If* X_1, X_2 *are two-sided ideals in* S, T *respectively, and* Y *is a sub-bimodule in* M *such that* $X_1M + MX_2 \subseteq Y$, *then* $\begin{pmatrix} X_1 & Y \\ 0 & X_2 \end{pmatrix}$ *is a two-sided ideal of A. Conversely, any two-sided ideal of A has this form.*

The following result easily follows from theorem 1.1.23:

Proposition 2.6.7. *Let* $A = \begin{pmatrix} S & M \\ 0 & T \end{pmatrix}$ *be a formal triangular matrix ring. Then A is right (resp. left) Noetherian (Artinian) if and only if S and T are both right (resp. left) Noetherian (Artinian), and M is a finitely generated right* T*-module (resp. left* S*-module).*

The Jacobson radical is one of the main tools in the investigation of the structure of rings. The next statements give the structure of the Jacobson radical of generalized matrix rings under some special conditions.

Lemma 2.6.8. *Let* $A = \begin{pmatrix} A_1 & X_{12} \\ X_{21} & A_2 \end{pmatrix}$ *be a generalized matrix ring, where* X_{ij} *is an* A_i-A_j*-bimodule, for* $i,j = 1,2$. *Suppose that* $X_{ij}X_{ji} \subseteq \mathrm{rad}(A_i)$. *Then*

$$\mathrm{rad}(A) = \begin{pmatrix} \mathcal{R}_1 & X_{12} \\ X_{21} & \mathcal{R}_2 \end{pmatrix} \tag{2.6.9}$$

where $\mathcal{R}_i = \mathrm{rad}(A_i)$ *for* $i,j = 1,2$.

Proof. Let $1 = e_1 + e_2$ and $e_iAe_i = A_i$, $e_iAe_j = X_{ij}$, for $i,j = 1,2$, and $\mathcal{R} = \mathrm{rad}(A)$. By proposition 1.1.19, $\mathcal{R}_i = e_i\mathcal{R}e_i$ for $i = 1,2$. Let

$$L = \begin{pmatrix} \mathcal{R}_1 & X_{12} \\ X_{21} & \mathcal{R}_2 \end{pmatrix}.$$

Since, by assumption, $X_{ij}X_{ji} \subseteq \mathrm{rad}(A_i)$, L is a two-sided ideal in A.

Note that

$$1 - u = \begin{pmatrix} 1 - r_1 & x \\ y & 1 - r_2 \end{pmatrix}$$

is invertible for any $u \in L$.

One has

$$\begin{pmatrix} 1 - r_1 & x \\ y & 1 - r_2 \end{pmatrix} \begin{pmatrix} 1 & -(1 - r_1)^{-1}x \\ 0 & 1 \end{pmatrix} = \begin{pmatrix} 1 - r_1 & 0 \\ y & 1 - r_2' \end{pmatrix},$$

where $r_2' = r_2 + y(1 - r_1)^{-1}x \in \mathcal{R}_2$. Furthermore,

$$\begin{pmatrix} 1 & 0 \\ -(1 - r_1)^{-1}y & 1 \end{pmatrix} \begin{pmatrix} 1 - r_1 & 0 \\ y & 1 - r_2' \end{pmatrix} = \begin{pmatrix} 1 - r_1 & 0 \\ 0 & 1 - r_2' \end{pmatrix}.$$

Since the latter matrix is clearly invertible, so is $1 - u$.

By proposition 1.1.18, $L \subseteq \mathrm{rad}\, A$, and by proposition 1.1.19, $\mathrm{rad}\, A \subseteq L$. Thus, $\mathrm{rad}\, A = L$, as required. □

Theorem 2.6.10. *Let*

$$A = \begin{pmatrix} A_1 & X_{12} & \cdots & X_{1n} \\ X_{21} & A_2 & \cdots & X_{2n} \\ \vdots & \vdots & \ddots & \vdots \\ X_{n1} & X_{n2} & \cdots & A_n \end{pmatrix} \tag{2.6.11}$$

be a generalized matrix ring, where X_{ij} *is an* A_i*-*A_j*-bimodule* $i = 1,2,\dots,n$. *Suppose that* $X_{ij}X_{ji} \subseteq \mathrm{rad}(A_i)$ *for* $i = 1,2,\dots,n;\ i \neq j$. *Then*

$$\mathrm{rad}(A) = \begin{pmatrix} \mathcal{R}_1 & X_{12} & \cdots & X_{1n} \\ X_{21} & \mathcal{R}_2 & \cdots & X_{2n} \\ \vdots & \vdots & \ddots & \vdots \\ X_{n1} & X_{n2} & \cdots & \mathcal{R}_n \end{pmatrix} \tag{2.6.12}$$

where $\mathcal{R}_i = \mathrm{rad}(A_i)$ *for* $i = 1,2,\dots,n$.

Proof. The theorem follows by induction on n using lemma 2.6.8. □

Corollary 2.6.13. *Let*

$$A = \begin{pmatrix} A_1 & X_{12} & \cdots & X_{1n} \\ 0 & A_2 & \cdots & X_{2n} \\ \vdots & \vdots & \ddots & \vdots \\ 0 & 0 & \cdots & A_n \end{pmatrix} \tag{2.6.14}$$

be a generalized triangular matrix ring, where X_{ij} *is an* A_i*-*A_j*-bimodule, for* $i,j = 1,2,\dots,n$. *Then*

$$\mathrm{rad}(A) = \begin{pmatrix} \mathcal{R}_1 & X_{12} & \cdots & X_{1n} \\ 0 & \mathcal{R}_2 & \cdots & X_{2n} \\ \vdots & \vdots & \ddots & \vdots \\ 0 & 0 & \cdots & \mathcal{R}_n \end{pmatrix} \tag{2.6.15}$$

where $\mathcal{R}_i = \mathrm{rad}(A_i)$, *for* $i = 1,2,\dots,n$.

Corollary 2.6.16. *Let*

$$A = \begin{pmatrix} A_1 & X_{12} & \cdots & X_{1n} \\ 0 & A_2 & \cdots & X_{2n} \\ \vdots & \vdots & \ddots & \vdots \\ 0 & 0 & \cdots & A_n \end{pmatrix} \tag{2.6.17}$$

be a generalized triangular matrix ring, where X_{ij} *is an* A_i*-*A_j*-bimodule, for* $i,j = 1,2,\dots,n$ *with Jacobson radical* R (2.6.15). *Then*

1. $A/R \cong A_1/\mathrm{rad}(A_1) \times \cdots \times A_n/\mathrm{rad}(A_n)$.
2. *Idempotents in* A *can be lifted modulo the radical* R *if and only if idempotents in* A_i *can be lifted modulo the Jacobson radical* $\mathrm{rad}(A_i)$ *for all* $i = 1,\dots,n$.
3. A *is semilocal if and only if the* A_i *are semilocal for all* $i = 1,\dots,n$.
4. A *is a semiperfect ring if and only if* A_i *are semiperfect for all* $i = 1,\dots,n$..
5. A *is a semiprimary ring if and only if* A_i *are semiprimary for all* $i = 1,\dots,n$..

Proof. Statements 1 and 2 follow immediately from corollary 2.6.13.

Statement 3 follows from statement 1. Statement 4 is a consequence of statements 2 and 3.

Suppose that A is semiprimary. Then A/R is semisimple and R is nilpotent. From statement 2 it follows that the $A_i/\mathrm{rad}(A_i)$ are semisimple for all $i = 1,\dots,n$. From (2.6.15) it follows that the $\mathcal{R}_i = \mathrm{rad}(A_i)$ are also nilpotent. □

2.7 *G*-graded Rings

In [146] graded rings were considered, which are the particular case of G-graded rings where G is the additive monoid of nonnegative integer numbers. This section considers the general case of G-graded rings when G is an arbitrary groupoid, that is a universal algebra with one binary operation.

Definition 2.7.1. Let G be a groupoid with its binary operation thought of as multiplication. An associative ring A is said to be **G-graded** (or **graded of type** G) if it can be represented as a direct sum of additive subgroups A_g:

$$A = \bigoplus_{g \in G} A_g$$

such that $A_g A_h \subseteq A_{gh}$ for all $g, h \in G$.

In particular, if $G = \mathbf{Z}$ is the additive Abelian group, then a ring A is $\mathbf{Z}$-graded if $A = \bigoplus\limits_{n \in \mathbf{Z}} A_n$, where A_i is an additive subgroup of A such that $A_i A_j \subseteq A_{i+j}$ for all $i, j \in \mathbf{Z}$. In this case A_0 is a subring of A, $1 \in A_0$, and A_i is an A_0-module for all $i \in \mathbf{Z}$.

If $G = \mathbf{N} \cup \{0\}$ is the additive monoid of nonnegative integer numbers, then A is called a $\mathbf{N}$-graded ring if $A = A_0 \oplus A_1 \oplus \cdots \oplus A_n \oplus \cdots$, where A_i is an additive subgroup of A such that $A_i A_j \subseteq A_{i+j}$ for all $i, j \in G$. In this case A_0 is also a subring of A, $1 \in A_0$, and A_i is an A_0-module for all $i \in \mathbf{N}$.

A G-graded ring A is called **strongly G-graded** if $A_g A_h = A_{gh}$ for all $g, h \in G$.

Groupoid-graded rings include as a special cases many other ring constructions: polynomial and skew polynomial rings, direct and semidirect product of rings, matrix and generalized matrix rings, group and semigroup rings, tensor algebras, crossed products, path algebras, incidence rings and various other ones considered in this chapter.

Examples 2.7.2.

1. Let A be a ring, and let G be a monoid. Then the monoid ring $A[G]$, defined by (2.2.1), is a G-graded ring with components $A_g = Ag$, where $g \in G$. In particular, the polynomial ring $B = A[x] = A[G]$ over A, where $G = \langle x \rangle \simeq \mathbf{N} \cup \{0\}$, is a $\mathbf{N}$-graded ring with components $B_i = Ax^i$, monomials of degree $i > 0$ and $B_0 = A$. In this case $A_i A_j \subseteq A_{i+j}$. If A is a domain, then $A[x]$ is a strongly G-graded ring.
2. Let $B = A[x_1, x_2, \dots, x_n]$ be the polynomial ring in n variables over an associative ring A, defined by (2.3.7). Then B is a $\mathbf{N}$-graded ring with

components B_k homogeneous polynomials of degree $k > 0$, and $B_0 = A$. In this case $B_iB_j \subseteq B_{i+j}$. If A is a domain, then B is a strongly $\mathbf{N}$-graded ring.

3. Let $B = A[[x]]$ be the ring of formal power series over an associative ring A. If G is as example 1, then B is a $\mathbf{N}$-graded ring with components $B_i = Ax^i$, monomials of degree $i > 0$ and $B_0 = A$. In this case $A_iA_j \subseteq A_{i+j}$. If A is a domain, then $A[x]$ is a strongly $\mathbf{N}$-graded ring.
4. Let $B = A[[x_i]]_{i \in I}$ be the formal power series ring over an associative ring A (here I is finite or infinite countable set of indices). Then B is a $\mathbf{N}$-graded ring with components B_i, homogeneous polynomials of degree $k > 0$ and $B_0 = A$. In this case $B_iB_j \subseteq B_{i+j}$. If A is a domain, then $B = A[x_i]_{i \in I}$ is a strongly $\mathbf{N}$-graded ring.
5. The skew polynomial ring $B = A[x, \sigma]$ is also a $\mathbf{N}$-graded ring with components $B_i = Ax^i$ for $i > 0$ and $B_0 = A$.
6. The Laurent formal series ring $B = A[[x, x^{-1}]]$ and the skew Laurent formal series ring $B = A[[x, x^{-1}; \sigma]]$ are $\mathbf{Z}$-graded rings with components $B_i = Ax^i$ for $i \in \mathbf{Z}$ and $B_0 = A$.
7. Let G be a monoid with the unity element e. The skew group ring $B = A[G, \sigma]$ is a G-graded ring with components $B_g = Ag$ for all $e \neq g \in G$ and $B_e = A$.

The components A_g are called **homogeneous components** (or g-**components**), and the elements of A_g are called **homogeneous of degree** g of the ring A. An element $a \in A$ has a unique decomposition as $a = \sum_{g \in G} a_g$, where $a_g \in A$ for all $g \in G$, and where all but finitely many of the a_g's are zero.

The set

$$\sup(a) = \{g \in G : a_g \neq 0\}$$

is called the **support** of $a \in A$ in G. The set

$$\sup(A) = \{g \in G : A_g \neq 0\}$$

is called the **support** of a G-graded ring A. In the case $\sup(A)$ is a finite set we will write $\sup(A) < \infty$ and A is said to be a G-**graded ring with finite support**.

If G is a group (semigroup), then A is called a **group-graded (semigroup-graded) ring**.

Lemma 2.7.3. *Let G be a group with neutral element e, and let A be a ring with identity* 1. *If A is a G-graded ring then A_e is a subring of A, $1 \in A_e$, and A_g is an A_e-module for all $g \in G$.*

Proof. As $A_eA_e \subseteq A_e$, A_e is closed under multiplication and thus it is a subring of A.

To see that $1 \in A_e$ write $1 = \sum_{g \in G} x_g$, where $x_g \in A_g$ and all but finitely many of the x_g's are zero. Then for all $h \in G$,

$$x_h = 1 \cdot x_h = \sum_{g \in G} x_g x_h.$$

Comparing degrees of the x_h it is easy to see that $x_h = x_e x_h$ for all $h \in G$. Therefore

$$x_e = 1 \cdot x_h = \sum_{g \in G} x_g x_e = \sum_{g \in G} x_g = 1.$$

Hence $1 = x_e \in A_g$.

The last statement follows from the fact that $A_e A_g \subseteq A_{eg} = A_g$ for all $g \in G$. □

Let G be a group with identity e, and let A be a ring with identity 1. If A is a G-graded ring then an A_e-module N gives rise to an A-module $M = N \otimes_{A_e} A$. The decomposition of A as a direct sum of its A_e-submodules A_g leads to a decomposition of M as a direct sum of its A_e-submodules $M_g \cong N \otimes_{A_e} A_g$ for $g \in G$. This decomposition makes M a G-graded A-module, i.e.,

$$A_g M_h \subseteq M_{gh}$$

for all $g, h \in G$.

There exist graded A-modules which are not induced by A_e-modules. In general one has the following definition.

Definition 2.7.4. Let A be a G-graded ring. An A-module $M = \bigoplus_{g \in G} M_g$ is called a G-**graded module** if $A_g M_h \subseteq M_{gh}$ for all $g, h \in G$.

The components M_g are called **homogeneous components** (or g-**components**) of a G-graded module M, and non-zero elements of M_g are called **homogeneous elements of degree** g of M.

A submodule N of a G-graded module M is said to be **homogeneous** if it can be decomposed into a direct sum of subgroups N_g such that $N_g \subseteq M_g$ for all $g \in G$.

If N is a homogeneous submodule of a G-graded module M then the quotient module $\bar{M} = M/N$ is also a G-graded module, i.e., $\bar{M} = \sum_{g \in G} \bar{M}_g$, where $\bar{M}_g$ is the image of the submodule M_g under the natural homomorphism $M \longrightarrow M/N$, $\bar{M}_g \cong M_g/N_g$.

Let A be a G-graded ring. The largest homogeneous ideal contained in $\mathrm{rad}(A)$ is called the **homogeneous radical** of A and denoted by $\mathcal{R}_G(A)$.

If $G = \mathbf{Z}_2 = \{0,1\}$, then a G-graded algebra is called a **superalgebra**. A **Lie superalgebra** is a superalgebra $L = L_0 \oplus L_1$ such that for all $a,b \in \mathbf{Z}_2$, $x \in L_a$, $y \in L_b$, $z \in L$ there are two identities:

$$[x,y] = -(-1)^{ab}[y,x],$$

$$[[x,y],z] = [x,[y,z]] - (-1)^{ab}[y,[x,z]].$$

Every associative algebra with direct sum decomposition $A = A_0 \oplus A_1$ gives a Lie superalgebra by setting

$$[x,y] = xy - (-1)^{ab}yx,$$

for all $a,b \in \mathbf{Z}_2$, $x \in A_a$, $y \in A_b$.

2.8 Notes and References

The crossed products are first arose in the study of finite dimensional division algebras and central simple algebras. Now they have found the important applications in many different fields of algebra, such as the study of infinite group algebras, group-graded rings and the Galois theory of noncommutative rings. Many publications and books are devoted to study the theory of the crossed products and their applications. First of all it should be distinguished two remarkable books of D.S. Passman [260], [261]. In these books the reader can also find a comprehensive reference on these topics.

O. Ore [256] was the first who systematically studied the skew polynomial rings. He also first showed in [256] that the left-hand division algorithm holds in the skew polynomial ring over a division ring.

D. Hilbert [149] was the first to consider skew Laurent series rings with $\sigma(x) = x^2$ to show the existence of noncommutative ordered division ring.

The general concept of a groupoid-graded ring was first considered in 1960 by G. Schiffels in [282]. Group graded rings and modules and their properties were studied by E.C. Dade [63], Cohen and S.Montgomery [57], M.E. Harris [143], Cohen and Montgomery [56], A. Jensen and Jøndrup [171], A.V. Kelarev and Okniński [185], [186].

CHAPTER 3

Valuation Rings

At first the theory of valuation rings was connected only with commutative fields. Discrete valuation domains are, excepting only fields, the simplest class of rings. Nevertheless they play an important role in algebra and algebraic geometry.

There is also a noncommutative side to this theory. In the noncommutative case there are different generalizations of valuation rings. The first generalization for valuation rings of division rings was obtained by Schilling [283], who introduced the class of invariant valuation rings and systematically studied them in [284].

Another significant contribution in noncommutative valuation rings was made by N.I. Dubrovin who introduced a more general concept of a valuation ring for simple Artinian rings and proved a number of nontrivial properties about them [73], [71], [72]. These rings are now called Dubrovin valuation rings. Dubrovin valuation rings have found a large number of applications.

In this chapter we present and briefly discuss many of the basic results for valuation rings and discrete valuation rings.

Section 3.1 is devoted to valuation rings of fields. In this section we give the main properties of these rings and describe different equivalent definitions of them. Section 3.2 presents the basic results about discrete valuation domains. We give the structure of these rings and a lot of equivalent definitions. In section 3.3 we describe noncommutative invariant valuation rings of division rings, where we give the main properties of these rings and their equivalent definitions. Some examples of noncommutative non-discretely-valued valuation rings are presented in 3.4. The main properties and structure of noncommutative discrete valuation rings are presented in section 3.5. In section 3.6 we briefly discuss total valuation rings which are more general class than invariant valuation rings. Section 3.7 is devoted to other type of valuation rings with zero divisors allowed. We consider some valuation rings of commutative rings with zero divisor and the Dubrovin valuation rings. Finally, in section 3.8 we consider the approximation theorems for noncommutative valuation rings and some corollaries from these theorems for noncommutative discrete valuation rings.

This chapter may be considered as a short introduction to the theory of valuation rings. More information about valuation rings of division rings can be found in [284], and about Dubrovin valuation rings, semihereditary and Prüfer orders in simple Artinian rings can be found in the book [233].

The first two sections of this chapter primarily deal with the commutative case. In particular 'a field' means a commutative field and an integral domain is a commutative ring without zero divisors so that it always has a field of fractions.

3.1 Valuation Domains

Consider first the important notion of a totally ordered group which plays a main role in the theory of valuations.

Definition 3.1.1. A group G (with operation written by $+$) is said to be **totally ordered** (or **linearly ordered**) if there is a binary order relation $\geq$ in G which satisfies the following axioms:

i. Either $\alpha \geq \beta$ or $\beta \geq \alpha$ (totality);
ii. If $\alpha \geq \beta$ and $\beta \geq \alpha$ then $\alpha = \beta$ (antisymmetry);
iii. If $\alpha \geq \beta$ and $\beta \geq \gamma$ then $\alpha \geq \gamma$ (transitivity);
iv. If $\alpha \geq \beta$ then $\gamma + \alpha \geq \gamma + \beta$ and $\alpha + \gamma \geq \beta + \gamma$ (translation-invariance) for all $\alpha, \beta, \gamma \in G$.

We take $\alpha \leq \beta$ to mean the same thing as $\beta \geq \alpha$, while $\alpha > \beta$ means that $\alpha \geq \beta$ but $\alpha \neq \beta$.

An element $\alpha \in G$ is said to be **positive** (**strictly positive**) if $\alpha \geq 0$ ($\alpha > 0$), where 0 is the identity element of G. The set of all positive elements of G is called the **positive cone** of G and denoted G^+. The set G^+ has the following properties:

1. $0 \in G^+$;
2. $G^+ + G^+ \subseteq G^+$;
3. $G^+ \cap (-G^+) = 0$;
4. $g + G^+ - g \subseteq G^+$ for each element $g \in G$;
5. $G^+ \cup (-G^+) = G$.

Conversely, if in a group G there is a set G^+ satisfying conditions (1)–(5), then G can be made a totally ordered group with G^+ as the set of positive elements.

Examples 3.1.2.

1. The typical example of a totally ordered group is $\mathbf{Z}^n$, the n-fold direct product of the integers $\mathbf{Z}$, where the group operation is componentwise addition and the order relation is the lexicographic ordering.
2. The similar example of a totally ordered group is $\mathbf{R}^n$, a direct product of the reals $\mathbf{R}$, where the group operation is componentwise addition and the order relation is again the lexicographic ordering.

A group G is called **orderable**, also called **totally orderable**, if a total order can be defined on it making it a totally ordered group.

The direct product, the complete direct product and the free product of totally orderable groups are also orderable extending the orders of the factors.

Examples of non-Abelian totally orderable groups are torsion-free nilpotent groups, free groups, free solvable groups (see [193], [195], [194], [196]).

Definition 3.1.3. Let $(G, +, \geq)$ be a totally ordered group. Add to G a special symbol ∞, defined to be larger than any other element of G, and such that $x + \infty = \infty = \infty + \infty$ for all $x \in G$. A (G-valued) **valuation** on a field K is a surjective map $v : K \to G \cup \{\infty\}$ satisfying the following axioms:

1. $v(x) \leq \infty$
2. $v(x) = \infty$ if and only if $x = 0$;
3. $v(xy) = v(x) + v(y)$;
4. $v(x + y) \geq \min(v(x), v(y))$,

for all $x, y \in K$.

The image $v(K^*)$ of K^* in G is a subgroup of G and it is called a **valuation group** of a valuation v.

Remark 3.1.4.

1. In the particular case when the group G is discrete[1], in particular $G = \mathbf{Z}$, a valuation is sometimes called a **discrete valuation**.
2. If v is a valuation on a field K then $v(1) = v(1^2) = v(1) + v(1)$, which means that $v(1) = 0$, the neutral element of the additive group $(G, +)$.
3. If G is a totally ordered group, written as a multiplicative group, the statement (3) has the following form
   (3*) $v(xy) = v(x)v(y)$
   and $v(1) = e$, the neutral element of the multiplicative group G.

Using the properties of a valuation v we immediately obtain the following lemma.

Lemma 3.1.5. *Let $(G, +, \geq)$ be a totally ordered Abelian group, and let $v : K \to G \cup \{\infty\}$ be a valuation on a field K. Then $A = \{x \in K : v(x) \geq 0\}$ is a subring in K.*

Definition 3.1.6. A subring A of a field K is called a **valuation domain** of K if there is a totally ordered Abelian group G and a valuation $v : K \to G \cup \{\infty\}$ on the field K such that $A = \{x \in K : v(x) \geq 0\}$.

---

[1]Recall that a **discrete group** is a group equipped with discrete topology.

Note that if the group G is trivial, i.e., $G = 0$, then a valuation is also trivial, and in this case $A = K$ is a field.

Examples 3.1.7.

1. Let K be a field and $A = K[[x]]$. Then A is a valuation domain with a valuation v given by $v(f(x)) = o(f(x)) = n$, where $f(x) = \sum_{i=0}^{\infty} a_i x^i$, and n is the least integer for which $a_n \neq 0$.
2. Let $A = \mathbf{Z}_{(p)}$ be a set of all rationals of the form a/b, where the chosen p is prime, $a, b \in \mathbf{Z}$, and $\gcd(b,p) = 1$. Then A is a valuation domain with $v(p^n a/b) = n$ if $\gcd(a,p) = 1$.
3. Let F be the field of formal Laurent series over a field K. Any element of F has the form $b = \sum_{i=r}^{\infty} a_i x^i$ with $a_i \in K$ for $i \in \mathbf{Z}$. Then F has a valuation defined by $v(b) = r$, where r is the least integer for which $a_r \neq 0$. The corresponding valuation domain is the formal power series ring $K[[x]]$. So this is really the same example as the first one.

Define the following subset of a field K with valuation v:

$$U = \{x \in K \ : \ v(x) = 0\} \tag{3.1.8}$$

It is easy to verify, that U is a subgroup in $K^* = K \setminus \{0\}$ which is called the **group of (valuation) units**.

Lemma 3.1.9. *Let A be the valuation domain of a field K with respect to a valuation v. Then $U(A) = U = \{x \in K^* \ : \ v(x) = 0\}$.*

Proof. Let $u \in U(A)$, then there is an element $w \in U(A)$ such that $uw = 1$. Therefore $0 = v(uw) = v(u) + v(w)$, hence $v(u) = v(w) = 0$, by property 3 of G^+, since $v(u) \geq 0$ and $v(w) \geq 0$. So $U(A) \subseteq U$.

Conversely, suppose $u \in U$, i.e., $u \in K^*$ and $v(u) = 0$. Then $u^{-1} \in K^*$ and $v(u^{-1}) = -v(u) = 0$. Hence $u, u^{-1} \in A$, which means that $u \in U(A)$. □

In the commutative case there is another equivalent definition of a valuation domain, which is given by the following statement.

Proposition 3.1.10. *The following statements are equivalent:*

1. *A is the valuation domain of a field K.*
2. *A is an integral domain with a field of fractions K, and for every $x \in K$ either $x \in A$ or $x^{-1} \in A$, or both.*

Proof.

$1 \Longrightarrow 2$. Let A be the valuation domain of a field K with respect to a valuation v and an additive group G. Obviously, A is an integral domain, since $A \subset K$. Suppose $x \in K$ and $x \notin A$, i.e., $v(x) < 0$. Then $1 = xx^{-1}$, and $0 = v(1) = v(xx^{-1}) =$

$v(x)+v(x^{-1})$, which implies that $x^{-1} \in A$. In this case $x = 1 \cdot y^{-1}$ where $y = x^{-1} \in A$, so K is the field of fractions of A.

$2 \Longrightarrow 1$. Let $U(A)$ be the group of units of A. Then $U(A)$ is a subgroup of the multiplicative group K^* of K. Consider the quotient group $G = K^*/U(A)$ written as a multiplicative group. Then there is a natural projection $v : K \to G \cup \{\infty\}$ given by $v(x) = U(A)x$ for any $x \in K^*$ and $v(0) = \infty$. For any unit element $u \in U(A)$ we have $v(u) = U(A)u = U(A)$, the neutral element of G. Obviously, v is a surjective mapping. The group G can be turned into a totally ordered group by defining $v(x) \geq v(y)$ iff $xy^{-1} \in A$ for any $v(x), v(y) \in G$. This is well-defined, that is, it does not depend on the choice of representatives x and y.

We now show that the relation $\geq$ thus defined is a total order on G.

i. Since, by assumption, $xy^{-1} \in A$ or $yx^{-1} = (xy^{-1})^{-1} \in A$, $v(x) \geq v(y)$ or $v(y) \geq v(x)$.
ii. Let $v(x) \geq v(y)$ and $v(y) \geq v(x)$. Then $xy^{-1} \in A$ and $yx^{-1} \in A$. Therefore $xy^{-1}, yx^{-1} \in U(A)$. Now we have $x = (xy^{-1})y \in U(A)y$, and $y = (yx^{-1})x \in U(A)x$. Therefore $U(A)x = U(A)y$, i.e., $v(x) = v(y)$.
iii. Let $v(x) \geq v(y)$ and $v(y) \geq v(z)$. Then $xy^{-1} \in A$ and $yz^{-1} \in A$. Therefore $xz^{-1} = (xy^{-1})(yz^{-1}) \in A$, i.e., $v(x) \geq v(z)$.
iv. Let $v(x) \geq v(y)$, that is $xy^{-1} \in A$. Then $xz(yz)^{-1} = xy^{-1} \in A$, which means that $v(xz) \geq v(yz)$, or $v(x)v(z) \geq v(y)v(z)$. This is where commutativity comes in.

Thus, G is a totally ordered group written as a multiplicative group.

Now we show that v is a valuation on K. We need only to verify property (4) from definition 3.1.3. Let $x, y \in K^*$ and $x + y \neq 0$. Assume that $v(x) \geq v(y)$, that is $xy^{-1} \in A$. We must to show that $v(x + y) \geq \min(v(x), v(y)) = v(y)$, or $(x + y)y^{-1} \in A$. But $(x + y)y^{-1} = xy^{-1} + 1 \in A$, since $1, xy^{-1} \in A$. □

The basic properties of a valuation domain are given by the following proposition.

Proposition 3.1.11. *Let A be a valuation domain of a field K. Then:*

1. *If B is a ring such that* $A \subseteq B \subset K$ *then B is a valuation domain.*
2. *A is a local ring.*
3. *A is an integrally closed ring.*
4. *The ideals of A are linearly ordered by inclusion, i.e., A is a uniserial ring.*

Proof.

1. This follows immediately from proposition 3.1.10.
2. Let A be the valuation domain of a field K with a valuation v, and $U(A)$ the group of units of A. One shows that the set

$$M = \{x \in A : v(x) > 0\} = A \setminus U(A) \qquad (3.1.12)$$

of all non-units of A is an ideal. Let $x, y \in M$. Suppose that $v(x) \geq v(y) = \alpha > 0$. Then $v(x + y) \geq \min(v(x), v(y)) = \alpha > 0$. Therefore $x + y \in M$.

If $x \in M$ and $a \in A$ then $v(xa) = v(x) + v(a) > 0$, which implies $xa \in M$. Analogously, $ax \in M$. So M is a two-sided ideal of A, which means that A is a local rings, by proposition 1.9.1.

3. Assume that $\alpha \in K$ is a root of some monic polynomial $f(x) \in A[x]$, where $f(x) = x^n + c_{n-1}x^{n-1} + \cdots + c_1x + c_0$ with $c_i \in A$. Then

$$\alpha^n + c_{n-1}\alpha^{n-1} + \cdots + c_1\alpha + c_0 = 0. \tag{3.1.13}$$

Assume that $\alpha \notin A$. Then $\alpha^{-1} \in A$, since A is a valuation domain, and from (3.1.13) we obtain

$$\alpha = -c_{n-1} - c_{n-2}\alpha^{-1} - \cdots - c_1\alpha^{-(n-2)} - c_0\alpha^{-(n-1)} \in A.$$

A contradiction.

4. Let I and J be ideals of A. It has to be shown that either $I \subseteq J$ or $J \subseteq I$. Suppose that I is not contained in J. Choose some element $a \in I$ and $a \notin J$. Then $a \neq 0$. Let b be an arbitrary non-zero element of J. Then $ab^{-1} \in A$ or $ba^{-1} \in A$. Suppose that $ab^{-1} \in A$, then $a = (ab^{-1})b \in J$, a contradiction. Therefore $ba^{-1} \in A$, and so $b = (ba^{-1})a \in I$, i.e., $J \subseteq I$.

This completes the proof of the theorem. □

Note that in the commutative case for $a, b \in A$ we write $a|b$ if $b = ac$ for some $c \in A$.

The following statement gives several other equivalent definitions of a valuation domain:

Theorem 3.1.14. *Let A be an integral domain. Then the following statements are equivalent:*

1. *A is a valuation domain.*
2. *For any $a, b \in A$ $a|b$ or $b|a$, which is equivalent to the statement that the principal ideals of A are linearly ordered by inclusion.*
3. *All ideals of A are linearly ordered by inclusion, i.e., A is a uniserial ring.*
4. *A is a local ring and any finitely generated ideal of A is principal.*

Proof.

$1 \Longrightarrow 2$. Let $I = (a)$, $J = (b)$ be principal ideals of A. By proposition 3.1.10, $b^{-1}a \in A$ or $(b^{-1}a)^{-1} = a^{-1}b \in A$. If $b^{-1}a \in A$, then $a = b(b^{-1}a) \in A$, and so $I \subseteq J$. If $a^{-1}b \in A$ then $b = a(a^{-1}b) \in A$, and so $J \subseteq I$.

$2 \Longrightarrow 3$. Let I, J be ideals of A. Suppose that I is not contained in J. Choose some non-zero element $x \in I \setminus J$. Let y be any element of J. Since $x \notin J$, $x \notin (y)$, and hence $(x) \nsubseteq (y)$. Therefore, by assumption, $(y) \subseteq (x) \subseteq (I)$. It follows that $J \subseteq I$.

$3 \Longrightarrow 4$. Since any uniserial ring is a local ring, we need only prove that any finitely generated ideal I of A is principal. The proof will be done by induction on the number of generators. Assume that I is generated by two elements a and b. Since

A is a uniserial ring, $(a) \subseteq (b)$ or $(b) \subseteq (a)$, which implies that $I = (b)$ or $I = (a)$. Induction is now straightforward.

$4 \Longrightarrow 1$. Suppose that A is a local ring with unique maximal ideal M, and any finitely generated ideal of A is principal. Let I be an ideal of A generated by two elements $a, b \in A$. Then I is a principal ideal, by assumption. Therefore I/IM is a one-dimensional vector space over the field $F = A/M$. Hence the elements $\bar{a} = a + IM, \bar{b} = b + IM$ are linearly dependent in K, i.e., there exist elements $u, v \in A$ such that $au + bv \in IM$, and either u or v does not belong to M. Furthermore one can find elements $x, y \in M$ such that $au + bv = ax + by$. Hence $a(u - x) = b(y - v)$. Assume that $u \in U(A)$, then $u - x \in U(A)$ as well, and so $ab^{-1} = (y - v)(u - x)^{-1} \in A$. Analogously, if $v \in U(A)$, then $ba^{-1} \in A$. Thus, A is a valuation domain, by proposition 3.1.10. □

Remark 3.1.15. Note that condition 3 in theorem 3.1.14 was considered by W. Krull as the definition of a valuation domain in [200], where he showed the equivalence of conditions 1 and 3 for an integral domain.

Prüfer domains were first introduced in 1932 by H. Prüfer in [267]. Following this definition an integral domain is called a **Prüfer domain** if every finitely generated non-zero ideal of A is invertible. Since for any integral domain any ideal is invertible if and only if it is projective, any integral domain is a Prüfer domain if and only if it is semihereditary.

Corollary 3.1.16. *Any valuation domain is a Prüfer domain.*

Proof. By theorem 3.1.14, a valuation domain A is a local ring which any finitely generated ideal is principal, and so is projective. Then from theorem 1.8.9 it follows that A is a Prüfer domain. □

Proposition 3.1.17. *Let A be a valuation domain. If P is a prime ideal of A then A/P and A_P are also valuation domains.*

Proof. Let A be a valuation domain. Then A_P and A/P are integral domains. By theorem 3.1.14, A is a uniserial ring, so the same is true for rings A/P and A_P. Therefore, again by theorem 3.1.14, A/P and A_P are valuation domains. □

Note that corollary 3.1.16 has no converse in the general case. In 1936 W.Krull proved the following theorem.

Theorem 3.1.18. (W. Krull [202]) *Let A be an integral domain with a field of fractions K. Then A is a Prüfer domain if and only if A_M is a valuation domain for any maximal ideal M of A.*

Proof. Let A be a Prüfer domain. Then any finitely generated ideal of A is projective, and so invertible, by [146, proposition 8.3.2]. Assume that M is a maximal ideal of A and $P = \langle p_1, p_2, \ldots, p_n \rangle$, where $p_i = a_i/s_i$ for $i \in \{1, 2, \ldots, n\}$, is a finitely generated ideal of A_M. Then $P = I_M$, where $I = \langle a_1, a_2, \ldots, a_n \rangle$ is finitely generated ideal of

A. Therefore I is an invertible ideal, and so is $P = I_M$, i.e., there is a fractional ideal P^{-1} such that $PP^{-1} = A_M$ and there is an element $s \in A_M$ such that $sP^{-1} \subseteq A_M$. Since M is a maximal ideal of an integral domain, A_M is a local domain with a unique maximal ideal which is consisted of all non-units of A_M. So there exist elements $q_1, q_2, \ldots, q_n \in P^{-1}$ such that $1 = \sum_{i=1}^{n} p_i q_i$ and $p_i q_i \in A_M$ for all $i = 1, 2, \ldots, n$. Therefore there is an i such that $p_i q_i$ is invertible in A_M. Then $P = (p_i)$, and from theorem 3.1.14 it follows that A_M is a valuation domain.

Conversely, suppose that A_M is a valuation domain for any maximal ideal M of A. Let I be a finitely generated ideal of A, and set $I^{-1} = \{x \in K : xI \subset A\}$, a fractional ideal. Assume that I is not invertible, i.e., $II^{-1} = J \neq A$. Then there exists a maximal ideal M such that $J \subseteq M$. Therefore $J_M = I_M I_M^{-1}$. The ideal I_M is a finitely generated ideal of A_M as I is a finitely generated ideal of A. Since A_M is a valuation ring, I_M is a principal ideal in an integral domain A_M. So I_M is invertible, and $J_M = A_M$, which means that $J \not\subseteq M$. A contradiction. □

Remark 3.1.19. Note that in 1949 L. Fuchs introduced a class of rings which he called arithmetical. Following to L. Fuchs [134] a commutative ring A is **arithmetical** if the ideals of the local ring A_M are totally ordered by inclusion for all maximal ideals M of A. Therefore theorem 3.1.18 states that an integral domain A is Prüfer if and only if it is arithmetical.

In his books [134] L. Fuchs considered commutative rings whose ideals are totally ordered by inclusion, i.e., uniserial rings. He called these rings valuation rings or chain rings. Note that by theorem 3.1.14 any uniserial integral domain is a valuation domain in the sense of definition 3.1.6. But if a ring is uniserial which is not an integral domain it is a valuation ring in the sense of Fuchs but not a valuation domain in the sense of definition 3.1.6.

Next we describe some basic facts about commutative uniserial rings.

Proposition 3.1.20. *Let A be a commutative uniserial ring with a unique maximal ideal M. Then*

1. *The only non-zero principal ideal that can possibly be prime is M.*
2. *If I is a proper ideal of A then either $I^n = 0$ for some $n \in \mathbf{N}$ or $J = \bigcap_{n=1}^{\infty} I^n$ is a prime ideal of A.*

Proof.

1. Let $I = tA \subset A$ with $t \in I$ be a principal prime ideal of A. Suppose $I \neq M$, there exists an $a \in M$ and $a \notin I$. Since A is a uniserial ring, $tA \subset aA$. Therefore there is an element $b \in A$ such that $t = ab \in I$. Since I is a prime ideal and $a \notin I$, $b \in I$, i.e., $b = tc$ for some $c \in A$. So $t = ab = tac$ which implies $t(1 - ac) = 0$, where $ac \in M$. Since A is a local ring with the Jacobson radical M, $1 - ac \in U(A)$. So $t = 0$, i.e., $I = 0$.

2. Assume that $J = \bigcap_{i=1}^{\infty} I^n$ and $I^n \neq 0$ for all $n > 0$. Suppose $J \neq 0$, $xy \in J$ for some non-zero $x, y \in A$ and $x, y \notin J$. Then $x, y \in M \setminus J$, and we can assume that $x, y \notin U(A)$, since A is a local ring. Since A is a uniserial ring and $(x) \nsubseteq J$ and $(y) \nsubseteq J$, there are $n, m > 0$ such that $I^n \subseteq (x)$ and $I^m \subseteq (y)$. Therefore $I^{n+m} \subseteq (xy)$, and we have a sequence of inclusions $I^{n+m} \subseteq (xy) \subseteq J \subseteq I^{n+m}$. So $(xy) = I^{n+m} = J$. Then $J^2 = I^{2(n+m)} \subseteq I^{n+m} = J$. Denote $xy = z$. Then $z \neq 0$ and $z \notin U(A)$. Since $J = zA = J^2 = z^2 A$, there is an element $a \in A$ such that $z = z^2 a$, which implies that $z(1 - za) = 0$. But $1 - za \in U(A)$, since A is local. So $z = 0$. A contradiction! □

The description of Artinian and Noetherian commutative uniserial rings are given by the following lemma.

Lemma 3.1.21.

1. *A commutative uniserial ring A is Artinian if and only if it is a field or a local principal ideal ring whose all ideals form a finite composition series*

$$A \supset tA \supset t^2 A \supset \cdots \supset t^n A = 0 \tag{3.1.22}$$

   for some $t \in A$ and $n > 0$.
2. *A commutative uniserial ring A is Noetherian, but not Artinian, if and only if it is a local principal ideal ring whose all ideals form an infinite composition series*

$$A \supset tA \supset t^2 A \supset \cdots \supset t^n A \supset \cdots \tag{3.1.23}$$

   for some $t \in A$.

Proof.

1. Let A be a commutative uniserial ring. If it has no maximal ideal then A is a field. Otherwise A is a local ring with the unique maximal ideal $M \neq 0$, which is the Jacobson radical of A, and any finitely generated ideal is principal, which follows from the proof of theorem 3.1.14. If A is an Artinian ring then M is nilpotent, i.e., $M^n = 0$ for some $n > 0$, by Hopkins theorem 1.1.21. Since any Artinian ring is Noetherian, M is a finitely generated ideal, and so principal, i.e., $M = tA = At$ for some $t \in A$. Then $M^i = t^i A = At^i$ for any $i > 0$. Moreover, from the Nakayama lemma 1.1.20 it follows that M^{i+1} is strictly contained in M^i for $0 < i < n - 1$. Since A is a uniserial ring, M^i / M^{i+1} is a simple module. Therefore we have exactly one composition series (3.1.22).
   The conversely statement is obvious.
2. The proof is similar to the previous case. □

If A is a commutative uniserial ring which is an integral domain, then it is a valuation domain by theorem 3.1.14. In this case we obtain the following corollary.

Corollary 3.1.24.

1. *A valuation domain is Artinian if and only if it is a field.*
2. *A valuation domain, which is not a field, is a Noetherian ring if and only if it is a local principal ideal domain. In this case there exists exactly one composition series* (3.1.23).

Thus any valuation domain A is a Prüfer ring, and if A is also a Noetherian ring then A is a principal ideal domain.

The following example shows existence of non-Noetherian valuation domains. It provides an instance of a non-discretely valued valuation ring which maximal ideal is not principal.

Example 3.1.25. Fractional Laurent formal power series.

To start with take a (base) field K and consider the Abelian group AFL_K of all formal sums of the form:

$$f = f(t) = \sum_{r \in \mathbf{Q}} a_r t^r, \quad a_r \in K. \tag{3.1.26}$$

The addition is 'componentwise', i.e., if we have an other element

$$g(t) = \sum_{r \in \mathbf{Q}} b_r t^r,\ b_r \in K \tag{3.1.27}$$

then

$$(f + g)(t) = \sum_{r \in \mathbf{Q}} (a_r + b_r) t^r. \tag{3.1.28}$$

So abstractly the Abelian group AFL_K is isomorphic to the countable direct product of copies of (the underlying Abelian group of) K.

The usual kind of formula for multiplying power series does not work on all of AFL_K. To make it work we introduce two restrictions:

$$\text{(i) There is an } r_0 \in \mathbf{Q} \text{ such that } a_r = 0 \text{ for all } r < r_0. \tag{3.1.29}$$

$$\text{(ii) There is an } n \in \mathbf{Q} \text{ such that } a_r = 0 \text{ for all } r \text{ with denominator } > n. \tag{3.1.30}$$

I.e., writing $r = p/q$, with $\gcd(p,q) = 1$, then $a_r = 0$ if $q > n$. Write

$$FL_K = \{f \in AFL_K \ :\ \text{conditions (3.1.29) and (3.1.30) hold for } f\} \tag{3.1.31}$$

This is an Abelian subgroup of AFL_K. Now define a multiplication on FL_K as follows: for f and g as above, both satisfying (3.1.29) and (3.1.30),

$$fg = fg(t) = \sum_{r \in \mathbf{Q}} c_r t^r, \tag{3.1.32}$$

where $c_r = \sum_{r_1+r_2=r} a_{r_1}b_{r_2}$. The two conditions (3.1.29) and (3.1.30) see to it that in fact the second sum in (3.1.32) has only finitely many non-zero terms.

This turns FL_K into a field which we will call the field of fractional formal Laurent series. The FL in the notation FL_K serves as mnemonic for 'fractional Laurent'. That it is in fact a field can either be proved directly in much the same way as is used when dealing with normal Laurent series or by observing that the field of Laurent series $K((t^{1/n}))$ is naturally a subset of AFL_K and that, thanks to (3.1.29) and (3.1.30), for each $f \in FL_K$ there is an n such that $f \in K((t^{1/n})) \subset FL_K \subset AFL_K$. So effectively the ring FL_K is a union of fields and hence itself is a field. Phrased more technically

$$FL_K = \varinjlim K((t^{1/n}))$$

where the direct limit is taken over the directed system given by the embeddings

$$K((t^{1/n})) \longrightarrow K((t^{1/mn})),\ t^{1/n} \mapsto t^{m/mn} = (t^{1/mn})^m. \tag{3.1.33}$$

Now define a valuation on FL_K as follows:

$$v : FL_K \longrightarrow \mathbf{Q} \cup \{\infty\},\ f = f(t) = \sum_{r\in\mathbf{Q}} a_r t^r \mapsto \begin{cases} \infty & \text{if } f = 0 \\ r_0 & \text{otherwise} \end{cases} \tag{3.1.34}$$

where r_0 is the largest rational number such that $a_r = 0$ for $r < r_0$. The valuation ring for this valuation consists of all the f for which $a_r = 0$ for $r < 0$. The maximal ideal M consists of all the f for which $a_r = 0$ for $r \leq 0$:

$$M = \{f \in FL_K \ :\ a_r = 0,\ \text{for}\ r \leq 0\}.$$

Obviously this maximal ideal M is not principal. Using somewhat loose notation one can write the valuation ring as

$$K[[t, t^{1/2}, t^{1/3}, \ldots, t^{1/n}, \ldots]] \tag{3.1.35}$$

More precisely it is the direct limit description

$$\varinjlim K[[t^{1/n}]],$$

where the various formal power series rings embed in each other according to the recipe (3.1.33) above.

Every $0 \leq r \in \mathbf{Q}$ defines two ideals, viz.,

$$I_r = \{f \in FL_K \ :\ a_s = 0\ \text{for}\ s < r\}$$

$$I_{r^+} = \{f \in FL_K \ :\ a_s = 0\ \text{for}\ s \leq r\}$$

The first of these ideals is principle with generating element t^r; the second one is not even finitely generated. Note that I_0 is the valuation ring itself and I_{0^+} is its maximal ideal. Besides (0) they are all the ideals of the valuation ring. So the ideals are linear (totally) ordered (as should be).

Note also that $M^2 = M$ and more generally $\bigcap_n M^n = M$. Indeed,

$$t^{1/n} = (t^{1/mn})^n \text{ for all } m, n \in \mathbf{N}$$

and as M is generated by the $t^{1/n}$ this suffices to show that $M^n = M$ for all $n \in \mathbf{N}$.

3.2 Discrete Valuation Domains

If in the definition of a valuation domain we have a totally ordered group G which is isomorphic to the integers $\mathbf{Z}$ under addition we obtain the particular case of a discrete valuation domain. This kind of rings was considered in [146, section 8.4].

Definition 3.2.1. A subring A of a field K is called a **discrete valuation domain** if there is a (discrete) valuation $v : K \to \mathbf{Z} \cup \{\infty\}$ such that

$$A = \{x \in K \ : \ v(x) \geq 0\}. \tag{3.2.2}$$

All rings considered in the examples 3.1.7 are discrete valuation domains.

Remark 3.2.3. Since in definition 3.2.1 $v : K \longrightarrow \mathbf{Z} \cup \{\infty\}$ is a surjective map, a discrete valuation domain A is not a field.

We formulate the basic properties of a discrete valuation domain in the following proposition.

Proposition 3.2.4. *Let A be a discrete valuation domain of a field K with respect to a valuation v, and $t \in A$ a fixed non-zero element with $v(t) = 1$. Then*

1. *A is a local ring with non-zero unique maximal ideal*
$$M = \{x \in A \ : \ v(x) > 0\} \tag{3.2.5}$$
which has the form $M = tA = (t)$.
2. *Any non-zero element $x \in A$ has the form $x = t^n u$, for some $u \in U(A)$, and $n \in \mathbf{Z}$, $n \geq 0$. Any element $y \in K^*$ has the form $y = t^n u$ for some $u \in U(A)$, and $n \in \mathbf{Z}$.*
3. *Any non-zero ideal I of A is principal and has the form $I = t^n A = M^n$ for some $n \in \mathbf{Z}$, $n \geq 0$, i.e., A is a principal ideal domain (abbreviated,* PID*).*
4. *The only non-zero prime ideal of A is M.*
5. *A is a Noetherian hereditary ring.*
6. *$\bigcap_{i=0}^{\infty} M^i = 0$, where M is the unique maximal ideal of A (as defined above).*
7. *A is integrally closed.*

Proof.

1. Since a discrete valuation domain A is a particular case of a valuation domain, A is a local ring with the unique maximal ideal $M = \{x \in A : v(x) > 0\}$ which is the Jacobson radical of A. Let $t \in A$ with $v(t) = 1$. Then $t \in M$ and $ta \in M$ for any $a \in A$, since $v(ta) = v(t) + v(a) \geq 1$. Therefore $(t) \subseteq M$. Let $x \in M$ with $v(x) = n > 0$. Then $v(xt^{-n}) = v(x) - n = 0$. Therefore from lemma 3.1.9 it follows that $xt^{-n} = u \in U(A)$ and $x = t^n u$. So, $M \subseteq (t)$. Thus, $M = (t) = tA \neq 0$ (and also A is not a field).
2. As in the previous case we find that any element $x \in A$ has the form $x = t^n u$ with $u \in U(A)$ and $n \geq 0$. Let $y \in K^*$. Since K is the field of fractions of A, $y = ab^{-1}$ with $a, b \in A$. Let $a = t^n u$ and $b = t^m w$ with $u, w \in U(A)$ and $n, m \geq 0$. Then $y = ab^{-1} = (t^n u)(t^m w)^{-1} = t^{n-m} uw^{-1}$, where $uw^{-1} \in U(A)$. (Note that this uses commutativity.)
3. Let I be a non-zero ideal of A. Choose in I an element x with minimal value $v(x) = n$ (if there are more than one such elements we can choose one arbitrarily). Then $x = t^n u$ with $u \in U(A)$. Therefore $t^n A = (t^n) \subseteq I$. Let $y \in I$, then $y = t^m w$ with $m \geq n$. So $v(yt^{-n}) \geq 0$, hence $yt^{-n} \in A$ and $y \in (t^n)$. Therefore $I = (t^n) = t^n A = M^n$.
4. This follows immediately from statements 1 and 3, since each non-zero prime ideal I of a commutative PID A is maximal, so $I = M$.
5. This follows immediately from statement 3, since each PID is Noetherian and hereditary.
6. Since A is a local PID which is not a field, $M \neq 0$ is the Jacobson radical of A. Suppose $J = \bigcap_{i=0}^{\infty} M^i \neq 0$. Then there exists $0 \neq x \in J$ with $v(x) = n$, i.e., $x = t^n u \in M^n$, for some $u \in U(A)$ and $n \in \mathbf{Z}$, $n \geq 0$. However $x \notin M^{n+1}$. Otherwise $x = t^{n+1} w$ with $w \in U(A)$ and so $t^n u = t^{n+1} w$, which implies $u = tw \notin U(A)$. This contradiction shows that $J = 0$.
7. This follows from proposition 3.1.11. □

In addition to definition 3.2.1 there are other equivalent definitions of a discrete valuation domain which are given in the following statement.

Proposition 3.2.6. *The following statements are equivalent.*

1. *A is a discrete valuation domain.*
2. *A is a principal ideal domain with a unique non-zero prime ideal.*
3. *A is a local principal ideal domain which is not a field.*
4. *A is a Noetherian valuation domain.*
5. *A is a Noetherian local domain whose maximal ideal is non-zero and principal.*

Proof. That statement 1 implies each of the other properties was proved above.

$2 \Longleftrightarrow 3$. This is obvious since any prime ideal of a commutative PID is maximal.

$3 \Longrightarrow 4$. A is a Noetherian ring, since A is a PID, by statement (3). Since A is also local, A is a valuation domain by theorem 3.1.14.

$4 \Longrightarrow 5$. This follows from theorem 3.1.14, since any ideal of a Noetherian ring is finitely generated.

$5 \Longrightarrow 1$. Let A be a Noetherian local domain which maximal ideal $M \neq 0$, and $M = (t)$. Note that $M^n \neq M^{n+1}$ for any $n \geq 0$. Otherwise, by the Nakayama lemma 1.1.20, $M^n = 0$, and $t^n = 0$. Since A is a domain, $t = 0$. A contradiction.

We now prove, as an intermediate step, that $J = \bigcap_{i=0}^{\infty} M^i = 0$. Let $x \in J$ and $x \neq 0$. Then for suitable $a_i \in A$

$$x = a_0 = a_1 t = a_2 t^2 = \cdots = a_n t^n = \cdots$$

This gives a chain of principal ideals $(a_1) \subset (a_2) \subset \cdots$ which must be stabilize because A is Noetherian. So $(a_n) = (a_{n+1})$ for some $n > 0$, and $a_{n+1} = a_n b$ for some $b \in A$. On the other hand, $a_n t^n = a_{n+1} t^{n+1}$. So $a_n = a_{n+1} t$, since A is a domain. Thus, $a_{n+1} = a_n b = a_{n+1} t b$, that is $tb \in U(A)$. So $t \in U(A)$, which is not the case. Hence, indeed, $\bigcap_{i=0}^{\infty} M^i = 0$.

Show that any non-zero element $x \in A$ has a unique representation in the form $x = t^n u$, where $u \in U(A)$ and $n \geq 0$. Let $x \notin U(A)$, then $x \in M$. Since $\bigcap_{i=0}^{\infty} M^i = 0$, there is an integer $n > 0$ such that $x \in M^n$ and $x \notin M^{n+1}$. Then $x = t^n u$ and $u \notin M$. Therefore $u \in U(A)$.

Since A is an integral domain, it has a field of fractions K and any element of K^* has the form ab^{-1}, where $a, b \in A$. If $a = t^n u$ and $b = t^m w$ with $u, w \in U(A)$ and $n, m \geq 0$, then $d = t^{n-m}\varepsilon$, where $n - m \in \mathbf{Z}$ and $\varepsilon \in U(A)$. If we set $v(d) = v(t^{n-m}\varepsilon) = n - m \in \mathbf{Z}$, we obtain a valuation on K with the discrete valuation domain A.

This finishes the proof of the proposition. □

3.3 Invariant Valuation Rings of Division Rings

Consider now the case of noncommutative rings. In this case there are three (at least) different generalizations of the concept of a valuation ring:

i. An invariant valuation ring (definitions 3.3.1 and 3.3.7);
ii. A total valuation ring (definition 3.6.1);
iii. A Dubrovin valuation ring (definition 3.7.9).

The terminology of the last two is a bit misleading in that the definitions make no mention of a valuation function. However there is already a large amount of publications on Dubrovin valuation rings; so the phrase is likely here to stay.

This section is about (not necessarily discrete) invariant valuation rings.

Historically the first generalization was proposed in 1945 by Schilling [283], who extended the concept of a valuation on a field to that of a valuation on a division ring.

Definition 3.3.1. Let G be a totally ordered group (written additively) with order relation $\geq$. Add to G a special symbol ∞ such that $x + \infty = \infty + x = \infty$ for all $x \in G$. Let D be a division ring. A **valuation** on D is a surjective map $v : D \to G \cup \{\infty\}$ which satisfies the following:

1. $v(x) \leq \infty$
2. $v(x) = \infty$ if and only if $x = 0$
3. $v(xy) = v(x) + v(y)$
4. $v(x + y) \geq \min(v(x), v(y))$
   for all $x, y \in D$.

Remark 3.3.2. Note that these requirements are exactly the same as those in definition 3.1.3. The only difference is that D is allowed to be noncommutative.

If D is a field then from condition 3 it follows immediately that D can admit only valuations with Abelian groups G.

Remark 3.3.3. Let D be a division ring with a valuation v and the multiplicative group D^*. Write

$$U = \{u \in D^* \; : \; v(u) = 0\}. \tag{3.3.4}$$

If $u_1, u_2 \in U$ then $v(u_1u_2) = v(u_1) + v(u_2) = 0$, and $v(u_2u_1) = v(u_2) + v(u_1) = 0$, which means that $u_1u_2 \in U$ and $u_2u_1 \in U$. Let 1 be the identity of D. Then $v(1) = v(1^2) = v(1) + v(1)$ implies that $1 \in U$. If $u \in U$ then $0 = v(1) = v(uu^{-1}) = v(u) + v(u^{-1}) = v(u^{-1})$, i.e., $u^{-1} \in U$. Thus U is a subgroup of D^* which is called the **group of valuation units**. Let $x \in D^*$. Then $v(xux^{-1}) = v(x) + v(u) + v(x^{-1}) = v(x) + v(x^{-1}) = v(xx^{-1}) = 0$ for any $u \in U$. Thus, U is an invariant subgroup of D^* which is equal to $\mathrm{Ker}(v)$. Therefore $D^*/U \cong G$.

Proposition 3.3.5. Let $(G, +, \geq)$ be a totally ordered group, and let $v : D \to G \cup \{\infty\}$ be a valuation on a division ring D. Then

$$A = \{x \in D \; : \; v(x) \geq 0\} \tag{3.3.6}$$

is a subring of D.

Proof. Let $x, y \in A$, then $v(x), v(y) \geq 0$. Therefore $v(xy) = v(x) + v(y) \geq 0$ and $v(x+y) \geq \min(v(x), v(y)) \geq 0$, which means that $xy \in A$ and $x + y \in A$. Moreover, $v(-x) = v((-1) \cdot x) = v(-1) + v(x) = v(1) + v(x) = v(x) \geq 0$ for any $x \in A$. □

Definition 3.3.7. A subring A of a division ring D is called an **invariant valuation ring**[2] of D if there is a totally ordered group G and a valuation $v : D \to G \cup \{\infty\}$ on D such that $A = \{x \in D \; : \; v(x) \geq 0\}$.

---

[2]This ring is often called a **noncommutative valuation ring**, or simply, a **valuation ring**.

Lemma 3.3.8. *Let A be an invariant valuation ring of a division ring D with respect to a valuation v. Then $U(A)$ is equal to U, the group of valuation units of D, i.e., $U(A) = U = \{x \in D : v(x) = 0\}$.*

Proof. Let $u \in U(A)$, then there is an element $w \in U(A)$ such that $uw = 1$. Therefore $0 = v(uw) = v(u) + v(w)$, hence $v(u) = v(w) = 0$, since $v(u) \geq 0$ and $v(w) \geq 0$.

Conversely, suppose $u \in D$ and $v(u) = 0$. Then $u^{-1} \in D^*$ and $v(u^{-1}) = -v(u) = 0$. Hence $u, u^{-1} \in A$, which means that $u \in U(A)$. □

For any invariant valuation ring A associated to the valuation v let

$$M = \{x \in D : v(x) > 0\} = A \setminus U(A) \tag{3.3.9}$$

the set of all non-units of A.

Lemma 3.3.10. *An invariant valuation ring A of a division ring D with respect to a valuation v is a local domain with a unique non-zero maximal ideal $M = A \setminus U(A)$.*

Proof. Let $x, y \in M$ and $a \in A$. Then

1. $v(x + y) = \min(v(x), v(y)) > 0$, that is, $x + y \in M$;
2. $v(xa) = v(x) + v(a) > 0$ and $v(ax) = v(a) + v(x) > 0$, that is, $ax, xa \in M$.

Thus, M is an ideal of A. In order to show that M is a maximal two-sided ideal in A, suppose that $\mathcal{I}$ is a two-sided ideal in A such that $M \subset \mathcal{I} \subseteq A$. Since $M = A \setminus U$ there is a unit $u \in \mathcal{I}$ such that $v(u) = v(u^{-1}) = 0$ and $u^{-1} \in A$. Consequently, $1 = uu^{-1} \in \mathcal{I}$. Thus, $\mathcal{I} = A$, i.e., M is a maximal ideal in A. Since $U(A) = U$, $M = A \setminus U(A)$ consists of all non-units of A. Therefore A is a local ring by proposition 1.9.1, and M is the unique maximal ideal in A. □

Lemma 3.3.11 (O.F.G. Schilling [283]). *If A is an invariant valuation ring of a division ring D with respect to a valuation v, then both A and M are invariant subsets of D^*, that is, $dAd^{-1} = A$ and $dMd^{-1} = M$ for any $d \in D^*$.*

Proof. Suppose that $dAd^{-1} \neq A$ for some $d \in D^*$. Then there is an element $x = dyd^{-1} \in dAd^{-1}$ with $y \in A$ and $x \notin A$. Therefore $v(x) < 0$ and $v(y) \geq 0$. On the other hand, $y = d^{-1}xd$, and so $v(y) = v(d^{-1}) + v(x) + v(d) < v(d^{-1}) + v(d) = v(1) = 0$, since G is a totally ordered group. This contradiction shows that $dAd^{-1} = A$ for any $d \in D^*$.

Suppose that $dMd^{-1} \neq M$ for some $d \in D^*$. Then there is an element $x = dyd^{-1} \in dMd^{-1}$ with $y \in M$ and $x \notin M$. Since $dMd^{-1} \subset dAd^{-1} = A$ and $M = A \setminus U(A) = M \setminus U$, $x \in A \setminus M = U$. By remark 3.3.3, U is an invariant subgroup of D^*, and so $y = d^{-1}xd \in U$. Consequently, $y \in M \cap U = \emptyset$. This contradiction shows that M is an invariant subset of D^*. □

The following theorem gives an equivalent definition of an invariant valuation ring which is similar to that for valuation domains.

Theorem 3.3.12 (O.F.G.Schilling [283]). *Let A be a subring of a division ring D. Then the following are equivalent:*

1. *A is an invariant valuation ring with respect to some valuation v on D.*
2. *A is an invariant subring of D^*, and for any element $x \in D^*$ either $x \in A$ or $x^{-1} \in A$.*

Proof.

$1 \Longrightarrow 2$. A is an invariant subring of D^*, by lemma 3.3.11. Suppose $x \in D^*$ and $x \notin A$, which means that $v(x) < 0$. Then $0 = v(1) = v(xx^{-1}) = v(x) + v(x^{-1})$, hence $v(x^{-1}) = -v(x) \geq 0$. Thus $x^{-1} \in A$.

$2 \Longrightarrow 1$. Suppose that A is an invariant subring of a division ring D^* with group of units $U(A)$. Let $u \in U(A)$ and $d \in D^*$. Then $x = dud^{-1} \in A$ and $x^{-1} = du^{-1}d^{-1} \in A$. Therefore $x, x^{-1} \in U(A)$, i.e., $U(A)$ is an invariant subgroup of D^*.

Let $M = A \setminus U(A)$. We show that this set is also invariant in D^*. Let $d \in D^*$. Assume that $dMd^{-1} \neq M$. This means that there exists an element $x = dyd^{-1} \in dMd^{-1}$ with $y \in M$ and $x \notin M$. Note that $x \in A$, since A is an invariant subring in D^*. Therefore $x \in U(A)$ and $y = d^{-1}xd \in U(A)$, since $U(A)$ is invariant in D^*. So $y \in M \cap U(A) = \emptyset$. This contradiction shows that M is invariant in D^*.

Since $U(A)$ is an invariant subgroup in D^*, one can consider the factor group $G = D^*/U(A)$ as a multiplicative group and define a natural map $v : D \to G \cup \{\infty\}$ by $v(d) = dU(A) = U(A)d$ for any $d \in D^*$ and $v(0) = \infty$. Obviously, $v(du) = v(ud)$ for all $u \in U(A)$ and $d \in D^*$. We set $v(u) = 0$ for any $u \in U(A)$. Then v is a surjective map with $\mathrm{Ker}(v) = U(A)$. It only remains introduce a total order on G assuming that $v(x) \leq \infty$ for all $x \in D$. Let $a, b \in D^*$. By assumption, either $a^{-1}b \in A$ or $b^{-1}a \in A$. Suppose $a^{-1}b \in A$, then $a(a^{-1}b)a^{-1} = ba^{-1} \in A$, since A is an invariant ring in D^*. We use this fact to order the group G. We set $v(a) > v(b)$ in the case $ab^{-1} \in M$ (and $b^{-1}a \in M$). In this way G turns out to be totally ordered. Now observe that v is a valuation of D with valuation ring A. Indeed,

1. $v(x) \leq \infty$ for all $x \in D$;
2. $v(x) = \infty$ if and only if $x = 0$;
3. v is a surjective map;
4. $v(d) = 0$ if and only if $d \in U(A)$;
5. $v(ab) = v(a)v(b)$.
6. Let $a, b \in D^*$ and $a + b \neq 0$. Assume that $v(a) > v(b)$ in G. This means that $ab^{-1} \in M$ or $ab^{-1} \in U(A)$. In both cases $ab^{-1} + 1 \in A$, since $1 = 1^{-1} \in A$. Since $(a + b)b^{-1} = ab^{-1} + 1 \in A$, $v(a + b) \geq v(b) = \min(v(a), v(b))$. If $a + b = 0$ then $v(a + b) = \infty$ and we also have that $v(a + b) \geq v(b) = \min(v(a), v(b))$. □

This theorem suggests the possibility to introduce the equivalent definition for an invariant valuation ring of a division ring D without mention of valuation on D.

Definition 3.3.13. A subring A of a division ring D is called an **invariant valuation ring** if it satisfies the following two conditions:

1. For every $x \in D^*$, $x \in A$ or $x^{-1} \in A$.
2. For every $d \in D^*$, $dAd^{-1} = A$.

Theorem 3.3.12 states that any invariant valuation ring is a total valuation ring, but not conversely. Note that in the case of (commutative) integral domains these notions for valuation rings are equivalent to the notion of a (commutative) valuation domain.

Lemma 3.3.14 (O.F.G. Schilling [283]). *Let A be an invariant valuation ring of a division ring D with respect to a valuation v, and $a,b \in A$. Then the following statements are equivalent:*

1. $a = bc_1$ with $c_1 \in A$;
2. $a = c_2b$ with $c_2 \in A$;
3. $v(a) \geq v(b)$.

Proof.
1,2 $\Longrightarrow$ 3. Suppose $a = bc_1 = c_2b$ with $c_1, c_2 \in A$. Then $v(a) = v(b) + v(c_1) = v(c_2) + v(b) \geq v(b)$ by condition (4) of definition 3.1.3.

3 $\Longrightarrow$ 1,2. Suppose $v(a) \geq v(b)$ and $b \neq 0$. Then $v(ab^{-1}) \geq 0$ and $v(b^{-1}a) \geq 0$, that is, $ab^{-1} \in A$ and $b^{-1}a \in A$. Thus, $a = b(b^{-1}a) = bc_1$ and $a = (ab^{-1})b = c_2b$.

Suppose $v(a) \geq v(b)$ and $b = 0$. Then $v(b) = \infty$ and so $v(a) = \infty$, hence $a = 0$. This means that a is again both a left and a right multiple of b. □

The next proposition gives the basic properties of invariant valuation rings.

Proposition 3.3.15. *Let A be an invariant valuation ring of a division ring D with respect to a valuation v. Then*

1. *Each ideal of A is two-sided.* [3]
2. *A is an Ore domain which classical ring of fractions is a division ring D.*
3. *Any finitely generated ideal of A is principal.*

---

[3]A ring A is called a **left (right) duo ring** if every left (right) ideal is two-sided. A **duo ring** means both a left and a right duo ring. Thus, this lemma says that any invariant valuation ring is a duo ring. Note that in literature these rings are also called **invariant** or **subcommutative.**

Proof.

1. Suppose that I is a left ideal of A, that is $AI \subseteq I$. Since $1 \in A$, $AI = I$. Let $x = \sum_{i=1}^{n} y_i a_i$ be an arbitrary element of the set IA, where $y_i \in I$, $a_i \in A$. Then $v(y_i a_i) = v(y_i)+v(a_i) \geq v(y_i)$. Consequently, by lemma 3.3.14, $y_i a_i = b_i y_i$ for some $b_i \in A$. Therefore $x = \sum_{i=1}^{n} b_i y_i \in AI = I$. Thus I is a right ideal.
2. Let $I = xA$, then $AI = AxA = xA$, since I is a two-sided ideal. Analogously, $Ax = AxA$. Therefore $Ax = xA$. This means that A satisfies the right and left Ore conditions. Since A is a domain[4], A has a left and right classical ring of fractions which is a division ring $F = Q_{cl}(A)$, by [146, corollary 9.1.3]. Let $x \in F$, then $x = ab^{-1}$, where $a, b \in A \subset D$. Therefore $F \subseteq D$. On the other hand, if $d \in D^* \setminus A$ then, by theorem 3.3.12, $x = d^{-1} \in A$, which implies $d = 1 \cdot x^{-1} \in F$. So $F = D$.
3. Let $I = a_1A + a_2A + \cdots + a_nA$, where $a_i \in A$. Since A is a valuation ring then we can choose among the elements $a_1, a_2, \ldots, a_n$ an element with minimal value. Without loss of generality we can consider that $v(a_i) \geq v(a_1)$ for all i. Then, by lemma 3.3.14, this means that $a_iA \subseteq a_1A$. So $I = a_1A$. □

The following theorem gives other equivalent definitions of a (noncommutative) invariant valuation ring.

Theorem 3.3.16 (Gilmer R. [93, p.161]). *Let A be a subring of a division ring D which is its classical ring of fractions, and let A be invariant in D. Then the following are equivalent:*

1. *A is an invariant valuation ring of D with respect to some valuation v.*
2. *The set of right (left) principal ideals of A is linearly ordered by inclusion.*
3. *A is a uniserial ring.*

Proof.

$1 \Longrightarrow 2$. Let $a, b \in A$ and $v(a) \geq v(b)$. Then from lemma 3.3.14 it follows that $a \in bA$ and $a \in Ab$. Therefore $aA \subseteq bA$ and $Aa \subseteq Ab$.

$2 \Longrightarrow 3$. Let I and J be right ideals of A. Suppose that I is not contained in J. Choose some non-zero element $x \in I \setminus J$. Let y be any element of J. Since $x \notin J$, $x \notin yA$, and so $xA \not\subseteq yA$. Therefore, by assumption, $yA \subseteq xA \subseteq I$. It follows that $J \subset I$.

$3 \Longrightarrow 1$. By assumption, A is a subring of a division ring D which is its classical ring of fractions. Let $x \in D^*$. Then $x = ab^{-1}$ for some non-zero $a, b \in A$. Since A is a uniserial ring, $Aa \subset Ab$ or $Ab \subset Aa$. If $Aa \subset Ab$ then $a = rb$ for some $r \in A$. Therefore $x = ab^{-1} = rbb^{-1} = r \in A$. If $Ab \subset Aa$ then $b = sa$ with $s \in A$.

---

[4]Recall that A is called a **domain** if it does not have divisors of zero.

Therefore $x^{-1} = ba^{-1} = saa^{-1} = s \in A$. Since A is invariant in D by assumption, A is a valuation ring, by theorem 3.3.12. □

Proposition 3.3.17 (Pirtle [263]). *Any invariant valuation ring of a division ring is integrally closed.*

Proof. Since all ideals in invariant valuation rings are two-sided the proof is like the commutative case. Let A be a valuation ring of a division ring D. Let $x \in D^*$ be integral over A, i.e., $x^n + a_{n-1}x^{n-1} + \cdots + a_1x + a_0 = 0$ with $a_i \in A$. If $x \notin A$ then $x^{-1} \in A$. Then $x^n = -a_{n-1}x^{n-1} - \cdots - a_1x - a_0$ and $x = a_{n-1} + \cdots + a_1(x^{-1})^{n-2} + a_0(x^{-1})^{n-1} \in A$. A contradiction. □

Recall that a ring is semihereditary if every finitely generated ideal is projective.

Proposition 3.3.18. *An invariant valuation ring of a division ring is semihereditary.*

Proof. Since A is a domain, each of its elements is regular[5]. Hence any principal ideal of A is free. So the statement follows immediately from proposition 3.3.15(3). □

3.4 Examples of Non-commutative Non-discretely-valued Valuation Rings

This section is concerned with three families of noncommutative non-discretely-valued valuation rings (where valuation ring is to be understood as 'invariant valuation ring', i.e., valuation ring according to definitions 3.3.1 and 3.3.7).

The first family is $\mathbf{Q}$-valued, the second and third are $\mathbf{Z}[n^{-1}]$-valued, where $\mathbf{Z}[n^{-1}]$ is the subgroup of $\mathbf{Q}$ of all rational numbers that can be written in the form $\frac{a}{n^b}$ for suitable $b \in \mathbf{N}$, $a \in \mathbf{Z}$. Note that this is a non-discrete group.

The first two families come from divisions rings of twisted fractional formal Laurent power series over a field of formal Laurent power series over a field; the third family consists of twisted fractional formal Laurent power series over finite fields; and more generally over a nontrivial finite degree Galois extension L/K.

Let k be a field and σ an automorphism of k. The division ring (skew field) $k((t;\sigma))$ consists of all power series

$$f(t) = \sum_{i=m}^{\infty} a_i t^i, \quad \text{where } m \in \mathbf{Z}, a_i \in k \tag{3.4.1}$$

[5]Recall that an element x of a ring A is **regular** if $ax \neq 0$ and $xa \neq 0$ for any non-zero element $a \in A$.

with addition defined componentwise and multiplication defined by the distributive law and the twist

$$ta = \sigma(a)t, \text{ which implies } t^{-1}a = \sigma^{-1}(a)t^{-1}. \tag{3.4.2}$$

This, in turn, explains why one needs an automorphism and not just an endomorphism (which would be enough to define twisted power series rings).

Within the context of valuation rings twisted formal Laurent power series rings are just about the example of noncommutative division rings. So, in order to find examples of noncommutative non-discretely-valued valuation rings, it is natural to try to see of one can make sense of something like (3.4.2) in the context of fractional Laurent formal power series rings.

Example 3.4.3. **Q**-valued case.

For this case one finds rapidly that something like a field with coherent system of automorphisms $(\sigma_n)_{n\in\mathbf{N}}$ is needed. Here 'coherence' means that

$$(\sigma_{mn})^m = (\sigma_n) \quad \text{for all } m, n \in \mathbf{N}. \tag{3.4.4}$$

These are not so easy to come by, but they do exist.

Let k be a field. A **coherent system of roots** in k is by definition a sequence of elements $(a_n)_{n\in\mathbf{N}}$, $a_n \in k$, such that

$$(a_{mn})^m = a_n, \quad n, m \in \mathbf{N} \tag{3.4.5}$$

Two trivial examples are $a_n = 0$ for all $n \in \mathbf{N}$ and $a_n = 1$ for all $n \in \mathbf{N}$. All others are called nontrivial. For a nontrivial coherent systems of roots one has $a_n \neq 0$ for all n and $a_n \neq 1$ for infinitely many n.

Here are two examples of nontrivial coherent root systems.

For the first one take $k = \mathbf{R}$, the real numbers, and a strictly positive element $a = a_1 \in \mathbf{R}^+$. For each $n \in \mathbf{N}$ let a_n be the unique strictly positive real number such that $(a_n)^n = a$. This defines a coherent system of roots in $\mathbf{R}$.

For a second example take $k = \mathbf{C}$, the complex numbers, and consider the roots of unity $\xi_n = e^{2\pi i/n}$. This is the first n-th root of unity one encounters when walking counter clockwise along the unit circle in the complex plane starting from the point $1 + 0i$. Then $(\xi_n)_{n\in\mathbf{N}}$ is a coherent system of roots.

Combining these two ideas one finds a nontrivial coherent system of roots starting with $a_1 = c$ for any non-zero complex number c. Indeed, write c in polar coordinates as

$$c = re^{i\varphi}, \ 0 < r \in \mathbf{R}, \ 0 < \varphi \leq 2\pi$$

and take $a_n = \sqrt[n]{r}e^{i\varphi/n}$, where $\sqrt[n]{r}$ is as in the first example.

The next step is to construct a field with a coherent system of automorphisms. Start with a field k with a nontrivial coherent systems of roots $(a_n)_{n\in\mathbf{N}}$. Now take the field of formal Laurent series $K = k((t))$. On K consider the automorphisms σ_n

determined by $t \mapsto a_n t$. This defines $(K, (\sigma_n))$, a field with a coherent system of automorphisms.

Further consider the field of twisted Laurent series $K((T^{1/n}; \sigma_n))$ where the 'twist' is the commutation relation

$$T^{1/n} b = \sigma_n(b) T^{1/n}, \quad b \in K. \tag{3.4.6}$$

There are embeddings

$$i_{n,mn} : K((T^{1/n}; \sigma_n)) \longrightarrow K((T^{1/mn}; \sigma_{mn})) \tag{3.4.7}$$

given by

$$T^{1/n} \mapsto (T^{1/mn})^m$$

and these embeddings are compatible with the twists (3.4.6). Indeed,

$$i_{n,mn}(T^{1/n} b) = (T^{1/mn})^m b = (\sigma_{mn})^m (b) (T^{1/mn})^m =$$

$$= \sigma_n(b) (T^{1/mn})^m = i_{n,mn}(\sigma_n(b) T^{1/n})$$

where the coherence of the automorphism system $(\sigma_n)_{n \in \mathbf{N}}$ is used for the one-but-last equality sign.

Thus the embeddings (3.4.7) form a directed system of twisted power series algebras and one can take the direct limit

$$TFL(K; (\sigma_n)_{n \in \mathbf{N}}) = \varinjlim K((t^{1/n}; \sigma_n)). \tag{3.4.8}$$

Here TFL is a mnemonic for 'twisted fractional Laurent'. The elements of $TFL(K; (\sigma_n)_{n \in \mathbf{N}})$ are fractional power series

$$f(T) = \sum_{r \geq r_0, r \in \mathbf{Q}} b_r T^r, \quad b_r \in K \text{ for a certain } r_0 = r_0(f) \in \mathbf{Q} \tag{3.4.9}$$

with the extra property that there is an $n \in \mathbf{N}$ such that $b_r = 0$ unless r can be written as a fraction which has n as its denominator. That is because f must 'come from' a $K((T^{1/n}; \sigma_n))$ for some n.

Now define a valuation on the division ring $TFL(K; (\sigma_n)_{n \in \mathbf{N}})$ by

$$v(f(T)) = s \tag{3.4.10}$$

if s is the largest rational number such that $b_r = 0$ for $r < s$; and $v(0) = \infty$.

Such an s exists because of the extra property just mentioned. This defines a valuation on $TFL(K; (\sigma_n)_{n \in \mathbf{N}})$ with valuation group $\mathbf{Q}$. Its valuation ring

$$ATFL(K; (\sigma_n)_{n \in \mathbf{N}}) = \{f \in TFL(K; (\sigma_n)_{n \in \mathbf{N}}) \ : \ v(f) \geq 0\} \tag{3.4.11}$$

is an example of a noncommutative non-discretely-valued valuation ring.

Every $0 \leq r \in \mathbf{Q}$ gives rise to two ideals in this valuation ring, viz.,

$$I_r = \{f \in TFL(K; (\sigma_n)_{n \in \mathbf{N}}) \;:\; v(f) \geq r\}$$

which is a principal ideal (generated by T^r) and an ideal

$$I_{r^+} = \{f \in TFL(K; (\sigma_n)_{n \in \mathbf{N}}) \;:\; v(f) > r\}$$

which is not finitely generated. Besides the zero ideal these are all ideals of this valuation ring. Indeed let I be an ideal. Consider

$$r = \inf\{v(f) \;:\; f \in I\}.$$

Then it is not difficult to show that $I = I_r$ if the infimum r is actually attained and $I = I_{r^+}$ if that is not the case. The simple proof of this relies mostly on the fact that the elements f with valuation zero are invertible.

The ideal I_0 is the valuation ring itself and $I_{0^+} = M$ is its maximal ideal. One has $I_{0^+} = M = M^2 = M^3 = \cdots$. More generally,

$$I_{r^+} I_{s^+} = I_{(r+s)^+} = I_r I_{s^+} = I_{r^+} I_s, \quad I_r I_s = I_{r+s}$$

Example 3.4.12. $\mathbf{Z}[n^{-1}]$-valued case.

The second family of noncommutative non-discretely-valued valuation ring is $\mathbf{Z}[n^{-1}]$-valued, which is less elegant than $\mathbf{Q}$-valued, but slightly easier to construct. The construction uses the notion of a string of roots.

Choose a natural number $n \geq 2$. Then an n-**string of roots** in a field k is a sequence of elements $a_0, a_1, a_2, \ldots$ such that

$$(a_m)^n = a_{m-1} \text{ for all } m \in \mathbf{N} \tag{3.4.13}$$

Such a string is called nontrivial if it does not consist of all zeros or all 1's. Such strings are easily obtained. For instance if k is algebraically closed, take an arbitrary $a_0 \neq 0, 1$. Let a_1 be a solution of $X^n = a_0$; a_2 a solution of $X^n = a_1$; etc.

There is also at least one class of other examples. Let n be a prime number p and consider a finite field $\mathbf{F}$ of characteristic p. Then the Frobenius morphism $x \mapsto x^p$ is an automorphism of the field $\mathbf{F}$ so that for every $a_0 \neq 0, 1$ there is a unique p-string of roots starting with a_0. This string is periodic. In the case $\mathbf{F} = \mathbf{F}_p$ it is actually constant.

An n-**string of automorphisms** of a field K is a sequence of field automorphisms of K, $\sigma_0, \sigma_1, \sigma_2, \ldots$ such that $(\sigma_m)^n = \sigma_{m-1}$ for all $m \in \mathbf{N}$.

Given a field k with an n-string of roots consider the field of ordinary (non-twisted) Laurent series $K = k((t))$. This field can be equipped with an n-string of automorphisms by defining $\sigma_n(t) = a_n t$.

In turn given K, consider the twisted Laurent power series skew fields

$$K((T^{n^{-i}}; \sigma_i)), \quad T^{n^{-i}} b = \sigma_i(b) T^{n^{-i}} \tag{3.4.14}$$

There are twist preserving (commutation relation preserving) embeddings

$$K((T^{n^{-i}};\sigma_i)) \longrightarrow K((T^{n^{-(i+1)}};\sigma_{i+1})), \quad T^{n^{-i}} \mapsto (T^{n^{-(i+1)}})^n \tag{3.4.15}$$

That these embeddings are compatible with the commutation relations (3.4.14) is an easy calculation very similar to the one done above for the previous family of examples.

Thus the embeddings (3.3.15) define a (linearly ordered) directed system of twisted Laurent series fields and one can take the direct limit

$$TFL(K,n,(\sigma_i)) = \varinjlim K((T^{n^{-i}};\sigma_i))$$

which is a noncommutative division ring. From here on in the things go pretty much the same as in the first family of examples described above. The details are just a bit different and so a short outline will be given. The formulations will be different in the hope that will make things a bit clearer.

Consider all power series of the form

$$f(t) = \sum_{r \in \mathbf{Z}[n^{-1}]} b_r T^r, \quad b_r \in K. \tag{3.4.16}$$

If a power series (3.4.16) is actually in the direct limit $TFL(K,n,(\sigma_i))$ it has two additional properties:

$$\text{(1) There is an } r_0 \in \mathbf{Z}[n^{-1}] \text{ such that } b_r = 0 \text{ for } r < r_0 \tag{3.4.17}$$

and

$$\text{(2) There is an } i \in \mathbf{N} \text{ such that } b_r = 0 \text{ if } r \text{ cannot be written as a fraction } r = \frac{j}{n^i} \text{ with } j \in \mathbf{Z} \tag{3.4.18}$$

Both conditions come from the fact that an f in the direct limit must come from some element (the same one really) in a $K((T^{n^{-i}};\sigma_i))$.

Conversely, if f as in (3.4.16) satisfies these two conditions (3.4.17) and (3.4.18), it is in the direct limit.

There is a valuation:

$$v : TFL(K,n,(\sigma_i)) \longrightarrow \mathbf{Z}[n^{-1}] \cup \{\infty\}$$

defined as follows. Let f be like in (3.4.16) and satisfy (3.4.17) and (3.4.18). Then (3.4.17) says that the set of r for which $b_r \neq 0$ is bounded below and (3.4.18) says that for some fixed i this set lies in the discrete subset $\{jn^{-1} : j \in \mathbf{Z}\} \subset \mathbf{Q}$. Thus, if $f \neq 0$, there is a largest r in $\mathbf{Z}[n^{-1}]$ such that $b_s = 0$ for all $s < r$. This r is the valuation $v(f)$ of f. And of course $v(0) = \infty$.

The situation with respect to the valuation ring, its maximal ideal and other ideals is now exactly as in the first family of example above, except that now the only relevant r lie in the dense additive sub-monoid

$$\{r \in \mathbf{Z}[n^{-1}] : r \geq 0\}$$

of the nonnegative rationals.

Example 3.4.19.

The third family of examples is actually simpler than the first two. The starting point this time is a pair (K,γ) consisting of a field K and an automorphism $\gamma : K \longrightarrow K$ that is of finite order and not the identity. For instance if K/K_0 is a nontrivial Galois extension any element of the Galois group $\mathrm{Gal}(K/K_0)$ will do for γ provided it is not the identity. Let the order of γ be $(n-1)$ (writing $n-1$ here instead of n saves typing, ink and paper later). A nice and interesting case is the field $\mathbf{F}_{p^{n-1}}$, $n > 2$, the field of p^{n-1} elements where p is a prime number. Then the Galois group $\mathrm{Gal}(\mathbf{F}_{p^n}/\mathbf{F}_{p^{n-1}})$ is cyclic of order $(n-1)$ with as canonical generator the Frobenius automorphism, which is raising to the power p. For a completely explicit instance take $p = 3$ and consider the equation $X^2 = 2$ over $\mathbf{F}_3$. This has no solution in $\mathbf{F}_3$. So adjoining a root of the equation gives a field of nine elements $\{a + bx \ : a, b \in \mathbf{F}_3\}$. The Frobenius automorphism γ is raising to the power 3. And indeed

$$(a + bx)^3 \equiv a^3 + b^3x^3 \equiv a + 2bx \bmod(3)$$

$$(a + 2bx)^3 \equiv a^3 + 8b^3x^3 \equiv a + 2bx^3 \equiv a + 4bx \equiv a + bx \bmod(3)$$

so that γ is indeed of order 2.

Again fractional Laurent power series are to be used. This time consider the twisted Laurent series fields

$$K((T^{n^{-i}};\gamma)), \quad T^{n^{-i}}b = \gamma(b)T^{n^{-i}}. \tag{3.4.20}$$

(where the same γ is used for all i). This time consider the embeddings

$$K((T^{n^{-i}};\gamma)) \longrightarrow K((T^{n^{-(i+1)}};\gamma)), \quad T^{n^{-i}} \mapsto (T^{n^{-(i+1)}})^n$$

Because $\gamma^n = \gamma$ these embeddings are compatible with the commutation relations written on the right sided of (3.4.18) and one can take the direct limit

$$TFL(K,\gamma) = \varinjlim K((T^{n^{-i}};\gamma). \tag{3.4.21}$$

This gives a noncommutative division ring which carries a valuation with as valuation group again $\mathbf{Z}[n^{-1}]$. The valuation is defined exactly as in example 3.4.12. From here on everything goes like in the family of examples.

Remark 3.4.22. The four families of non-discrete valuation rings A constructed in examples 3.1.25, 3.4.3, 3.4.12 and 3.4.19 with value groups $\mathbf{Q}$ and $\mathbf{Z}[n^{-1}]$, $n \geq 2$, all have the property that $\bigcap_{n=1}^{\infty} M^n = M$, where M is a maximal ideal of A. This is not an accident.

Indeed let D be a commutative or noncommutative division ring with a valuation $v : D \longrightarrow \mathbf{Z}[n^{-1}] \cup \{\infty\}$, and the valuation ring A with maximal ideal M. Let $x \in M$ have the valuation jn^i, $j \in \mathbf{N}$, $i \in \mathbf{Z}$. Because v is surjective and $M = \{x \in D \ : \ v(x) > 0\}$, there is an $y \in M$ of valuation n^{i-1}. Then $y^{jn} \in M^{jn} \subset M^2$ and

$v(y^{jn}) = jn^i = v(x)$. Then $y^{-jn}x \in U(A) \subset A$ and $x = y^{jn}(y^{-jn}x) \in M^2A = M^2$. So $M^2 = M$ and that suffices.

The proof for the case of a **Q**-valued valuation is just about the same. (Only a bit easier.)

3.5 Discrete Valuation Rings of Division Rings

As in the commutative case one can introduce the notion of a discrete valuation ring of a noncommutative division ring.

Definition 3.5.1. A subring A of a division ring D is called a (noncommutative) **discrete valuation ring** (DVR for short) if there is a (discrete) valuation $v : D \to \mathbf{Z}\bigcup\{\infty\}$ of D such that

$$A = \{x \in D \ : \ v(x) \geq 0\}.$$

As in the commutative case a discrete valuation ring A is not a division ring, since $v : D \to \mathbf{Z}\bigcup\{\infty\}$ is a surjective map by definition. From lemma 3.3.8 it follows that the group of units $U(A)$ is equal to the group of valuation units

$$U = \{x \in D \ : v(x) = 0\}.$$

Example 3.5.2.
A main example of a noncommutative discrete valuation ring is a skew power series ring $K[[x,\sigma]]$ with $xa = \sigma(a)x$ for any $a \in K$, where K is a field and $\sigma : K \to K$ is a nontrivial automorphism of K.

Here are some of the basic properties of a discrete valuation ring.

Proposition 3.5.3. *Let A be a (noncommutative) discrete valuation ring of a division ring D with respect to a valuation v. Let t be a fixed element of A with $v(t) = 1$. Then*

1. *A is a local domain with a unique non-zero maximal ideal*

$$M = \{x \in A \ : \ v(x) > 0\}.$$

2. *Any non-zero element $x \in A$ has unique representations in the form $x = t^n u = wt^n$, for some $u, w \in U(A)$, and $n \in \mathbf{Z}$, $n \geq 0$. Any element $y \in D^*$ can be written in the form $y = t^n u = wt^n$ for some $u, w \in U(A)$, and $n \in \mathbf{Z}$.*
3. *Any one-sided ideal I of A is a two-sided ideal and has the form $I = t^n A = At^n$ for some $n \in \mathbf{Z}$, $n \geq 0$; in particular A is a principal ideal ring[6], and moreover, $M = tA = At$, and $I = M^n = t^n A = At^n$.*
4. *$\bigcap\limits_{i=0}^{\infty} M^i = 0$, where M is the unique maximal ideal of A.*

---

[6]Recall that a ring A is called a **principal ideal ring** if each one-sided ideal of A is principal.

5. *A is a Noetherian uniserial ring.*
6. *A is a hereditary ring.*

Proof.

1. Since a discrete valuation ring A is a particular case of an invariant valuation ring, this statement follows from lemma 3.3.10.
2. Let t be a fixed element of A with $v(t) = 1$, and $x \in A$ with $v(x) = n \geq 0$. Then $t \in M$, and $v(xt^{-n}) = v(x) - n = 0 = v(t^{-n}x)$. Therefore from lemma 3.3.8 it follows that $xt^{-n} = u \in U(A)$ and $t^{-n}x = u_1 \in U(A)$. So $x = ut^n = t^n u_1$.
   Let $y \in D^*$. Since D is the classical ring of fractions of A, by proposition 3.3.15(2), any element $y \in D^*$ can be represented in the form $y = ab^{-1}$ with $a, b \in A$. Let $a = t^n u$ and $b = t^m w$ with $u, w \in U(A)$ and $n, m \geq 0$. Then $y = (t^n u)(t^m w)^{-1} = t^{n-m} u_1 w_1 = u_2 w_2 t^{n-m}$, where $u_1 w_1, u_2 w_2 \in U(A)$ and $n - m \in \mathbf{Z}$.
3. Since A is an invariant valuation ring, any one-sided ideal of A is two-sided. Let I be an ideal of A. Choose in I an element x with minimal value $v(x) = n$ (if there are more than one such elements we can choose an arbitrarily one). Then $x = t^n u = wt^n$ with $u, w \in U(A)$. Therefore $t^n A \subseteq I$ and $At^n \subseteq I$. Let $y \in I$, then $y = t^m w$ with $m \geq n$. So $v(t^{-n}y) = m - n \geq 0$, hence $t^{-n}y \in A$ and $y \in t^n A$. Therefore $I = t^n A$. Analogously, $I = At^n$. In particular, since $t \in M$, $M = tA = At$, and $M^n = t^n A = At^n = I$.
4. Assume that $N = \bigcap_{i=0}^{\infty} M^i \neq 0$. Let x be non-zero element of N with $v(x) = n \geq 0$. Then $x = t^n u \in M^n$ with $u \in U(A)$. Since $x \in N$, $x \in M^{n+1}$. Therefore $x = t^{n+1} w$ with $w = U(A)$. So $t^n u = t^{n+1} w$. Since A is a domain, $u = tw \in M$. A contradiction. Thus $N = 0$.
5. This follows immediately from statement 3 and theorem 3.3.16.
6. This follows from the fact that A is a principal ideal domain and any principal ideal over a domain is free. □

In addition to definition 3.5.1 there are other equivalent definitions of a discrete valuation domain. These are given by the following statements.

Proposition 3.5.4. *The following statements for a ring A are equivalent.*

1. *A is a discrete valuation ring.*
2. *A is a local ring with non-zero maximal ideal M of the form $M = tA = At$, where $t \in A$ is a non-nilpotent element, and $\bigcap_{i=0}^{\infty} M^i = 0$.*

Proof. 1 $\Longrightarrow$ 2. From proposition 3.5.3 it follows that A is a local ring with non-zero maximal ideal M of the form $M = tA = At$, where $t \in M$ with $v(t) = 1$, and $\bigcap_{i=0}^{\infty} M^i = 0$. Since A is a domain, t is a non-nilpotent element.

2 $\Longrightarrow$ 1. Since $M = tA = At$, it is easy to show directly that $M^n = t^n A = At^n$.

We now show that any non-zero element $x \in A$ has a unique representation in the form $x = t^n u = wt^n$, where $u, w \in U(A)$ and $n \geq 0$. Let $x \notin U(A)$, then $x \in M$. Since $\bigcap_{i=0}^{\infty} M^i = 0$, there is an integer $n > 0$ such that $x \in M^n$ and $x \notin M^{n+1}$. Then $x = t^n u$ and $u \notin M$. Therefore $u \in U(A)$. Analogously, $x = wt^n$.

The ring A is domain. Otherwise there are elements $x, y \in A$ such that $xy = 0$. Let $x = t^n u$, $y = t^m w$ and $ut^m = t^m u_1$ with $u, w, u_1 \in U(A)$. Then $xy = t^{n+m} u_1 w = 0$, and so $t^{n+m} = 0$, which is not the case, since t is a non-nilpotent element.

It now follows that A is a right and left Ore domain (see definition 1.11.1). Let x, y be non-zero elements of A. Suppose $x = t^n u$, $y = t^m w$, $ut^m = t^m u_1$ and $wt^n = t^n w_1$ with $u, w, u_1, w_1 \in U(A)$. Then $xy = t^n ut^m w = t^n t^m u_1 w = t^m t^n u_1 w = t^m ww^{-1} t^n u_1 w = yx_1$, where $x_1 = w^{-1} t^n u_1 w \in A$. Analogously, $yx = xy_1$, where $y_1 = u^{-1} t^m w_1 u$. This shows that A satisfies the right and the left Ore conditions. So A has a classical ring of fractions D. Any element of D^* can be represented in the form $d = ab^{-1}$, where $a, b \in A$. If $a = t^n u$ and $b = t^m w$ with $u, w \in U(A)$ and $n, m \geq 0$, then $d = t^{n-m} \varepsilon$, where $n - m \in \mathbf{Z}$ and $\varepsilon \in U(A)$. Now set $v(d) = v(t^{n-m} \varepsilon) = n - m \in \mathbf{Z}$ to obtain a valuation on D^* with discrete valuation ring A.

This finishes the proof of the proposition. □

Proposition 3.5.5. *The following statements for a ring A are equivalent.*

1. *A is a discrete valuation ring.*
2. *A is a local principal ideal domain which is not a division ring.*
3. *A is a Noetherian local ring with non-zero maximal ideal which is two-sided and principal.*
4. *A is a right (left) Noetherian local ring with non-zero maximal ideal M of the form $M = tA = At$ for a non-nilpotent element $t \in A$.*

Proof. That statement 1 implies each of the other properties was proved above. The implications $2 \Longrightarrow 3$ and $3 \Longrightarrow 4$ are trivial.

$4 \Longrightarrow 1$. Let A be a Noetherian local ring which maximal ideal $M \neq 0$, and $M = tA = At$. Note that $M^n \neq M^{n+1}$ for any $n \geq 0$. Otherwise, by the Nakayama lemma, $M^n = 0$, and $t^n = 0$, which is not the case, since t is a non-nilpotent element.

We now prove that $\bigcap_{i=0}^{\infty} M^i = 0$. Otherwise there is a non-zero element $x \in \bigcap_{i=0}^{\infty} M^i$. Then for suitable $a_i \in A$

$$x = a_0 = a_1 t = a_2 t^2 = \cdots = a_n t^n = \cdots$$

with every $a_i \notin U(A)$. Now $a_i - a_{i+1} t \notin U(A)$, for otherwise $a_i t^i = a_{i+1} t^{i+1}$ would imply that $t^i = 0$, i.e., that t is nilpotent, which is not the case. So $a_i - a_{i+1} t \in M = At = tA$ and we have an ascending chain of right principal ideals $a_1 A \subset a_2 A \subset \cdots$ which must be stabilize because A is Noetherian, i.e., there is a number $n > 0$ such that $a_n A = a_{n+1} A$. Then $a_{n+1} = a_n b$ for some $b \in A$. Hence $a_n t^n = a_{n+1} t^{n+1} = a_n b t^{n+1}$. As $M = At = tA$, $bt^{n+1} = t^{n+1} c$ for some $c \in A$. Then $a_n t^n = a_n t^{n+1} c$

and $a_n t^n(1 - tc) = 0$. As $tc \in M$, $1 - tc \in U(A)$. So that $x = a_n t^n = 0$. This contradiction shows that $\bigcap_{i=0}^{\infty} M^i = 0$. Now we can apply proposition 3.5.4. □

Proposition 3.5.6. *The following statements for a ring A are equivalent.*

1. *A is a discrete valuation domain.*
2. *A is a Noetherian non-Artinian uniserial ring.*
3. *A is a Noetherian invariant valuation ring.*

Proof. The implications $1 \Longrightarrow 2$ and $1 \Longrightarrow 3$ were proved in proposition 3.5.3.

$2 \Longrightarrow 1$. Let A be a uniserial Noetherian but non-Artinian ring. Then, the unique maximal ideal M of A is the Jacobson radical of A, $M \neq 0$ and M^n/M^{n+1} is a simple A-module. So we have a strictly descending chain of ideals

$$A \supset M \supset M^2 \supset \cdots \supset M^n \supset \cdots \tag{3.5.7}$$

Hence M is not nilpotent, otherwise (3.5.7) is a composition series for M and so A is an Artinian ring, which is not the case. Choose an element $t \in M \setminus M^2$. Since A is uniserial $M^2 \subset tA \subseteq M$. Hence $M = tA$, since M/M^2 is a simple module. Analogously, $M = At$. This gives exactly statement 4 of proposition 3.5.5.

$3 \Longrightarrow 1$. Let A be a Noetherian invariant valuation ring. Then any ideal of A is finitely generated, hence it is principal by proposition 3.3.15(3). Thus, A is local principal ideal domain which is not a division ring. So we can apply proposition 3.5.5. □

3.6 Total Valuation Rings

Theorem 3.3.12 suggests the possibility to introduce another kind of generalization for an invariant valuation ring of a division ring.

Definition 3.6.1. A subring A of a division ring D is called a **total valuation ring** of D if $x \in A$ or $x^{-1} \in A$ for each $x \in D^*$.

From this definition it immediately follows that a total valuation ring A of a division ring D is an order in D and D is its classical ring of fractions. Therefore A is a prime Goldie ring.

The following theorem gives an equivalent characterization of total valuation rings of division rings.

Theorem 3.6.2 (R.B. Warfield [320, theorem 4.9]). *The following properties are equivalent for a ring A:*

i. *A is uniserial and semihereditary.*
ii. *A is a total valuation ring of a division ring.*

Proof.

i $\Longrightarrow$ ii. Since A is a uniserial and semihereditary ring, any its finitely generated right (left) ideal I is principal and free, i.e., $I \simeq A$. Therefore any non-zero element $x \in A$ is regular, i.e., A is a domain. So A is an order in a division ring D which is a classical ring of fractions of A. Let $x = ab^{-1} \in D$ with $a, b \in A$. Since A is uniserial, $aA \subseteq bA$ or $bA \subseteq aA$. If $Aa \subset Ab$ then $a = rb$ for some $r \in A$. Then $x = ab^{-1} = rbb^{-1} = r \in A$. If $Ab \subset Aa$ then $b = sa$ with $s \in A$. Then $x^{-1} = ba^{-1} = saa^{-1} = s \in A$.

ii $\Longrightarrow$ i. Let A be a total valuation ring. Then it is a subring in a division ring D and $x \in A$ or $x^{-1} \in A$ for every $x \in D^*$. Suppose $a, b \in A$. Let $x = ab^{-1}$. If $x \in A$ then $a = xb$, and so $Aa \subseteq Ab$. If $x^{-1} = ba^{-1} \in A$, then $b = x^{-1}a$ and so $Ab \subseteq Aa$. Analogously, for all $a, b \in A$, $aA \subseteq bA$ or $bA \subseteq aA$. Thus the set of right and left principal ideals of A is linearly ordered by inclusion.

Let I and J be right ideals of A. Suppose that I is not contained in J. Choose some non-zero element $x \in I \setminus J$. Let y be any element of J. Since $x \notin J$, $x \notin yA$, and so $xA \not\subseteq yA$. Therefore, by assumption, $yA \subseteq xA \subseteq I$. It follows that $J \subset I$, i.e., A is a right uniserial ring. In a similar way we can prove that A is a left uniserial ring. Therefore A is a uniserial ring.

Since A is a domain, any principal right (left) ideal of A is isomorphic to A, i.e., it is projective. As all finitely generated right (left) ideals of a right (left) uniserial ring are right (left) principal, and so they are projective, A is a right (left) semihereditary ring. □

From theorem 3.3.12 it follows that every invariant valuation ring is also a total valuation ring. However not all total invariant rings are invariant valuation rings. In the commutative case these concepts are the same. But in the noncommutative case there do exists total valuation rings which are not invariant. The first such example was constructed by J. Gräter in [120].

Example 3.6.3 (J. Gräter [121]). Example of the total valuation ring which is not an invariant valuation ring.

Let K be a commutative field, and let L be a finite cyclic Galois extension of K with Galois group $\langle\sigma\rangle$. Let W be a valuation ring of L. Consider a skew Laurent series ring $D = L((x, \sigma))$ which consists of all power series

$$f(x) = \sum_{i=m}^{\infty} a_i x^i, \quad \text{where } m \in \mathbf{Z}, a_i \in L$$

with addition defined componentwise and multiplication defined by the distributive law and the twist $xa = \sigma(a)x$, which implies that $x^{-1}a = \sigma^{-1}(a)x^{-1}$. Then D is a division ring by proposition 2.5.10. Consider two rings:

$$B_1 = L[[x, \sigma]]$$

and

$$B_2 = \{a_0 + a_1x + a_2x^2 + \cdots \mid a_0 \in W;\ a_1, a_2, \ldots \in L\}.$$

From the proof of proposition 2.5.10 it follows that B_1 and B_2 are total valuation rings. Since σ is an automorphism of L, $xB_1 = B_1x$, i.e., B_1 is also an invariant valuation ring. However if $\sigma(W) \neq W$ then B_2 is not an invariant valuation ring.

Theorem 3.6.4 [235, Corollary 1.3]. *Let B be a total valuation ring of a division ring D. The set Ω of all subrings of D containing B is totally ordered by inclusion.*

Proof. Let B be a total valuation ring, A a subring of a division ring D and $B \subseteq A \in \Omega$. If $a \in D$ and $a \notin A$ then $a \notin B$, hence $a^{-1} \in B$. So any subring $A \in \Omega$ is a total valuation ring of D.

Suppose A_1 and A_2 are subrings of D and $A_1, A_2 \in \Omega$. Note that $A_iB \subseteq A_iA_i \subseteq A_i$ and $BA_i \subseteq A_iA_i \subseteq A_i$ for $i = 1,2$. Suppose that $A_1 \nsubseteq A_2$. Choose an element $a \in A_1$ such that $a \notin A_2$. Let $b \in A_2$, $b \neq 0$. Since B is a total valuation ring in D, $b^{-1}a \in B$ or $a^{-1}b \in B$ and $b^{-1} \in B$. If $b^{-1}a \in B$ then $a = b(b^{-1}a) \in A_2$ contrary to the choice of a. Hence $a^{-1}b \in B$ and therefore $b = a(a^{-1}b) \in A_1$. So $A_2 \subset A_1$. □

One of the generalizations of total valuation rings was provided in [29] by Brungs and Gräter:

Definition 3.6.5. Let D be a skew field and W_v a totally ordered set with largest element ϑ. A mapping $v : D \to W_v$ is called a **valuation** of D if for all $x, y, z \in D$ the following conditions hold:

i. $v(x) = \vartheta$ if and only if $x = 0$;
ii. $v(x + y) \geq \min\{v(x), v(y)\}$;
iii. $v(x) \geq v(y)$ implies $v(zx) \geq v(zy)$.

Note that if in the definition above the condition (iii) is changed by

iv. $v(xy) = v(x) + v(y)$
provided $W_v \setminus \vartheta = G$ is a totally ordered group (written additively) we obtain the definition 3.3.1 of a valuation first proposed by O.F.G. Schilling in [283].

The valuation in definition 3.6.5 defines

$$B_v = \{x \in D : v(x) \geq v(1)\}$$

which is a valuation ring of D associated to v. Moreover, B_v is a total valuation ring in the sense of definition 3.6.1, i.e., $x \notin B_v$ implies $x^{-1} \in B_v$ for each $x \in D$.

Conversely, for each total valuation subring B of a skew field D, one can define a valuation in the sense of definition 3.6.5. For this end one can construct a totally ordered set W by the following way. Let $W = \{xB : x \in D\}$ and define the order relation $xB \leq yB$ if and only if $xB \supseteq yB$. Then $0 = 0 \cdot B$ is the largest element in respect to this order. Setting $v(x) = xB$ we obtain the corresponding valuation which satisfies all condition of definition 3.6.5.

3.7 Other Kinds of Valuation Rings with Zero Divisors

During the last 40 year the theory of valuation rings with zero divisors was developed. In this section we give a short characteristic of these rings and present some their properties.

First the notion of a valuation ring for commutative rings with zero divisors was considered by M.E. Manis in 1969 in [231]. These rings are now called Manis valuation rings after him.

Definition 3.7.1 [231]. A **valuation** v on a commutative ring K is a surjective map from K to a totally ordered additive group G, called the **valuation group**, together with symbol ∞ (such that $\infty > \gamma$ for all $\gamma \in G$ and $\infty + \gamma = \gamma + \infty = \infty$) with the following properties:

i. $v(ab) = v(a) + v(b)$ for all $a, b \in K$;
ii. $v(a + b) \geq \min(v(a), v(b))$ for all $a, b \in K$;
iii. $v(1) = 0$, $v(0) = \infty$.

Note that he requirement (iii) in this definition is added to avoid the trivial valuations.

With this valuation one can construct a ring

$$A = A_v = \{x \in K \ : \ v(x) \geq 0\} \tag{3.7.2}$$

which is called a **Manis valuation ring** with respect to a valuation v, and the ideal

$$P = P_v = \{x \in K \ : \ v(x) > 0\}$$

which is called the **prime ideal** of v. The pair A and P is denoted by (A, P) and called a **Manis valuation pair**.

There were obtained other equivalent definitions of the notion of a Manis valuation ring which are given in the following theorem:

Theorem 3.7.3. ([231]) *Let A be a subring of a commutative ring K, and let P be a prime ideal of A. Then the following statements are equivalent*:

1. *(A, P) is a Manis valuation pair of K, i.e., there is a valuation v on K such that*
$$A = \{x \in K \ : \ v(x) \geq 0\}$$
and
$$P = \{x \in K \ : \ v(x) > 0\}.$$
2. *For all $x \in K \setminus A$ there exists $y \in P$ such that $xy \in A \setminus P$.*
3. *If B is a subring of K such that $A \subseteq B$ and M is a prime ideal of B with $M \cap A = P$, then $A = B$.*

As was showed in corollary 3.1.16 any valuation domain is a Prüfer domain. The notion of Prüfer domain has been generalized to a commutative ring with zero divisors in many different ways. There are at least twenty two different characterization of Prüfer domains collected in [19, Theorem 1.1], many of which may be generalized in several various ways.

An ideal I of a commutative ring A is called **regular** if it contains a regular element, i.e., element which is not a zero divisor of A.

M. Griffin [123] defined a **Prüfer ring** as a commutative ring which every regular finitely generated ideal is invertible. First these rings appeared in the literature under α-rings in [31]. Some results on Prüfer rings can be found in [31], [229], [123]. The most full survey on Prüfer rings and their generalizations are given by S.Glaz and Ryan Schwarz in [98].

There exist Manis valuation rings which are not Prüfer rings (see [155]). One of these interesting example was provided by R. Gilmer in [95].

The connection of Manis valuation rings with Prüfer rings was studied by Monte B. Boisen and Max D. Larsen. They obtained the following result:

Theorem 3.7.4 [23, theorem 2.3]. *Let A be a ring and let P be a prime ideal of A. The following conditions are equivalent*:

1. *(A,P) is a Manis valuation ring and A is a Prüfer ring.*
2. *A is a Prüfer ring and P is the unique regular maximal ideal of A.*
3. *(A,P) is a Manis valuation ring and P is the unique regular maximal ideal of A.*

Definition 3.7.5 [232], [19]. A commutative ring A is called a **Marot ring** if every regular ideal of A is generated by its set of regular elements.

The following theorem gives the connection of Manis valuation rings with Prüfer rings that are Marot rings.

Theorem 3.7.6. (See [232, theorem 2.1], and also [19, theorem 2.10].) *Let A be a Marot ring. Then the following statements are equivalent*:

1. *A is a Manis valuation ring and a Prüfer ring.*
2. *A is a Manis valuation ring.*
3. *For each regular element of Q, the classical ring of fractions of A, either $x \in A$ or $x^{-1} \in A$.*

Note that any commutative uniserial ring is a Marot ring, and so a Manis valuation ring and a Prüfer ring. An example of a commutative ring which is not uniserial but it is a Manis valuation ring and a Prüfer ring was provided in [155].

The notion of a Manis valuation is possible to extend to the noncommutative case.

Definition 3.7.7. Let A be a ring. A **valuation** on A with values in a totally ordered group G is a surjective map $v : A \to G \cup \{\infty\}$ subject to the conditions:

i. $v(ab) = v(a) + v(b)$ for all $a, b \in K$;
ii. $v(a + b) \geq \min(v(a), v(b))$ for all $a, b \in K$;
iii. $v(1) = 0$, $v(0) = \infty$.

The set

$$\mathrm{Ker}(v) = \{x \in A \ : \ v(x) = \infty\} \tag{3.7.8}$$

is a proper ideal of A and from (i) it follows that $A/\mathrm{Ker}(v)$ is a domain.

The main progress in the general theory of valuation rings has been made by N. Dubrovin in 1982 [71], [72] when he introduced the new type of valuation rings. These rings, which are now called Dubrovin valuation rings, are defined not only for division rings but also for simple Artinian rings, especially for central simple algebras. They have significantly better extension properties than the classical valuation rings considered above.

Definition 3.7.9. Let S be a simple Artinian ring. A subring A of S is called a **Dubrovin valuation ring** if there is an ideal M of A such that

1. A/M is a simple Artinian ring;
2. For each $s \in S \setminus A$ there are $a_1, a_2 \in A$ such that $sa_1 \in A \setminus M$ and $a_2 s \in A \setminus M$.

Lemma 3.7.10 [234, Lemma 1.4.2]. *If A is a Dubrovin valuation ring with ideal M as in definition* 3.7.9, *then M is the Jacobson radical of A. In particular, M is a unique maximal ideal in A.*

Proof. Let A be a Dubrovin valuation ring with Jacobson radical $J(A)$ in a simple Artinian ring S. Since the Jacobson radical of a simple Artinian ring A/M is equal to zero, $J(A) \subseteq M$. Therefore it suffices to prove that $M \subseteq J(A)$. Let $m \in M$. By proposition 1.1.18 one only needs to show that $1 + m \in U(A)$.

We write $X = \mathrm{l.ann}_S(1 + m)$ and $Y = \mathrm{r.ann}_S(1 + m)$. Let $x \in X \subset S$. If $x \notin A$ then there exists an element $b \in A$ such that $bx \in A \setminus M$. On the other hand $bx = -bxm \in M$, since M is an ideal of A. A contradiction. Therefore $x = -xm \in M$ for all $x \in X$ and all $m \in M$. In a similar way one can show that $y = -my \in M$ for all $y \in X$ and all $m \in M$. Thus $X, Y \subset M$.

Since XY is an ideal of the simple Artinian ring S, $XY = 0$ or $XY = S$.

Suppose that $XY = S$. Then $1 = \sum_i x_i y_i$ for some $x_i \in X$ and $y_i \in Y$. On the other hand, $0 = \sum_i x_i(1+m)y_i = \sum_i x_i y_i + \sum_i x_i m y_i = 1 + \sum_i x_i m y_i$. Since $X, Y \subset M$, this implies that $1 \in M$. This contradiction shows that $XY = 0$.

We write $X^{(n)} = \text{l.ann}_S((1+m)^n) \subset S$ for $n \in \mathbf{Z}$. Since S is a Noetherian ring, the ascending chain of ideals

$$X = X^{(1)} \subseteq X^{(2)} \subseteq X^{(3)} \subseteq \cdots \subseteq X^{(n)} \subseteq \cdots$$

must terminate, i.e., there exists n such that $X^{(n)} = X^{(n+1)} = \cdots$. Since $(1+m)^n = 1 + m_1$ for some $m_1 \in M$, we may assume that $n = 1$, i.e., $X = X^{(1)} = X^{(2)} = \cdots$, because $\text{r.ann}_S(1+m_1) = 0$ implies $X = 0$.

Write $S(1+m) = Se$ for some idempotent $e \in S$. Then

$$Y = \text{r.ann}_S(S(1+m)) = (1-e)S.$$

Therefore

$$\text{l.ann}_S(Y) = Se = S(1+m).$$

Since $XY = 0$, $X \subseteq \text{l.ann}_S(Y) = S(1+m)$. Therefore for any $x \in X$ there exists $s \in S$ such that $x = s(1+m)$. Then $0 = x(1+m) = s(1+m)^2$, which implies $s \in X^{(2)} = X$, i.e., $x = s(1+m) = 0$. Thus $X = 0$, i.e., $1+m$ is a left regular element in S. Therefore $1+m$ is invertible in S.

Now we show that $1+m \in U(A)$. Suppose that $(1+m)^{-1} \in S \setminus A$. Therefore $(1+m)^{-1} - 1 \in S \setminus A$ as well. Since A is a Dubrovin valuation ring, there exists an element $a \in A$ such that $t = [(1+m)^{-1} - 1]a \in A \setminus M$. Therefore $(1+m)t = [1-(1+m)]a = -ma$ which implies $t = -m(t+a) \in M$. This contradiction shows that $(1+m)^{-1} \in A$, i.e., $1+m \in U(A)$. Thus $M \subseteq J(A)$. So that $M = J(A)$. □

From this lemma one can give the equivalent definition of a Dubrovin valuation ring.

Definition 3.7.11. Let S be a simple Artinian ring. A subring A of S, with Jacobson radical $J(A)$ is called a **Dubrovin valuation ring** if

1. $A/J(A)$ is a simple Artinian ring;
2. For each $s \in S \setminus A$ there are $a_1, a_2 \in A$ such that $sa_1 \in A \setminus J(A)$ and $a_2 s \in A \setminus J(A)$.

Example 3.7.12.

Every total valuation ring A of a division ring D is a Dubrovin valuation ring of D.

Indeed, by theorem 3.6.2, A is a uniserial semihereditary ring. Therefore A is a local ring, i.e., $A/J(A)$ is a division ring, where $J(A)$ is the Jacobson radical of A. Let $s \in S \setminus A$. Then $s^{-1} \in A$. Therefore $ss^{-1} = s^{-1}s = 1 \in A \setminus J(A)$. Thus A is a Dubrovin valuation ring of S.

So we obtain the following chain of classes of valuation rings:

$$\begin{pmatrix}\text{Invariant}\\ \text{valuation}\\ \text{rings}\end{pmatrix} \subsetneq \begin{pmatrix}\text{Total}\\ \text{valuation}\\ \text{rings}\end{pmatrix} \subsetneq \begin{pmatrix}\text{Dubrovin}\\ \text{valuation}\\ \text{rings}\end{pmatrix}$$

Note that the class of Dubrovin valuation rings is much wider than the class of total valuation rings. Profound applications were found. More information about Dubrovin valuation rings can be found in the book [233] which presents a good many main results about these rings with complete proofs.

Let A be a semisimple ring. Then we write $d(A) = m$ for the number of all mutually non-isomorphic simple right A-modules $U_i = e_i A$ where $e_i^2 = e_i \in A$ for $i = 1, 2, \ldots, m$. If M is a finitely generated right module over a semisimple ring A then it is decomposed into a direct sum of simple right A-modules

$$M = U_{n_1} \oplus U_{n_2} \oplus \cdots \oplus U_{n_d}, \tag{3.7.13}$$

where each U_{n_i} is one of $U_1, U_2, \ldots, U_m$. By the Krull-Schmidt theorem the number of direct summands in this decomposition is constant and do not depend on the chosen decomposition of M into simple right A-modules. We write $d_A(M) = d$ the number of all direct summands in the decomposition (3.7.13) of a module M. If $d = d_A(M) \le d_A(A) = m$ then $M = e_{i_1} A \oplus \cdots \oplus e_{i_d} A$ with all different idempotents e_i. Therefore $M = xA$ where $x = e_{i_1} + \cdots + e_{i_d}$ and so M is a cyclic A-module. If $d_A(M) \ge d_A(A)$ then $M \simeq A \oplus N$ and $d_A(M) = d_A(A) + d_A(N)$.

Lemma 3.7.14. *Let A be a Dubrovin valuation ring of a simple Artinian ring S. Then*

1. *If $q \in S$ and*
$$qA \cap A \subseteq J(A) \tag{3.7.15}$$
then $q \in A$ and so $q \in J(A)$;
2. *If I is a proper ideal of S then $I \subseteq J(A)$.*

Proof.

1. Suppose that $q \in S \setminus A$. Then there exists $0 \neq x \in A$ such that $qx \in A \setminus J(A)$. But from (3.7.15) it follows that $qx \in qA \cap A \subseteq J(A)$. This contradiction shows that $q \in A$, and (3.7.15) implies that $q \in J(A)$.
2. Since I is a proper ideal of S, $I \cap A$ is a proper ideal of A, and so $I \cap A = IA \cap A \subseteq J(A)$. Hence $I \subseteq J(A)$. □

Proposition 3.7.16 [234, Lemma 1.4.3]. *Let A be a Dubrovin valuation ring of a simple Artinian ring S. Then every finitely generated right A-submodule M of S is cyclic.*

Proof. Let A be a Dubrovin valuation ring with Jacobson radical R and $\bar{A} = A/R$. Consider a finitely generated right A-module M of S and let $\bar{M} = M/MR$ which is a finitely generated right $\bar{A}$-module. First we will prove that $\bar{M}$ is a cyclic $\bar{A}$-module. As said above it suffices to show that $d_{\bar{A}}(M/MR) \le d_{\bar{A}}(\bar{A})$.

Suppose that the opposite holds, i.e., $d_{\bar{A}}(M/MR) > d_{\bar{A}}(\bar{A})$. Then $M/MR \simeq \bar{A} \oplus N$ for some $\bar{A}$-module N. Therefore there exist elements $x, y \in M$ such that

$$(xA + yA + MR)/MR = (xA + MR)/MR \oplus (yA + MR)/MR \tag{3.7.17}$$

with $y \in M \setminus MR$ and

$$(xA + MR)/MR \simeq \bar{A}.$$

Since

$$(xA + MR)/MR \simeq xA/(xA \cap MR) \simeq A/R \tag{3.7.18}$$

and $xR \subseteq xA \cap MR$, we obtain that

$$xA \cap MR = xR. \tag{3.7.19}$$

Thus (3.7.18) and (3.7.19) yield $xA/xR \simeq \bar{A}$ which implies $\text{r.ann}_A(x) \subseteq R$. We now show that $\text{r.ann}_S(x) = \text{r.ann}_A(x)$. Suppose that $y \in \text{r.ann}_S(x) \subseteq S$, i.e., $xy = 0$, and $y \notin A$. Since A is a Dubrovin valuation ring, there exists $b \in A$ such that $yb \in A \setminus R$. Then $x(yb) = (xy)b = 0$ which implies $yb \in \text{r.ann}_A(x) \subseteq R$. This contradiction shows that $\text{r.ann}_S(x) = \text{r.ann}_A(x) \subseteq R$.

Suppose that $I = \text{r.ann}_S(x) \neq 0$. Then it contains an idempotent $e^2 = e$. Therefore $e(1 - e) = 0$. Since $e \in I \subset R$, $1 - e$ is an invertible element in A, so that $e = 0$. Thus $\text{r.ann}_S(x) = 0$, which implies that x is invertible in S, by proposition 1.1.17. Therefore (3.7.19) can be rewritten in the form:

$$A \cap x^{-1}MR = R.$$

Hence $x^{-1}MR \subseteq R$ by lemma 3.7.14(2). Since $x \in M$, we obtain $MR = xR$. By assumption $xA \cap yA \subseteq MR = xR$ and so $x^{-1}yA \cap A \subseteq R$. Again by lemma 3.7.14(1) this implies that $y \in xR = MR$, a contradiction. Hence $d_{\bar{A}}(M/MR) \leq d_{\bar{A}}(\bar{A})$.

So M/MR is a cyclic $\bar{A}$-module, and hence M is a cyclic A-module by Nakayama's lemma. □

Corollary 3.7.20. *Every right (left) finitely generated ideal of a Dubrovin valuation ring is principal.*

Proposition 3.7.21. *If A is a Dubrovin valuation ring of a simple Artinian ring S then A is an order in S.*

Proof. Let A be a Dubrovin valuation ring of a simple Artinian ring S. Let $q \in S \setminus A$. Then by proposition 3.7.16 the finitely generated right A-submodule $A + qA$ of S is cyclic, i.e., there exists an element $s \in S$ such that $A + qA = sA$. In particular, there exist elements $t, r \in A$ such that $1 = st$ and $q = sr$. Therefore $q = t^{-1}r$ where $r \in A$ and t is regular in A.

We now show that any regular element $x \in A$ is invertible in S, or equivalently that $\text{l.ann}_S(x) = 0$, by proposition 1.1.17. Suppose that $yx = 0$ for some $0 \neq y \in S \setminus A$. Since A is a Dubrovin valuation ring of S, there exists $a \in A$ such that $0 \neq ay \in A \setminus J(A)$. Then $ayx = 0$ which contradicts $\text{l.ann}_A(x) = 0$. Therefore any regular element of A is invertible in S. So A is a left order in S. In a similar way we can prove that A is a right order in S. Thus A is an order in S. □

Proposition 3.7.22. *Let A be a Dubrovin valuation ring of a simple Artinian ring S. Then A is a prime Goldie ring.*

Proof. By proposition 3.7.21 A is an order in a simple Artinian ring S. Then the result follows immediately from theorem 1.11.4. □

Definition 3.7.23. A ring A is said to be a **right (left) Bézout ring** if every finitely generated right (left) ideal of A is principal. A ring which is both right and left Bézout is called a **Bézout ring**.

Theorem 3.7.24 [234, Theorem 1.4.7]. *Let A be a Dubrovin valuation ring of a simple Artinian ring S. Then*:

1. *A is a Bézout order in S with a unique maximal ideal.*
2. *A is a semihereditary order in S with a unique maximal ideal.*

Proof.

1. Follows from corollary 3.7.20 and proposition 3.7.21.
2. Is a special case of (1). □

Definition 3.7.25. Let A be a subring in a ring S. Then A is called a **right n-chain ring in** S if for any $n+1$ elements $x_0, x_1, \dots, x_n \in S$ there is i such that

$$x_i \in \sum_{k \neq i} x_k A.$$

A right n-chain ring in itself is called a **right n-chain ring**. A **left n-chain ring** is defined similarly. A ring which is both a right and left n-chain ring is called an **n-chain ring**.

Note that 1-chain ring is exactly an uniserial ring (or a chain ring).

Lemma 3.7.26. [234, Lemma 1.4.6]. *Let A be a ring with Jacobson radical $J(A) = 0$. Then A is (right) Artinian if and only if A is a (right) n-chain ring for some $n \in \mathbf{N}$.*

Proof. Suppose that A is a right Artinian ring with $J(A) = 0$. Then A is a semisimple ring, by theorem 1.1.24. Therefore, by the Wedderburn-Artin theorem 1.1.6, $A \simeq M_{n_1}(D_1) \oplus M_{n_2}(D_2) \oplus \cdots \oplus M_{n_k}(D_k)$, where D_i is a division ring for $i = 1, 2, \dots, k$. So A has $n = n_1 + n_2 + \cdots + n_k$ direct summands into the decomposition into simple right A-modules, i.e., $d_A(A) = n$. Therefore for any $n+1$ different elements $a_0, a_1, \dots, a_n \in A$ the ascending chain of right ideals $a_0A \subseteq a_0A + a_1A \subseteq \cdots \subseteq a_0A + a_1A + \cdots + a_nA$ has length at most n, therefore there is an integer k such that $a_0A + a_1A + \cdots + a_kA = a_0A + a_1A + \cdots + a_kA + a_{k+1}A$, which implies that $a_{k+1} \in a_0A + a_1A + \cdots + a_kA$.

Conversely, suppose that A is a right n-chain ring for some $n \in \mathbf{N}$ and $J(A) = 0$. We will show that for any set of different maximal right ideals there is a finite subset which intersection is zero. Consider a set $N_1, N_2, \dots, N_k, \dots$ of maximal right ideals of A. Since $J(A) = 0$ then $\bigcap_{i=1}^{\infty} N_i = 0$. Consider the ascending chain of right ideals of A:

$$A \supset N_1 \supset N_1 \cap N_2 \supset \cdots \supset N_1 \cap N_2 \cap \cdots \cap N_k \supset \cdots \tag{3.7.27}$$

Since $N_i \cap N_j = 0$ for $i \neq j$,

$$A/(N_1 \cap N_2 \cap \cdots \cap N_k) \simeq A/N_1 \oplus A/N_2 \oplus \cdots \oplus A/N_k \simeq S_1 \oplus S_2 \oplus \cdots \oplus S_k,$$

where each $A/N_i \simeq S_i$ is a simple right A-module.

Consider natural epimorphisms $\varphi_i : A \to A/N_i$. Select some non-zero elements $\bar{a}_1, \bar{a}_2, \dots, \bar{a}_k \in A/N_i$, and their preimages $a_1, a_2, \dots, a_k \in A$, i.e., $\varphi(a_i) = \bar{a}_i$. Clearly, $a_i \notin \sum_{i \neq j} a_j A$ since $A/N_i \bigcap \bigoplus_{i \neq j} A/N_j = 0$. Therefore using that $J(A) = 0$ and extending the chain of right ideals (3.7.27), we obtain that for some large k, $N_1 \cap \cdots \cap N_k = 0$.

Suppose that we have an ascending chain of right ideals of A:

$$A \supset I_1 \supset I_2 \supset \cdots I_n \supset \cdots \tag{3.7.28}$$

Let N_k be a maximal ideal of A such that $I_k \subset N_k$. By proving above there is a some integer m such that $I_m = I_1 \cap I_2 \cap \cdots \cap I_m \subseteq N_1 \cap N_2 \cap \cdots \cap N_m = 0$, i.e., the chain (3.7.28) stabilizes. Thus A is a right Artinian ring. □

Theorem 3.7.29. [233, Lemma 5.9]. *Let A be an n-chain ring in a simple Artinian ring S for some n with unique maximal ideal. If $n = d(A/J(A))$, where $J(A)$ is the Jacobson radical of A, then A is a Dubrovin valuation ring of S.*

Proof. Since A is an n-chain ring, so is $\bar{A} = A/J(A)$. Then by lemma 3.7.26 $\bar{A}$ is an Artinian ring. Thus $\bar{A}$ is a simple Artinian ring, since A has a unique maximal ideal.

Let $n = d(\bar{A})$ and $\bar{e}_1, \bar{e}_2, \dots, \bar{e}_n$ be all mutually orthogonal primitive idempotents of $\bar{A}$. Select $e_1, e_2, \dots, e_n \in A$ such that $\varphi(e_i) = \bar{e}_i$, $(i = 1, 2, \dots, n)$, where $\varphi : A \to A/J(A)$ is a natural epimorphism.

For $x \in S$ it is possible two cases:

1. $xA + J(A) \subseteq \sum_{i}^{n} e_i A + J(A)$;
2. There exists an integer i such that $e_i A + J(A) \subseteq xA + \sum_{j \neq i}^{n} e_j A + J(A)$.

Suppose that $x \in S \setminus A$. Then it is possible only the second case and so

$$e_i = xa + \sum_{j \neq i}^{n} e_j b_j + r,$$

where $a, b_j \in A$ and $r \in J(A)$. Hence $xa = e_i - \sum_{j \neq i}^{n} e_j b_j - r \in A \setminus J(A)$. In a similar way one can show that there exists an $b \in A$ such that $bx \in A \setminus J(A)$. So A is a Dubrovin valuation ring of S. □

Theorem 3.7.30. [234, Theorem 1.4.7]. *Let A be a Bézout order in a simple Artinian ring S with unique maximal ideal. Then A is an n-chain ring in S for some n with* $n = d(A/J(A))$, *where* $J(A)$ *is the Jacobson radical of A.*

Proof. Suppose A is a Bézout order in S and $d_{\bar{A}}(\bar{A}) = n$, where $\bar{A} = A/J(A)$. Let $x_0, x_1, x_2, \ldots, x_n \in S$. Then

$$x_0 A + x_1 A + \cdots + x_n A = yA$$

for some $y \in S$. Since $d(yA/yJ(A)) \leq d_{\bar{A}}(\bar{A}) = n$, there is an integer i such that

$$yA = \sum_{j \neq i} x_j A + yJ(A)$$

Then by Nakayama's lemma $yA = \sum_{j \neq i} x_j A$ and so $x_i \in \sum_{j \neq i} x_j A$, i.e., A is a right n-chain ring in S. In a similar way one can show that A is a left n-chain ring in S. Thus A is an n-chain ring in S. □

Remark 3.7.31. In section 8.3 we will show the equivalent conditions for a ring to be a Dubrovin valuation ring connected with Bézout orders, semihereditary orders and n-chain rings.

3.8 Approximation Theorems for Valuation Rings

When working with several valuation rings of a given division ring, the Approximation theorem is a basic tool which is frequently used. In this section we consider the approximation theorem for noncommutative valuation rings and some corollaries from this theorem for noncommutative DVR's.

In the general algebra a number of approximation theorems are well known, which are different generalizations of the famous Chinese Remainder Theorem obtained by Chinese mathematicians for solving a finite system of congruence in integers.

Among these generalizations in the theory of commutative fields the CRT for ideals of principal ideal domains is the most known one (See [146, Theorem 7.6.2]. Another forms of approximation theorems are the general Ribenboim's approximation theorem [271, p.136] and the Krull approximation theorem for independent valuations.

Definition 3.8.1. Two valuation rings B_1 and B_2 of a field K are called **independent** if their least overing in K is K itself. Two valuations v and w on a field K are said to be **independent** if their valuation rings B_v and B_w are independent.

Two valuation rings B_1 and B_2 of a field K are called **incomparable** if $B_1 \not\subseteq B_2$ and $B_2 \not\subseteq B_1$.

Theorem 3.8.2 (Approximation Theorem). *Let v_i $(1 \leq i \leq n)$ be a pairwise independent valuations on a field K with value groups G_i, and let $\gamma_i \in G_i$, $a_i \in K$ $(1 \leq i \leq n)$. Then there exists an element $x \in K$ such that*

$$v_i(x - a_i) \geq \gamma_i \tag{3.8.3}$$

for each $i \in \{1,2,\dots,n\}$.

Corollary 3.8.4. *Let v_i $(1 \leq i \leq n)$ be a pairwise independent valuations on a field K with value groups G_i, and let $\gamma_i \in G_i$ $(1 \leq i \leq n)$. Then there exists an element $x \in K$ such that*

$$v_i(x) = \gamma_i \tag{3.8.5}$$

for each $i \in \{1,2,\dots,n\}$.

The full proofs of theorem 3.8.2 and Corollary 3.8.4 can be found in [24, Chapter 6].

Note that analogous form of the approximation theorem and its corollary for R-Prüfer rings[7] with zero divisors, were proved by M. Griffin in [124]. In his theory M. Griffin used the valuation theory of M.E. Manis.

In the case of valuations on a division ring D J. Gräter in [119] extends Ribenboim's general theorem to the case of a finite number of total valuation rings of D.

In his paper [119] J.Gräter investigated under which circumstance the general Ribenboim theorem holds for valuations on division rings. For this aim he introduced the special class of valuations and locally invariant rings.

Recall that a (two-sided) ideal I of a ring A is called prime if $I_1 I_2 \subseteq I$ implies that either $I_1 \subseteq I$ or $I_2 \subseteq I$ for any (two-sided) ideals I_1, I_2 of A. If A is a commutative ring this condition is equivalent to the following: $xy \in I$ implies that either $x \in I$ or $y \in I$ for any $x, y \in A$.

A (two-sided) ideal I of a ring A is called **completely prime** if $xy \in I$ implies that either $x \in I$ or $y \in I$ for any $x, y \in A$. So that in the commutative case the notions of a prime ideal and a completely prime ideal are the same.

Definition 3.8.6 [119]. Let A be a total valuation ring of a division ring D with the maximal ideal M, and $P(x)$ the minimal completely prime ideal of A containing $x \in M$. The ring A is called **locally invariant** if

$$xP(x) = P(x)x$$

---

[7]A subring B of a ring A is called an R-**Prüfer ring** with regard to a set of fixing elements R of A if $B_{[P]}$ is a valuation ring for every maximal ideal P of B. Here $B_{[P]} = \{b \in R \mid db \in B \text{ for some } d \in B \backslash P\}$.

for all $x \in M$.

The example of a locally invariant ring is an invariant valuation ring in sense of definition 3.3.7.

The following theorem provides the characterization of locally invariant valuation rings which we give here without proof.

Theorem 3.8.7. ([119], [20], [28, Theorem 3.1].) *A chain domain A is locally invariant if and only if any one of the following equivalent conditions is satisfied:*

1. $\bigcap_n a^n A$ *is a two-sided ideal for any* $a \in A$.
2. $Aa^2 \subseteq aA$ *for any* $a \in A$.
3. $\sqrt{I} = \{x \in A | \exists n = n(x) \in \mathbf{N}, x^n \in I\}$ *is a completely prime ideal for every right ideal I in A.*
4. *Between any two prime ideals of A there is another two-sided ideal of A.*
5. $xP(x) = P(x)x$ *for any* $x \in R$ *and* $P(x)$ *the minimal completely prime ideal of R containing x.*

From this theorem it follows immediately the following statement:

Corollary 3.8.8. *Any discrete valuation ring O of a division ring D (in sense of definition 3.5.1) is locally invariant.*

In [119, Satz 4.1] J.Gräter proved the approximation theorem for total valuation rings of a division ring when they are locally invariant. We give here this theorem following to [28]. We need some definitions which were introduced in this paper.

Let B_1, B_2 be valuation rings of a division ring D. Then there exists a minimal valuation ring $B_{1,2}$ of D containing both B_1 and B_2. A right ideal I_1 of B_1 and a right ideal I_2 of B_2 are said to be **compatible** if $I_1 B_{1,2} = I_2 B_{1,2}$. If in addition, $a_1, a_2 \in D$ then a_1, a_2, I_1, I_2 are called **compatible** if $a_1 - a_2 \in I_1 B_{1,2} = I_2 B_{1,2}$.

Theorem 3.8.9 (Approximation Theorem). [28, Theorem 3.2]. *Let* $B_1, \ldots, B_n$ *be invariant valuation rings of a division ring D. Assume that* B_i *is locally invariant for* $i = 1, \ldots, n-1$. *Let* $a_i \in D$ *for* $i = 1, \ldots, n$ *and let* I_i *be a right ideal of* B_i *such that* a_i, a_j, I_i, I_j *are compatible for every* $i, j \in \{1, \ldots, n\}$. *Then there exists an* $x \in D$ *with*

$$x - a_i \in I_i \tag{3.8.10}$$

for $i = 1, \ldots, n$.

Note that this theorem holds for invariant valuation rings in finite dimensional division algebras.

We show now that this theorem holds also for discrete valuation rings. Let B_1, B_2 be discrete valuation rings of a division ring D. By [146, Theorem 10.2.5], any discrete valuation ring O of D is a maximal (under inclusion) order in D. Therefore if $B_1 \neq B_2$ then $B_{1,2} = D$. Therefore $I_1 B_{1,2} = I_2 B_{1,2}$ for each ideal I_1 of B_1 and each ideal I_2 of B_2, and so a_1, a_2, I_1, I_2 are compatible for all elements $a_1, a_2 \in D$ and any ideal I_1 of B_1, any ideal I_2 of B_2. Taking into account that any discrete valuation ring is locally invariant from theorem 3.8.9 we immediately obtain the following corollary which is the approximation theorem for discrete valuation rings.

Corollary 3.8.11. *Let $O_1, \dots, O_n$ be distinct discrete valuation rings of a division ring D. Then for any elements $a_i \in D$ and ideals I_i of O_i, for $i = 1, \dots n$, there is an element $x \in D$ such that*

$$x - a_i \in I_i \tag{3.8.12}$$

for $i = 1, \dots n$.

Proposition 3.8.12. *Let O_1, O_2 be distinct discrete valuation rings of a division ring D with maximal ideals M_1 and M_2. Then*

1. *$D = O_1 + O_2$.*
2. *There is an element $\pi_1 \in M_1 \setminus M_1^2$ and $\pi_1 \in O_2^*$. Analogously there is an element $\pi_2 \in M_2 \setminus M_2^2$ and $\pi_2 \in O_1^*$. So that $M_1 = \pi_1 O_1$, and $M_2 = \pi_2 O_2$.*
3. *$D^* = O_1^* \cdot O_2^* = O_2^* \cdot O_1^*$.*

Proof.

1. Put $I_1 = O_1$, $I_2 = O_2$, $a_1 = d \in D$, $a_2 = 0$ in corollary 3.8.11. Then $\exists x \in D$ that $x - d \in O_1$ and $x \in O_2$, i.e., $x - d = a$, $x = b$. So $d = b - a$, where $b \in O_2$, $a \in O_1$.
2. Putting $I_1 = M_1^2$, $I_2 = M_2$, $a_1 \in M_1 \setminus M_1^2$, $a_2 = 1$ in corollary 3.8.11, we obtain an element $x = \pi_1 \in M_1 \setminus M_1^2$ and $\pi_1 \in O_2^*$. Analogously we obtain an element $\pi_2 \in M_2 \setminus M_2^2$ and $\pi_2 \in O_1^*$.
3. It is obvious that $O_1^* \cdot O_2^* \subseteq D^*$ and $O_2^* \cdot O_1^* \subseteq D^*$. On the other hand, any element $d \in D^*$ can be written in the form $d = \pi_1^n u$ where $\pi_1 \in (M_1 \setminus M_1^2) \cap O_2^*)$, by statement 2 and theorem 3.5.3(3). Therefore $D^* \subseteq O_2^* \cdot O_1^*$. Analogously, $D^* \subseteq O_1^* \cdot O_2^*$. Thus, $D^* = O_2^* \cdot O_1^* = O_1^* \cdot O_2^*$. □

Lemma 3.8.13. *Let O_1, O_2 be distinct discrete valuation rings of a division ring D, and $A = O_1 \cap O_2$. Then the group of units $U(A)$ is invariant in D.*

Proof. Let $u \in U(A)$, then $u^{-1} \in A$. Therefore $u^{-1} \in O_1$ and $u^{-1} \in O_2$, and so $U(A) \subseteq U(O_1) \cap U(O_2)$.

On the other hand, if $x \in U(O_1) \cap U(O_2)$ then $x \in U(O_1)$ and $x \in U(O_1)$, so $x^{-1} \in O_1$ and $x^{-1} \in O_2$, that is, $x^{-1} \in A$ which implies $x \in U(A)$. Therefore $U(O_1) \cap U(O_2) \subseteq U(A)$. So,

$$U(A) = U(O_1) \cap U(O_2). \tag{3.8.14}$$

Let $u \in U(A)$ and $d \in D^*$. Since $U(O_1)$, $U(O_2)$ are invariant in D, then from (3.8.14) $dud^{-1} \in U(O_1)$ and $dud^{-1} \in U(O_2)$. And so $dud^{-1} \in U(A)$, that is, $dU(A)d^{-1} \subseteq U(A)$ for any $d \in D^*$. Let $dud^{-1} = u_1$, then $u = d^{-1}u_1 d \in U(A)$, that is, $U(A) \subseteq d^{-1}U(A)d$ for any $d \in D^*$. Therefore $dU(A)d^{-1} = U(A)$ for any $d \in D^*$, that is, $U(A)$ is invariant in D. □

Proposition 3.8.15. *Let O_1, O_2 be distinct discrete valuation rings of a division ring D, and $A = O_1 \cap O_2$. Then there exists an element $x \in D$ such that $x^n \notin A$ and $x^{-n} \notin A$ for any $n \in N$.*

Proof. Suppose that A is a totally valuation ring of D. Then by theorem 3.6.4 subrings of D containing A must be totally ordered by inclusion. Since $O_1 \not\subseteq O_2$ and $O_2 \not\subseteq O_1$ we have a contradiction.

Therefore A is not a total valuation ring and so by definition there exists and element $x \in D$ such that $x \notin A$ and $x^{-1} \notin A$.

Suppose there exists $n \in N$ such that $x^n \in A$ or $x^{-n} \in A$. Since $x^n \in A$, $x^n \in O_1$ and $x^n \in O_2$. Therefore $v_1(x^n) = nv_1(x) \geq 0$, which implies that $v_1(x) \geq 0$, that is $x \in O_1$. Analogously, $x \in O_2$, that is, $x \in A$. A contradiction! In the same way we can show that $x^{-n} \notin A$ for any $n \in N$. □

In [245] P.J. Morandi extends Gräter's approximation theorem to the case of Dubrovin valuation rings of a finite-dimensional central simple K-algebra S.

Definition 3.8.16. Two Dubrovin valuation rings B_1, B_2 of a finite-dimensional central simple algebra S are said to be **independent** (resp. **incomparable**) if the least overring in S of B_1 and B_2 is S itself (resp. $B_1 \not\subseteq B_2$ and $B_2 \not\subseteq B_1$).

We will write $Z(A)$ for the center of a ring A. If B_i and B_j are Dubrovin valuation rings in S, then B_{ij} will denote the least overring in S of B_i and B_j, $\overline{B_{ij}} = B_{ij}/J(B_{ij})$, $\tilde{B}_i = B_i/J(B_{ij})$ where $J(B_{ij})$ is the Jacobson radical of B_{ij}.

Theorem 3.8.17. (**Approximation theorem.**) [245, Theorem 2.3]. *Let S be a finite-dimensional central simple algebra and let $B_1, B_2, \ldots, B_n$ be pairwise incomparable Dubrovin valuation rings of S. Suppose for all $i \neq j$, $Z(\tilde{B}_i)$ and $Z(\tilde{B}_j)$ are independent in $Z(\overline{B_{ij}})$. If I_i is a right ideal of B_i with $I_iB_{ij} = I_jB_{ij}$ and $s_i \in S$ with $s_i - s_j \in I_iB_{ij}$ then there is an $x \in S$ with*

$$x \equiv s_i \bmod I_i$$

for all $i = 1, 2, \ldots, n$.

3.9 Notes and References

The theory of valuations and valuation rings has its beginning in the early 20th century. The concepts of valuations on fields and valuation domains first were introduced in 1932 by W. Krull in his famous paper [201]. In this paper a valuation ring is defined as an integral domain whose ideals are totally ordered by inclusion, i.e., as a commutative uniserial domain. He also showed the connection between the concepts of valuation domains and valuation rings of fields. He first proved the equivalence (1) and (3) in theorem 3.1.14, and proved the basic existence theorem:

Theorem 3.9.1. *Given a field K and a totally ordered Abelian group G, there exists a valuation domain A with field of fractions isomorphic to K and with a value group order-isomorphic to G.*

Prüfer domains were defined in 1932 by H.Püfer [267]. In 1936 W. Krull in [202] named these rings Prüfer rings in honor of H.Prüfer. He also proved many equivalent definitions of Prüfer domains. The most significant investigations connected with Prüfer domains were obtained by R. Gilmer and collected in his famous book on commutative rings [93].

The theory of valuations for division rings was developed by O.F.G.Shilling in [283] and systematically presented in his book [284]. If we consider invariant valuation rings of division rings as introduced by Schilling one obtains that any invariant valuation ring is a semihereditary ring. So semihereditary rings can be considered as some generalization of Prüfer domains for noncommutative rings.

The notion of valuation rings for commutative rings with divisors of zero was considered in 1969 by M.E. Manis in [231]. These rings are called Manis valuation rings after him.

The theory of noncommutative valuations and valuations rings has evolved substantially during the last 30 years. The main progress in this theory has been made by N.I. Dubrovin who first introduced the new type of valuation rings which are now called Dubrovin valuation rings. This class of rings is defined not only for division rings but also for simple Artinian rings. Dubrovin valuation rings can be considered as a noncommutative analogue of commutative valuation rings in the sense of W. Krull [201]. In noncommutative valuation theory it was shown that a Dubrovin valuation ring of a simple Artinian ring S is exactly a local semihereditary order of S. So semihereditary orders in general can be considered as a global version of Dubrovin valuation rings.

Commutative fully (linearly) ordered groups were considered first in 1907 by H.Hahn in the paper [137]. The interesting book [134] is devoted to the study of properties of partially ordered algebraic systems. Fully ordered groups were considered and studied in [193].

CHAPTER 4

Homological Dimensions of Rings and Modules

There is a discussion of homological dimensions of rings and modules in [146, chapter 6] and [147, chapter 4]. This chapter considers some other questions connected with these notions.

For the reader's convenience Section 4.1 summarizes the basic concepts and results on projective and injective dimensions of modules, and on the global dimensions of rings. Concepts and results on flat dimensions of modules and weak dimensions of rings are presented in Section 4.2. In Section 4.3 we collect some useful results for calculating global homological dimensions of various classes of rings and present some examples.

The homological characterization of some classes of rings, such as semisimple, right Noetherian, right hereditary, right semihereditary, semiperfect, right perfect and quasi-Frobenius rings are considered in Section 4.4.

Duality over Noetherian rings, which is given by the covariant functor $^* = \mathrm{Hom}_A(-, A)$, was considered in [147, Section 4.10]. For an arbitrary ring A this functor induces a duality between the full subcategories of finitely generated projective right A-modules and finitely generated projective left A-modules. The main properties of this functor and of torsion-free modules, and the relationships between them are studied in Section 4.5.

The basic properties of flat modules were considered in [146, Section 5.4], and in [147, Section 5.1]. Further properties of flat modules are studied in Section 4.6. In particular, the main theorem of this section, which was proved by S.U. Chase [41], gives equivalent conditions for a direct product of any family of flat modules to be flat. As the corollary of this theorem we obtain the homological characterizations of semihereditary rings, which were proved by S.U.Chase in [41].

Section 4.7 gives necessary and sufficient conditions under which a formal triangular matrix ring is right (left) hereditary.

4.1 Projective and Injective Dimension

In this section we recall the basic concepts and results on projective and injective dimension.

Definition 4.1.1. Let A be a ring and X a right A-module. One says that the **projective dimension** of X is equal to n and writes $\text{proj.dim}_A X = n$ if there is a projective resolution[1] of length n:

$$0 \longrightarrow P_n \xrightarrow{d_n} \dots \xrightarrow{d_2} P_1 \xrightarrow{d_1} P_0 \xrightarrow{\varepsilon} X \longrightarrow 0 \tag{4.1.2}$$

and there is no shorter one.

Set $\text{proj.dim}_A X = \infty$ if there is no finite length resolution.

From this definition it follows that $\text{proj.dim}_A X = 0$ if and only if X is projective.

Theorem 4.1.3. (See [146, proposition 6.5.4].) *The following conditions are equivalent for a right A-module X*:

1. $\text{proj.dim}_A X \leq n$;
2. $\text{Ext}_A^k(X,Y) = 0$ *for all right modules Y and all* $k \geq n+1$;
3. $\text{Ext}_A^{n+1}(X,Y) = 0$ *for all right modules Y*;
4. $\text{Ker}(d_{n-1})$ *is a projective right A-module for any projective resolution* (4.1.2) *of X*.

(For definition of $\text{Ext}_A^k(-,-)$ see Section 1.7.)

If A is a right Noetherian ring then one can say some more about finitely generated A-modules.

Proposition 4.1.4. (See [147, proposition 5.1.3].) *Let A be a right Noetherian ring. Then every finitely generated flat right A-module is projective.*

Definition 4.1.5. Let A be a ring and X a right A-module. One says that the **injective dimension** of X is equal to n and writes $\text{inj.dim}_A X = n$ if there is a injective resolution of length n:

$$0 \longrightarrow X \xrightarrow{i} Q_0 \xrightarrow{d_0} \dots \longrightarrow Q_{n-1} \xrightarrow{d_{n-1}} Q_n \longrightarrow 0 \tag{4.1.6}$$

and there is no shorter one.

Theorem 4.1.7 (M. Auslander [9]). (See [147, theorem 5.1.11, corollary 5.1.12].) *Let A be an arbitrary ring and M a right A-module. Then* $\text{inj.dim}_A M \leq n$ *if and only if* $\text{Ext}_A^{n+1}(A/\mathcal{I},M) = 0$ *for any right ideal $\mathcal{I}$ of A. In particular, a right A-module B is injective if and only if* $\text{Ext}_A^1(A/\mathcal{I},B) = 0$ *for any right ideal $\mathcal{I}$ of A.*

[1]"projective resolution" here means that the P_i are projective and the sequence is exact.

The following statement is dual to theorem 4.1.3.

Theorem 4.1.8. (See [146, proposition 6.5.8].) *The following conditions are equivalent for a right A-module X:*

1. $\text{inj.dim}_A X \le n$;
2. $\text{Ext}_A^k(X,Y) = 0$ *for all modules Y and* $k \ge n+1$;
3. $\text{Ext}_A^{n+1}(X,Y) = 0$ *for all modules Y*;
4. $\text{Im}(d_{n-1})$ *is an injective module for any injective resolution* (4.1.6) *of X*.

Definition 4.1.9. The **right projective global dimension** of a ring A, abbreviated as r.proj.gl.dim A, is defined as follows:

$$\text{r.proj.gl.dim}\, A = \sup\{\text{proj.dim}_A M : M \in \mathbf{M}_A\} \tag{4.1.10}$$

Analogously one can introduce the **left projective global dimension** of A:

$$\text{l.proj.gl.dim}\, A = \sup\{\text{proj.dim}_A M : M \in {}_A\mathbf{M}\} \tag{4.1.11}$$

Theorem 4.1.3 immediately implies:

Proposition 4.1.12. r.proj.gl.dim$A \le n$ *if and only if* $\text{Ext}_A^{n+1}(X,Y) = 0$ *for all right A-modules X and Y.*

Dually one can define right injective global dimension of a ring.

Definition 4.1.13. The **right injective global dimension** of a ring A, abbreviated as r.inj.gl.dim A, is defined as follows:

$$\text{r.inj.gl.dim} A = \sup\{\text{inj.dim}_A M : M \in \mathbf{M}_A\} \tag{4.1.14}$$

Analogously, one can introduce the **left injective global dimension** of A:

$$\text{l.inj.gl.dim} A = \sup\{\text{inj.dim}_A M : M \in {}_A\mathbf{M}\} \tag{4.1.15}$$

Theorem 4.1.8 immediately implies:

Proposition 4.1.16. r.inj.gl.dim$A \le n$ *if and only if* $\text{Ext}_A^{n+1}(X,Y) = 0$ *for all right A-modules X and Y.*

Comparing propositions 4.1.12 and 4.1.16 there results:

Theorem 4.1.17. *For any ring A*

$$\text{r.proj.gl.dim} A = \text{r.inj.gl.dim} A.$$

In view of this theorem one defines the **right global dimension** of a ring A, abbreviated as r.gl.dim A, as the common value of r.proj.gl.dim A and r.inj.gl.dim A. Considering left A-modules one can analogously define the **left global dimension** of a ring A.

Theorem 4.1.18 (M. Auslander [9]). (See also [147, proposition 5.1.13].) *For any ring A,*

$$\text{r.gl.dim}\, A = \sup\{r.\text{proj.dim}_A\,(A/\mathcal{I}) : \mathcal{I} \textit{ is a right ideal of } A\}. \tag{4.1.19}$$

Theorem 4.1.20 (M. Auslander [9]). (See also [147, proposition 5.1.16].) *If A is a Noetherian ring, then*

$$\text{r.gl.dim}\, A = \text{l.gl.dim}\, A.$$

4.2 Flat and Weak Dimension

This section collects basic concepts and results on flat and weak dimension as presented in [147, Section 5.1].

Definition 4.2.1. A right A-module X has **flat dimension** n which is written $\text{w.dim}_A X = n$ if there is an exact sequence

$$0 \longrightarrow F_n \xrightarrow{d_n} F_{n-1} \xrightarrow{d_{n-1}} \dots \xrightarrow{d_2} F_1 \xrightarrow{d_1} F_0 \xrightarrow{\varepsilon} X \longrightarrow 0, \tag{4.2.2}$$

where all F_i are flat and there is no shorter such sequence.

Set $\text{w.dim}_A M = \infty$ if there is no n with $\text{w.dim}_A X \leq n$.

Theorem 4.2.3. (See [147, proposition 5.1.5].) *The following conditions are equivalent for a right A-module X*:

1. $\text{w.dim}_A X \leq n$;
2. $\text{Tor}_k^A(X,Y) = 0$ *for all left A-modules Y and all* $k \geq n+1$;
3. $\text{Tor}_{n+1}^A(X,Y) = 0$ *for all left A-modules Y*;
4. $\text{Ker}\,(d_{n-1})$ *is a flat A-module for any projective resolution* (4.2.2) *of X*.

(For definition of $\text{Tor}_k^A(-,-)$ see Section 1.6.)

Definition 4.2.4. The **right weak dimension** of a ring A, abbreviated as r.w.dim A, is defined as follows:

$$\text{r.w.dim}\, A = \sup\{\text{w.dim}_A M : M \in \mathbf{Mod}_r\, A\} \tag{4.2.5}$$

Analogously one can introduce the **left weak dimension** of A:

$$\text{l.w.dim}\, A = \sup\{\text{w.dim}_A M : M \in \mathbf{Mod}_l\, A\} \tag{4.2.6}$$

Theorem 4.2.3 immediately implies:

Proposition 4.2.7. (See [147, corollary 5.1.6].) *Given a ring A* r.w.dim $A \leq n$ *if and only if* $\text{Tor}_{n+1}^A(X,Y) = 0$ *for all right A-modules X and all left A-modules Y.*

Theorem 4.2.8. (See [147, theorem 5.1.7].) *For any ring A*

$$\text{r.w.dim}\, A = \text{l.w.dim}\, A.$$

Definition 4.2.9. The common value of r.w.dim A and l.w.dim A is called the **weak dimension** of a ring A and written as w.dim A.

Proposition 4.2.10. (See [147, proposition 5.1.9].) *For any ring A,*

$$\text{w.dim}\, A = \sup\{\text{r.w.dim}_A\, (A/\mathcal{I})\ :\ \mathcal{I}\ \textit{is a right ideal of}\ A\} =$$

$$= \sup\{\text{l.w.dim}_A\, (A/\mathcal{I})\ :\ \mathcal{I}\ \textit{is a left ideal of}\ A\}$$

As a corollary from this proposition one obtains the following result.

Proposition 4.2.11. *For any ring A the following conditions are equivalent*:

1. *Every right ideal of A is flat.*
2. *Every submodule of a flat right A-module is flat.*

Proof.
$2 \Longrightarrow 1$. This is trivial, since any right ideal of A is a submodule of A.
$1 \Longrightarrow 2$. Let I be a right ideal of A. Consider an exact sequence

$$0 \to I \longrightarrow A \longrightarrow A/I \to 0 \qquad (4.2.12).$$

Since (4.2.12) is a flat resolution of A/I and I is flat, r.w.dim$(A/I) \leq 1$. Then by proposition 4.2.10 w.dim $A \leq 1$. Therefore if N is a submodule of a flat right module M then r.w.dim$(M/N) \leq 1$. Consider an exact sequence $0 \to N \longrightarrow M \longrightarrow M/N \to 0$. This sequence is a flat resolution for M/N, since M is flat and r.w.dim$(M/N) \leq 1$. This means that N is flat, by theorem 4.2.3. □

For Noetherian rings M.Auslander obtained the following results.

Proposition 4.2.13. (See [147, proposition 5.1.14].) *If A is a right Noetherian ring then*

$$\text{w.dim}_A\, X = \text{r.proj.dim}_A\, X$$

for any finitely generated right A-module X.

Theorem 4.2.14. (Auslander [9, theorem 4]) (See [147, theorem 5.1.15].)

1. *If A is a right Noetherian ring then*

$$\text{w.dim}\, A = \text{r.gl.dim}\, A.$$

2. *If A is a left Noetherian ring then*

$$\text{w.dim}\, A = \text{l.gl.dim}\, A.$$

Corollary 4.2.15. *If A is a Noetherian ring then*

$$\text{l.gl.dim}\, A = \text{w.dim}\, A = \text{r.gl.dim}\, A \tag{4.2.16}$$

This common value of global homological dimension of a Noetherian ring A is called the **global dimension** of the ring A and denoted by gl.dim A.

4.3 Some Examples

In this section we consider some useful results for calculating global homological dimensions of some important classes of rings.

First we consider the semiprimary rings. The most important results about global homological dimensions of these rings were obtained by M. Auslander in [9].

Recall that a ring A with Jacobson radical R is **semiprimary** if A/R is a semisimple ring and R is nilpotent. An important example of semiprimary rings is a Artinian ring.

Lemma 4.3.1 (M. Auslander [9, Lemma 6]). *Let A be a semiprimary ring with Jacobson radical R, and F a left (or right) exact functor for all right A-modules. If $F(X) = 0$ for each right A-module X such that $XR = 0$, then $T = 0$.*

Proof. Suppose that $F \neq 0$. Then there is an A-module M such that $F(M) \neq 0$. Since R is nilpotent, there is minimal integer n such that $R^n = 0$, so that $F(MR^n) = 0$. So that there is the maximal integer k such that $F(MR^k) \neq 0$. Obviously $k < n - 1$ by assumption of the statement. Consider an exact sequence

$$0 \longrightarrow MR^{k+1} \longrightarrow MR^k \longrightarrow MR^k/MR^{k+1} \longrightarrow 0$$

and apply the functor F to this sequence. Then $F(MR^{k+1}) = 0$ in due of the choice of k, and $F(MR^k/MR^{k+1}) = 0$ as $(MR^k/MR^{k+1})R = 0$. Since F is a right (or left) exact functor $F(MR^k) = 0$ as well, which contradicts to the minimality of k. □

Proposition 4.3.2 (M. Auslander [9, Proposition 7]). *Let A be a semiprimary ring with Jacobson radical R, and $B = A/R$. Then for each left A-module M the following conditions are equivalent*:

1. $\text{Tor}_n^A(B, M) = 0$
2. $\text{Tor}_n^A(U, M) = 0$ *for every simple right A-module U*
3. $\text{w.dim}_A(M) < n$.
4. $\text{Ext}_A^n(M, B) = 0$
5. $\text{Ext}_A^n(M, U) = 0$ *for every simple right A-module U*
6. $\text{proj.dim}_A(M) < n$.

Proof.

1 $\Longrightarrow$ 2. Let $U \neq 0$ be a simple A-module, and $B = A/R$. Then $UR = 0$ since R is nilpotent, so that U can be considered as a simple B-module. Since B is a semisimple

ring, U is a direct summand of B. Then $\mathrm{Tor}_n^A(U,M) = 0$ as a direct summand of $\mathrm{Tor}_n^A(B,M) = 0$ since Tor_n^A is commutes with direct sum in both variables.

$2 \Longrightarrow 3$. Let X be a right A-module such that $XR = 0$. Then X can be considered as an B-module. Since B is a semisimple ring, X is a direct sum of simple modules. Taking into account that Tor_n^A is commutes with direct sum in both variables $\mathrm{Tor}_n^A(X,M) = 0$ for each A-module M. Thus $\mathrm{Tor}_n^A(X,M) = 0$ for each right A-module X such that $XR = 0$. Now since the functor Tor_n^A is right exact on both variables and A is a semiprimary ring, we can apply lemma 4.3.1. It yields that $\mathrm{Tor}_n^A(Y,M) = 0$ for each A-module Y, i.e., $\mathrm{r.w.dim}_A(M) < n$.

$3 \Longrightarrow 1$. This follows as a particular case from theorem 4.2.3.

$4 \Longrightarrow 5$. The proof is similar to the case $1 \Longrightarrow 2$ by substituting the functor Ext_A^n for the functor Tor_n^A.

$5 \Longrightarrow 6$. Let X be a left A-module such that $RX = 0$. Then X is a module over a semisimple ring B and so it is a direct sum of simple modules B-modules X_i, i.e., $X = \bigoplus\limits_i X_i$, which is a submodule of $\prod\limits_i X_i$. So that $\mathrm{Ext}_A^n(M, \bigoplus\limits_i X_i) = 0$ as a direct summand of $\mathrm{Ext}_A^n(M, \prod\limits_i X_i) = \prod\limits_i \mathrm{Ext}_A^n(M, X_i) = 0$. Thus $\mathrm{Ext}_n^A(M,X) = 0$ for each left A-module X such that $RX = 0$. Now since the functor Ext_A^n is left exact on both variables and A is a semiprimary ring, we can apply lemma 4.3.1. It yields that $\mathrm{Ext}_A^n(M,Y) = 0$ for each left A-module Y, i.e., $\mathrm{proj.dim}_A(M) < n$.

$6 \Longrightarrow 4$. This follows as a particular case from theorem 4.1.3.

$6 \Longrightarrow 1$. This follows from [147, Theorem 5.1.8] according which $\mathrm{r.w.dim}_A(M) \le \mathrm{r.proj.dim}_A(M)$.

Corollary 4.3.3 (M. Auslander [9]). *If A is a semiprimary ring and M is a right (left) A-module then*

$$\mathrm{w.dim}_A(M) = \mathrm{proj.dim}_A(M) \tag{4.3.4}$$

and

$$\mathrm{l.gl.dim}\, A = \mathrm{w.dim}\, A = \mathrm{r.gl.dim}\, A \tag{4.3.5}$$

Similar as in the case of Noetherian rings, this common value of global homological dimension of a semiprimary ring A is called a **global dimension** of the ring A and denoted by $\mathrm{gl.dim}\, A$.

Proposition 4.3.6. *Let A be a semiprimary ring with Jacobson radical R and $B = A/R$, then*

$$\mathrm{gl.dim}\, A = \mathrm{r.proj.dim}_A(B) = \mathrm{r.proj.dim}_A R + 1 =$$

$= \sup\{\mathrm{r.proj.dim}_A\, U : U \in \mathbf{Mod}_r\, A,\}$, *where U are simple right A-modules.*

$$\mathrm{gl.dim}\, A = \mathrm{l.proj.dim}_A(B) = \mathrm{l.proj.dim}\, R + 1 =$$

$= \sup\{\mathrm{l.proj.dim}_A\, V : V \in \mathbf{Mod}_l\, A,\}$, *where V are simple left A-modules.*

Proof.

By definition, gl.dim $A \geq$ r.proj.dim$_A(B)$. On the other hand, from proposition 4.3.2 r.proj.dim $M \leq$ r.proj.dim$_A(B)$. Therefore gl.dim $A \leq$ r.proj.dim$_A(B)$. So that gl.dim A = r.proj.dim$_A(B)$.

Suppose that $R \neq 0$ and consider an exact sequence

$$0 \longrightarrow R \longrightarrow A \longrightarrow A/R = B \longrightarrow 0$$

Then by [146, Proposition 6.5.1] 1 + r.proj.dim$_A(R)$ = r.proj.dim$_A(B)$ = gl.dim A.

Since $B = A/R$ is semisimple, $B \simeq \bigoplus_{i=1}^{n} U_i$, where U_i are simple right A-modules. Moreover, for any simple A-module U there exists i such that $U \simeq U_i$. So that sup{r.proj.dim$_A$ U = gl.dim A where U ranges over all simple right A-modules. □

Example 4.3.7.

Let D be a division ring and

$$A = \begin{pmatrix} D & D & D & D \\ 0 & D & 0 & D \\ 0 & 0 & D & D \\ 0 & 0 & 0 & D \end{pmatrix} \tag{4.3.8}$$

which is obviously an Artinian ring. There are 3 principal right A-modules $P_1 = \begin{pmatrix} D & D & D & D \end{pmatrix}$, $P_2 = \begin{pmatrix} 0 & D & 0 & D \end{pmatrix}$, $P_3 = \begin{pmatrix} 0 & 0 & D & D \end{pmatrix}$, $P_4 = \begin{pmatrix} 0 & 0 & 0 & D \end{pmatrix}$, and corresponding simple A-modules $U_i = P_i/P_iR$ for $i = 1, 2, 3$. Since A is an Artinian ring, one can use proposition 4.3.6.

The Jacobson radical of A is

$$R = \begin{pmatrix} 0 & D & D & D \\ 0 & 0 & 0 & D \\ 0 & 0 & 0 & D \\ 0 & 0 & 0 & 0 \end{pmatrix}.$$

Therefore $R = P_1R \oplus P_2R \oplus P_3R$ and $P_2R = P_3R = P_4$ are projective right A-modules. So that one need only to compute r.proj.dim$A(P_1R)$. Since the right A-module $P_1R = \begin{pmatrix} 0 & D & D & D \end{pmatrix}$ is not projective, r.proj.dim$_A R \geq 1$. Therefore gl.dim $A \geq 2$, by proposition 4.3.6.

To compute the projective cover of P_1R we note that $P_1R/P_1R^2 = \begin{pmatrix} 0 & D & D & 0 \end{pmatrix} \simeq U_2 \oplus U_3$, so that $P(P_1R) = P_2 \oplus P_3$ (See [146, p.240]). Therefore we get a projective resolution of P_1R:

$$0 \longrightarrow P_4 \longrightarrow P_2 \oplus P_3 \xrightarrow{\varphi} P_1R \longrightarrow 0,$$

where Ker$(\varphi) \simeq P_4$. Thus r.proj.dim $R = 1$ and so gl.dim $A = 2$.

Example 4.3.9.

Let $A = T_n(D)$ be a ring of upper triangular matrices with entries from a division ring D. This ring is Artinian hereditary and so gl.dim $A = 1$. Consider a ring $B = A/R^2$, where R is the Jacobson radical of A. Then B is an Artinian ring with Jacobson radical $J = R/R^2$ and $J^2 = 0$. There are exactly n principal right B-modules $P_i = e_i B$ corresponding to simple right B-modules $U_i = P_i/P_iJ$ $(i = 1,2,\dots,n)$. Then $P_iJ = U_{i+1}$ for $i = 1,2,\dots,n-1$ and $P_nJ = 0$. Since $J^2 = 0$, the projective cover $P(P_iJ) = P_{i+1}$ for $i = 1,2,\dots,n$ and $P(P_nR) = 0$. Therefore one obtains the projective resolution of the simple module U_i:

$$0 \longrightarrow P_n \longrightarrow P_{n-1} \longrightarrow \cdots \longrightarrow P_{i+1} \longrightarrow P_i \longrightarrow U_i \longrightarrow 0$$

for $i = 1,2,\dots,n$. So that $\text{proj.dim}_B(U_i) = n - i$ for $0 < i \leq n$. Therefore by proposition 4.3.6 gl.dim $B = \sup\{\text{proj.dim}_B(U_i)\} = n - 1$.

For example, if $A = T_4(D)$ and

$$B = \begin{pmatrix} D & D & 0 & 0 \\ 0 & D & D & 0 \\ 0 & 0 & D & D \\ 0 & 0 & 0 & D \end{pmatrix} \tag{4.3.10}$$

then gl.dim $B = 3$.

Example 4.3.11. (A.W.Chatters and Hajarnavis [47, Example 4.2]).

Let K be a 4 dimensional extension of a field $\mathbf{Q}$. Consider a ring

$$A = \begin{pmatrix} K & K & K & K & K \\ 0 & \mathbf{Q} & K & K & K \\ 0 & 0 & \mathbf{Q} & K & K \\ 0 & 0 & 0 & K & K \\ 0 & 0 & 0 & 0 & K \end{pmatrix} \tag{4.3.12}$$

which is an Artinian ring with 5 principal right A-modules $P_i = e_{ii}A$ for $i = 1,\dots,5$. By proposition 4.3.6 we need only to compute r.proj.dim$_A$ R, where R is the Jacobson radical of A:

$$R = \begin{pmatrix} 0 & K & K & K & K \\ 0 & 0 & K & K & K \\ 0 & 0 & 0 & K & K \\ 0 & 0 & 0 & 0 & K \\ 0 & 0 & 0 & 0 & 0 \end{pmatrix}. \tag{4.3.13}$$

Let $\{x_1,x_2,x_3,x_4\}$ be a basis of K over $\mathbf{Q}$. Define right A-modules $X_i = \begin{pmatrix} 0 & 0 & x_i\mathbf{Q} & K & K \end{pmatrix} \simeq P_3$ for $i = 1,2,3,4$, $X = \begin{pmatrix} 0 & 0 & K & K & K \end{pmatrix}$, and $Y = \begin{pmatrix} 0 & 0 & 0 & K & K \end{pmatrix} \simeq P_4$. Then X_i, Y are projective right A-modules. We claim that $X_1 + X_2$ is not projective. Otherwise the exact sequence

$$0 \longrightarrow Y \longrightarrow X_1 \oplus X_2 \longrightarrow X_1 + X_2 \longrightarrow 0 \tag{4.3.14}$$

splits. Therefore there is an epimorphism $f : X_1 \oplus X_2 \longrightarrow Y$.

$X_1 \oplus X_2$ is generated by $a = (x_1e_{13}, 0)$ and $b = (0, x_2e_{13})$. Now $Ye_{33} = 0$ so that $f(a) = f(b) = 0$ which is a contradiction.

Since Y is essential in X_2, X_2/Y cannot be projective. It follows that $\text{r.proj.dim}_A(X_2/Y) = 1$. So that the exact sequence

$$0 \longrightarrow X_1 \longrightarrow X_1 + X_2 \longrightarrow X_2/Y \longrightarrow 0$$

yields $\text{r.proj.dim}(X_1 + X_2) = 1$. Proceeding with similar arguments, we obtain $\text{r.proj.dim}_A X = 1$.

Now define $Z_i = \begin{pmatrix} 0 & x_iQ & K & K & K \end{pmatrix}$, $Z = \begin{pmatrix} 0 & K & K & K & K \end{pmatrix}$. Since $Z_i \simeq P_2$ is projective, Z_i/X is not projective, and $\text{r.proj.dim}_A X = 1$ we have $\text{r.proj.dim}\, Z_i/X = 1$. Hence the exact sequence

$$0 \longrightarrow Z_1 \longrightarrow Z_1 + Z_2 \longrightarrow Z_2/X \longrightarrow 0$$

gives $\text{r.proj.dim}\,(Z_1 + Z_2) = 2$. Continuing we obtain $\text{r.proj.dim}_A Z = 2$. Since R is a direct sum of Z and a right A-module of smaller projective dimension, we get that $\text{r.proj.dim}_A R = 2$. So that applying proposition 4.3.6 one obtains $\text{r.gl.dim}\, A = 3$.

The next class of rings which global dimensions were studying by many authors are tiled orders (See e.g., [192], [278], [302], and also [147, Chapter 6]).

Example 4.3.15 (G. Michler [241]).

Let O be a discrete valuation ring with unique maximal ideal $M = \pi O$ where $\pi \in M \setminus M^2$ is a prime element of O.

Consider a ring

$$A = \begin{pmatrix} O & \pi O & \pi O \\ \pi O & O & \pi O \\ O & O & O \end{pmatrix}, \tag{4.3.16}$$

which a prime Noetherian semimaximal ring by [146, Theorem 14.5.2].

There are 3 principal right A-modules $P_1 = \begin{pmatrix} O & \pi O & \pi O \end{pmatrix}$, $P_2 = \begin{pmatrix} \pi O & O & \pi O \end{pmatrix}$, $P_3 = \begin{pmatrix} O & O & O \end{pmatrix}$.

The Jacobson radical of A has the form:

$$R = \begin{pmatrix} \pi O & \pi O & \pi O \\ \pi O & \pi O & \pi O \\ O & O & \pi O \end{pmatrix}, \tag{4.3.17}$$

and so we have 3 simple right A-modules $U_1 = P_1/P_1R = \begin{pmatrix} O/M & 0 & 0 \end{pmatrix}$, $U_2 = P_2/P_2R = \begin{pmatrix} 0 & O/M & 0 \end{pmatrix}$, $U_3 = P_3/P_3R = \begin{pmatrix} 0 & 0 & O/M \end{pmatrix}$. Therefore $P_1R = P_2R \simeq P_3$ are projective A-modules, and so to compute the right projective dimension of R we need only to compute the projective dimension of P_3.

Since $P_3R/P_3R^2 = \begin{pmatrix} O/\pi O & O/\pi O & 0 \end{pmatrix} \simeq U_1 \oplus U_2$ and $P_3R^2 = \begin{pmatrix} \pi O & \pi O & \pi O \end{pmatrix} \simeq P_3$, the projective cover of $P(P_3R)$ is $P_1 \oplus P_2$ and we have a projective resolution for P_3R:

$$0 \longrightarrow P_3 \longrightarrow P_1 \oplus P_2 \longrightarrow P_3R \longrightarrow 0.$$

This implies that proj.dim $P_3R = 1$ and so proj.dim $R = 1$. Therefore

$$\text{gl.dim}\, A \geq \text{proj.dim}\, R + 1 = 2 \tag{4.3.18}$$

by proposition 4.2.10 and corollary 4.2.15.

On the other hand A is a tiled order in $M_3(D)$, where D is a classical ring of fractions of O (See [147, Section 6.6]). The quiver $Q(A)$ of A has the following form:

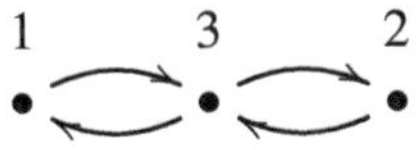

So that $Q(A)$ has no loops and $w(A) = 1 < 2$. Therefore gl.dim $A < \infty$ by [147, Theorem 6.10.12]. Thus applying [147, Theorem 6.10.8] we obtain that gl.dim $A \leq 2$. So that taking into account inequality (4.3.18) we obtain that gl.dim $A = 2$.

The class of rings which global dimension is also well-studied are quasi-Frobenius rings. Recall that a ring is called quasi-Frobenius if it is right (left) Noetherian and left (right) self-injective (See e.g. [147, Chapter 4]).

It is well known the theorem about quasi-Frobenius rings (see [147, Corollary 4.12.20]) which states that the global dimension of a quasi-Frobenius ring, which is not semisimple, is equal to ∞.

The easiest example of quasi-Frobenius rings is $\mathbf{Z}/n\mathbf{Z}$ for any $n > 0$. Since this ring is not semisimple gl.dim $\mathbf{Z}/n\mathbf{Z} = \infty$.

Example 4.3.19.

Let O be a discrete valuation ring with unique maximal ideal M, and $H_n(O)$ is a ring of the form (1.10.4). Let J be the Jacobson radical of $H_n(O)$. Then the ring $H_n(O)/J^m$ for $m > 2$ is a quasi-Frobenius ring which is not semisimple (See [147, p.176]). Therefore gl.dim $H_n(O)/J^m = \infty$.

At the end of this section we give for completeness the following important results for polynomial rings. (The proofs of these theorems can be find, e.g., in [262]).

Theorem 4.3.20. *If D a nontrivial ring with* gl.dim $D = n$ *then*

$$\text{gl.dim}\, D[x] = n + 1. \tag{4.3.21}$$

Theorem 4.3.22. *If D a nontrivial ring with* gl.dim $D = n$ *then*

$$\text{gl.dim}\, D[x_1, x_2, \ldots, x_k] = n + k. \tag{4.3.23}$$

As an immediate corollary of this theorem one obtains the famous Hilbert's Syzygies Theorem:

Theorem 4.3.24. *Let F be a field and $x_1, x_2, \ldots, x_k$ indeterminates. Then*

$$\text{gl.dim}\, F[x_1, x_2, \ldots, x_k] = k. \tag{4.3.25}$$

4.4 Homological Characterization of Some Classes of Rings

A large number of rings can be characterized uniquely by their homological properties. This section collects the main properties of rings considered in [146] and [147].

We start with a nice class of rings which are semisimple rings. Recall that a ring is **semisimple** if it is a direct sum of finite number of simple right (left) ideals. The famous Wedderburn-Artin theorem 1.1.6 says that every semisimple ring A is isomorphic to a finite direct sum of full matrix rings over division rings. There are a lot of other equivalent conditions which characterize semisimple rings, some of them are given in the following theorem.

Theorem 4.4.1. *The following conditions are equivalent for a ring A:*

1. *A is semisimple.*
2. *A is isomorphic to a finite direct sum of full matrix rings over division rings.*
3. *Any A-module is semisimple.*
4. *Any A-module is projective.*
5. *Any A-module is injective.*
6. *All simple A-modules are projective.*
7. *All cyclic A-module are injective.*
8. r.gl.dim A = l.gl.dim $A = 0$.
9. *A is right Artinian and semiprimitive*[2].
10. *A is left Artinian and semiprimitive.*

Proof. This theorem is an immediate consequence of theorems 1.1.6, 1.1.7, 1.1.24, 1.5.9 and [146, Proposition 6.6.6(1)]. □

The next important class of rings is formed Noetherian rings. Noetherian rings can be characterized by the following properties:

Theorem 4.4.2. *The following conditions are equivalent for a ring A:*

1. *A is right Noetherian.*
2. *Every right ideal in A is finitely generated.*
3. *Every non-empty set of right ideals in A has a maximal element.*
4. *Every direct sum of injective right A-modules is injective.*

Proof. These equivalences immediately follow from propositions [146, proposition 3.1.5] and [146, theorem 5.2.12]. □

---

[2]A ring is called **semiprimitive** if its Jacobson radical is equal to zero.

An A-module M is said to be **finitely presented** (f.p.) if the following equivalent conditions hold:

1. There is an exact sequence
$$A^m \longrightarrow A^n \longrightarrow M \to 0,$$
for some $m,n \in N$.
2. M is finitely generated and there is an epimorphism $\varphi : P \longrightarrow M$ with P projective, such that $\mathrm{Ker}(\varphi)$ is a finitely generated module.

Proposition 4.4.3. *Any finitely generated projective module is finitely presented.*

Proof. Let P be a finitely generated projective A-module, then there is an exact sequence:
$$0 \to X \longrightarrow A^n \longrightarrow P \to 0.$$
Since P is projective, X is a direct summand of A^n, so it is finitely generated, which shows that P is finitely presented. □

Lemma 4.4.4. *Let M be a finitely presented A-module, and let $\alpha : X \longrightarrow M$ be an epimorphism. If X is a finitely generated A-module, then so is* $\mathrm{Ker}(\alpha)$.

Proof. Since X is a finitely generated A-module, there is an epimorphism $\beta : A^k \to X$. Let $\alpha\beta$ be a composition of epimorphisms β and $\alpha : X \longrightarrow M$, which induces a homomorphism $\gamma : \mathrm{Ker}(\alpha\beta) \to \mathrm{Ker}(\alpha)$. Then one has the following commutative diagram:

$$\begin{array}{ccccccccc}
 & & & & 0 & & 0 & & \\
 & & & & \downarrow & & \downarrow & & \\
 & & & & \mathrm{Ker}(\alpha\beta) & \xrightarrow{\gamma} & \mathrm{Ker}(\alpha) & & \\
 & & & & \downarrow{\scriptstyle \iota_2} & & \downarrow{\scriptstyle \iota_1} & & \\
0 & \longrightarrow & \mathrm{Ker}(\beta) & \longrightarrow & A^k & \xrightarrow{\beta} & X & \longrightarrow & 0 \\
 & & & & \downarrow{\scriptstyle \alpha\beta} & & \downarrow{\scriptstyle \alpha} & & \\
 & & & & M & = & M & & \\
 & & & & \downarrow & & \downarrow & & \\
 & & & & 0 & & 0 & &
\end{array}$$

with exit row and columns. Since ι_1 and ι_2 are monomorphisms and β is an epimorphism, the commutativity of this diagram implies that γ is also an epimorphism. Since $\mathrm{Ker}(\alpha\beta)$ is a finitely generated A-module, $\mathrm{Ker}(\alpha)$ is f.g. as well. □

Noetherian rings can be characterized in terms of finitely presented modules by the follows.

Proposition 4.4.5. *A ring A is right Noetherian if and only if every finitely generated right A-module is finitely presented.*

Proof.

1. Let A be a right Noetherian ring, and M a finitely generated right A-module. Take a surjection $\varphi : A^n \longrightarrow M$. Since A^n is a Noetherian module, $\mathrm{Ker}(\varphi)$ is finitely generated, so M is finitely presented.
2. Conversely, suppose that all finitely generated A-modules are finitely presented. Let I be a right ideal in A. Consider the exact sequence

$$0 \to I \longrightarrow A \longrightarrow A/I \to 0.$$

Since A/I is a cyclic A-module, it is finitely presented, by assumption. Then from lemma 4.4.4 it follows that I is finitely generated, therefore A is a right Noetherian ring. □

Semisimple rings are also very simple as regards homological properties. These are rings whose global homological dimension is equal to zero. In this series of rings with finite global homological dimension the next rings after semisimple ones are hereditary rings. Recall that a ring A is said to be **right hereditary** if each of its right ideals is projective. For the convenience of the reader we also list several main properties of right hereditary rings which where proved in [146] and [147].

Proposition 4.4.6. *If a ring A is right hereditary then*

1. *Any submodule of a free A-module is isomorphic to a direct sum of right ideals of A* ([146, theorem 5.5.1]);
2. *Every submodule of a projective right A-module is projective* ([146, corollary 5.5.2]);
3. *A right A-module P is projective if and only if it is embedded into a free right A-module* ([146, corollary 5.5.5]);
4. *For any non-zero idempotent $e^2 = e \in A$ the ring eAe is also right hereditary* ([146, proposition 5.5.7]);
5. *Any non-zero homomorphism $\varphi : P_1 \to P_2$ of indecomposable non-zero projective right A-modules is a monomorphism* ([146, lemma 5.5.8]).

In particular, for a principal ideal domain[3] one has the following proposition:

---

[3]An integral domain A is called a **principal ideal domain** if each of its ideals is principal.

Proposition 4.4.7. *If A is a principal ideal domain then*

1. *Every submodule of a free A-module is free and every projective A-module is free* ([146, corollary 5.5.3]);
2. *Every projective A-module is free* ([146, corollary 5.5.4]).

The following theorem yields a number of equivalent definitions for a right hereditary ring:

Theorem 4.4.8. *The following conditions are equivalent for a ring A:*

1. *A Is a right hereditary ring, i.e., every right ideal of A is projective.*
2. *Any submodule of a projective right A-module is projective.*
3. *Any quotient module of an injective right A-module is injective.*
4. r.gl.dim $A \leqslant 1$.
5. $\mathrm{Ext}^2_A(X,Y) = 0$ *for all right A-modules* X,Y.
6. $\mathrm{Ext}^k_A(X,Y) = 0$ *for all* $k > 1$ *and all right A-modules* X,Y.

Proof. This theorem is the immediate corollary of [146, theorem 5.5.6] and [146, proposition 6.6.6]. □

A ring A is said to be **right semihereditary** if each its finitely generated right ideal is projective. Right semihereditary rings can be also characterized by the following equivalent conditions:

Proposition 4.4.9. *The following conditions are equivalent for a ring A:*

1. *A is right semihereditary.*
2. *Every finitely generated submodule of a right projective A-module is projective.*

Proof. This is exactly what is proved in [146, corollary 5.5.10]. □

Undoubtedly other interesting classes of rings are semiperfect and perfect rings. Recall that a ring A with Jacobson radical R is called **semilocal** if A/R is a right Artinian ring. A ring A is called **semiperfect** if A is semilocal and idempotents can be lifted modulo R. The following statement gives a characterization of semiperfect rings.

Theorem 4.4.10 (H.Bass [15]). *The following conditions are equivalent for a ring A:*

1. *A is semiperfect.*
2. *A can be decomposed into a direct sum of right ideals each of which has exactly one maximal submodule.*
3. *The identity* $1 \in A$ *can be decomposed into a sum of a finite number of pairwise orthogonal local idempotents.*

4. *Any finitely generated right A-module has a projective cover.*
5. *Any cyclic right A-module has a projective cover.*

Proof. This theorem is an immediate consequence of theorem 1.9.3. □

Recall that a ring A with Jacobson radical R is **right (left) perfect** if A/R is semisimple and R is right (left) T-nilpotent[4]. A ring A is called **right (left) socular** if every non-zero right (left) A-module has a non-zero socle[5].

The following theorem yields the characterization of right perfect rings.

Theorem 4.4.11 (H. Bass [15]). *The following conditions are equivalent for a ring A:*

1. *A is right perfect.*
2. *Every right A-module has a projective cover.*
3. *Every flat right A-module is projective.*
4. *A satisfies the descending chain condition on principal left ideals.*
5. *A is semilocal and left socular.*

Proof. This theorem is an immediate consequence of theorems 1.9.8 and [147, theorem 4.8.2]. □

In [147, Section 5.1] a famous result of M. Auslander was proved. It says that for a two-sided Noetherian ring being right hereditary implies being left hereditary. The analogous result is also true for perfect rings as formulated the next proposition.

Proposition 4.4.12 (H.Bass [15]). *Let A be a right perfect ring. Then for any right A-module M,* $\text{w.dim}_A(M) = \text{proj.dim}_A(M)$. *In particular,* $\text{r.gl.dim}A = \text{w.dim}A \leq \text{l.gl.dim}A$. *If A is a right and left perfect ring, a strong equality holds, i.e.,* $\text{r.gl.dim}A = \text{l.gl.dim}A$.

Proof. Let A be a right perfect ring. Then by theorem 4.4.11 every flat right A-module is projective. Since $\text{w.dim}_A M \leq \text{r.proj.dim}_A M$ always holds, it remains to show that $\text{w.dim}_A M \geq \text{r.proj.dim}_A M$, i.e., if $\text{Tor}^A_{n+1}(M,Y) = 0$ for all left A-modules Y, then $\text{Ext}^{n+1}_A(M,B) = 0$ for all right A-modules B and all $n \geq 0$. Since A is right perfect, by theorem 1.9.8, for any right A-module M there exists a projective resolution, constructed by linking projective covers,

$$\ldots \longrightarrow P_n \xrightarrow{d_n} P_{n-1} \xrightarrow{d_{n-1}} \ldots \longrightarrow P_1 \xrightarrow{d_1} P_0 \longrightarrow M \longrightarrow 0$$

---

[4]See Section 1.9.

[5]Recall that the **socle** of module M is the sum of all simple submodules of M. See Section 1.5.

where each d_n is a minimal epimorphism[6] onto the kernel of d_{n-1}. Since, by [146, lemma 10.4.1], $d_n(P_n) \subset P_{n-1}R$, where $R = \text{rad}A$ is the Jacobson radical of A, the sequence

$$\ldots \longrightarrow P_n \otimes_A \bar{A} \longrightarrow P_{n-1} \otimes_A \bar{A} \longrightarrow \ldots \longrightarrow P_0 \otimes_A \bar{A} \longrightarrow M \otimes_A \bar{A} \longrightarrow 0$$

is also exact, where $\bar{A} = A/R$. Since $X \otimes_A \bar{A} \simeq X/XR$, there results that $\text{Tor}_n^A(M, \bar{A}) \simeq P_n/P_nR$. Hence, if $\text{w.dim}_A(M) < n$, $\text{Tor}_n^A(M, \bar{A}) = 0$ so $P_n = 0$ and $\text{proj.dim}_A M < n$. □

Corollary 4.4.13. *A right hereditary perfect ring is left hereditary.*

Recall that a ring A is called **right self-injective** if the right regular module A_A is injective. A **left self-injective ring** is defined analogously. (See [147, Section 4.12].)

A ring A is called a **quasi-Frobenius** ring (or a **QF-ring**) if it is a self-injective Artinian ring. For any subset S of a ring A denote

$$\text{l.ann}_A(S) = \{x \in A \;:\; xs = 0 \;\text{ for all }\; s \in S\}$$

which is a left annihilator of S. Analogously,

$$\text{r.ann}_A(S) = \{x \in A \;:\; sx = 0 \;\text{ for all }\; s \in S\}$$

is a right annihilator of S.

Let A be a semiperfect ring with Jacobson radical R and $A_A = P_1^{n_1} \oplus \cdots \oplus P_s^{n_s}$ (${}_AA = Q_1^{n_1} \oplus \cdots \oplus Q_s^{n_s}$) the canonical decomposition of A into a direct sum of right (left) principal modules. Let ν be a permutation on the set $\{1,2,\ldots,s\}$. One says that a semiperfect ring A admits a **Nakayama permutation** ν if the following conditions hold:

1. $\text{soc}P_k = P_{\nu(k)}/P_{\nu(k)}R$,
2. $\text{soc}P_{\nu(k)} = Q_k/RQ_k$,
   for each $k \in \{1,2,\ldots,s\}$.

The following theorem gives equivalent characterizations of QF-rings.

Theorem 4.4.14. *The following conditions are equivalent for a ring A:*

1. *A is a quasi-Frobenius ring.*
2. *A is a right Noetherian and right self-injective ring.*
3. *A is a left Noetherian and left self-injective ring.*
4. *A is an Artinian ring and satisfies the following double annihilator conditions:*

$$(a)\quad \text{r.ann}_A(\text{l.ann}_A(H)) = H \;\textit{for any right ideal}\; H$$

$$(b)\quad \text{l.ann}_A(\text{r.ann}_A(L)) = L \;\textit{for any left ideal}\; L.$$

---

[6]A homomorphism $A \to B$ is called **minimal** if $\text{Ker}(f)$ is small in A.

5. *A is an Artinian ring and admits the Nakayama permutation.*
6. *Any projective A-module is injective.*
7. *Any injective A-module is projective.*

Proof. This theorem is the immediate consequence of [147, theorems 4.2.14, 4.12.17, 4.12.19]. □

The following important result about homological dimension of QF-rings was obtained by M.Auslander:

Theorem 4.4.15 (M.Auslander [9, Propsition 15]). (See [147, corollary 4.12.20].) *If A is a QF-ring, then* $\mathrm{gl.\,dim}\,A = 0$ *or* $\mathrm{gl.\,dim}\,A = \infty$.

4.5 Torsion-less Modules

In [147, Section 4.10] a **duality functor** $*$ for Noetherian rings was considered. This functor can also be defined and used for arbitrary rings.

Let A be a ring and X a right A-module. Then

$$X^* = \mathrm{Hom}_A(X, A_A)$$

is called the **dual module** to X. Analogously if Y is a left A-module, then

$$Y^* = \mathrm{Hom}_A(Y, {}_AA)$$

is called the **dual module** to Y. If $N \to M$ is a homomorphism of right A-modules then a map $f^* : M^* \to N^*$ is an A-homomorphism of left A-modules and the homomorphism f^* is called **dual** to f.

If M is a right A-module, then its dual module M^* is a left A-module. The left A-module structure on M^* is given by $af(m) = f(ma)$ for $f \in M^*$, $a \in A$, $m \in M$. The module M^* itself has a dual module M^{**}, which is a right A-module.

Suppose $m \in M$ and $f \in M^* = \mathrm{Hom}_A(M, A)$. Define a mapping

$$\varphi_m : M^* \longrightarrow A$$

by $\varphi_m(f) = f(m)$. Obviously,

$$\varphi_m(f_1 + f_2) = \varphi_m(f_1) + \varphi_m(f_2).$$

For any $a \in A$ there results $\varphi_m(af) = af(m) = a\varphi_m(f)$. Thus φ_m is an A-homomorphism, i.e., $\varphi_m \in M^{**}$.

Consider the mapping

$$\delta_M : M \longrightarrow M^{**} \tag{4.5.1}$$

defined by

$$\delta_M(m)(f) = f(m) \tag{4.5.2}$$

for $m \in M$ and $f \in M^*$. It is easy to verify that δ_M is an A-homomorphism. If $g : M \longrightarrow N$ is a homomorphism of A-modules, then it induces a dual homomorphism $g^* : N^* \longrightarrow M^*$, which in turn induces a homomorphism $g^{**} : M^{**} \longrightarrow N^{**}$. It is easy to check that the diagram

$$\begin{array}{ccc} M & \xrightarrow{g} & N \\ {\scriptstyle \delta_M}\downarrow & & \downarrow{\scriptstyle \delta_N} \\ M^{**} & \xrightarrow[g^{**}]{} & N^{**} \end{array}$$

Diag. 4.5.3.

is commutative. Therefore δ_M is a natural transformation.

A module M is called **reflexive** if δ_M is an isomorphism. It is called **torsion-less** (or **semi-reflexive**) if δ_M is a monomorphism.

The following lemma gives a useful equivalent characterization of a torsion-less module, which is formulated in terms (4.5.2).

Lemma 4.5.4. *Let M be a right A-module. Then A is torsion-less if and only if for each non-zero element $m \in M$ there exists an $f \in M^*$ such that $f(m) \neq 0$.*

Another equivalent characterization is given by the following proposition.

Proposition 4.5.5. *An A-module M is torsion-less if and only if it can be embedded into a direct product of a number of copies of A.*

Proof.

1. Suppose an A-module M is embedded into a direct product A^I of I copies of A, i.e., there exists a monomorphism $\varphi : M \longrightarrow A^I$. Write $a_i = \pi_i \varphi(m)$ for any $i \in I$, where $\pi_i : A^I \longrightarrow A$ is the natural projection given by $\pi_i(\ldots, a_j, \ldots, a_i, \ldots) = a_i$. Since φ is a monomorphism, for any m there exists an index i such that $a_i \neq 0$. Then $f = \pi_i \varphi : M \longrightarrow A$ is a homomorphism such that $f(m) \neq 0$, and so, by lemma 4.5.4, M is a torsion-less module.
2. Conversely, assume that M is a torsion-less right A-module. Take an exact sequence $F \longrightarrow M^* \longrightarrow 0$, where F is a free left A-module which is isomorphic to a direct sum of copies of the left regular A-module ${}_AA$. An application of the duality functor gives an exact sequence $0 \longrightarrow M^{**} \longrightarrow F^*$, where F^* is a free right A-module. Since $\delta_M : M \longrightarrow M^{**}$ is a monomorphism, one obtains the required embedding. □

Corollary 4.5.6.

1. *Any submodule of a torsion-less module is torsion-less.*

2. *Any submodule of a free module is torsion-less. In particular, any submodule of a projective module as well as any ideal is torsion-less.*
3. *Any direct product of torsion-less modules is torsion-less.*

Let A be an integral domain, and M an A-module. The set

$$t(M) = \{m \in M \ : \ \exists x \in A, x \neq 0, mx = 0\}$$

is called the **set of torsion elements** of M. The module M is called **torsion-free** if $t(M) = 0$, and M is called **torsion** if $t(M) = M$. Torsion and torsion-free modules play an important role in studying modules over integral domains. The following proposition gives the connection between the concepts of torsion-free modules and torsion-less modules over an integral domain.

Proposition 4.5.7. *Let A be an integral domain. Then any torsion-less A-module M is torsion-free. A finitely generated torsion-free A-module M is torsion-less.*

Proof.

1. Assume that an A-module M is torsion-less. Let $m \in M$ and $a \in A$ with $m \neq 0$ and $a \neq 0$. In view of lemma 4.5.4 there is a homomorphism $f \in M^*$ such that $f(m) \neq 0$. Then $f(m)a \neq 0$, since A is an integral domain. Therefore $f(m)a = f(ma) \neq 0$, and so $ma \neq 0$, i.e., M is torsion-free.
2. Assume that the finitely generated A-module M is torsion-free. Let $\{m_1, m_2, \ldots, m_n\}$ be a set of generators of M. In this set one can choose a maximal subset $\{m_1, m_2, \ldots, m_k\}$ (renaming the m_i if necessary) which is linearly independent with respect to A. Then $N = m_1A + m_2A + \ldots + m_kA$ is a free A-module with the basis $\{m_1, m_2, \ldots, m_k\}$. We show that there exists an element $x \in A$ such that $Mx \subseteq N$.

Consider the set of elements $\{m_1, \ldots, m_k, m_{k+1}\}$. Because of maximality there are $a_1, \ldots, a_k, a_{k+1} \in A$ such that

$$m_1a_1 + \cdots + m_ka_k + m_{k+1}a_{k+1} = 0$$

with $a_{k+1} \neq 0$. Then $m_ia_{k+1} \in M$ for all $i = 1, \ldots, k+1$. Next look at the set of elements $\{m_1, \ldots, m_k, m_{k+2}a_{k+1}\}$. Then there is a relation

$$m_1b_1 + \cdots + m_kb_k + m_{k+2}a_{k+1}b_{k+2} = 0$$

with $a_{k+1}b_{k+2} \neq 0$. Then $m_ia_{k+1}b_{k+2} \in N$ for all $i = 1, \ldots, k+2$. Further look at the set of elements $\{m_1, \ldots, m_k, m_{k+3}a_{k+1}b_{k+2}\}$ to find a $c_{k+3} \in A$ such that $a_{k+1}b_{k+2}c_{k+3} \neq 0$ and $m_ia_{k+1}b_{k+2}c_{k+3} \in N$ for all $i = 1, \ldots, k+3$. This process terminates because n is finite, giving the desired element $x \in A$.

Let $\varphi_i : N \longrightarrow A$ be the A-homomorphism defined by

$$\varphi_i(m_1a_1 + m_2a_2 + \ldots + m_ka_k) = a_i$$

for $i = 1, 2, \dots, k$.

Let $m \in M$ and $m \neq 0$. Then $mx = m_1a_1 + m_2a_2 + \ldots + m_ka_k$, where all $a_i \in A$. Since M is torsion-free, $t(M) = 0$, i.e., $mx \neq 0$ for all $m \in M$ and $x \in A$. Therefore there exists an index i such that $a_i \neq 0$. Define a homomorphism $f \in M^*$ by $f(m) = \varphi_i(mx)$. Then $f(m) = a_i \neq 0$, and so M is torsion-less by lemma 4.5.1. □

Remark 4.5.8. In general, a torsion-free module need not be torsion-less. For example, if $A = \mathbf{Z}$ is a ring of integers, then the A-module $M = \mathbf{Q}$, the rings of rational numbers, is torsion-free, but it is not torsion-less because the only morphism of Abelian groups $\mathbf{Q} \longrightarrow \mathbf{Z}$ is zero.

It is interesting to know when a module M is torsion-less or reflexive. In [147, Section 4.12] it was shown that any finitely generated torsion-less module is reflexive and that for a quasi-Frobenius ring A any finitely generated A-module is reflexive. The following proposition states that every finitely generated projective module, and, in particular, any free module with a finite free basis is reflexive for an arbitrary ring A.

Proposition 4.5.9. *Let P be a finitely generated projective right A-module. Then the dual module P^* is a finitely generated projective left A-module, and $P^{**} \simeq P$. Moreover, for each right A-module M:*

$$\mathrm{Hom}_A(P, M) \simeq M \otimes_A P^*. \tag{4.5.10}$$

Proof. The first statement is [147, lemma 4.10.1] and the second statement is [147, proposition 4.10.4]. So it remains to prove the last statement, i.e., the (functorial) equality (4.5.10). Since P is a finitely generated projective right A-module, by theorem 1.5.5, there is a system of elements $\{p_1, p_2, \dots, p_n\}$ in P and a system of homomorphisms $\{\varphi_1, \varphi_2, \dots, \varphi_n\}$ in P^* such that any element $p \in P$ can be written in the form

$$p = \sum_{i=1}^{n} \varphi_i(p) \cdot p_i.$$

Given this one can construct a map $\alpha : \mathrm{Hom}_A(P, M) \longrightarrow M \otimes_A P^*$ by setting $\alpha(h) = \sum\limits_{i=1}^{n} h(p_i) \otimes \varphi_i$, and a map $\beta : M \otimes_A P^* \longrightarrow \mathrm{Hom}_A(P, M)$ given by $\beta(m \otimes \varphi)(p) = m \cdot \varphi(p)$. It easy to verify that these maps are well-defined and inverse to one another. □

Lemma 4.5.11. *If A is a semiperfect ring with Jacobson radical R and P is a principal right A-module, then the dual P^* is a principal left A-module. Further, if $P/PR \simeq U$ then $P^*/RP^* \simeq V$, where V is dual to the module U over the ring A/R.*

Proof. Let P be a principal right A-module, then, by proposition 4.5.9, P^* is a finitely generated projective left A-module. Suppose that $P^* = X \oplus Y$, then $P^{**} = X^* \oplus Y^*$ and

$P^{**} \simeq P$. Therefore $X^* = 0$ or $Y^* = 0$, so $X = 0$ or $Y = 0$, i.e., P^* is indecomposable. So P^* is a principal left A-module.

The last statement follows from [147, lemma 4.11.4]. □

An epimorphism $g : M \longrightarrow N$ is called a **split epimorphism** if there exists a homomorphism $g_1 : N \longrightarrow M$ such that $gg_1 = 1_N$. A monomorphism $f : M \longrightarrow N$ is called a **split monomorphism** if there is a homomorphism $f_1 : N \longrightarrow M$ such that $f_1 f = 1_M$.

Let $f : N \longrightarrow M$ be a homomorphism of right A-modules. Then there is the natural map $f^* : M^* \longrightarrow N^*$ defined by the formula $f^*(\varphi) = \varphi f$ for $\varphi \in M^*$. It is easy to show that f^* is an A-homomorphism of left A-modules. The homomorphism f^* is called **dual** to f. Note the following simple facts.

Lemma 4.5.12. *Let A be a ring, and M a right A-module. Then a homomorphism $\delta_{M^*} : M^* \longrightarrow M^{***}$ is a split monomorphism.*

Proof. Consider two homomorphisms: $\delta_{M^*} : M^* \longrightarrow M^{***}$ and $(\delta_M)^* : M^{***} \longrightarrow M^*$. We show that $(\delta_M)^*\delta_{M^*} = 1_{M^*}$. Indeed, let $f \in M^*$, then by definition for any $m \in M$:

$$(\delta_M)^*(\delta_{M^*}(f))(m) = (\delta_M)^*(f)(\delta_M(m)) = \delta_M(m)(f) = f(m),$$

so that $(\delta_M)^*(\delta_{M^*}(f)) = f$, and so $(\delta_M)^*\delta_{M^*} = 1_{M^*}$. □

Corollary 4.5.13. *Let M be a right A-module. Then the left A-module M^* is always torsion-less.*

Lemma 4.5.14. *Let A be a ring, and M, N right A-modules. Then a homomorphism $g : M \longrightarrow N$ is a split epimorphism (monomorphism) if and only if $g^* : N^* \longrightarrow M^*$ is a split monomorphism (epimorphism).*

Proof. If a homomorphism $g : M \longrightarrow N$ is a split epimorphism, then there exists a homomorphism $g_1 : N \longrightarrow M$ such that $gg_1 = 1_N$. Then $(gg_1)^* = g_1^* g^* = 1_{N^{**}}$, and so, by definition, g^* is a split monomorphism.

The dual statement is proved analogously. □

Proposition 4.5.15. *Let A be a Noetherian ring, and M a finitely generated A-module such that $M^* = 0$. Then there is a finitely generated A-module N such that $M \simeq \mathrm{Ext}_A^1(N, A)$.*

Proof. Suppose $M^* = 0$ and consider a projective resolution of M:

$$P_1 \xrightarrow{\varphi} P_0 \longrightarrow M \longrightarrow 0$$

with finitely generated projective modules P_0 and P_1. Setting $N = \mathrm{Coker}\varphi^*$, one has an exact sequence

$$0 \longrightarrow M^* = 0 \longrightarrow P_0^* \longrightarrow P_1^* \longrightarrow N \longrightarrow 0.$$

Since P_i is a finitely generated projective module, P_i^* is also projective and $P_i^{**} \simeq P_i$ for $i = 0, 1$, by proposition 4.5.9. This yields the exact sequence

$$0 \longrightarrow N^* \longrightarrow P_1 \xrightarrow{\varphi} P_0 \longrightarrow \mathrm{Ext}_A^1(N,A) \longrightarrow \mathrm{Ext}_A^1(P_1^*,A) = 0,$$

which shows that $M = \mathrm{Coker}\varphi \simeq \mathrm{Ext}_A^1(N,A)$. □

Let N be a submodule of M. Like in [147, Section 4.1], introduce the following notations:

$$N^{\perp} = \{f \in M^* \ : \ f(N) = 0\} \tag{4.5.16}$$

$$N^{\perp\perp} = \{m \in M \ : \ \varphi(m) = 0 \text{ for } \forall\varphi \in N^{\perp}\} \tag{4.5.17}$$

It is easy to check that $N^{\perp}$ is a submodule of M^*, and that $N^{\perp\perp}$ is a submodule of M. □

Theorem 4.5.18. *Let $N \subseteq M$, and let $N^{\perp}$ be defined by* (4.5.16), *then $N^{\perp} \simeq (M/N)^*$. Furthermore, M/N is torsion-less if and only if $N^{\perp\perp} = N$.*

Proof. Consider the exact sequence

$$0 \longrightarrow N \longrightarrow M \xrightarrow{f} M/N \longrightarrow 0$$

and the dual exact sequence

$$0 \longrightarrow (M/N)^* \xrightarrow{f^*} M^* \longrightarrow N^*.$$

Therefore $(M/N)^*$ can be identified with $N^{\perp}$, which is clearly the kernel of the map $M^* \longrightarrow N^*$ induced by the inclusion $N \hookrightarrow M$.

Assume that M/N is torsion-less. Let $\psi \in (M/N)^*$, then ψ is induced by a map $\varphi \in M^*$ such that $\varphi(N) = 0$, i.e., $\varphi \in N^{\perp}$. Let $m \in N^{\perp\perp} \subseteq M$, then, by definition,

$$\delta_{M/N}(m + N)(\psi) = \psi(m + N) = \varphi(m) = 0.$$

Since M/N is torsion-less, so that $\delta_{M/N}$ is a monomorphism, one has $m + N = N$, i.e., $N^{\perp\perp}/N = 0$ and so $N^{\perp\perp} = N$.

Conversely, suppose that $N^{\perp\perp} = N$. Let $x = m + N \in M/N$, $x \neq 0$, i.e., $m \notin N$ and so $m \notin N^{\perp\perp}$. Consequently, there is a $g \in N^{\perp}$ such that $g(m) \neq 0$. Since $N^{\perp} \simeq (M/N)^*$, there is a $\varphi \in (M/N)^*$, such that $\varphi(x) \neq 0$. Therefore, by lemma 4.5.4, M/N is torsion-less. □

For the case of right (left) Noetherian rings one can obtain deeper results. Recall that for a right Noetherian ring A the dual to a finitely generated left A-module is finitely generated, by [147, lemma 4.10.2], and the dual to a finitely generated projective left A-module is finitely generated and projective, by [147, lemma 4.10.1].

Note that the converse statement to corollary 4.5.6(2) which would say that each torsion-less module can be embedded into a projective module, is not true in the general case. But the next statement shows that this is true for finitely generated modules over a right Noetherian ring.

Proposition 4.5.19. *Let A be a right Noetherian ring, and M a finitely generated left A-module. Then M is torsion-less if and only if it is isomorphic to a submodule of a finitely generated projective left A-module.*

Proof. Assume that M is a torsion-less left A-module. Consider an epimorphism $F \longrightarrow M \longrightarrow 0$, where F is a finitely generated free left A-module. Then one has a monomorphism $0 \longrightarrow M^* \longrightarrow F^*$. Since A is right Noetherian and F^* is finitely generated, by proposition 4.5.9, M^* is also finitely generated. So there is a finitely generated free right A-module G with an epimorphism $G \longrightarrow M^* \longrightarrow 0$. Then $0 \longrightarrow M^{**} \longrightarrow G^*$ embeds M^{**} into a finitely generated free module. Since M is torsion-less, δ_M is a monomorphism. Therefore the composition of monomorphisms $M \hookrightarrow M^{**} \hookrightarrow G^*$ is a monomorphism. □

Consider a projective resolution $\mathcal{P}$ of a module K:

$$\cdots \longrightarrow P_{n+1} \xrightarrow{f_{n+1}} P_n \xrightarrow{f_n} P_{n-1} \longrightarrow \cdots \longrightarrow P_1 \xrightarrow{f_1} P_0 \xrightarrow{\varepsilon} K \longrightarrow 0$$

where each P_i is projective. Recall that every finitely generated A-module over a right Noetherian ring has a finitely generated projective resolution, i.e., a projective resolution where each P_i is finitely generated. Any projective resolution $\mathcal{P}$ can be cut at f_n to yield

$$0 \longrightarrow M \longrightarrow P_n \xrightarrow{f_n} P_{n-1} \longrightarrow \cdots \longrightarrow P_1 \xrightarrow{f_1} P_0 \xrightarrow{\varepsilon} K \longrightarrow 0 \qquad (4.5.20)$$

The module $M = \operatorname{Ker} f_n = \operatorname{Coker} f_{n+1}$ is called the n-th **syzygy** of K.

In particular a right A-module M is called a 0-th **syzygy** (or simply **syzygy**) if it is isomorphic to a submodule of a projective module.

Using this terminology proposition 4.5.19 can be rewritten in the following equivalent form:

Proposition 4.5.21. *Let A be a right Noetherian ring, and let M be a finitely generated left A-module. Then M is torsion-less if and only if it is a 0-th syzygy.*

Theorem 4.5.22. *Let A be a Noetherian ring, and M be a finitely generated right A-module. Then M is a first syzygy if and only if there is a left A-module N such that $M \simeq N^*$.*

Proof.

1. Suppose M is a first syzygy. Then there is an exact sequence

$$0 \longrightarrow M \longrightarrow P_1 \longrightarrow P_0$$

where P_1 and P_0 are finitely generated projective A-modules. Let $M^{\perp} \subset P_1^*$ and $M^{\perp\perp} \subset P_1$ be defined as above. Consider $\delta_{P_1} : P_1 \longrightarrow P_1^{**}$. Then

$$M^{\perp\perp} = \{p \in P_1 \ : \ \delta_{P_1}(M^{\perp}) = 0\}.$$

Therefore there is an exact sequence

$$0 \longrightarrow M^{\perp\perp} \longrightarrow P_1 \xrightarrow{f} (M^{\perp})^*,$$

where f is the composition of δ_{P_1} and a homomorphism ξ induced by the inclusion $M^{\perp} \hookrightarrow P_1^*$. So there is a commutative diagram:

$$\begin{array}{ccccccc} 0 & \longrightarrow & M^{\perp\perp} & \longrightarrow & P_1 & \longrightarrow & (M^{\perp})^* \\ & & & & {\scriptstyle \delta_{P_1}}\Big\downarrow{\scriptstyle \simeq} & & \Big\| \\ 0 & \longrightarrow & (P_1^*/M^{\perp})^* & \longrightarrow & P_1^{**} & \xrightarrow{\xi} & (M^{\perp})^* \end{array}$$

where $\mathrm{Ker}\xi = (P_1^*/M^{\perp})^*$, as indicated. Thus $M^{\perp\perp} \simeq (P_1^*/M^{\perp})^*$. Since $P_1/M \subset P_0$, P_1/M is a 0-th syzygy, and so it is torsion-less. Therefore, by theorem 4.5.18, $M^{\perp\perp} = M$. Thus $M \simeq N^*$, where $N \simeq P_1^*/M^{\perp}$.

2. Suppose $M = N^*$. Then there is an exact sequence

$$0 \longrightarrow K \longrightarrow P_1 \longrightarrow N \longrightarrow 0$$

with a finitely generated projective module P_1, and there is the dual exact sequence

$$0 \longrightarrow N^* \longrightarrow P_1^* \longrightarrow K^*.$$

By corollary 4.5.13, K^* is torsion-less, and so it is a 0-th syzygy, by proposition 4.5.21. Thus, N^* is a first syzygy. □

Theorem 4.5.23. *Let A be a Noetherian ring, and M a finitely generated torsion-less right A-module. Then there exists a torsion-less left A-module N such that there are exact sequences:*

$$0 \longrightarrow M \xrightarrow{\delta_M} M^{**} \longrightarrow \mathrm{Ext}_A^1(N, A) \longrightarrow 0 \tag{4.5.24}$$

$$0 \longrightarrow N \xrightarrow{\delta_N} N^{**} \longrightarrow \mathrm{Ext}_A^1(M, A) \longrightarrow 0. \tag{4.5.25}$$

Proof. Consider an exact sequence

$$0 \longrightarrow K \longrightarrow P \longrightarrow M \longrightarrow 0$$

with P projective, and the dual exact sequence

$$0 \longrightarrow M^* \longrightarrow P^* \xrightarrow{\psi} N \longrightarrow 0, \tag{4.5.26}$$

where $N = P^*/M^*$. Since $M = P/K$, from theorem 4.5.18 it follows that $M^* \simeq K^{\perp}$, and so there is an exact sequence

$$0 \longrightarrow K^{\perp} \longrightarrow P^* \xrightarrow{\psi} N \longrightarrow 0.$$

Apply the duality functor * to the sequence (4.5.26). Since P^* is projective, there results an exact sequence

$$0 \longrightarrow N^* \xrightarrow{\psi^*} P^{**} \longrightarrow M^{**} \longrightarrow \mathrm{Ext}_A^1(N,A) \longrightarrow 0. \qquad (4.5.27)$$

Let $K^{\perp\perp} = \{p \in P : \varphi(p) = 0 \text{ for } \forall\varphi \in K^{\perp}\}$, then $K^{\perp\perp} = K$, since P/K is torsion-less, and $\mathrm{Im}\psi^* = \delta_P(K^{\perp\perp})$. Therefore one obtains the following commutative diagram:

$$\begin{array}{ccccccccc}
0 & \longrightarrow & K^{\perp\perp} & \longrightarrow & P & \longrightarrow & M & \longrightarrow & 0 \\
 & & \simeq\big\downarrow\delta_K & & \delta_P\big\downarrow\simeq & & \delta_M\big\downarrow & & \\
0 & \longrightarrow & N^* & \longrightarrow & P^{**} & \longrightarrow & M^{**} & &
\end{array}$$

Since M is torsion-less, δ_M is a monomorphism. Thus $M \simeq P^{**}/N^*$. Combining this with (4.5.27), there results a commutative diagram

$$\begin{array}{ccccccccc}
0 & \longrightarrow & P^{**}/N^* & \longrightarrow & M^{**} & \longrightarrow & \mathrm{Ext}_A^1(N,A) & \longrightarrow & 0 \\
 & & \simeq\big\downarrow & & \big\| & & \big\| & & \\
0 & \longrightarrow & M & \xrightarrow[\delta_M]{} & M^{**} & \longrightarrow & \mathrm{Ext}_A^1(N,A) & \longrightarrow & 0
\end{array}$$

Diag. 4.5.28.

with exact rows. Hence one obtains an exact sequence (4.5.24). Applying the same reasoning to the sequence (4.5.26), there results the exact sequence (4.5.25). □

4.6 Flat Modules and Coherent Rings

The concept of weak dimensions of modules was considered in Section 4.2 (See also [147, Section 5.1]). From the definition of a flat module, proposition 1.5.15 and proposition 4.2.3 one obtains the following equivalent conditions for a module to be flat:

Proposition 4.6.1. *The following conditions are equivalent for a right A-module M:*

1. *M is flat.*
2. $\mathrm{w.dim}_A M \leq 1$.
3. *For each monomorphism of right A-modules $N \longrightarrow M$, the natural group homomorphism $N \otimes_A X \longrightarrow M \otimes_A X$ is a monomorphism for any left A-module X.*
4. $\mathrm{Tor}_k^A(M,X) = 0$ *for all left A-module X and all $k \geq 2$.*
5. *The canonical map $M \otimes_A I \longrightarrow M \otimes_A A \cong M$ is a monomorphism for any left ideal I in A.*

6. *The canonical map $M \otimes_A I \longrightarrow M \otimes_A A \cong M$ is a monomorphism for any finitely generated left ideal I in A.*
7. *The natural map $M \otimes_A I \longrightarrow MI$ is an isomorphism of Abelian groups for any finitely generated left ideal I in A.*

While semisimple rings and hereditary rings are defined unique by their global (projective) dimension, for semihereditary rings there is only the following property in connection with weak dimension:

Theorem 4.6.2. *If A is a right semihereditary ring then*

$$\text{w.dim}\, A \leq 1.$$

Proof. Let A be a right semihereditary ring, and M a right A-module. Consider an exact sequence

$$0 \longrightarrow N \longrightarrow P \longrightarrow M \longrightarrow 0$$

with P projective. For any left A-module X, by theorem 1.6.6(3), $\mathrm{Tor}_{n+1}^A(X,M) \simeq \mathrm{Tor}_n^A(X,N)$. Since from proposition 1.4.10(2) it follows that the functor Tor_n^A commutes with direct limits, it suffices to prove that $\mathrm{Tor}_{n+1}^A(X,L) = 0$ for any finitely generated submodule $L \subset N$. But any such submodule is projective, since A is a semihereditary ring and $L \subset N \subset P$. Therefore $\mathrm{Tor}_n^A(X,L) = 0$ for all $n \geq 1$, and so $\mathrm{Tor}_{n+1}^A(X,M) = 0$ for all $n \geq 1$. This means, by proposition 4.2.3, that $\text{w.dim}\, A \leq 1$. □

As a consequence of this theorem and proposition 4.2.11 one obtains the following results, which are right-left symmetric.

Corollary 4.6.3. *The following conditions are equivalent for any ring A:*

1. $\text{w.dim}\, A \leq 1$.
2. *All right ideals of A are flat.*
3. *All left ideals of A are flat.*
4. *Any submodule of a flat right A-module is flat.*
5. *Any submodule of a flat left A-module is flat.*
In particular, if A is a semihereditary ring then the conditions 2–5 are equivalent.

Note that the converse statements of theorem 4.6.2 and corollary 4.6.3 do not hold. That is, there are rings with $\text{w.dim}\, A \leq 1$ which are not semihereditary. An example of such a ring was obtained by S.Glaz in [97].

Example 4.6.4 (S. Glaz [97, Example 3.1.2]).

Let $\mathbf{Q}$ be the field of rational numbers, and let $\mathbf{Q}[x]$ be the ring of polynomials in one variable over $\mathbf{Q}$. Let A be the subring of $\prod \mathbf{Q}[x]$, the infinite product of copies of $\mathbf{Q}[x]$, generated by the sequence $(x, 0, x^2, 0, x^3, 0\ldots)$, and all sequences that eventually consist of constants.

Then the ring A has weak dimension w.dim $A \leq 1$, but it is not semihereditary.

So semihereditary rings are not defined uniquely by their flatness property.

Lemma 4.6.5. *Any finitely generated right ideal of a right semihereditary ring is finitely presented.*

Proof. Since any finitely generated right ideal $\mathcal{I}$ of a right semihereditary ring A is projective, the statement follows immediately from proposition 4.4.3. □

Theorem 1.5.15 gives criteria for a module to be flat. The following theorem provides another flatness test.

Theorem 4.6.6. *Let $0 \longrightarrow X \longrightarrow P \longrightarrow M \longrightarrow 0$ be an exact sequence of right A-modules, where P is projective. Then the following statements are equivalent:*

1. *M is a flat module.*
2. *For any $x \in X \subset P$ there is a $\theta \in \mathrm{Hom}_A(P, X)$ with $\theta(x) = x$.*
3. *For any finite set of elements $x_1, x_2, \ldots, x_n \in X$ there is a $\theta \in \mathrm{Hom}_A(P, X)$ with $\theta(x_i) = x_i$ for all i.*

Proof.
$1 \Longrightarrow 2$. First observe that if P' is a projective module then M is flat if and only if $M \oplus P'$ is flat. This follows directly from definition ([146, page 131]).

Choose P' such that $F = P \oplus P'$ is free, and replace the exact sequence $0 \longrightarrow X \longrightarrow P \longrightarrow M \longrightarrow 0$ by the exact sequence

$$0 \longrightarrow X \longrightarrow F \longrightarrow M \oplus P' \longrightarrow 0.$$

Choose a basis $\{p_i \in P \ : i \in I\}$ for F and let $\varphi_i : F \longrightarrow A$ be the corresponding family of (coordinate) homomorphisms.

An $x \in X$ can be written in the form $x = p_{i_1}a_1 + p_{i_2}a_2 + \ldots + p_{i_m}a_m$, where $a_r = \varphi_r(x) \in A$. Let $\mathcal{I}$ be the left ideal $\mathcal{I} = Aa_1 + Aa_2 + \ldots + Aa_m$. Since $M \oplus P'$ is flat, $x \in X \cap F\mathcal{I} = X F$. (It is the equality $X \cap F\mathcal{I} = XF$ that is guaranteed by the flatness of $M \oplus P'$; see proposition 1.5.15). So $x = \sum x_j c_j$, where $x_j \in X$ and $c_j \in \mathcal{I}$. Write $c_j = \sum\limits_i b_{ij}a_i$, so that $x = \sum\limits_i x_i' a_i$, where $x_i' = \sum\limits_j x_j b_{ij}$.

Define $\theta' : F \longrightarrow X$ by $\theta'(p_{i_k}) = x_k'$ for $k = 1, \ldots, m$ and $\theta'(p_j) = 0$ for all other p's. Then

$$\theta'(x) = \theta'(\sum_{k=1}^{m} p_{i_k}a_k) = \sum_{k=1}^{m} \theta'(p_{i_k}a_k = \sum_{k=1}^{m} x_k' a_k = x.$$

Let θ be the restriction of θ' to $P \subset F$. Then because $x \in X \subset P$, $\theta(x) = x$. This finishes the proof of $1 \Longrightarrow 2$.

Incidentally, write $p_i = (p_i', p_i'') \in P \oplus P'$ and let φ_i' be the restriction of φ_i to $P \subset F$. Then the $\{p_i' \ : \ i \in I\}$ and $\{\varphi_i' \ : \ i \in I\}$ form a family of elements and a family of homomorphisms as in the Kaplansky theorem 1.5.5. This also illustrates that the Kaplansky family of elements is usually not linearly independent. Otherwise P would be free.

$2 \Longrightarrow 1$. We aim to apply the flatness criterion 1.5.15. So it is to be shown that $X \cap P\mathcal{I} = X\mathcal{I}$ for all finitely generated left ideals $\mathcal{I} \subset A$.

So let $x \in X \cap P\mathcal{I}$, where $\mathcal{I}$ is a left ideal in A. Then x can be written $x = p_1a_1 + p_2a_2 + \ldots + p_ma_m$, where $p_i \in P$, $a_i \in A$. Let $\theta \in \mathrm{Hom}_A(P,X)$ be such that $\theta(x) = x$. Then $x = \theta(x) = \theta(p_1)a_1 + \theta(p_2)a_2 + \ldots + \theta(p_m)a_m \in X\mathcal{I}$ finishing the proof $2 \Longrightarrow 1$.

$2 \Longrightarrow 3$. This is proved by induction on n. Let $x_1,x_2,\ldots,x_n \in X$. If $n = 1$, then the existence of θ follows from statement 2. Assume that $n > 1$ and statement 3 holds for all $k < n$. Let $\theta_n : P \longrightarrow X$ be a homomorphism such that $\theta_n(x_n) = x_n$. Let $y_i = x_i - \theta_n(x_i)$ for $i = 1,2,\ldots,n-1$. By induction hypothesis there exists a homomorphism θ' such that $\theta'(y_i) = y_i$ for $i = 1,2,\ldots,n-1$. Now define $\theta = \theta' + \theta_n - \theta'\theta_n \in \mathrm{Hom}_A(P,X)$. Note that this makes sense because $\theta_n(p) \in X \subset P$. Then

$$\theta(x_n) = \theta'(x_n) + \theta_n(x_n) - \theta'\theta_n(x_n) = \theta'(x_n) + x_n - \theta'x_n = x_n,$$

$$\theta(x_i) = \theta'(x_i) + \theta_n(x_i) - \theta'\theta_n(x_i) = \theta'(x_i) + (x_i - y_i) - \theta'(x_i - y_i) =$$
$$= x_i - y_i + \theta'(y_i) = x_i$$

for $i = 1,2,\ldots,n-1$. So θ is a homomorphism as required.

$3 \Longrightarrow 2$ follows by taking $n = 1$. □

Remark 4.6.7. From theorem 4.6.6 there immediately follows the following theorem which was first proved by O. Villamayor and was stated by S.U. Chase in his paper [41].

Theorem 4.6.8 (O. Villamayor [41]). *Let $0 \longrightarrow X \longrightarrow F \longrightarrow P \longrightarrow 0$ be an exact sequence of right A-modules, where F is free with a basis $\{e_i : i \in I\}$. Then the following statements are equivalent:*

1. *P is a flat module.*
2. *For any $x \in X$ there is a $\theta \in \mathrm{Hom}_A(F,X)$ with $\theta(x) = x$.*
3. *For any $x_1,x_2,\ldots,x_n \in X$ there is a $\theta \in \mathrm{Hom}_A(F,X)$ with $\theta(x_i) = x_i$ for all i.*

Proposition 4.6.9. *If an A-module M is flat and has a projective cover then M is projective.*

Proof. Let P be a projective cover of M. Consider an exact sequence

$$0 \longrightarrow X \longrightarrow P \xrightarrow{\varphi} M \longrightarrow 0$$

where $X = \mathrm{Ker}(\varphi)$ is a small module in P. Let $x \in X$ be an arbitrary element. Consider a cyclic A-module $N = (x) \subseteq X$. By theorem 4.6.6, there is a homomorphism $\theta : P \longrightarrow X$ such that $\theta(x) = x$. Therefore $\theta(N) = N$ and so $(1-\theta)N = 0$. Consider the homomorphism $1-\theta : P \longrightarrow X$. Since $\theta(P) = \mathrm{Im}(\theta) \subseteq X$, $\theta(P)$ is small in P. Then from the equality $P = \theta(P)+(1-\theta)(P)$ it follows that $P = (1-\theta)(P)$, i.e., $1-\theta$ is an epimorphism. Since P is projective, $\mathrm{Ker}(1-\theta)$ is a direct summand of P. Let $p \in \mathrm{Ker}(1-\theta)$, then $(1-\theta)p = 0$, hence $\theta(p) = p$, which means that $p \in X$. Thus $N \subseteq \mathrm{Ker}(1-\theta) \subseteq X$. Since X is small in P, $\mathrm{Ker}(1-\theta)$ is small in P as well. But $\mathrm{Ker}(1-\theta)$ is a direct summand of P. Therefore $\mathrm{Ker}(1-\theta) = 0$, and so $N = 0$. Since x is an arbitrary element of X, $X = 0$, and $M \simeq P$ is projective. □

Proposition 4.6.10. *If A is a semiperfect ring then any finitely generated flat A-module M is projective.*

Proof. Since A is semiperfect and M is a finitely generated A-module, M has a projective cover. Then the flat A-module M is projective, by proposition 4.6.9. □

Proposition 4.6.11. *If A is a semiperfect ring then A is right semihereditary if and only if A is left semihereditary.*

Proof. Suppose A is a semiperfect right semihereditary ring. Let $\mathcal{I}$ be a finitely generated left ideal in A. By corollary 4.6.3, $\mathcal{I}$ is flat. Therefore $\mathcal{I}$ is projective, by proposition 4.6.10. Thus A is left semihereditary. □

Since any right serial ring is semiperfect, from this proposition there follows immediately

Corollary 4.6.12. *If A is a right serial ring then A is right semihereditary if and only if A is left semihereditary.*

The last part of this section is devoted to proving results of S.U.Chase characterizing rings for which all direct products of any family of flat modules are flat.

If $\{P_i\}_{i\in I}$ is a family of flat A-modules for some index set I, their direct product $P = \prod\limits_{i\in I} P_i$ need not be flat in a general case. S.U. Chase in [42] showed in what case the direct product of flat modules is flat, and gave some applications of this result to characterizations of semihereditary rings and Prüfer rings. The proofs of these results given here follow K.R. Goodearl [109].

First consider compatibility between the tensor product and the direct product of modules. Let $\{X_i\}_{i\in I}$ be a family of right A-modules for some index set I, and M a left A-module over the same ring A. For any $j \in I$, the canonical projection $\pi_j : \prod\limits_{i\in I} X_i \longrightarrow X_j$ induces a map $\varphi_j : (\prod\limits_{i\in I} X_i) \otimes_A M \longrightarrow X_j \otimes_A M$. These maps together induce a map $\varepsilon : (\prod\limits_{i\in I} X_i) \otimes_A M \longrightarrow \prod\limits_{i\in I} (X_i \otimes_A M)$ which is called the "natural map" from $(\prod\limits_{i\in I} X_i) \otimes_A M$ into $\prod\limits_{i\in I} (X_i \otimes_A M)$.

If there are only a finite number of non-zero modules in the family $\{X_i\}_{i\in I}$, then ε is an isomorphism. But in general ε need not be an isomorphism.

In a special case when all $X_i = A$ one can identify $X_i \otimes_A M$ with M in the usual way and there results the natural map:

$$\delta : A^I \otimes_A M \longrightarrow M^I$$

defined by $(\delta(a \otimes m))_i = a_i m$ for all $i \in I$, $a \in A^I$, and $m \in M$.

Proposition 4.6.13. *The following conditions are equivalent for any left A-module M:*

1. *M Is finitely generated.*
2. *The natural map $(\prod_{i \in I} X_i) \otimes_A M \longrightarrow \prod_{i \in I} (X_i \otimes_A M)$ is surjective for all families $\{X_i\}_{i \in I}$ of right A-modules.*
3. *The natural map $A^I \otimes M \longrightarrow M^I$ is surjective for any index set I.*

Proof.
$1 \Longrightarrow 2$. Let M be a finitely generated left A-module with generators $m_1, m_2, \dots, m_n$. Given any element $u \in \prod_{i \in I} (X_i \otimes_A M)$ one can write $u_i = x_{1i} \otimes m_1 + x_{2i} \otimes m_2 + \dots + x_{ni} \otimes m_n$ for each $i \in I$ and for suitable elements $x_{1i}, x_{2i}, \dots, x_{ni} \in X_i$. For each j, elements x_{ji} can be considered to be the components of an element $x_j \in \prod_i X_i$, and the element $x_1 \otimes m_1 + x_2 \otimes m_2 + \dots + x_n \otimes m_n \in (\prod_{i \in I} X_i) \otimes_A M$ maps onto the element x via the natural map.
$2 \Longrightarrow 3$. This is clear, since statement 3 is a special case of statement 2.
$3 \Longrightarrow 1$. Take the set M for the index set I. Then the natural map $\delta : A^M \otimes M \longrightarrow M^M$ is surjective. Define $t \in M^M$ by setting $t_m = m$ for all $m \in M$. Writing $t = \delta(\sum_{i=1}^{n} a_i \otimes m_i)$ for $a_i \in A^M$ and $m_i \in M$, one has

$$m = t_m = \sum_{i=1}^{n} (\delta(a_i \otimes m_i))_m = \sum_{i=1}^{n} (a_i)_m m_i,$$

so the elements $m_1, m_2, \dots, m_n$ generate M. □

Proposition 4.6.14. *The following conditions are equivalent for any left A-module M:*

1. *M Is finitely presented.*
2. *The natural map $(\prod_{i \in I} X_i) \otimes_A M \longrightarrow \prod_{i \in I} (X_i \otimes_A M)$ is bijective for all families $\{X_i\}_{i \in I}$ of right A-modules.*
3. *The natural map $A^I \otimes M \longrightarrow M^I$ is bijective for any index set I.*

Proof.
$1 \Longrightarrow 2$. Let $\{X_i\}_{i \in I}$ be a family of right A-modules. Since M is a finitely presented left A-module, there is an exact sequence $A^m \longrightarrow A^n \longrightarrow M \longrightarrow 0$. Consider the following commutative diagram with exact rows:

$$\begin{array}{ccccccc}
(\prod_{i\in I} X_i)\otimes_A A^m & \longrightarrow & (\prod_{i\in I} X_i)\otimes_A A^n & \longrightarrow & (\prod_{i\in I} X_i)\otimes_A M & \longrightarrow & 0\\
\downarrow{\scriptstyle\alpha} & & \downarrow{\scriptstyle\beta} & & \downarrow{\scriptstyle\gamma} & & \\
\prod_{i\in I}(X_i\otimes_A A^m) & \longrightarrow & \prod_{i\in I}(X_i\otimes_A A^n) & \longrightarrow & \prod_{i\in I}(X_i\otimes_A M) & \longrightarrow & 0
\end{array}$$

Diag. 4.6.15.

Since α and β are bijective, γ is also bijective, by corollary 1.2.9.

$2 \Longrightarrow 3$. Obvious.

$3 \Longrightarrow 1$. Proposition 4.6.13 gives that M is finitely generated. So there is an exact sequence $0 \longrightarrow K \longrightarrow F \longrightarrow M \longrightarrow 0$ of left A-modules, where F is finitely generated and free. For any index set I consider the following commutative diagram:

$$\begin{array}{ccccccccc}
 & & A^I\otimes K & \longrightarrow & A^I\otimes F & \longrightarrow & A^I\otimes M & \longrightarrow & 0\\
 & & \downarrow{\scriptstyle\alpha} & & \downarrow{\scriptstyle\beta} & & \downarrow{\scriptstyle\gamma} & & \\
0 & \longrightarrow & K^I & \longrightarrow & F^I & \longrightarrow & M^I & \longrightarrow & 0
\end{array}$$

Diag. 4.6.16.

By assumption γ is bijective, and β is surjective, by proposition 4.6.13. So it follows from diagram 4.6.16 that α is surjective. Then from proposition 4.6.13 it follows that K is finitely generated, as required. □

Theorem 4.6.17 (S.U. Chase [41]). *The following statements are equivalent for any ring A:*

1. *The direct product of any family of flat right A-modules is flat.*
2. *The direct product of any family of copies of A is flat as a right A-module.*
3. *Any finitely generated left ideal in A is finitely presented.*

Proof.

$1 \Longrightarrow 2$. By proposition 1.5.14(1) as A_A is flat.

$2 \Longrightarrow 3$. Let $\mathcal{I}$ be a finitely generated left ideal in A. Given any index set I, from condition (2) and proposition 4.6.1(5) it follows that $\psi : A^I \otimes \mathcal{I} \longrightarrow \mathcal{I}^I$ is injective. Since $\mathcal{I}$ is finitely generated, ψ is surjective, by proposition 4.6.13. Hence proposition 4.6.14 shows that $\mathcal{I}$ is finitely presented.

$3 \Longrightarrow 1$. Let $\{X_i\}_{i\in I}$ be a family of flat right A-modules, and let $\mathcal{I}$ be a finitely generated left ideal in A. Consider the following commutative diagram:

$$\begin{array}{ccc}
(\prod\limits_{i\in I} X_i)\otimes_A \mathcal{I} & \xrightarrow{\ \varphi\ } & (\prod\limits_{i\in I} X_i)\mathcal{I} \\
\Big\downarrow{\scriptstyle\alpha} & & \Big\downarrow{\scriptstyle\beta} \\
\prod\limits_{i\in I}(X_i\otimes_A \mathcal{I}) & \xrightarrow[\ \psi\ ]{} & \prod\limits_{i\in I}(X_i\mathcal{I})
\end{array}$$

Diag. 4.6.18.

Since $\mathcal{I}$ is finitely generated, $\mathcal{I}$ is finitely presented by assumption. So from proposition 4.6.14 it follows that α is bijective. Since each X_i is flat, the maps $X_i \otimes_A \mathcal{I} \longrightarrow X_i\mathcal{I}$ are all bijective, by one of the flatness tests (theorem 1.5.15(4)). Hence the map ψ is bijective. Then from diagram 4.6.18 it follows that φ is injective. Since the natural map φ is surjective, it follows again from the same flatness test that $\prod\limits_{i\in I} X_i$ is flat, as required. □

Definition 4.6.19. A ring A which satisfies the equivalent conditions of theorem 4.6.17 is called a **left coherent ring**. A **right coherent ring** is defined analogously.

Examples 4.6.20.

1. Any right Noetherian ring is a right coherent, which follows from proposition 4.4.5.
2. Any right semihereditary ring is a right coherent, which follows from lemma 4.6.5.

Theorem 4.6.21. *Let for an A-module M there exist an exact sequence*

$$0 \longrightarrow K \longrightarrow P \longrightarrow M \longrightarrow 0, \tag{4.6.22}$$

where P is a projective A-module and K is a finitely generated A-module. Then M is flat if and only if it is projective.

Proof. Since any projective module is flat, by proposition 1.5.14(2), it remains only to prove that if M is flat and there exists an exact sequence (4.6.22) then M is projective.

Consider an exact sequence (4.6.22), where P is a projective A-module and K is a finitely generated A-module with generators $c_1, c_2, \dots, c_n$. Then, by theorem 4.6.6, there exists $\varphi \in \mathrm{Hom}_A(P, K)$ such that $\varphi(c_i) = c_i$ for all i. Therefore φ is an identity on K, and so this sequence is split, which means that M is projective as a direct summand of the projective module P. □

Definition 4.6.23. An A-module M is called **finitely related** if there exists an exact sequence $0 \longrightarrow K \longrightarrow F \longrightarrow M \longrightarrow 0$, where F is a free A-module of arbitrary rank and K is a finitely generated A-module.

Obviously, any finitely presented A-module is finitely related. As an immediately corollary of theorem 4.6.21 we obtain the following theorem.

Theorem 4.6.24 (S.U. Chase [41]). *Let M be a finitely related A-module. Then M is flat if and only if it is projective. In particular, any finitely presented flat A-module is projective.*

Theorem 4.6.25. *Let A be a right coherent ring. Then if A is left semihereditary then A is right semihereditary.*

Proof. Let a ring A be right coherent and left semihereditary, and I a finitely generated right ideal in A. Then I is a finitely presented right A-module. Since A is left semihereditary, from corollary 4.6.3 it follows that I is flat, and so, by theorem 4.6.24, I is projective. So, A is right semihereditary. □

Since any right Noetherian is right coherent, from theorem 4.6.25 one obtains the following result.

Corollary 4.6.26. *Let A be a right Noetherian ring. Then if A is left semihereditary then it is right semihereditary.*

Theorem 4.6.27 (S.U. Chase [41]). *The following conditions are equivalent for a ring A:*

1. *A Is a right semihereditary ring.*
2. $\text{w.dim}\, A \leq 1$ *and A is a right coherent ring.*
3. *All torsion-less left A-modules are flat.*

Proof.

$1 \Longrightarrow 2$. Let A be a right semihereditary ring. Then $\text{w.dim}\, A \leq 1$, by theorem 4.6.2. Since any finitely generated right ideal in A is projective, it is flat and finitely presented, by lemma 4.6.5. So, by theorem 4.6.17, the direct product of an arbitrary family of copies of A is flat as a left A-module, i.e., A is a right coherent ring.

$2 \Longrightarrow 1$. Let I be a finitely generated right ideal in A. Then I is flat, since $\text{w.dim}\, A \leq 1$, and it is finitely presented, because A is right coherent, see theorem 4.6.17. So I is projective, by theorem 4.6.24. Thus, A is right semihereditary.

$2 \Longrightarrow 3$. Let M be a torsion-less left A-module. By proposition 4.5.5, M can be embedded into a direct product of copies of A, which is a flat left A-module, by assumption. Since $\text{w.dim}\, A \leq 1$, any submodule of a flat A-module is flat. So, M is flat.

$3 \Longrightarrow 2$. Let I be a left ideal in A. Since any ideal is torsion-less, I is flat, by assumption. Then $\text{w.dim}\, A \leq 1$, by proposition 4.2.10. Since any direct product of an arbitrary family of copies of A is torsion-less, and by assumption any torsion-less left A-module is flat, one obtains that A is a right coherent ring. □

4.7 Modules over Formal Triangular Matrix Rings

This section considers ideals and modules over formal triangular matrix rings (See Section 2.6 for the definition of 'formal triangular matrix ring'). It also gives necessary and sufficient conditions under which a formal triangular matrix ring is right (left) hereditary.

Proposition 4.7.1. *Let* $I = \begin{pmatrix} I_1 & N \\ 0 & I_2 \end{pmatrix}$ *be a right ideal of a formal triangular matrix ring* $A = \begin{pmatrix} S & M \\ 0 & T \end{pmatrix}$. *Then* I *is projective if and only if the following conditions hold:*

i. I_1 *is projective as a right* S*-module*;
ii. I_2 *and* N/I_1M *are projective as right* T*-modules*;
iii. *The map* $f : I_1 \otimes_S M \longrightarrow M$ *is a monomorphism.*

Proof.

1. Suppose that $I = \begin{pmatrix} I_1 & N \\ 0 & I_2 \end{pmatrix}$ is a projective right ideal of the ring $A = \begin{pmatrix} S & M \\ 0 & T \end{pmatrix}$. Let $e_1 = \begin{pmatrix} 1 & 0 \\ 0 & 0 \end{pmatrix}$, $e_2 = \begin{pmatrix} 0 & 0 \\ 0 & 1 \end{pmatrix}$. Then $H = e_1A = \begin{pmatrix} S & M \\ 0 & 0 \end{pmatrix}$ is a two-sided ideal of A, $A/H \cong T$, and $IH = \begin{pmatrix} I_1 & I_1M \\ 0 & 0 \end{pmatrix}$. Therefore I_2 and N/I_1M are projective right T-modules. This uses special cases of the observation that if $A \longrightarrow B$ is a homomorphism of rings and P a projective right A-module then $P \otimes_A B$ is a projective right B-module. This is easy because P if and only if P is a direct summand of a free module.
Analogously, $L = Ae_2 = \begin{pmatrix} 0 & M \\ 0 & T \end{pmatrix}$ is a two-sided ideal of A, $A/L \cong S$, and $IL = \begin{pmatrix} 0 & N \\ 0 & I_2 \end{pmatrix}$. Therefore $I/IL \cong I_1$ is a projective right S-module.
Consider the following left ideal of A: $E = \begin{pmatrix} 0 & M \\ 0 & 0 \end{pmatrix}$. Since I is a projective A-module, it is flat. So one has a monomorphism $\varphi : I \otimes_A E \longrightarrow I$. So the composition of the monomorphism φ and the natural inclusion $I \longrightarrow A$ yields a monomorphism $I \otimes_A E \longrightarrow A$, which by restriction gives a natural monomorphism $e_1Ie_1 \otimes_{e_1Ae_1} e_1Ee_2 \longrightarrow e_1Ae_2$, i.e., $I_1 \otimes_S M \longrightarrow M$ is a monomorphism, as required.
2. Conversely, suppose that I_1 is projective as a right S-module, I_2 and N/I_1M are projective as right T-modules, and the map $f : I_1 \otimes_S M \longrightarrow M$ is a monomorphism. Now $H = e_1A = \begin{pmatrix} S & M \\ 0 & 0 \end{pmatrix}$ is a two-sided ideal of A, $A/H \cong T$ and $IH = \begin{pmatrix} I_1 & I_1M \\ 0 & 0 \end{pmatrix}$. Therefore $I/IH \cong \begin{pmatrix} N/I_1M \\ I_2 \end{pmatrix}$ is a right projective A/H-module. Since $H = e_1A$, $A/H \cong e_2A$ is a projective A-module. So I/IH is also projective as an A-module.
Since I_1 is a projective right S-module, it is a direct summand of a free module $\bigoplus_\alpha S^\alpha$. Therefore there exists a right S-module K such that $I_1 \oplus K \cong \bigoplus_\alpha S^\alpha$.

Consequently,

$$\begin{pmatrix} I_1 & I_1 \otimes_S M \\ 0 & 0 \end{pmatrix} \oplus \begin{pmatrix} K & K \otimes_S M \\ 0 & 0 \end{pmatrix} = \begin{pmatrix} I_1 \oplus K & (I_1 \oplus K) \otimes_S M \\ 0 & 0 \end{pmatrix} =$$

$$\begin{pmatrix} \bigoplus\limits_{\alpha} S^{\alpha} & (\bigoplus\limits_{\alpha} S^{\alpha}) \otimes_S M \\ 0 & 0 \end{pmatrix} = \bigoplus_{\alpha} \begin{pmatrix} S & S \otimes_S M \\ 0 & 0 \end{pmatrix} \cong \bigoplus_{\alpha} H$$

is a projective A-module. Therefore $\begin{pmatrix} I_1 & I_1 \otimes_S M \\ 0 & 0 \end{pmatrix}$ is also a projective A-module. Since $f : I_1 \otimes_S M \longrightarrow M$ is a monomorphism, $IH \cong \begin{pmatrix} I_1 & I_1 \otimes_S M \\ 0 & 0 \end{pmatrix}$. So IH is also a projective A-module.

Therefore I/IH and IH are both projective A-modules, so that I is projective as well, as required. □

Lemma 4.7.2. *Let $L \subseteq H$ be A-modules for an arbitrary ring A. If every submodule of H, which contains L or is contained in L, is projective, then every submodule of H is projective.*

Proof. Let K be any submodule of H. Then it follows from the assumptions that $K + L$ and $K \cap L$ are both projective. Therefore there is an exact split sequence

$$0 \longrightarrow K \cap L \longrightarrow K \oplus L \longrightarrow K + L \longrightarrow 0,$$

from which it follows that $K \oplus L$ is projective, and so K is also projective. □

Theorem 4.7.3. *A ring $A = \begin{pmatrix} S & M \\ 0 & T \end{pmatrix}$ is right hereditary if and only if the following conditions hold:*

a. *S and T are both right hereditary;*
b. *M is a flat left S-module;*
c. *M/KM is a projective right T-module for any right ideal K of S.*

Proof.

1. Suppose that A is a right hereditary ring. Let $1 = e_1 + e_2$, where $e_1 = \begin{pmatrix} 1 & 0 \\ 0 & 0 \end{pmatrix}$, $e_2 = \begin{pmatrix} 0 & 0 \\ 0 & 1 \end{pmatrix}$, be the obvious decomposition of the identity $1 \in A$ into a sum of orthogonal idempotents. Then $S \cong e_1Ae_1$ and $T \cong e_2Ae_2$. Therefore both S and T are right hereditary, by proposition 4.4.6(4).

   Let K be a right ideal of S. Then $\mathcal{J} = \begin{pmatrix} K & M \\ 0 & 0 \end{pmatrix}$ is a right ideal of A. Therefore $\mathcal{J}$ is projective and by proposition 4.7.1, M/KM is a projective right T-module and $K \otimes_S M \longrightarrow M$ is a monomorphism. Hence M is a flat left S-module, by proposition 4.6.1(5).

2. Conversely, assume that (a), (b) and (c) hold. Consider the following two right ideals of A: $L = \begin{pmatrix} 0 & M \\ 0 & 0 \end{pmatrix}$ and $H = \begin{pmatrix} S & M \\ 0 & 0 \end{pmatrix}$. Let $\mathcal{I} = \begin{pmatrix} 0 & N \\ 0 & 0 \end{pmatrix} \subseteq L$ be an ideal in A, then N is a right T-submodule of M. Since M_T is projective and T is right hereditary, by assumption, N is also a projective right T-module, by proposition 4.4.6(2). Therefore it follows from proposition 4.7.1, that $\mathcal{I}$ is a projective A-module. Now consider a right ideal $\mathcal{I} \subseteq H$ such that $L \subseteq \mathcal{I}$. In this case $\mathcal{I} = \begin{pmatrix} I_1 & M \\ 0 & 0 \end{pmatrix}$, where I_1 is a right ideal in S. Then M/I_1M is projective, by assumption (c). I_1 is a projective right ideal of S, since S is right hereditary; and $I_1 \otimes_S M \longrightarrow S \otimes_S M \cong M$ is a monomorphism, since M is a flat left S-module. Therefore it follows from proposition 4.7.1 that $\mathcal{I}$ is a projective ideal. So, by lemma 4.7.2, any ideal of H is projective.
   Now consider any ideal $\mathcal{I}$ of A which contains H. In this case $\mathcal{I} = \begin{pmatrix} S & M \\ 0 & I_2 \end{pmatrix}$, where I_2 is a right ideal of T. Then I_2 is projective, since T is right hereditary; $M/SM = 0$ and $S \otimes_S M \longrightarrow M$ is an isomorphism. So $\mathcal{I}$ is projective, by proposition 4.7.1. Thus, all ideals of A which contains H and which are contained in H are projective. Therefore it follows from lemma 4.7.2 that all right ideals of A are projective, i.e., A is right hereditary. □

We now describe modules over a formal triangular ring $A = \begin{pmatrix} S & M \\ 0 & T \end{pmatrix}$. Given a right S-module X, a right T-module Y and a right T-module homomorphism $f : X \otimes_S M \longrightarrow Y$ one constructs a right A-module B as follows. The elements of this module B are pairs $(x, y) \in X \oplus Y$ with coordinatewise addition and the operation of multiplication on the right by elements of A defined by the rule:

$$(x, y) \begin{pmatrix} s & m \\ 0 & t \end{pmatrix} = (xs, f(x \otimes m) + yt) \tag{4.7.4}$$

for any element $\begin{pmatrix} s & m \\ 0 & t \end{pmatrix} \in A$. It is easy to check that these operations make B into a right A-module. We will write this module B as a triple (X, Y, f), where $X \in \mathbf{Mod}_r S$, $Y \in \mathbf{Mod}_r T$ and $f : X \otimes_S M \longrightarrow Y$ is a homomorphism in $\mathbf{Mod}_r T$.

Given two right A-modules $B = (X, Y, f)$ and $B_1 = (X_1, Y_1, f_1)$ a morphism $\varphi : B \longrightarrow B_1$ is a pair (α, β) where $\alpha : X \longrightarrow X_1$, $\beta : Y \longrightarrow Y_1$ are homomorphisms of modules such that the diagram is commutative.

$$\begin{array}{ccc} X \otimes_S M & \xrightarrow{f} & Y \\ {\scriptstyle \alpha\otimes 1}\downarrow & & \downarrow{\scriptstyle \beta} \\ X_1 \otimes_S M & \xrightarrow{f_1} & Y_1 \end{array}$$

Diag. 4.7.5.

It is clear that $\varphi = (\alpha, \beta)$ is injective (surjective) if and only if α and β are both injective (surjective).

Analogously one can consider left A-modules which can be constructed as follows. Given a left S-module X, a left T-module Y and a left S-module homomorphism $f : M \otimes_T Y \longrightarrow X$ construct a left A-module C as follows. The elements of this module C are vectors $\begin{pmatrix} x \\ y \end{pmatrix}$ such that $\begin{pmatrix} x \\ y \end{pmatrix}^T \in X \oplus Y$ with coordinatewise addition and the operation of multiplication on the left by elements of A defined by the rule:

$$\begin{pmatrix} s & m \\ 0 & t \end{pmatrix} \begin{pmatrix} x \\ y \end{pmatrix} = \begin{pmatrix} sx + f(m \otimes y) \\ ty \end{pmatrix} \tag{4.7.6}$$

for any element $\begin{pmatrix} s & m \\ 0 & t \end{pmatrix} \in A$.

It is easy to check that these operations make C into a left A-module. We will write this module C as a triple (X, Y, f), where $X \in \mathbf{Mod}_l S$, $Y \in \mathbf{Mod}_l T$ and $f : M \otimes_T Y \longrightarrow X$ is a homomorphism in $\mathbf{Mod}_l S$.

Given two left A-modules $C = (X, Y, f)$ and $C_1 = (X_1, Y_1, f_1)$ a morphism $\psi : B \longrightarrow B_1$ is a pair (α, β) where $\alpha : X \longrightarrow X_1$, $\beta : Y \longrightarrow Y_1$ are homomorphisms of modules such that the diagram

$$\begin{array}{ccc} M \otimes_T Y & \xrightarrow{\ f\ } & X \\ {\scriptstyle 1\otimes\beta}\downarrow & & \downarrow{\scriptstyle \alpha} \\ M \otimes_T Y_1 & \xrightarrow{\ f_1\ } & X_1 \end{array}$$

Diag. 4.7.7.

is commutative.

Conversely, all right and left A-modules can be written in the form described above, which is proved by the following theorem which is stated without proof.

Theorem 4.7.8 (Palmer [257]). *Let* $A = \begin{pmatrix} S & M \\ 0 & T \end{pmatrix}$ *be a formal triangular matrix ring. Let* Ω *be a category which objects are triples* (X, Y, f), *where* $X \in \mathbf{Mod}_l S, Y \in \mathbf{Mod}_l T$ *and* $f \in \mathrm{Hom}_T(X \otimes_S M, Y)$, *and morphisms of this category be morphisms from* $B = (X, Y, f)$ *to* $B_1 = (X_1, Y_1, f_1)$ *defined by pairs* (α, β) *where* $\alpha \in \mathrm{Hom}_S(X, X_1)$, $\beta \in \mathrm{Hom}_T(Y, Y_1)$ *such that the diagram* (4.7.5) *is commutative. Then the category* Ω *and the category of right A-modules are equivalent.*

There is a similar theorem for left modules.

Using the construction of right and left modules one can write right and left ideals as follows. Any right ideal I in $A = \begin{pmatrix} S & M \\ 0 & T \end{pmatrix}$ has the form $I = \begin{pmatrix} I_1 & N \\ 0 & I_2 \end{pmatrix}$, where I_1 is a right ideal in S, I_2 is a right ideal in T, N is a sub-bimodule in M and $I_1M \subset N$. This ideal can be also considered as a right A-module. Set $X = \begin{pmatrix} I_1 \\ 0 \end{pmatrix}$, $Y = \begin{pmatrix} N \\ I_2 \end{pmatrix}$. Then X is a right S-module, Y is a right T-module, and inclusion $I_1M \subset N$ induces a homomorphism $f : I_1 \otimes_S M \longrightarrow N$. So the operation of multiplication given by (4.7.4) for right ideals can be written in the following form:

$$(x, y)\begin{pmatrix} s & m \\ 0 & t \end{pmatrix} = \begin{pmatrix} i_1 & n \\ 0 & i_2 \end{pmatrix}\begin{pmatrix} s & m \\ 0 & t \end{pmatrix} = \begin{pmatrix} i_1 s & f(i_1 \otimes m) + nt \\ 0 & i_2 t \end{pmatrix}. \tag{4.7.9}$$

Similarly, consider a left ideal $J = \begin{pmatrix} J_1 & K \\ 0 & J_2 \end{pmatrix}$ in $A = \begin{pmatrix} S & M \\ 0 & T \end{pmatrix}$, where J_1 is a left ideal in S, J_2 is a left ideal in T, K is a sub-bimodule in M and $MJ_2 \subset K$. This ideal can be considered as a left A-module, if we set $X = \begin{pmatrix} J_1 & N \end{pmatrix}$, $Y = \begin{pmatrix} 0 & J_2 \end{pmatrix}$. Then X is a left S-module, Y is a left T-module, and the inclusion $MJ_2 \subseteq K$ induces a homomorphism $f : M \otimes_T J_2 \longrightarrow K$. So the operation of multiplication given by (4.7.6) in this case can be written in the following form:

$$\begin{pmatrix} s & m \\ 0 & t \end{pmatrix}\begin{pmatrix} x \\ y \end{pmatrix} = \begin{pmatrix} s & m \\ 0 & t \end{pmatrix}\begin{pmatrix} j_1 & k \\ 0 & j_2 \end{pmatrix} = \begin{pmatrix} sj_1 & sk + f(m \otimes j_2) \\ 0 & tj_2 \end{pmatrix} \tag{4.7.10}$$

Proposition 4.7.11. *Let* $A = \begin{pmatrix} S & M \\ 0 & D \end{pmatrix}$ *be a formal triangular matrix ring. Then* A *is a right (left) perfect ring if and only if* S *and* D *are both right (left) perfect rings*[7].

Proof. Suppose A is a right perfect ring with Jacobson radical R, and $J(S)$, $J(D)$ are the Jacobson radicals of rings S and D, respectively. Then A/R is semisimple, and so $S/J(S)$ and $D/J(D)$ are also semisimple, by corollary 2.6.16. Assume that one has two sequences $s_1, s_2, \ldots, s_n, \ldots$ of elements of $J(S)$ and $d_1, d_2, \ldots, d_n, \ldots$ of elements of $J(D)$. Put $a_i = \begin{pmatrix} s_i & m_i \\ 0 & d_i \end{pmatrix}$ for any chosen sequence $m_1, m_2, \ldots, m_n, \ldots$. From the assumption it follows that R is T-nilpotent, i.e., for any sequence $a_1, a_2, \ldots, a_n, \ldots$ of elements of R there exists a number n such that $a_n a_{n-1} \cdots a_1 = 0$. Since $a_n a_{n-1} \cdots a_1 = \begin{pmatrix} s_n s_{n-1} \cdots s_1 & x \\ 0 & d_n d_{n-1} \cdots d_1 \end{pmatrix}$, one obtains that $s_n s_{n-1} \cdots s_1 = 0$ and $d_n d_{n-1} \cdots d_1 = 0$, i.e., $J(S)$ and $J(D)$ are both T-nilpotent. So, S and D are both right perfect.

Conversely, suppose that S and D are both right perfect. Then $S/J(S)$ and $D/J(D)$ are both semisimple, and so is A/R. Let $B = (X, Y, f)$ be a right A-module and $BR = B$. Then from (4.7.4) it follows that $XJ(S) = X$ and $f(X \otimes_S M) + YJ(D) = Y$. Since by assumption $J(S)$ and $J(D)$ are both T-nilpotent, from theorem 1.9.7 it follows that $X = 0$, therefore $f(X \otimes_S M) = 0$, and so $Y = 0$, i.e., $B = 0$. By the same theorem it follows that R is T-nilpotent. Therefore A is right perfect. □

---

[7]For the notion of a right (left) perfect ring see Section 1.9 (an references given there).

Criteria for modules over formal triangular matrix rings to be flat are given by the following theorem which was obtained by R.M. Fossum, P.A. Griffith and I. Reiten in [86].

Proposition 4.7.12 [86]. *Let* $A = \begin{pmatrix} S & M \\ 0 & T \end{pmatrix}$ *be a formal triangular matrix ring, and* $B = (X, Y, f)$ *a right A-module. Then B is flat if and only if the following conditions are satisfied:*

1. Y *is a flat right* T*-module*;
2. $\mathrm{Coker} f$ *is a flat right* T*-module*;
3. f *is monomorphism.*

The following theorem gives a global characterization of formal triangular matrix rings.

Theorem 4.7.13 (Palmér, Roos [258], [259]). *Let* $A = \begin{pmatrix} S & M \\ 0 & T \end{pmatrix}$ *be a formal triangular matrix ring. The following conditions are equivalent*:

1. $\mathrm{r.gl.dim}\, A \leq n$;
2. $\mathrm{r.gl.dim}\, S \leq n$; $\mathrm{r.gl.dim}\, T \leq n$; $R^n\mathrm{Hom}_T(* \otimes_S M, -) = 0$;
3. $\mathrm{r.gl.dim}\, S \leq n$; $\mathrm{r.gl.dim}\, T \leq n$; $R^n\mathrm{Hom}_T(S/I \otimes_S M, -) = 0$, *for all right ideals* I *of* S.

4.8 Notes and References

The study in algebra (ring theory) of homological dimensions started in the second half of the 1950's with among others papers by Maurice Auslander, David A. Buchshbaum, Gerhard P. Hochschild, Samuel Eilenberg, M. Nagata, T.Nakayama and the influential Chicago lecture notes by Irving Kaplansky entitled "Homological dimension of rings and modules", i.e., virtually the same title as the one of this chapter.

Direct products of flat modules were studied by S.U.Chase. Theorem 4.6.8 was proved by O. Villamayor (unpublished) and published by S.U.Chase (see [41], proposition 2.2). Theorems 4.6.17, 4.6.24 and 4.6.27 were proved by S.U. Chase in [41]. Propositions 4.6.9 and 4.6.10 are due to H.Lenzing [219]. The proof of theorem 4.6.27 in this book follows to K.R.Goodearl [109].

The construction of formal triangular matrix rings was first considered by S.U.Chase in [41] for the construction of an example of a ring which is left hereditary but not right hereditary. The structure of right ideals over formal triangular matrix rings was completely described in [109]. Note that theorem 4.7.3 is a particular case of the more general theorem 4.7.13 obtained by I.Palmér and Jan-Erik Roos (See [258], theorem 1 and [259], theorem 5). The proof of theorem 4.7.3 in this book follows to K.R. Goodearl [109].

Formal triangular matrix rings and modules over them were also considered by A.Haghany and K.Varadarajan in [135] and [136]. Modules over formal triangular matrix rings were earlier studied by I. Palmér and J.-E. Roos in [258], [259], [257].

CHAPTER 5

Goldie and Krull Dimensions of Rings and Modules

This chapter consists of a short introduction to the theory of uniform, Goldie and Krull dimensions of rings and modules. Uniform modules and their main properties are considered in Section 5.1, where the uniform dimension of modules is also introduced and studied. Modules of finite Goldie dimension as defined by A. Goldie are considered in Section 5.2.

The notions of singular and nonsingular modules are introduced in Section 5.3. Some of the main properties of such modules are also studied there as well as some properties of nonsingular rings. In Section 5.4 the results of this theory are applied to prove a theorem which gives equivalent conditions for a ring being a Goldie ring. This theorem encompasses the famous Goldie theorem which was proved in [100], [101], [102] (see also [146, Section 9.3]). This well-known theorem gives necessary and sufficient conditions for a ring to have a semisimple classical quotient ring.[1] In 1966 L.Small generalized this theorem and described Noetherian rings which have Artinian classical rings of quotients [292]. A variant of Small's theorem without the Noetherian hypothesis was obtained by R.B.Warfield, Jr. [321]. In Section 5.5 the proofs of these results are given.

Originally Krull dimension was introduced for commutative Noetherian rings by W. Krull. There are a few different generalizations of this concept for the case of noncommutative rings. One of them, the classical Krull dimension introduced by G.Krause in [197], is considered in Section 5.6. The more important generalization, the concept of Krull dimension, is considered in Section 5.7. The module-theoretic form of this notion in the general case for noncommutative rings was introduced by R.Rentschler and P. Gabriel in 1967 [270] and later by G. Krause [197]. Note that not all modules have a Krull dimension, but each Noetherian module does. The Krull dimension of any Artinian module is equal to 0. So in some sense the Krull dimension of a module can be considered as a measure which shows how far

[1]Recall that in this book a semisimple ring means always Artinian semisimple.

the module is from being Artinian. In Section 5.8 we consider some relationships between the concepts of classical Krull dimension and Krull dimension.

5.1 Uniform Modules and Uniform Dimension

The notion of dimension is very important in the study of vector spaces (linear algebra). Uniform dimension, introduced by Alfred W. Goldie, [101], [102] (and hence also called Goldie dimension[2]), is a successful attempt to define a similar notion for modules. It is based on the idea of a uniform module. Uniform dimension generalizes some, but not all aspects of the notion of dimension of a vector space. It is also referred to as simply the dimension of a module or the rank of a module (besides 'Goldie dimension'). It should not be confused with the idea of the reduced rank of a module (also due to Goldie). Reduced rank is discussed in Section 5.5 below.

Definition 5.1.1. A non-zero module M is called **uniform** if the intersection of any two non-zero submodules of M is non-zero.

Example 5.1.2.

1. Any simple module is uniform.
2. Any uniserial module is uniform.

There are some other equivalent definitions of uniform modules. One of them which will be considered in this section is connected with the concept of essential modules. Recall that a submodule N of a module M is called **essential** in M if it has a non-zero intersection with every non-zero submodule of M (See [146, Section 5.3]). In this case M is called an **essential extension** of N, and we write $N \subseteq_e M$.

The next statement states the connection between concepts of uniform and essential modules.

Proposition 5.1.3. *A module M is uniform if and only if every non-zero submodule of M is essential in M.*

Proof. Let M be a uniform module and N a non-zero submodule of M. If X is any non-zero submodule of M then $N \cap X \neq 0$, i.e., $N \subseteq_e M$.

Conversely, if $N \subseteq_e M$ then for any non-zero submodule X of M one has $X \cap N \neq 0$, i.e., M is uniform. □

---

[2]This fits with the notion of 'right Goldie ring', See corollary 5.1.20 below.

Note that all non-zero submodules and all essential extensions of uniform modules are uniform. Any uniform module is obviously indecomposable. Over a semisimple ring the concepts of simple, uniform and indecomposable modules coincide. In the general case there is the following chain of strong inclusions:

$$\text{simple modules } \subsetneq \text{ uniform modules } \subsetneq \text{ indecomposable modules}$$

The following examples show that this chain is really strict.

Example 5.1.4.

1. Let $A = \mathbf{Z}$, then $\mathbf{Z}$, $\mathbf{Q}$ and $\mathbf{Z}/p^n\mathbf{Z}$ (where $n \geq 2$ and p is prime) are all uniform, but not simple $\mathbf{Z}$-modules.
2. Let $A = \mathbf{Q}[x, y]$ be the commutative $\mathbf{Q}$-algebra, that is defined by the relations $x^2 = y^2 = xy = 0$. Then the right regular module A_A is indecomposable, but it is not uniform, because it contains the direct sum $\mathbf{Q}x \oplus \mathbf{Q}y$ of two non-zero ideals.

Proposition 5.1.5. *If a module M is a finite direct sum of n uniform modules, then M does not contain any direct sum of $n + 1$ non-zero submodules.*

Proof. If $n = 0$, then $M = 0$. If $n = 1$, then M is a uniform module and so it can not be a direct sum of two or more modules. Therefore one can consider that $n \geq 2$. The proposition will be proved by induction on the number n. Suppose that the statement is true for $n - 1$ uniform modules.

Let $M = M_1 \oplus \ldots \oplus M_n$, where each M_i is a uniform module. Suppose M contains a direct sum $N_1 \oplus \ldots \oplus N_n \oplus N_{n+1}$ of $n+1$ non-zero submodules. Set $N = N_1 \oplus \ldots \oplus N_n$. Suppose $N \cap M_i \neq 0$ for all i. Since M_i is uniform, the submodule $N \cap M_i \subset M_i$ is essential in M_i for all i by proposition 5.1.3. So $N \cap M_i \cap X \neq 0$ for any submodule X of M and all $i = 1, 2, \ldots, n$. Therefore the module $N = N \cap M = (N \cap M_1) \oplus \ldots \oplus (N \cap M_n)$ is a non-zero and $N \cap X = (N \cap M_1 \cap X) \oplus \ldots \oplus (N \cap M_n \cap X) \neq 0$ for any submodule X of M, i.e., N essential in $M_1 \oplus \ldots \oplus M_n = M$. This is impossible because $N \cap N_{n+1} = 0$.

Therefore there exists an index i such that $N \cap M_i = 0$. Without loss of generality one can assume that $i = 1$, i.e., $N \cap M_1 = 0$. Consider the natural projection $\varphi : M \to M_2 \oplus \ldots \oplus M_n$ with $\operatorname{Ker} \varphi = M_1$. Since $N \cap \operatorname{Ker} \varphi = 0$, N embeds in $M_2 \oplus \ldots \oplus M_n$, i.e., $N \simeq \varphi(N)$. Whence $M_2 \oplus \ldots \oplus M_n$ contains a direct sum of n non-zero submodules $\varphi(N_1) \oplus \ldots \oplus \varphi(N_n)$, that contradicts the induction hypothesis.

Therefore M does not contain a direct sum of $n + 1$ submodules. □

Corollary 5.1.6. *Let $M = \bigoplus_{i=1}^{m} M_i = \bigoplus_{j=1}^{n} N_j$ be two decompositions of a module M into a direct sum of uniform modules, then $m = n$.*

Definition 5.1.7. The **uniform dimension**[3] of a module M is the largest integer n such that M contains a direct sum $M_1 \oplus M_2 \oplus \ldots \oplus M_n$ of non-zero submodules. It is denoted $\text{u.dim}\, M = n$. If no such maximum exists, then $\text{u.dim}\, M = \infty$. By assumption, $\text{u.dim}\, M = 0$ if and only if $M = 0$.

From this definition it follows that $\text{u.dim}\, M = 1$ if and only if M is a uniform module.

Remark 5.1.8. If A is a field K, and M is a finite dimensional vector space over K, then $\text{u.dim}\, M = \dim_K M$. Thus the concept of uniform dimension is some generalization of the dimension of a finitely dimension vector space over a field.

The following proposition gives some elementary properties of the notion of uniform dimension.

Proposition 5.1.9. *Let N be a submodule of M. Then*

1. $\text{u.dim}\, N \leq \text{u.dim}\, M$; *moreover, if* $N \subseteq_e M$ *then* $\text{u.dim}\, N = \text{u.dim}\, M$.
2. *If* $\text{u.dim}\, N = \text{u.dim}\, M < \infty$ *then* $N \subseteq_e M$.
3. *If the composition length of M is equal to k, then* $\text{u.dim}\, M \leq k$.
4. *If M is the direct sum of k simple modules then* $\text{u.dim}\, M = k$.

Proof.

1. The statement $\text{u.dim}\, N \leq \text{u.dim}\, M$ is obvious. Suppose $N \subseteq_e M$, $\text{u.dim}\, M = n$, and $M_1 \oplus \ldots \oplus M_n \subseteq M$, where all M_i are non-zero. Since $N \subseteq_e M$, $X_i = N \cap M_i \neq 0$ for all $i = 1, \ldots, n$. Then $X_1 \oplus \ldots \oplus X_n \subseteq N$, whence $\text{u.dim}\, N \geq n = \text{u.dim}\, M$. Taking into account that $\text{u.dim}\, N \leq \text{u.dim}\, M$ one obtains $\text{u.dim}\, N = \text{u.dim}\, M$.
2. Suppose that $\text{u.dim}\, N = \text{u.dim}\, M$, and N is not essential in M. Then there exists a non-zero submodule X of M such that $N \cap X = 0$. Then $N \oplus X \subseteq M$, whence

$$\text{u.dim}\, N + \text{u.dim}\, X \leq \text{u.dim}\,(N \oplus X) \leq \text{u.dim}\, M = \text{u.dim}\, N.$$

   Therefore $\text{u.dim}\, X = 0$, whence $X = 0$. A contradiction.
3. Suppose $\text{u.dim}\, M = n$, then there exists a direct sum of n non-zero submodules $M_1 \oplus \ldots \oplus M_n \subset M$ and so there is the following series of submodules of M

$$0 \subset M_1 \subset M_1 \oplus M_2 \subset \ldots \subset M_1 \oplus \ldots \oplus M_n \subset M,$$

   which has length n.

---

[3]Note, that another name of this dimension is the **Goldie dimension** after A.W.Goldie who first introduced this notion in [102] to prove his famous theorem. He denoted this dimension with $\dim M = n$.

4. Since any simple module is indecomposable, and M is a direct sum of k simple modules, $\mathrm{u.dim}\, M \geq k$. On the other hand M is an Artinian semisimple module with composition length equal to k. Therefore from the previous statement $\mathrm{u.dim}\, M \leq k$. So $\mathrm{u.dim}\, M = k$. □

Corollary 5.1.10. *If a module M is either Artinian or Noetherian, then* $\mathrm{u.dim}\, M < \infty$. *In particular, if A is a right Noetherian ring, then*

$$\mathrm{u.dim}\, A_A < \infty.$$

Corollary 5.1.11. *Suppose a module M has uniform dimension n. If $f : M \to M$ is a monomorphism, then* $\mathrm{Im} f$ *is an essential submodule in M.*

Proof. Since $\mathrm{Im} f$ is isomorphic to M, $\mathrm{u.dim}(\mathrm{Im} f) = \mathrm{u.dim}\, M$. Therefore the statement follows from proposition 5.1.9(2). □

Proposition 5.1.12. *A module M has uniform dimension n if and only if there exists an essential submodule $N \subseteq M$ which is a direct sum of n uniform submodules.*

Proof. Let N be an essential submodule of M and $N = N_1 \oplus \ldots \oplus N_n$, where all N_i are uniform submodules. Suppose M contains a submodule X which is a direct sum of $n+1$ non-zero submodules $X = X_1 \oplus \ldots \oplus X_n \oplus X_{n+1}$. Since $N \subseteq_e M$, $Y_i = N \cap X_i \neq 0$ for each $i = 1, \ldots, n+1$, and $Y_1 \oplus Y_2 \oplus \ldots \oplus Y_{n+1} \subset N$, which contradicts proposition 5.1.5. Therefore, $\mathrm{u.dim}\, M = n$.

Conversely, let $\mathrm{u.dim}\, M = n$, i.e., there exists n non-zero submodules $M_1, \ldots, M_n$ such that their direct sum is a submodule of M. Suppose M_1 is not uniform, then there exists two submodules X, Y of M_1 such that $X \cap Y = 0$. In this case M contains a direct sum of $n+1$ non-zero submodules $X \oplus Y \oplus M_2 \oplus \ldots \oplus M_n$. Therefore M_1 is a uniform submodule of M. Similarly all submodules M_i are uniform. If $N = M_1 \oplus \ldots \oplus M_n$ is not an essential submodule of M, then there exists a non-zero submodule $M_{n+1} \subset M$ such that $M_{n+1} \cap N = 0$. In this case M would contain the direct sum $M_1 \oplus \ldots \oplus M_n \oplus M_{n+1}$ of $n+1$ non-zero submodules. Therefore $N = M_1 \oplus \ldots \oplus M_n$ is an essential submodule of M. □

Definition 5.1.13. Following A.W. Goldie [102] a module M is called **finite-dimensional** if it contains no infinite direct sum of non-zero submodules. A ring A will be called **right finite dimensional** (in sense of Goldie) if it contains no infinite sum of non-zero right ideals of A, i.e., the right regular module A_A is finite-dimensional. Analogously one can define a **left finite-dimensional ring**.

Lemma 5.1.14. *Let M be a finite-dimensional module. Then any non-zero submodule $N \subseteq M$ contains a uniform submodule.*

Proof. If the statement is false then M is not uniform and it contains a direct sum $X_1 \oplus Y_1$ of two non-zero submodules. Then again Y_1 is not uniform and it contains a direct sum $X_2 \oplus Y_2$ of two non-zero submodules. Continuing this process we obtain that M contains an infinite direct sum $X_1 \oplus X_2 \oplus X_3 \oplus \ldots$ of non-zero submodules. A contradiction. □

Lemma 5.1.15. *Let M be a finite-dimensional module. Then for any submodule $N \subseteq M$ there exist a finite number (possibly zero) of uniform submodules $M_1, M_2, \ldots, M_n$ such that the direct sum $N \oplus M_1 \oplus M_2 \oplus \ldots \oplus M_n$ is an essential submodule in M.*

Proof. Suppose that statement is not true. Therefore $N \subseteq M$ is not an essential submodule in M. Then there is a non-zero submodule $X_1 \subset M$ such that $N \cap X_1 = 0$. By lemma 5.1.14 X_1 contains a uniform submodule $Y_1 \subseteq X_1$. Since $N \cap Y_1 = 0$, $N_1 = N \oplus Y_1$ is a direct sum in M. By assumption N_1 is not essential in M. Therefore there is a non-zero submodule $X_2 \subset M$ such that $N_1 \cap X_2 = 0$. By lemma 5.1.14 X_2 contains a uniform submodule $Y_2 \subseteq X_2$. Since $N_1 \cap Y_2 = 0$, $N_2 = N_1 \oplus Y_2 = N \oplus Y_1 \oplus Y_2$ is a direct sum in M. Continuing this process we obtain an infinite direct sum $N \oplus \bigoplus_{i=1}^{\infty} Y_i$ of non-zero submodules. A contradiction. □

Proposition 5.1.16. *Let M be an A-module, then* u.dim $M < \infty$ *if and only if M is a finite-dimensional module.*

Proof. Obviously u.dim $M < \infty$ implies that M is a finite-dimensional module.

Conversely, assume that M is a finite-dimensional module. Then by lemma 5.1.15 there exist uniform submodules $M_1, M_2, \ldots, M_n$ such that their direct sum $M_1 \oplus M_2 \oplus \ldots \oplus M_n$ is an essential submodule in M. Then by proposition 5.1.12 u.dim $M = n$. □

Corollary 5.1.17. u.dim $M = \infty$ *if and only if M contains an infinite direct sum of non-zero submodules.*

Corollary 5.1.18. u.dim $M = \infty$ *if and only if M contains an essential submodule which is a direct sum of an infinite number of uniform submodules.*

Remark 5.1.19. Proposition 5.1.12 and corollary 5.1.18 give equivalent definitions of the uniform dimension of a module. One can say that a module M has the **uniform dimension** n if there is an essential submodule $N \subseteq M$ which is a direct sum of n uniform submodules. If no such integer n exists one says that the uniform dimension of M is equal to ∞.

Recall, See [146, Section 9.3], that A is a **right Goldie ring** if A satisfies the ascending chain condition on right annihilators and A contains no infinite direct sum of non-zero right ideals, i.e., the right regular module A_A is finite-dimensional.

From this definition and proposition 5.1.16 there immediately follows:

Corollary 5.1.20. *If A is a right Goldie ring, then* u.dim$(A_A) < \infty$.

Proposition 5.1.21. *Let M and N be A-modules, then*

$$\text{u.dim}(M \oplus N) = \text{u.dim}\, M + \text{u.dim}\, N. \tag{5.1.22}$$

In particular, if $\text{u.dim}\, M < \infty$ *and* $\text{u.dim}\, N < \infty$ *then* $\text{u.dim}\,(M \oplus N) < \infty$ *as well.*

Proof. If either $\text{u.dim}\, M$ or $\text{u.dim}\, N$ is infinite, then so is $\text{u.dim}\,(M \oplus N)$. Thus one can assume that $\text{u.dim}\, M$ and $\text{u.dim}\, N$ are both finite. By corollary 5.1.17, M and N contain no infinite direct sums of non-zero submodules. Then, by lemma 5.1.15, M and N contain uniform submodules U_i $(i = 1,2,\dots,m)$ and V_j $(j = 1,2,\dots,n)$ such that $U = U_1 \oplus \dots \oplus U_m$ is essential in M and $V = V_1 \oplus \dots \oplus V_n$ is essential in N. Moreover, by proposition 5.1.12, $\text{u.dim}\, M = m$ and $\text{u.dim}\, N = n$. Since $U \oplus V$ is essential in $M \oplus N$, and $U \oplus V$ is a direct sum of $m + n$ uniform submodules, from proposition 5.1.12 it follows that

$$\text{u.dim}\,(M \oplus N) = m + n = \text{u.dim}\, M + \text{u.dim}\, N.$$

□

Corollary 5.1.23. *Let M_i be A-modules for $i = 1,\dots,n$. Then*

$$\text{u.dim}(\bigoplus_{i=1}^{n} M_i) = \sum_{i=1}^{n} \text{u.dim}\, M_i. \tag{5.1.24}$$

In particular, if all $\text{u.dim}\, M_i < \infty$ *then* $\text{u.dim}(\bigoplus_{i=1}^{n} M_i) < \infty$ *as well.*

Proposition 5.1.25. *Suppose that A is a ring,* $\text{u.dim} A_A = m$*, and $S = M_n(A)$ is the ring of $n \times n$-matrices with entries from A. Then* $\text{u.dim} S_S = mn$.

Proof. Let $K = A^n$ and $S = M_n(A)$. Considering elements of K as row vectors, one can make K into a right S-module by means of the matrix multiplication. Let e_{ij} be the matrix in $M_n(A)$ with all entries other than the (i,j) entry being 0 and the (i,j) entry being 1. Since $S_S = e_{11}S \oplus e_{22}S \oplus \cdots \oplus e_{nn}S$, and $e_{ii}S \simeq K_S$ as S-modules, it follows from corollary 5.1.23 that $\text{u.dim} S_S = n(\text{u.dim} K_S)$.

Show that $\text{u.dim} K_S = \text{u.dim} A_A = m$. Let $\mathcal{I}$ be a right ideal in A. Define

$$X(\mathcal{I}) = \{(x_1, x_2, \dots, x_n) \in K \,:\, x_i \in \mathcal{I}\}$$

which is an S-submodule in K. Conversely, for any S-submodule $L \subseteq K$ define

$$Y(L) = \{y \in A \,:\, (0,\dots,y,0,\dots,0) \in L\}$$

which is a right ideal in A. It is easy to verify that X and Y are inverse one-to-one correspondences between right ideals in A and S-submodules in K. Indeed, for any right ideal $\mathcal{I}$ of A we have $YX(\mathcal{I}) = \mathcal{I}$. On the other hand, for any right S-module L we have $XY(L) = L$.

Moreover, both X and Y preserve direct sums. It follows that u.dimK_S = u.dimA_A = m. Since S_S is the direct sum of n copies of K_S, there results u.dim$S_S = mn$. □

If V is a finite-dimensional vector space over a field K, then any subspace $U \subset V$ is also finite-dimensional and there is a complementary subspace $W \subset V$ such that $V = U \oplus W$. This complement is a direct summand and has the important property that $V \cap W = 0$. In the case of modules over rings this is not true in general. As a substitute A.W. Goldie in [102] introduced the notion of a complementary submodule as follows.

Definition 5.1.26. Let N be a submodule of a module M. A submodule C of M is said to be a **complement** to N in M if C is maximal with respect to the property that $N \cap C = 0$. A submodule C of M is called a **complementary submodule** (or simply, **complement**) in M if there exists a submodule N of M such that C is a complement to N in M.

Note that for any submodule N of M a complement C to N is not necessarily unique. The existence of a complement to any submodule N of M follows from Zorn's lemma. To this end it suffices to consider the set of all submodules K of M with the property $K \cap N = 0$. This set is not empty and by Zorn's lemma it has a maximal element C.

An important relation between complements and essential modules is embodied in the following proposition.

Proposition 5.1.27. *Let C be a complementary submodule in M, and let N be a submodule of M such that $N \cap C = 0$. Then C is a complement to N if and only if $N \oplus C$ is an essential submodule in M.*

Proof. Assume that C is a complementary submodule in M which is a complement to a submodule N in M. Consider an arbitrary submodule K of M. If $K \subseteq C$, then obviously $K \cap (N \oplus C) \neq 0$. Since C is maximal with respect to the property that $C \cap N = 0$, $(K + C) \cap N \neq 0$. Let $0 \neq n \in (K + C) \cap N$, then $n = k + c$ for some $k \in K$ and $c \in C$. Hence $k = n - c \in N + C$. Since $N \cap C = 0$, this implies that $k \neq 0$, i.e., $K \cap (N + C) \neq 0$. So $N \oplus C \subseteq_e M$.

Conversely, assume that C is a complementary submodule in M, and N be a submodule of M such that $N \cap C = 0$. Suppose that $N \oplus C \subseteq_e M$. We will show that C is a complement to N in M. To this end we now show that C is maximal with respect to the property that $C \cap N = 0$. Since C is a complementary submodule in M, by definition 5.1.26 there exists a submodule K of M such that C is a complement to K in M. Assume that there is a submodule $D \subset M$ such that $C \subseteq D$ and $D \cap N = 0$. Then

$$(N \oplus C) \cap (D \cap K) = ((N \oplus C) \cap D) \cap K = C \cap K = 0.$$

Since $N \oplus C \subseteq_e M$, $D \cap K = 0$, whence $D = C$ because C is maximal with respect to the property that $K \cap C = 0$. This shows that C is a complement to N in M. □

Since for any submodule N of M its complement C exists, proposition 5.1.27 immediately implies the following:

Corollary 5.1.28. *Any submodule N in M is a direct summand of an essential submodule $N \oplus C$ in M, where C is a complement to N in M.*

Corollary 5.1.29. *Let C be a complement in M. If N is a complement to C in M (which always exists), then C is a complement to N in M.*

Proof. The proof follows immediately from the symmetry of proposition 5.1.27. □

Proposition 5.1.30. *Suppose that K, N are submodules in an A-module M and $K \subseteq N$.*

1. *If K is a complement in M then K is a complement in N.*
2. *If K is a complement in N and N is a complement in M, then K is a complement in M.*

Proof.

1. If K is a complement in M then, by definition, there exists a submodule K_1 such that K is a complement to K_1 in M, i.e., K_1 is a maximal submodule in M with respect to the property $K_1 \cap K = 0$. Obviously, $K \cap (K_1 \cap N) = 0$ and $K_1 \cap N \subseteq N$. We now show that K is a complement to $K_1 \cap N$ in N. Otherwise there exists a submodule $D \subseteq N$ such that $K_1 \cap N \subseteq D$ and $D \cap K = 0$. In this case $K_1 \subseteq D + K_1 \subseteq M$ and $(D + K_1) \cap K = (D \cap K) + (K_1 \cap K) = 0$. Since K_1 is maximal with respect to the property $K_1 \cap K = 0$, $D \subseteq K_1$, that is $D = K_1 \cap N$.
2. Suppose that K is a complement to a submodule K_1 in N and N is a complement to a submodule N_1 in M. We shall show that K is a complement to the submodule $K_1 \oplus N_1$ in M. Note that $K_1 \cap N_1 = 0$ because $K_1 \subset N$ and $N \cap N_1 = 0$. Obviously, $(K_1 \oplus N_1) \cap K = (K_1 \cap K) \oplus (N_1 \cap K) = 0$.

So it remains to show that $D \cap (K_1 \oplus N_1) \neq 0$ for any submodule $D \subset M$ such that $K \subsetneq D$.

Consider the chain of submodules

$$K \subseteq N \cap D \subseteq D \subseteq M.$$

Suppose $D_1 = N \cap D \neq K$. Then $K_1 \cap D_1 \neq 0$ since K is a complement to K_1 in N and $K \subsetneq D_1 \subset N$. Therefore

$$(K_1 \oplus N_1) \cap D_1 = (K_1 \cap D_1) \oplus (N_1 \cap D_1) \neq 0$$

and so $(K_1 \oplus N_1) \cap D \neq 0$ as well.

So one can assume that $K = N \cap D$ and there is a chain of submodules:

$$K = N \cap D \subsetneq D \subset M.$$

Therefore there exists a non-zero element $d \in D\backslash N$, because $D \neq N \cap D = K$. Since $N + dA \supsetneq N$ and N is a complement to N_1 in M, $N_1 \cap (N + dA) \neq 0$. This means that there exists $0 \neq n_1 \in N_1$ such that

$$n_1 = n + da \tag{5.1.31}$$

for some $n \in N$ and $a \in A$.

i. If $n \in K$ then $0 \neq n + da \in D$ and so $(K_1 \oplus N_1) \cap D \neq 0$.
ii. Suppose that $n \notin K$. Consider the submodule $K + nA \supsetneq K$. Since K is a complement to K_1 in N, $K_1 \cap (K + nA) \neq 0$, which implies that there exists $0 \neq k_1 \in K_1$ such that

$$k_1 = k + nb \tag{5.1.32}$$

for some $k \in K$ and $b \in A$. Then from (5.1.31) and (5.1.32) one has:

$$k_1 = k + (n_1 - da)b = k + n_1 b - dab,$$

or

$$k - dab = k_1 - n_1 b \in (K_1 \oplus N_1) \cap D$$

and $k_1 - n_1 b \neq 0$, since $k_1 \neq 0$, $n_1 \neq 0$ and $K_1 \cap N_1 = 0$. Thus, $(K_1 \oplus N_1) \cap D \neq 0$, as required. □

Proposition 5.1.33. *If* u.dim$M = n < \infty$, *then* $n \geq s$, *where* s *is the length of any strictly ascending chain of complements in* M.

Proof. Let u.dim$M = n < \infty$. Consider a strictly ascending chain of submodules of M:

$$K_0 \subsetneq K_1 \subsetneq K_2 \subsetneq \cdots \subsetneq K_s$$

where each K_i is a complement in M for $i = 0, 1, \dots, s$. Then K_i is a complement in K_{i+1} for $i = 0, 1, \dots, s-1$, by proposition 5.1.30. So there is a submodule $N_i \subseteq K_{i+1}$ such that K_i is a complement to N_i in K_{i+1}. Since $K_i \subsetneq K_{i+1}$, $N_i \neq 0$. Since $N_i \subseteq K_{i+1}$, $K_i \cap N_i = 0$ and $K_j \subset K_i$ for each $j < i$, there results that $N_i \cap N_s = 0$ for all $i \neq s$. So the direct sum $N_1 \oplus N_2 \oplus \cdots \oplus N_s$ is a submodule in M, which means that $n =$ u.dim$M \geq s$. □

Proposition 5.1.34 (A.W. Goldie). *The following conditions are equivalent for a module* M:

1. u.dim$M < \infty$;
2. *M satisfies a.c.c. on complements in M;*
3. *M satisfies d.c.c. on complements in M.*

Proof.
1 ⇒ 2 follows from proposition 5.1.33.
3 ⇒ 1. Suppose M contains an infinite direct sum of non-zero submodules $\bigoplus_{i=1}^{\infty} N_i \subseteq M$. Denote $X_n = \bigoplus_{i=n}^{\infty} N_i \subset M$ for each $n = 1, 2, \dots$. Let Y_n be a complement

submodule to X_n in M. Since $X_n \cap Y_n = 0$ and $X_n \subsetneq X_{n+1}$, $Y_{n+1} \subsetneq Y_n$ for each n. So there exists a strictly descending chain of non-zero submodules $M \supseteq Y_1 \supsetneq Y_2 \supsetneq \dots$ where each Y_n is a complement in M. This contradiction shows that u.dim$M < \infty$.

$2 \Rightarrow 3$. Suppose there exists an infinite strictly descending chain of submodules of M: $C_1 \supsetneq C_2 \supsetneq \cdots$, where each C_i is a complement in M. Let Y_n be a complement submodule to C_n in M. Then one obtains a strictly ascending chain of non-zero submodules $Y_1 \subsetneq Y_2 \subsetneq \dots \subset M$, where each Y_n is a complement in M. □

Recall that B is an **essential submodule** in M if it has a non-zero intersection with every non-zero submodule of M. In this case M is called an **essential extension** of B. We need the following simple lemma.

Lemma 5.1.35. *If $N \subseteq_e M$ and $K \subseteq M$ then $N \cap K \subseteq_e K$.* Since

Proof. Let P be a submodule of K. Then $K \cap P = P \subseteq K \subseteq M$. Since $N \subseteq_e M$, $N \cap P \neq 0$. Therefore $(N \cap K) \cap P = N \cap (K \cap P) = N \cap P \neq 0$, i.e., $N \cap K \subseteq_e K$. □

Definition 5.1.36. A submodule C is called **essentially closed** (or shortly, **closed**) in M if C has no proper essential extensions inside M. In other words, if $C \subseteq N \subset M$ and $C \subseteq_e N$ then $N = C$.

Proposition 5.1.37. *A submodule C of M is a complement in M if and only if C is closed in M.*

Proof.

1. Let C be a complement in M. Then there is a submodule N of M such that C is a complement to N. Assume that $C \subseteq K \subseteq M$ and K is an essential extension of C. Then $C \cap (K \cap N) = 0$ says that $K \cap N = 0$ as $C \cap K = K$. Since C is a complement to N, $C = K$, i.e., C is closed in M.
2. Conversely, assume that C is closed in M. Let N be a complement to C (which always exists). We now show that C is a complement to N in M. Let $K \subseteq M$ be maximal with respect to $C \subseteq K$ such that $K \cap N = 0$. Since N is a complement to C, $C \oplus N \subseteq_e M$, by proposition 5.1.27. Then $(C \oplus N) \cap K \subseteq_e K$, by lemma 5.1.35. Since $(C \oplus N) \cap K = (C \cap K) \oplus (N \cap K) = C$, we obtain that $C \subseteq_e K \subseteq M$. As C is closed in M, this means that $C = K$, and so C is a complement to N in M. □

Summarizing all results in this section there results the following main theorem which shows the equivalence of various different kinds of finiteness conditions for a module.

Theorem 5.1.38. *The following conditions are equivalent for a module M:*

1. *M is finite-dimensional.*
2. u.dim$M < \infty$.
3. *M satisfies a.c.c. on complements.*

4. *M satisfies d.c.c. on complements.*
5. *M satisfies a.c.c. on closed submodules.*
6. *M satisfies d.c.c. on closed submodules.*
7. *M contains no infinite direct sum of closed submodules.*

Corollary 5.1.39. *Any right Noetherian module is finite-dimensional. In particular, if A is a right Noetherian ring then the right regular module A_A is finite-dimensional, i.e., A is a finite-dimensional ring (in sense of definition 5.1.13).*

Proof. Suppose that a Noetherian module M is not finite-dimensional, then, by definition 5.1.17, it contains an infinite direct sum of non-zero submodules. So M contains an infinite sequence $M_1, M_2, \dots$ of different non-zero submodules. Then there is a strictly ascending infinite chain

$$M_1 \subset M_1 \oplus M_2 \subset M_1 \oplus M_2 \oplus M_3 \subset \dots$$

of submodules of M. A contradiction. □

Below we give some simple examples of right and left finite-dimensional rings.

Examples 5.1.40.

1. If D is a division ring then $\text{u.dim}D_D = \text{u.dim}_D D = 1$.
2. If A is a simple Artinian ring then A $\text{u.dim}A_A = \text{u.dim}_A A = 1$.
3. If A is a semisimple ring and $A = M_{n_1}(D_1)\oplus\cdots\oplus M_{n_k}(D_k)$, then $\text{u.dim}A_A = \text{u.dim}_A A = n_1 + \cdots + n_k$.
4. If A is a uniserial ring then $\text{u.dim}A_A = \text{u.dim}_A A = 1$.
5. If A is a serial ring and $A_A = P_1^{n_1} \oplus \cdots \oplus P_1^{n_1}$ is a decomposition into a direct sum of principal A-modules, then $\text{u.dim}A_A = n_1 + \cdots + n_k$.

5.2 Injective Uniform Modules

This section considers injective uniform modules and their connection with finite-dimensional modules.

Recall that a module $E(M)$ is called an **injective hull** of a module M if it is both an essential extension of M and an injective module.

Proposition 5.2.1. *A non-zero module M is uniform if and only if an injective hull $E(M)$ is indecomposable.*

Proof. Suppose that M is a non-zero uniform module and $E(M) = X \oplus Y$ for some submodules X and Y. Since M is uniform and $(X \cap M) \oplus (Y \cap M) = M$, either $X \cap M = 0$ or $Y \cap M = 0$. Whence either $X = 0$ or $Y = 0$ because M is essential in $E(M)$. Thus $E(M)$ is indecomposable.

Conversely, if M is not uniform it has non-zero submodules X and Y such that $X \cap Y = 0$. Then $E(M)$ has a non-zero submodule E which is an injective hull for

X, and $E \cap Y = 0$, because X is essential in E. Whence $E \neq E(M)$. Therefore E is a nontrivial direct summand of $E(M)$, so that $E(M)$ is not indecomposable. □

Theorem 5.2.2. *Let A be a ring, and M an injective right A-module. Then the following conditions are equivalent*:

1. *M is indecomposable.*
2. *M is uniform.*
3. *The endomorphism ring* $\mathrm{End}_A(M)$ *of M is local.*

Proof.

$1 \Longleftrightarrow 2$. This follows immediately from proposition 5.2.1.

$3 \Longrightarrow 1$. Since $\mathrm{End}_A(M)$ is a local ring, it does not contain nontrivial idempotents and so M is indecomposable.

$1 \Longrightarrow 3$. Show that all non-invertible elements of $\mathrm{End}_A(M)$ form its proper ideal. Let M be an indecomposable injective module and $\varphi \in \mathrm{End}_A(M)$. Suppose $\varphi \neq 0$ is a monomorphism, that is, $\mathrm{Ker}\,\varphi = 0$. Then $\mathrm{Im}\,\varphi \subseteq M$. Since $M \cong \mathrm{Im}\,\varphi$ and M is injective and indecomposable, $\mathrm{Im}\,\varphi = M$. So φ is an isomorphism.

Assume that $\varphi_1 \neq 0$ and $\varphi_2 \neq 0$ are not invertible in $\mathrm{End}_A(M)$, then by what was proved above $\mathrm{Ker}\,\varphi_1 \neq 0$ and $\mathrm{Ker}\,\varphi_2 \neq 0$. Since M is a uniform module, $\mathrm{Ker}\,\varphi_1 \cap \mathrm{Ker}\,\varphi_2 \neq 0$. Therefore $\mathrm{Ker}\,(\varphi_1 + \varphi_2) \neq 0$ as well, since $\mathrm{Ker}\,\varphi_1 \cap \mathrm{Ker}\,\varphi_2 \subseteq \mathrm{Ker}\,(\varphi_1 + \varphi_2)$, i.e., $\varphi_1 + \varphi_2$ is not invertible in $\mathrm{End}_A(M)$. Thus, $\mathrm{End}_A(M)$ is a local ring, by proposition 1.9.1. □

Definition 5.2.3. An A-module M is called **strongly indecomposable** if the endomorphism ring $\mathrm{End}_A(M)$ is local.

Obviously, each strongly indecomposable module is indecomposable. Any simple module M over a ring A is strongly indecomposable, since $\mathrm{End}_A(M)$ is a division ring and so it is a local ring. The field of rational numbers $\mathbf{Q}$ considering as a $\mathbf{Z}$ module is also strongly indecomposable.

Remark 5.2.4. From theorem 5.2.2 it follows that for injective modules the notions of indecomposability and strong indecomposability are the same.

Definition 5.2.5. A module M has **finite rank** if the injective hull $E(M)$ is a finite direct sum of indecomposable modules. Otherwise M has **infinite rank**.

The following theorem shows that in the finite case this notion coincides with the notion of the uniform dimension.

Theorem 5.2.6. *A module M has finite rank if and only if M has finite uniform dimension.*

Proof. Suppose M has finite uniform dimension. Then M has an essential submodule N which is a finite direct sum of uniform submodules $A_1, A_2, \ldots, A_n$. Since $N \subseteq_e M$ and $M \subseteq_e E(M)$, $N \subseteq_e E(M)$, by [146, Lemma 5.3.2]. Therefore from the definition of the injective hill it follows that $E(M) = E(N) = E(A_1 \oplus \ldots \oplus A_n) \simeq E(A_1) \oplus \ldots \oplus E(A_n)$, where the $E(A_i)$ are indecomposable, by proposition 5.2.1. Thus, M has finite rank.

Conversely, suppose that M has finite rank, that is, $E(M) = E_1 \oplus \ldots \oplus E_n$, where all E_i are non-zero and indecomposable modules. By theorem 5.2.2, all E_i are uniform modules. Each of submodules $M_i = M \cap E_i$ is non-zero because $M \subseteq_e E(M)$, and each M_i is uniform as a submodule of the uniform module E_i. Since E_i is uniform, each M_i is essential in E_i. Therefore $M_1 \oplus \ldots \oplus M_n \subseteq_e E_1 \oplus \ldots \oplus E_n = E(M)$, and so $M_1 \oplus \ldots \oplus M_n \subseteq_e M$. □

The following theorem gives the equivalent definition of the finite rank of a module.

Theorem 5.2.7 (A.W. Goldie). *A module M has finite rank if and only if M is finite-dimensional.*

Proof. Although this theorem immediately follows from theorem 5.2.6 and proposition 5.1.16, we give here the another proof which is due to A.W. Goldie [101].

1. Suppose that M has finite rank, then $E(M)$ is a direct sum of n uniform submodules for some integer n. Then, by proposition 5.1.5, $E(M)$ cannot contain a direct sum of more than n non-zero submodules, and hence neither can M.
2. Suppose $M \neq 0$ is a finite-dimensional module and $E(M)$ is not a finite direct sum of indecomposable submodules. Write $C_0 = E(M)$. Since C_0 is not indecomposable $C_0 = B_1 \oplus C_1$ for some non-zero submodules B_1 and C_1; moreover, B_1 and C_1 cannot both be finite direct sums of indecomposable submodules. Suppose C_1 is not a finite direct sum of indecomposable submodules. Repeating this argument with respect to C_1 yields $C_1 = B_2 \oplus C_2$ and C_2 is not a finite direct sum of indecomposable submodules. Continuing this process inductively we obtain a sequence of non-zero submodules $B_1, C_1, B_2, C_2, \ldots, B_n, C_n, \ldots$ such that each $C_{n-1} = B_n \oplus C_n$ with B_n non-zero and C_n not a finite direct sum of indecomposable submodules. Since $B_k \subseteq C_n$ whenever $k > n$, it follows that
$$B_n \bigcap \Big(\bigoplus_{k=n+1}^{\infty} B_k \Big) \subseteq B_n \cap C_n = 0 \qquad (5.2.8)$$
for all n, whence $B_1, B_2, \ldots$ is a sequence of disjoint submodules of $E(M)$. Since $B_1 \cap M$, $B_2 \cap M$, $\ldots$ is an infinite sequence of submodules of M and $M \subseteq_e E(M)$, $B_n \cap M \neq 0$ for each $n \geq 1$. Therefore M contains an infinite direct sum of non-zero modules. A contradiction. □

Corollary 5.2.9. *If A is a right Goldie ring then the right regular module A_A has finite rank.*

Proof. This statement immediately follows from theorem 5.2.7 and corollary 5.1.20. □

Corollary 5.2.10. *Let M be a finitely generated right module over a right Noetherian ring. Then M has finite rank, i.e., the injective hull $E(M)$ is a finite direct sum of indecomposable modules.*

Proof. Since by proposition 1.1.11 M is a Noetherian module, the statement follows immediately from corollary 5.1.39 and theorem 5.2.7. □

Proposition 5.2.11. *Every non-zero Noetherian module has a uniform submodule.*

Proof. By corollary 5.2.10 a Noetherian module contains no infinite direct sums of non-zero submodules. Therefore the statement follows immediately from lemma 5.1.14. □

Suppose that a module M has finite rank. Then $E(M)$ is a direct sum of n uniform submodules. From theorem 5.2.6 and proposition 5.1.5 it follows that the number n of summands appearing in a decomposition of $E(M)$ is an invariant of M. Indeed, as follows from theorems 5.2.6 and 5.2.7, this notion is the same as the uniform dimension of a module M if this dimension is finite. One can write a corollary of these theorems in the following form.

Proposition 5.2.12. *Let M be an A-module. Then the following conditions are equivalent:*

1. *The injective hull $E(M)$ is a finite direct sum of n indecomposable modules.*
2. *M has an essential submodule which is a direct sum of n uniform submodules.*
3. u.dim $M = n$, *i.e., M contains a direct sum of n non-zero submodules, but no direct sums of $n + 1$ non-zero submodules.*

Definition 5.2.13. A module M over a ring A is called **essentially finitely generated** if there exists a finitely generated submodule $N \subseteq M$ which is essential in M.

Lemma 5.2.14. *An A-module M has finite uniform dimension if and only if every submodule of M is essentially finitely generated.*

Proof. Assume that u.dim $M = n$. Let N be an arbitrary submodule of M. Then, by proposition 5.1.9, u.dim $N = k \leq n$. Therefore, by lemma 5.1.15, there exist uniform submodules $U_1, \ldots, U_k \subseteq N$ such that their direct sum $U = U_1 \oplus \ldots \oplus U_k$ is an essential submodule in N. Fixing non-zero elements $x_i \in U_i$, $i = 1, \ldots, k$, it follows that $X = \sum_{i=1}^{k} x_i A$ is an essential submodule in U. Since the property of being an essential submodule is transitive by [146, Lemma 5.3.2], it follows that $X \subseteq_e N$. Since X is obviously a finitely generated submodule in N, the module N is essentially finitely generated.

Conversely, assume that u.dim $M = \infty$. Then M contains an infinite direct sum of non-zero submodules $N = \sum_{i=1}^{\infty} M_i$. If X is a finitely generated submodule of N then it is contained in some finite direct sum $N' = M_1 \oplus \ldots \oplus M_k$. Then $X \cap M_{k+1} = 0$, that is, X is not essential in N. Therefore N is not essentially finitely generated. □

5.3 Nonsingular Modules and Rings

This section presents the notions of singular and nonsingular modules, which were introduced by R.E. Johnson in [175]. Some main properties of these modules are studied in this section. These modules are also closely connected with essential submodules.

The following proposition gives some useful properties of essential submodules.

Proposition 5.3.1.

1. *Let N_1, N_2, K_1, K_2 be submodules of a module M. If N_i is essential in K_i for $i = 1,2$, then $N_1 \cap N_2$ is essential in $K_1 \cap K_2$. In particular, if N_1, N_2 are both essential in M then $N_1 \cap N_2$ is essential in M.*
2. *Let X be an essential submodule of M and $f : N \to M$ a homomorphism of modules. If $N_1 = f^{-1}(X) = \{n \in N : f(n) \in X\}$, then N_1 is an essential submodule in N.*
3. *Let $\{N_i : i \in I\}$ and $\{K_i : i \in I\}$ be disjoint families of submodules of a module M. If each N_i is essential in K_i, then $\bigoplus_{i \in I} N_i$ is essential in $\bigoplus_{i \in I} K_i$.*
4. *If N is an essential submodule of an A-module of M then*

$$N(m) = \{x \in A : mx \in N\}$$

is an essential right ideal of A for any $m \in M$. In particular, if $N = \mathcal{I}$ is an essential right ideal of A then $\mathcal{I}(m)$ is an essential right ideal of A.

Proof.

1. Let X be an arbitrary submodule of $K_1 \cap K_2$. Since N_1 is an essential submodule in K_1, and X is a submodule of K_1, $X \cap N_1 \neq 0$. Since $X \cap N_1 \subset K_2$ and N_2 is essential in K_2, $X \cap N_1 \cap N_2 \neq 0$. Thus $N_1 \cap N_2$ is essential in $K_1 \cap K_2$.
2. It is easy to show that $N_1 = f^{-1}(X)$ is a submodule in N. Suppose Y is a non-zero submodule of N. If $f(Y) = 0$, then $Y \subset N_1$, whence $Y \cap N_1 \neq 0$, i.e., N_1 is essential in N. If $f(Y) \neq 0$, then $f(Y) \cap X \neq 0$, since X is essential in M. Hence $N_1 \cap X \neq 0$. So N_1 is essential in N.
3. The statement is proved by induction on the number of elements of the index set I. If I contains one index there is nothing to prove. Let $I = \{1,2\}$. Since $N_1 \cap N_2 = 0$, it follows from statement 1 that 0 is essential in $K_1 \cap K_2$, whence $K_1 \cap K_2 = 0$. Therefore there is a direct sum $K_1 \oplus K_2$. Consider the canonical projections $\varphi_i : K_1 \oplus K_2 \to K_i$, $i = 1,2$. Applying statement 2

one obtains that $N_1 \oplus K_2$ and $K_1 \oplus N_2$ are essential in $K_1 \oplus K_2$. Therefore $N_1 \oplus N_2 = (N_1 \oplus K_2) \cap (K_1 \oplus N_2)$ is essential in $K_1 \oplus K_2$, by condition 1. So the statement is true if $I = \{1,2\}$. By induction hypothesis applying the analogous arguments one can prove that the statement is true if I is a finite set.

Suppose that I has an infinite number of elements. Since any finite subset of submodules K_i form a direct sum, for any submodule $M \subset \bigoplus_{i \in I} K_i$ there is a finite subset $J \subset I$ such that $M \cap \bigoplus_{j \in J} K_j \neq 0$. Since, by proving above, $\bigoplus_{j \in J} K_j$ is an essential submodule in $\bigoplus_{j \in J} K_j$, one obtains $M \cap \bigoplus_{j \in J} N_j \neq 0$. Therefore $\bigoplus_{i \in I} N_i$ is essential in $\bigoplus_{i \in I} K_i$.

4. Let $N \subseteq_e M$ and $N(m) = \{x \in A \,:\, mx \in N\}$ for some $m \in M$. Obviously, $N(m)$ is a right ideal in A. We now show that $N(m)$ is essential in A. Let $\mathcal{I}$ be a non-zero right ideal in A. If $m\mathcal{I} = 0$ then $\mathcal{I} \subseteq N(m)$, so $N(m) \cap \mathcal{I} \neq 0$. If $m\mathcal{I} \neq 0$ then $N \cap m\mathcal{I} \neq 0$, since $N \subseteq_e M$, and so $N(m) \cap \mathcal{I} \neq 0$ as well. Therefore $N(m) \cap \mathcal{I} \neq 0$ for any non-zero right ideal $\mathcal{I}$ in A, which means that $N(m) \subseteq_e A$. □

Here are the main notions which will be needed to introduce singular modules. Let M be a right module over a ring A, and $m \in M$. Then

$$\text{r.ann}_A(m) = \{x \in A \,:\, mx = 0\}$$

is called the **right annihilator** of the element m. Obviously, $\text{r.ann}_A(m)$ is a right ideal in A. If $\text{r.ann}_A(m) \neq 0$ the element m is called **torsion**, otherwise m is called **torsion-free**. If A is a commutative ring and all elements of a module M are torsion, M is called a **torsion module**. If all elements of M are torsion-free, M is called a **torsion-free module**.

Definition 5.3.2. Let M be a right module over a ring (not necessary commutative) A. An element $m \in M$ is called **singular** if the right ideal $\text{r.ann}_A(m)$ is essential in A_A. The set of all singular elements of M is denoted by $\mathcal{Z}(M)$.

We can also consider the set

$$T = \{m \in M \,:\, m\mathcal{I} = 0 \text{ for some right ideal } \mathcal{I} \subseteq_e A\}.$$

It is easy to show that $\mathcal{Z}(M) = T$. Indeed, let $m \in \mathcal{Z}(M)$ and $\mathcal{I} = \text{r.ann}_A(m)$. Then $\mathcal{I} \subseteq_e A$ and $m\mathcal{I} = 0$, which implies that $m \in T$. So $\mathcal{Z}(M) \subseteq T$. On the other hand, let $m \in T$. Then there is a right ideal $\mathcal{I} \subseteq_e A$ such that $m\mathcal{I} = 0$, so that $\mathcal{I} \subseteq \text{r.ann}_A(m) \subseteq A$. From [146, Lemma 5.3.2] it follows that $\text{r.ann}_A(m) \subseteq_e A$, i.e., $m \in \mathcal{Z}(M)$. Therefore

$$\mathcal{Z}(M) = \{m \in M \,:\, \text{r.ann}_A(m) \subseteq_e A\} =$$

$$= \{m \in M \,:\, \exists\, \mathcal{I} \subseteq_e A \text{ such that } m\mathcal{I} = 0\}.$$

Note that from this equality it follows immediately that

$$\mathcal{Z}(\mathcal{Z}(M)) = \mathcal{Z}(M).$$

Lemma 5.3.3. *$\mathcal{Z}(M)$ is a submodule of M.*

Proof. Let $m_1, m_2 \in \mathcal{Z}(M)$, then $\text{r.ann}_A(m_1)$ and $\text{r.ann}_A(m_2)$ are essential in A_A. Then by proposition 5.3.1(1), $\text{r.ann}_A(m_1) \bigcap \text{r.ann}_A(m_2)$ is also essential in A_A. Since $\text{r.ann}_A(m_1) \bigcap \text{r.ann}_A(m_2) \subseteq \text{r.ann}_A(m_1+m_2)$, $\text{r.ann}_A(m_1+m_2) \subseteq_e A_A$ by [146, Lemma 5.3.2]. So $m_1 + m_2 \in \mathcal{Z}(M)$.

Let $m \in \mathcal{Z}(M)$, then there is an essential right ideal I in A such that $mI = 0$. For any $a \in A$ consider a right ideal $I(a) = \{x \in A : ax \in I\}$. Then $maI(a) \subseteq mI = 0$, which means that $ma \in \mathcal{Z}(M)$, since $I(a) \subseteq_e A$, by proposition 5.3.1(4).

Thus, $\mathcal{Z}(M)$ is a submodule of M. □

Corollary 5.3.4.

1. *$\mathcal{Z}(A_A)$ is a two-sided ideal in A.*
2. *$\mathcal{Z}(A_A) \neq A$.*

Proof.

1. This follows from lemma 5.3.3, taking into account that $\text{r.ann}_A(m) \subseteq \text{r.ann}_A(am)$ for any $a \in A$ and $m \in \mathcal{Z}(A_A)$.
2. Since $A \neq 0$, $\text{r.ann}_A(1) = 0$ cannot be essential in A. □

Definition 5.3.5. The submodule $\mathcal{Z}(M)$ is called the **singular submodule** of M. The ideal

$$\mathcal{Z}(A_A) = \{a \in A \ : \ aI = 0 \ \text{ for some right ideal } \ I \subseteq_e A\}$$

is called the **right singular ideal** of A. The **left singular ideal** of A can be defined similarly.

Recall that the **socle** of an A-module M is the sum of all simple A-submodules of M. In particular, the right socle $\text{soc}(A_A)$ of a ring A is the sum of all minimal right ideals of A. From [146, proposition 2.2.4], it follows that $\text{soc}_A(M) = M$ if and only if M is semisimple. The following proposition gives the equivalent definition of the socle of a module in terms of essential submodules.

Proposition 5.3.6. *The socle of a right A-module M is equal to the intersection of all essential submodules of M.*

Proof. Let $S = \text{soc}_A(M) = M_1 \oplus M_2 \oplus \ldots \oplus M_n$, where each M_i is a simple module. Suppose N is an essential A-submodule in M. Then $N \cap M_i \neq 0$ for all i. Since $N \cap M_i \subseteq M_i$ and M_i is simple, $N \cap M_i = M_i$. Therefore $M_i \subset N$ for all i. So $S = M_1 \oplus M_2 \oplus \ldots \oplus M_n \subset N$ for an arbitrary essential submodule in M. Therefore $S = \text{soc}_A(M)$ is contained in the intersection of all essential submodules of M.

Conversely, let T be the intersection of all essential submodules of M. It will now be shown that any submodule $N \subseteq T$ is a direct summand of T. Let N' be a complement to N in M. Then $N \cap N' = 0$ and $N \oplus N'$ is an essential submodule in M, by proposition 5.1.27. By the definition of T, $T \subset N \oplus N'$. So there is the following chain of inclusions:

$$N \subseteq T \subseteq N \oplus N'. \tag{5.3.7}$$

Let $0 \neq t \in T$. Then $t \in N \oplus N'$, i.e., $t = n + n_1$, where $n \in N$ and $n_1 \in N'$. Whence $n_1 = t - n \in T \cap N_1$. Therefore $T = N + (T \cap N_1)$. This sum is direct because $N \cap (T \cap N') = 0$. Thus, $T = N \oplus (T \cap N_1)$, that is, N is a direct summand of T. By [146, proposition 2.2.4], T is semisimple, i.e., it is a direct sum of simple submodules of M, so that $T \subseteq S$. This implies that $T = S$, as desired. □

The following proposition gives some basic properties of singular submodules.

Proposition 5.3.8. *Let M, N be right A-modules, and let* $\mathrm{soc}(A_A)$ *be the right socle of A. Then*

1. $\mathcal{Z}(M) \cdot \mathrm{soc}(A_A) = 0$.
2. *If $f : M \to N$ is an A-homomorphism, then $f(\mathcal{Z}(M)) \subseteq \mathcal{Z}(N)$.*
3. *If N is a submodule of M then $\mathcal{Z}(N) = N \cap \mathcal{Z}(M)$.*

Proof.

1. Since, by the definition, $\mathrm{r.ann}_A(m)$ is an essential right ideal of A for any $m \in \mathcal{Z}(M)$, $\mathrm{r.ann}_A(m) \supset \mathrm{soc}(A_A)$, by proposition 5.3.6. Hence $m \cdot \mathrm{soc}(A_A) = 0$ as required.
2. This follows from the fact that $\mathrm{r.ann}_A(m) \subset \mathrm{r.ann}_A(f(m))$ for any $m \in M$, because $f(m)a = f(ma)$.
3. This follows from the definition of singular submodules. □

Definition 5.3.9. A right A-module M is called **singular** if $\mathcal{Z}(M) = M$. A right A-module M is called **nonsingular** if $\mathcal{Z}(M) = 0$. A ring A is called **right nonsingular** if $\mathcal{Z}(A_A) = 0$. A **left nonsingular ring** is defined similarly. A ring which is right and left nonsingular is called a **nonsingular ring**.

Note that the only module M that is both singular and nonsingular is $M = 0$.

Examples 5.3.10.

1. Any simple Artinian ring is nonsingular. This follows immediately from corollary 5.3.4(1).
2. Any semisimple ring is nonsingular.
3. Let A be a commutative domain. Then any essential ideal in A is a non-zero ideal. Therefore a singular submodule of any A-module is just a torsion submodule. A nonsingular submodule of A is just a torsion-free A-module.

The next propositions give some basic properties of singular and nonsingular modules.

Proposition 5.3.11. *An A-module M is singular if and only if $M \simeq N/X$ for some A-module N and some essential submodule X of N.*

Proof. Suppose M is a singular right A-module. There is a free right A-module F such that $M \simeq F/X$ for some submodule X of F. Let $\{f_i : i \in I\}$ be a free basis of F. Since $F/X \simeq M$ is a singular module, for each $i \in I$, there is a right essential ideal $\mathcal{J}_i$ of A such that $f_i\mathcal{J}_i \subset X$. Then by proposition 5.3.1(3) $\bigoplus_{i \in I} f_i\mathcal{J}_i$ is an essential submodule in $\bigoplus_{i \in I} f_i A = F$, and so X is essential in F.

Conversely, suppose that $M \simeq N/X$, where X is an essential submodule of N. Let $m \in M$, then $m = n + X$, where $n \in N$ and $n \notin X$. It will now be shown that $\mathcal{J} = \{a \in A : na \in X\}$ is an essential right ideal in A. Let $y \in A \setminus \mathcal{J}$. Then $ny \notin X$. Since X is essential in N, by lemma 1.5.10, there exists a non-zero element $b \in A$ such that $(ny)b \in X$. Therefore $yb \in \mathcal{J}$ and so the right ideal $\mathcal{J}$ is essential in A. □

Proposition 5.3.12. *Let N be a submodule of a nonsingular module M. Then M/N is a singular module if and only if N is essential in M.*

Proof. Let N be an A-submodule of a nonsingular A-module M.

If N is essential in M, then M/N is a singular module, by proposition 5.3.11.

Conversely, suppose M/N is a singular module. Let X be an arbitrary submodule of M, and let $x \in X$. Then, since M/N is singular, there is a right essential ideal $\mathcal{I}$ in A such that $x\mathcal{I} \subset N$. Since M is nonsingular, $x\mathcal{I} \neq 0$ for otherwise $x \in \mathcal{Z}(M) = 0$. So that $N \cap X \neq 0$. Thus N is essential in M. □

Proposition 5.3.13.

1. *All submodules, quotient modules and sums of singular modules are singular.*
2. *All submodules, direct products, and essential extensions of nonsingular modules are nonsingular.*
3. *Let N be a submodule of a module M. If N and M/N are both nonsingular modules, then M is nonsingular.*

Proof.

1. If N is an A-submodule of a singular A-module M then by proposition 5.3.8(3) it follows that $\mathcal{Z}(N) = N \cap \mathcal{Z}(M) = N \cap M = N$. So N is singular. Let M be a singular A-module, and $N = M/X$. Consider the canonical projection $f : M \to M/X$. By proposition 5.3.8(2), $f(\mathcal{Z}(M)) \subseteq \mathcal{Z}(M/X)$. Since f is an epimorphism and M is singular, $M/X = f(M) = f(\mathcal{Z}(M)) \subseteq \mathcal{Z}(M/X)$. Hence $M/X = \mathcal{Z}(M/X)$, i.e., $N = M/X$ is singular.

Let $\{X_i : i \in I\}$ be a family of singular A-submodules of an A-module M. Then $\mathcal{Z}(X_i) = X_i$ is a submodule of $\mathcal{Z}(M)$ for each $i \in I$. Therefore $\sum_{i \in I} X_i \subset \mathcal{Z}(M)$, and so $\sum_{i \in I} X_i$ is singular.

2. If N is an A-submodule of a nonsingular A-module M then by proposition 5.3.8(3) it follows that $\mathcal{Z}(N) = N \cap \mathcal{Z}(M) = 0$. So N is nonsingular.
   Let M be an essential extension of a nonsingular submodule N. Then by proposition 5.3.8(3) $0 = \mathcal{Z}(N) = N \cap \mathcal{Z}(M)$. Since N is essential in M, $\mathcal{Z}(M) = 0$, i.e., M is nonsingular.
   Let $\{X_i : i \in I\}$ be a family of nonsingular A-submodules of A-module M. Then for each $j \in I$ one can consider the canonical projection $\pi_j : \prod_{i \in I} X_i \to X_j$. By proposition 5.3.8(2) $\pi_j(\mathcal{Z}(\prod X_i)) \subset \mathcal{Z}(X_j)$. Since $\mathcal{Z}(X_j) = 0$ for each $j \in I$, $\pi_j(\mathcal{Z}(\prod X_i)) = 0$ for all $j \in I$. Whence $\mathcal{Z}(\prod X_i) = 0$.
3. Let N be a submodule of a module M. Consider the canonical projection $\pi : M \to M/N$. Then by proposition 5.3.8(2) $\pi(\mathcal{Z}(M)) \subset \mathcal{Z}(M/N)$. Since $\mathcal{Z}(M/N) = 0$, it follows that $\pi(\mathcal{Z}(M)) = 0$. Whence $\mathcal{Z}(M) \subset \operatorname{Ker} \pi = N$. Then by proposition 5.3.8(3) $\mathcal{Z}(N) = N \cap \mathcal{Z}(M) = \mathcal{Z}(M)$. Since N is nonsingular, $\mathcal{Z}(M) = 0$. □

Proposition 5.3.14. *A right A-module N is nonsingular if and only if* $\operatorname{Hom}_A(M, N) = 0$ *for each singular right A-module M.*

Proof.

1. Let N be a nonsingular A-module, i.e., $\mathcal{Z}(N) = 0$. Suppose that M is a singular A-module, i.e., $\mathcal{Z}(M) = M$, and $f \in \operatorname{Hom}_A(M, N)$. In due of proposition 5.3.8(2), $f(M) = f(\mathcal{Z}(M)) \subseteq \mathcal{Z}(N) = 0$, which implies that $f = 0$. Therefore $\operatorname{Hom}_A(M, N) = 0$.
2. Conversely, suppose that $\operatorname{Hom}_A(M, N) = 0$ for each singular right A-module M. Then, in particular, $\operatorname{Hom}_A(\mathcal{Z}(N), N) = 0$. So the inclusion $\mathcal{Z}(N) \hookrightarrow N$ is zero, which implies that $\mathcal{Z}(N) = 0$. □

Proposition 5.3.15. *Let A be a right nonsingular ring. Then*

1. $\mathcal{Z}(M/\mathcal{Z}(M)) = 0$ *for all right A-modules M.*
2. *A right A-module M is singular if and only if* $\operatorname{Hom}_A(M, N) = 0$ *for all nonsingular right A-modules N.*

Proof.

1. Suppose that $\mathcal{Z}(M/\mathcal{Z}(M)) \neq 0$ for some A-module M and consider the natural map $\varphi : M \longrightarrow M/\mathcal{Z}(M)$. Let $m \in M$ be such that $0 \neq \varphi(m) \in \mathcal{Z}(M/\mathcal{Z}(M))$. Then there is an essential right ideal $\mathcal{I}$ in A such that $m\mathcal{I} \subseteq \mathcal{Z}(M)$. To prove that $\varphi(m) = 0$, i.e., $m \in \mathcal{Z}(M)$, it suffices to show that $\operatorname{r.ann}_A(m) \subseteq_e A_A$. Let K be a right ideal in A. Since $\mathcal{I} \subseteq_e A_A$ there is $0 \neq x \in K \cap \mathcal{I}$. Since $x \in \mathcal{I}$, $mx \in \mathcal{Z}(M)$, i.e., there is a right ideal $\mathcal{J}$ in A which is essential in A and $mx\mathcal{J} = 0$. Then $x\mathcal{J} \neq 0$ for otherwise

$0 \neq x \in \mathcal{Z}(A_A) = 0$, since A is right nonsingular. Therefore there exists a non-zero element $y \in \mathcal{J}$ such that $xy \neq 0$ and $mxy = 0$. This means that $0 \neq xy \in \text{r.ann}_A(m) \cap K$, i.e., $\text{r.ann}_A(m) \cap K \neq 0$ for any right ideal of A. Thus $\text{r.ann}_A(m) \subseteq_e A_A$. So that $\mathcal{Z}(M/\mathcal{Z}(M)) = 0$, as required.

2. Suppose M is singular. Then from proposition 5.3.14 it follows that $\text{Hom}_A(M,N) = 0$ for all nonsingular right A-modules N. Conversely, suppose that $\text{Hom}_A(M,N) = 0$ for all nonsingular right A-modules N. In view of statement 1 proved above, $M/\mathcal{Z}(M)$ is nonsingular. Hence the natural map $M \to M/\mathcal{Z}(M)$ must be zero, whence $\mathcal{Z}(M) = M$. □

Let S be a subset in a ring A. Recall that the **right annihilator** of S is a set

$$\text{r.ann}_A(S) = \{x \in A \ : \ sx = 0 \ \text{ for all } s \in S\} \tag{5.3.16}$$

In a similar way, a set

$$\text{l.ann}_A(S) = \{x \in A \ : \ xs = 0 \ \text{ for all } s \in S\} \tag{5.3.17}$$

is a **left annihilator** of S. It is easy to show that $\text{r.ann}_A(S)$ is a right ideal in A and $\text{l.ann}_A(S)$ is a left ideal in A.

Definition 5.3.18. A **right (left) annihilator** in a ring A is any right (left) ideal of A which equals the right (left) annihilator of some subset $S \subseteq A$.

Lemma 5.3.19. *A right ideal $\mathcal{I}$ of a ring A is a right annihilator if and only if*

$$\mathcal{I} = \text{r.ann}_A(\text{l.ann}_A(\mathcal{I})). \tag{5.3.20}$$

An analogous statement is true for left ideals.

Proof. Obviously, if it holds the equality (5.3.20) then a right ideal $\mathcal{I}$ is a right annihilator.

Conversely, let a right ideal $\mathcal{I}$ of a ring A be a right annihilator of some set $S \subseteq A$, i.e., $\mathcal{I} = \text{r.ann}_A(S)$. Then $S \subseteq \text{l.ann}_A(\mathcal{I})$, and so

$$\mathcal{I} \subseteq \text{r.ann}_A(\text{l.ann}_A(\mathcal{I})) = \text{r.ann}_A(S) = \mathcal{I},$$

which implies that $\mathcal{I} = \text{r.ann}_A(\text{l.ann}_A(\mathcal{I}))$. □

Proposition 5.3.21. (Mewborn-Winton [239].) *If A is a ring with a.c.c. on right annihilators then the right singular ideal $\mathcal{Z}(A_A)$ is nilpotent.*

Proof. Write $N = \mathcal{Z}(A_A)$ and consider the chain of inclusions

$$\text{r.ann}_A(N) \subseteq \text{r.ann}_A(N^2) \subseteq \cdots \subseteq \text{r.ann}_A(N^i) \subseteq \text{r.ann}_A(N^{i+1}) \subseteq \cdots$$

Since A has a.c.c on right annihilators, there is a number n such that $\text{r.ann}_A(N^n) = \text{r.ann}_A(N^{n+1})$. We claim that $N^{n+1} = 0$. Suppose $N^{n+1} \neq 0$. Then there is a non-zero element $a \in N$ such that $N^n a \neq 0$. We can choose an element $x \in A \setminus$

$\text{r.ann}_A(N^n)$ such that $\text{r.ann}_A(x)$ is maximal among all $\text{r.ann}_A(c)$ for which $N^n c \neq 0$. Let $b \in N$. Then $\text{r.ann}_A(b)$ is an essential right ideal in A, therefore $\text{r.ann}_A(b) \cap xA \neq 0$. So there exists a non-zero element $y \in A$ such that $xy \neq 0$ and $bxy = 0$. This means that $\text{r.ann}_A(bx) \subsetneq \text{r.ann}_A(x)$, which is not the case, since $\text{r.ann}_A(x)$ is maximal. So $N^n bx = 0$ for all $b \in N$, i.e., $x \in \text{r.ann}_A(N^{n+1}) = \text{r.ann}_A(N^n)$, which contradicts the choice of x. Therefore $N^{n+1} = 0$. □

Proposition 5.3.22. *If A is a semiprime ring with a.c.c. on right annihilators then A is right nonsingular.*

Proof. Since A is semiprime it has no non-zero nilpotent ideals. So from proposition 5.3.21 it follows that $\mathcal{Z}(A_A) = 0$. □

Proposition 5.3.23. *The following conditions are equivalent for a ring A:*

1. *A satisfies d.c.c. on right (left) annihilators.*
2. *A satisfies a.c.c. on left (right) annihilators.*

Proof. Suppose that A satisfies d.c.c. on right annihilators and consider an ascending chain of left annihilators: $\mathcal{J}_1 \subseteq \mathcal{J}_2 \subseteq \cdots$, where $\mathcal{J}_k = \text{l.ann}_A(\mathcal{I}_k)$ for any k. Taking right annihilators we obtain a descending chain of right annihilators $\text{r.ann}_A(\mathcal{J}_1) \supseteq \text{r.ann}_A(\mathcal{J}_2) \supseteq \cdots$. This chain stabilizes, by assumption, i.e., there is a number n such that $\text{r.ann}_A(\mathcal{J}_n) = \text{r.ann}_A(\mathcal{J}_{n+1})$. Since J_n are left ideals, taking again left annihilators we obtain $\mathcal{J}_n = \mathcal{J}_{n+1}$ by lemma 5.3.19, i.e., A satisfies a.c.c. on left annihilators.

The remaining cases can be dealt with similarly. □

Lemma 5.3.24. *Let A be a ring with a.c.c. on right annihilators. Then $\text{l.ann}_A(\mathcal{Z}({}_AA))$ is an essential left ideal in A.*

Proof. By proposition 5.3.23, A satisfies d.c.c. on left annihilators. Write $N = \mathcal{Z}({}_AA)$, the left singular ideal of A. Let $\mathcal{M}$ be the set of all finite subsets of N. By the same proposition 5.3.23, the set of all left annihilators of the subsets in $\mathcal{M}$ has a minimal element $\text{l.ann}_A(S)$ for some finite subset $S \in \mathcal{M}$. Since $\text{l.ann}_A(S \cup x) \subseteq \text{l.ann}_A(S)$ for any $x \in N$, $\text{l.ann}_A(S \cup x) = \text{l.ann}_A(S)$ by minimality. Therefore $\text{l.ann}_A(N) = \text{l.ann}_A(S)$. Suppose $S = \{x_1, x_2, \ldots, x_n\}$, where $x_i \in N$, then $\text{l.ann}_A(S) = \bigcap\limits_{i=1}^{n} \text{l.ann}_A(x_i)$. Since $x_i \in S \subseteq N$, each $\text{l.ann}_A(x_i)$ is an essential ideal, by definition of $\mathcal{Z}({}_AA)$. So $\text{l.ann}_A(N) = \text{l.ann}_A(S)$ is also an essential ideal as a intersection of a finite number of essential ideals, by proposition 5.3.1(1). □

Proposition 5.3.25. *If A is a semiprime ring with a.c.c. on right (left) annihilators then A is right and left nonsingular.*

Proof. By proposition 5.3.22, A is right nonsingular. Write $N = \mathcal{Z}({}_AA)$ and $K = N \cap \text{l.ann}_A(N)$. Then $K^2 = 0$, which means that $K = 0$, since A is semiprime. By lemma 5.3.24, $\mathrm{l.ann}_A(N)$ is an essential left ideal of A. Therefore $N = 0$, i.e., A is also left nonsingular. □

Since a right Goldie ring has a.c.c. on right annihilators then we immediately obtain the following corollary.

Corollary 5.3.26. *A semiprime right Goldie ring A is right and left nonsingular.*

Lemma 5.3.27. *Let an A-module M is an essential extension of an A-module N. Then for any $0 \neq m \in M$ there is a non-zero essential right ideal L in A such that $mL \subseteq N$.*

Proof. Since $N \subseteq_e M$, for any $0 \neq m \in M$ there is a non-zero element $a \in A$ such that $0 \neq ma \in N$ by lemma 1.5.10. Then $L = \{a \in A : ma \in N\}$ is as easy to verify form a non-zero right ideal in A and $mL \subseteq N$. We need only to show that $L \subseteq_e A$. Let I be a non-zero right ideal in A. If $mI = 0$ then $I \subseteq L$ and so $I \cap L \neq 0$. Suppose that $mI \neq 0$. Then $mI \cap N \neq 0$ since $N \subseteq_e M$. Therefore there exists a non-zero element $b \in I$ such that $mb \in N$, i.e., $b \in L$. So that $I \cap L \neq 0$. Thus $L \subseteq_e A$. □

Proposition 5.3.28. *Let A be a right nonsingular ring with finite right uniform dimension. Then A satisfies a.c.c. and d.c.c. on right annihilators.*

Proof. Let N, M be two right annihilators in A and $N \subseteq M$. Suppose that $N \subseteq_e M$. Let $0 \neq m \in M$. Then by lemma 5.3.27 there is a non-zero essential right ideal L in A such that $mL \subset N$. Therefore $\mathrm{l.ann}_A(N)mL = 0$. Since A is a right nonsingular ring and L is an essential ideal, $\mathrm{l.ann}_A(N)m = 0$. Therefore $m \in \mathrm{r.ann}_A(\mathrm{l.ann}_A(N)) = N$. Thus $N = M$.

So that if for two annihilators M, N in A we have a strict inclusion $N \subsetneq M$ then N is not essential in M. In this case there is a right ideal K in A such that $K \subseteq M$ and $K \cap N = 0$. Therefore any chain of distinct annihilators can be risen to a direct sum of non-zero ideals. Since A has a finite uniform dimension this means that any such chains of right annihilators must stabilize. Therefore A satisfies a.c.c. and d.c.c. on right annihilators. □

Proposition 5.3.29. *A semiprime right Goldie ring satisfies a.c.c. and d.c.c. on right annihilators.*

Proof. This follows immediately from corollary 5.1.20, corollary 5.3.26 and proposition 5.3.28. □

For semiprime rings there is an important relationship between complementary ideals and annihilators of ideals that shows the following lemma.

Lemma 5.3.30. *If $\mathcal{I}$ is an ideal in a semiprime ring A then*:

1. $\mathrm{r.ann}_A(\mathcal{I}) = \mathrm{l.ann}_A(\mathcal{I}) = \mathrm{ann}_A(\mathcal{I})$
2. $\mathrm{ann}_A(\mathcal{I})$ *is a unique complement ideal to $\mathcal{I}$ in A.*

Proof.

1. Let $X = \text{r.ann}_A(\mathcal{I}), Y = \text{l.ann}_A(\mathcal{I})$. Since $\mathcal{I}$ is a two-sided ideal, X and Y are both two-sided ideals in A and $\mathcal{I}X = Y\mathcal{I} = 0$. Then $(X\mathcal{I})^2 = X(\mathcal{I}X)\mathcal{I} = 0$, which implies $X\mathcal{I} = 0$, since A is a semiprime ring. So $X \subseteq Y$. Analogously, $Y \subseteq X$. Therefore $X = Y$.
2. Let $X = \text{ann}_A(\mathcal{I})$. Then $(\mathcal{I} \cap X)^2 \subseteq \mathcal{I}X = X\mathcal{I} = 0$ which implies $\mathcal{I} \cap X = 0$, since A is a semiprime ring. If C is an ideal in A such that $\mathcal{I} \cap C = 0$, then $\mathcal{I}C \subseteq \mathcal{I} \cap C = 0$ which implies $C \subseteq \text{ann}_A(\mathcal{I})$. So $\text{ann}_A(\mathcal{I})$ is a complement ideal to I in A which is obviously unique. □

5.4 Nonsingular Rings and Goldie Rings

Goldie rings and the famous Goldie theorem were considered in [146, Section 9.3]. This theorem states that a ring A has a classical right ring of fractions, which is a semisimple ring, if and only if A is a right Goldie ring. This section contains the proof of an extended version of this theorem which shows when rings are Goldie rings using the notions of Goldie dimension and nonsingular rings.

Recall that an element $x \in A$ is **right regular** (resp. **left regular**) if $xa \neq 0$ (resp $ax \neq 0$) for any non-zero element $a \in A$. Denote by $C_A^r(0)$ (resp. $C_A^l(0)$) the set of all right (resp. left) regular elements of A. An element $x \in A$ is called **regular** if it is right and left regular. Denote by $C_A(0)$ the set of all regular elements of A.

A ring Q with identity is a **classical right ring of fractions** (**classical right quotient ring**, or **classical right ring of quotients**) of a ring A if

1. $A \subset Q$;
2. Every regular element of A is invertible in Q;
3. Each element of Q has the form $q = ab^{-1}$, where $a, b \in A$ and b is a regular element in A.

In this case a ring A is called a **right order** in Q. Analogously one can define a **left order** in Q. If A is both a right and left order in Q then A is called an **order** in Q.

Lemma 5.4.1. *Let A be a right nonsingular ring with finite right uniform dimension. If d is a right regular element of A then d is a left regular element of A, as well, and the right ideal dA is essential in A.*

Proof. Since d is a right regular element of A, $dA \simeq A$. Since $dA \subseteq A$ and A has finite right uniform dimension, $dA \subseteq_e A$ by corollary 5.1.11. Therefore by definition of $\mathcal{Z}(A_A) = 0$ one obtain that $\text{l.ann}_A(dA) \subseteq \mathcal{Z}(A_A) = 0$. This implies $\text{l.ann}_A(d) = 0$, i.e., d is left regular in A. □

Corollary 5.4.2. *Let A be a semiprime right Goldie ring. Then*:

1. $C^r_A(0) = C_A(0)$.
2. *A principal right ideal xA is essential in A if and only if $x \in C_A(0)$.*

Proof.

1. This follows from lemma 5.4.1 taking into account that any semiprime right Goldie ring is right non-singular by corollary 5.4.26, and have finite right uniform dimension by corollary 5.1.20.
2. If x is a regular element then xA is an essential ideal in A, by lemma 5.4.1. Conversely, let xA be an essential ideal in A. Then x is a regular element, by [146, Corollary 9.3.4]. □

Lemma 5.4.3. *Let A be a semiprime right nonsingular ring with finite right uniform dimension. Then a right ideal of A is essential if and only if it contains a regular element of A.*

Proof. We now prove that any a non-zero essential right ideal $\mathcal{I}$ in A contains a regular element of A. Let's first show that $\mathcal{I}$ contains an element x such that $\text{r.ann}_A(x) \cap \mathcal{I} = 0$. Since A has finite uniform dimension, every right ideal in A contains a uniform right ideal by lemma 5.1.14. Let U be a uniform right ideal which is contained in $\mathcal{I}$.

Suppose $\mathcal{I} = U$. Since A is semiprime, $\mathcal{I}^2 \neq 0$. Therefore $xy \neq 0$ for some $x, y \in \mathcal{I}$. Let $B = \text{r.ann}_A(x) \cap \mathcal{I} \subseteq \mathcal{I}$, and suppose that $B \neq 0$. Since $\mathcal{I} = U$ is uniform, B is essential in I. Then B is also essential in A, by [146, lemma 5.3.2]. So, by proposition 5.3.1(4), $B(y) = \{a \in A \ : \ ya \in B\}$ is an essential right ideal in A. Since $xyB(y) = 0$ for $xy \neq 0$ and $\mathcal{Z}(A_A) = 0$, we get a contradiction. So that $B = 0$, i.e., there exists an element $x \in \mathcal{I}$ such that $\text{r.ann}_A(x) \cap \mathcal{I} = 0$.

Suppose $\mathcal{I} \neq U$ and set $U_1 = U \subsetneq \mathcal{I}$. Then by the previous argument one can choose a non-zero element $a_1 \in U_1$ such that $\text{r.ann}_A(a_1) \cap U_1 = 0$. Therefore $a_1 A \oplus V_1 \subseteq \mathcal{I}$, where $V_1 = \text{r.ann}_A(a_1) \cap \mathcal{I}$. If $V_1 \neq 0$, V_1 contains a uniform right ideal U_2 and one can choose an element $a_2 \in U_2$ such that $\text{r.ann}_A(a_2) \cap U_2 = 0$. Then

$$a_1 A \oplus a_2 A \oplus V_2 \subseteq \mathcal{I},$$

where

$$V_2 = \text{r.ann}_A(a_1) \cap \text{r.ann}_A(a_2) \cap \mathcal{I}.$$

Continuing this process we obtain that if $V_k \neq 0$ then

$$a_1 A \oplus a_2 A \oplus \ldots a_k A \oplus V_k \subseteq \mathcal{I}$$

for some $a_k \in \mathcal{I}$ and

$$V_k = V_{k-1} \cap \text{r.ann}_A(a_k).$$

So that if $V_k \neq 0$ for all k we can construct an infinite direct sum $\bigoplus_{i=1}^{\infty} a_i A$ of right ideals $a_i A \subset \mathcal{I}$. This is impossible because A has a finite right uniform dimension.

So that there exists an integer k such that $V_k = 0$, i.e.,

$$V_k = \text{r.ann}_A(a_1) \cap \text{r.ann}_A(a_2) \cap \ldots \cap \text{r.ann}_A(a_k) \cap \mathcal{I} = 0 \tag{5.4.4}$$

and

$$a_1 A \oplus a_2 A \oplus \ldots \oplus a_k A \subseteq \mathcal{I}. \tag{5.4.5}$$

Set $x = a_1 + a_2 + \ldots + a_k \in \mathcal{I}$. Then from (5.4.5) it follows that

$$\text{r.ann}_A(x) = \bigcap_{i=1}^{k} \text{r.ann}_A(a_i). \tag{5.4.6}$$

Therefore from (5.4.4) it follows that $\text{r.ann}_A(x) \cap \mathcal{I} = 0$. Since $\mathcal{I}$ is an essential ideal in A, this means that $\text{r.ann}_A(x) = 0$, i.e., x is a right regular element. By lemma 5.4.1, x is a regular element in A.

Conversely, if x is a regular element in $\mathcal{I}$, then xA is an essential ideal in A, by lemma 5.4.1. Whence $\mathcal{I}$ is an essential ideal in A. □

Corollary 5.4.7. *Let A be a semiprime right Goldie ring. Then a right ideal in A is essential if and only if it contains a regular element of A.*

Proof. This follows from corollary 5.3.26 and lemma 5.4.3. □

Lemma 5.4.8. *Let A be a right Ore ring with nonempty set $S = C_A(0)$ of all regular elements of A, and let Q be the classical right ring fractions of A. Then*

1. *If $\mathcal{I}$ is a right ideal of A then*

$$\mathcal{I}Q = \mathcal{I}S^{-1} = \{xs^{-1} \ : \ x \in \mathcal{I}, s \in S\}$$

is a right ideal of Q and

$$\mathcal{I}Q \cap A = \{a \in A \ : \ as \in \mathcal{I} \text{ for some } s \in S\}.$$

2. *If $\mathcal{J}$ is a right ideal of Q then $\mathcal{J} \cap A$ is a right ideal of A and*

$$\mathcal{J} = (\mathcal{J} \cap A)Q.$$

3. *If I_1, I_2 are right ideals of A then $I_1 \cap I_2 = 0$ implies $I_1 Q \cap I_2 Q = 0$.*

Proof. These statements follows immediately from [146, Lemma 9.1.5, Lemma 9.1.6].

Lemma 5.4.9. *Let a ring A have the classical right ring of fractions Q. If $\mathcal{J}$ is an essential right ideal in Q then $\mathcal{J} \cap A$ is an essential right ideal in A.*

Proof. By theorem 1.11.2 A is a right Ore ring. Suppose that there is a non-zero right ideal $\mathcal{I}$ in A such that $\mathcal{I} \cap (\mathcal{J} \cap A) = 0$. By lemma 5.4.8 IQ and $J = (J \cap A)Q$ are non-zero right ideals in Q. Therefore from this lemma it follows that $\mathcal{I}Q \cap \mathcal{J} = \mathcal{I}Q \cap (\mathcal{J} \cap A)Q = 0$, since $\mathcal{I} \cap (\mathcal{J} \cap A) = 0$. But $\mathcal{J}$ is an essential ideal in Q. Thus, the contradiction obtained shows that $\mathcal{J} \cap A$ is an essential ideal in A. □

Theorem 5.4.10 (A.W. Goldie [102]). *The following conditions on a ring A are equivalent:*

1. *A is a semiprime right Goldie ring.*
2. *A is a semiprime right nonsingular ring and* $\text{u.dim}(A_A) < \infty$.
3. *A has a classical right ring of fractions which is semisimple.*

Proof.
$1 \Longleftrightarrow 3$. This is the Goldie theorem 1.11.3.
$1 \Longrightarrow 2$. This follows from corollary 5.2.9 and corollary 5.3.26.
$2 \Longrightarrow 3$. We first show that A is a right Ore ring. Let $a, d \in A$, where d is a right regular element of A. Then the right ideal dA is essential in A and d is a regular element, by lemma 5.4.1. Therefore $X(a) = \{u \in A : au \in dA\}$ is an essential right ideal in A, by proposition 5.3.1(4). Then, by lemma 5.4.3, $X(a)$ contains a regular element $y \in A$. Therefore $ay = dx$ for some $x \in A$, which shows that A is a right Ore ring. So, by theorem 1.11.2, A has a classical right ring of fractions Q.

We now further show that Q is a semisimple ring. Suppose I is a non-zero essential right ideal in Q. Then, by lemma 5.4.9, $I \cap A$ is an essential right ideal in A. Thus, by lemma 5.4.3, $I \cap A$ contains a regular element of A. Since all regular elements of A are units in Q, $I = Q$. Therefore the only essential right ideal in Q is Q itself. Then, by proposition 5.3.6, $\text{soc}_Q(Q) = Q$, i.e., sum of all minimal right ideals of Q is equal to Q. So that, from [146, proposition 2.2.4] it follows that Q is semisimple. □

Proposition 5.4.11. *Let A be a right order in a ring Q. Then Q is flat as a left A-module.*

Proof. Let A be a right order in a ring Q. Then Q is its a classical right ring of fractions, and so A is a right Ore ring, by theorem 1.11.2.

By the flatness test 1.5.15, it suffices to prove that for any finitely generated right ideal $I \subseteq A$ the natural map $\varphi : I \otimes_A Q \to IQ$, given by $\varphi(x \otimes s) = xs$ is injective. Suppose $\varphi(u) = 0$ for $u \in I \otimes_A Q$. Let $u = \sum_{i=1}^{k} (x_k \otimes q_k)$ for some $x_k \in I$, $q_k \in Q$. Applying induction to [146, lemma 9.1.4], one can find elements $b_1, b_2, \dots, b_k \in A$ and a regular element $s \in C_A(0)$ such that $q_i = b_i s^{-1}$, whence

$$u = \sum_{i=1}^{k} x_i b_i \otimes s^{-1} = x \otimes s^{-1}.$$

Thus, any element $u \in I \otimes_A Q$ has the form $x \otimes s^{-1}$ for some $x \in I$ and $s \in C_A(0)$. Therefore if $\varphi(u) = 0$ then $\varphi(u) = xs^{-1} = 0$ in Q, and so $x = 0$. Therefore $u = x \otimes s^{-1} = 0$, which implies that φ is injective. □

Corollary 5.4.12.

1. *If A is a right order in a ring Q and I is a left ideal in A then*
   i. $IQ \simeq I \otimes_A Q$ *as Abelian groups.*
   ii. $Q/IQ \simeq A/I \otimes_A Q$ *as Abelian groups.*
2. *If A is a right and left order in a ring Q, then Q is flat as both a right and left A-module.*

Proof.

i. Since by theorem 5.4.11 Q is a flat left A-module, the statement follows from theorem 1.5.15.
ii. Consider the exact sequence $0 \longrightarrow I \longrightarrow A \longrightarrow A/I \longrightarrow 0$. Applying the functor $* \otimes_A Q$ to this sequence we obtain the exact sequence

$$0 \to I \otimes_A Q \longrightarrow A \otimes_A Q \longrightarrow A/I \otimes_A Q,$$

since Q is a flat left A-module. Then $A/I \otimes_A Q \simeq Q/IQ$ since $I \otimes_A Q \simeq IQ$. □

Let M be an A-module. Write

$$t(M) = \{m \in M \ : \ ma = 0 \text{ for some regular element } a \in A\},$$

for the set of torsion elements of a right A-module M.

Proposition 5.4.13. *Let A be a semiprime right Goldie ring with a semisimple classical right ring of fractions Q. Let M be a right A-module. Then*

$$\mathcal{Z}(M) = t(M) \tag{5.4.14}$$

is the set of torsion elements of M.

Proof. Let $m \in \mathcal{Z}(M)$, then $\text{r.ann}_A(m) \subseteq_e A$. Then from corollary 5.4.7 it follows that $\text{r.ann}_A(m)$ contains a regular element $a \in A$. Therefore $ma = 0$ and so $m \in t(M)$.

Conversely, let $m \in t(M)$. Then there exists a regular element $a \in A$ such that $ma = 0$. Therefore $a \in \text{r.ann}_A(m)$. By corollary 5.4.7 $\text{r.ann}_A(m) \subseteq_e A$, i.e., $m \in \mathcal{Z}(M)$. □

Recall that an A-module M over a commutative ring A is called **torsion** if all its elements are torsion. If all elements of M are torsion-free, M is called a **torsion-free module**.

From proposition 5.4.13 it follows that a singular module over a semiprime Goldie ring is an analog of a torsion module over a commutative ring. Therefore in this case torsion terminology is often used.

Definition 5.4.15. Let A be a semiprime right Goldie ring, and M a right A-module. The submodule $\mathcal{Z}(M)$ is called the **torsion submodule** of M. A module M is called a **torsion module** if and only if $\mathcal{Z}(M) = M$, and M is called a **torsion-free module** if and only if $\mathcal{Z}(M) = 0$.

Proposition 5.4.16. *Let U be a uniform right module over a semiprime right Goldie ring. Then U is either torsion or torsion-free.*

Proof. Suppose U is not torsion-free, i.e., $\mathcal{Z}(U) \neq 0$. Since $\mathcal{Z}(U) \subseteq U$ and U is uniform, $\mathcal{Z}(U)$ is essential in U by proposition 5.1.3. Therefore $U/\mathcal{Z}(U)$ is a singular module, i.e., $\mathcal{Z}(U/\mathcal{Z}(U)) = U/\mathcal{Z}(U)$, by proposition 5.3.11. On the other hand, $\mathcal{Z}(U/\mathcal{Z}(U) = 0$, by proposition 5.3.15, since A is a nonsingular ring. So, $U = \mathcal{Z}(U)$, i.e., U is a torsion module. □

Definition 5.4.17. A right A-module M is called **faithful** if

$$\mathrm{ann}_M(A) = \{m \in M \mid ma = 0 \text{ for all } a \in A\} = 0.$$

Proposition 5.4.18. *Let O be a discrete valuation ring of a division ring D. Then any non-zero right O-module M is torsion-free and faithful.*

Proof. Let M be a non-zero right O-module. From proposition 5.4.16 it follows that M is torsion or torsion-free. Suppose that M is torsion, i.e., $\mathcal{Z}(M) = M$. Since O is a semiprime Goldie ring, from corollary 5.4.7 and definition of $\mathcal{Z}(M)$ it follows that that for any element $m \in M$ there exists a regular element $x \in O$ such that $mx = 0$. Since x is a regular element in O, there exists an element $x^{-1} \in D$. Then $0 = mx(x^{-1}) = m$, i.e., $M = 0$. Therefore M is a torsion-free module, and so faithful. □

We now further we consider some useful properties of minimal prime ideals of semiprime right Goldie rings.

Definition 5.4.19. A **minimal prime ideal** in a ring A is any prime ideal of A which does not properly contain any other prime ideal.

Lemma 5.4.20. [115, Proposition 2.3] *Any prime ideal of a ring A contains a minimal prime ideal.*

Proof. Let X be a set of all prime ideals which are contained in a given prime ideal P of A. Suppose that $Y \subset X$ is a chain of prime ideals. We will show that each such a chain has a lower bound in X. Consider the set $Q = \bigcap Y$ which is an intersection of all such chains. Then Q is an ideal in A and $Q \subseteq P$. We claim that Q is a prime ideal in A.

Let $xAy \in Q$ for some $x, y \in A$. Suppose $x \notin Q$. Since Q is an intersection of all ideals containing in a prime ideal P, there is a chain Y and an ideal $P_1 \in Y$ such that $x \notin P_1$. Since $xAy \in Q \subseteq P_1$ and P_1 is a prime ideal, $y \in P_1$, by [146, Proposition 9.2.1(6)]. Let $P_2 \in Y$. Since Y is a chain, either $P_1 \subseteq P_2$ or $P_2 \subseteq P_1$. If $P_1 \subseteq P_2$ then $y \in P_2$. If $P_2 \subseteq P_1$ then $x \notin P_2$. Since $xAy \in Q \subset P_2$ and P_2 is a prime ideal, $y \in P_2$. So that any ideal contained in the chain Y contains the element y as well.

Let $\tilde{Y}$ be any other chain which contains the ideal P_1. Analogously as above if $\tilde{P} \in \tilde{Y}$ then either $\tilde{P} \subseteq P_1$ or $P_1 \subseteq \tilde{P}$. In each case $y \in \tilde{P}$. So that each ideal in each chain contains the element y, therefore $y \in Q$, which shows that Q is a prime ideal.

Thus $Q \in X$ is a lower bound for a set X. Since X is a partially ordered set we can apply Zorn's lemma. Therefore there is a minimal element in the set X, i.e., there is a minimal prime ideal contained in the prime ideal P. □

Proposition 5.4.21. *Let A be a semiprime right Goldie ring with classical right ring of fractions Q. Then there are only a finite number of distinct minimal prime ideals of A each of them is of the form $P_i = A \cap M_i$, where M_i are maximal ideals of Q, $i = 1,2,\dots,m$. Moreover, $P_1P_2\cdots P_m = 0$.*

Proof. We will give the proof of this statement following [115, Proposition 6.1].

Since the classical right ring of fractions Q of a semiprime right Goldie ring A is semisimple, by the Wedderburn-Artin theorem Q is isomorphic to a direct sum of finite number of full matrix rings over division rings. Therefore we can assume that $Q = \bigoplus_{k=1}^{m} Q_k$, where $Q_k \simeq M_{n_k}(D_k)$ with division rings D_k for $k = 1,2,\dots,m$. Then any $M_k = \bigoplus_{i \neq k} Q_i$ is a maximal ideal of Q and $Q_k \simeq Q/M_k$ for $k = 1,2,\dots,m$. Moreover, $M_1 \cap M_2 \cap \cdots \cap M_m = 0$.

We set $P_i = A \cap M_i$ for $i = 1,2,\dots,m$. Then P_i is a right ideal of A and $M_i = P_iQ$, by lemma 5.4.8(2). Consider a homomorphism $\varphi_i : A \to Q/M_i$ which is a composition of the inclusion $A \hookrightarrow Q$ and the natural map $\pi_i : Q \to Q/M_i$. Then $\mathrm{Ker}(\varphi_i) = A \cap M_i = P_i$. If a is a regular element of A, then a is invertible in Q and so $\pi_i(a)$ is invertible in Q/M_i. Let $0 \neq x \in Q/M_i$. Then there exists $0 \neq y \in Q$ such that $x = \pi_i(y)$. Since Q is the classical right ring of fractions of A, $y = ab^{-1}$, where $a,b \in A$ and b is a regular element of A. Then $x = \pi_i(ab^{-1}) = \varphi_i(a)[\varphi_i(b)]^{-1}$. Since $Q/M_i \simeq Q_i$ is an Artinian simple ring, any regular element of $\varphi_i(A) \subset Q/M_i$ is invertible in Q/M_i, by proposition 1.1.17. So that $\varphi_i(A) \simeq A/P_i$ is a right order in Q/M_i. Since Q/M_i is an Artinian simple ring, A/P_i is a prime right Goldie ring, by theorem 1.11.4. So that each P_i is a prime ideal of A.

Since $P_i = A \cap M_i$ and $M_1 \cap M_2 \cap \cdots \cap M_m = 0$, we obtain that $P_1P_2\cdots P_m \subseteq P_1 \cap P_2 \cap \cdots \cap P_m = 0$. Since $M_i = P_iQ$ and $M_i \cap M_j = 0$ for $i \neq j$, we obtain that $P_i \cap P_j = 0$ for all $i \neq j$, by lemma 5.4.8(3).

We claim that $S = \{P_1, P_2, \dots, P_m\}$ is precisely the set of minimal prime ideals of A. Suppose that P is a prime ideal of A. Then $0 = P_1P_2\cdots P_m \subseteq P$. Since P is prime, $P_i \subseteq P$ for some i. So that S contains all minimal prime ideals of A. Suppose that there exists $P_k \in S$ which is not minimal. Then there exists i such that $P_i \subseteq P \subseteq P_k$, but this impossible, since $P_i \cap P_k = 0$ for $i \neq k$. □

Lemma 5.4.22. *Let I be an essential right ideal of a prime right Goldie ring A and $a \in A$. Then $a + I = \{a + x \ : \ x \in I\}$ contains a regular element of A.*

Proof. Let $x \in I$. Consider a set $\mathcal{R}$ of right ideals $\mathrm{r.ann}_A(a + x)$. Since A is a right Goldie ring, A satisfies d.c.c. for right annihilators, by proposition 5.3.29.

Therefore the set $\mathcal{R}$ contains an element $\text{r.ann}_A(a+y)$ which is minimal with respect to inclusion, where $y \in I$. Write $c = a + y$. We will show that $c \in C_A(0)$. Consider a right ideal cA. If cA is an essential right ideal in A then $\text{r.ann}_A(c) = 0$ and so $\text{l.ann}_A(c) = 0$, by corollary 5.4.2. Therefore $c \in C_A(0)$. If cA is not an essential ideal then there is a non-zero right ideal B in A such that $B \cap cA = 0$. Since I is an essential ideal and $B \neq 0$, $B \cap I \neq 0$. Let $0 \neq b \in B \cap I$, then $c+b = a+y+b \in a+I$. Since $B \cap cA = 0$, $cA \cap bA = 0$ and so $\text{r.ann}_A(c + b) = \text{r.ann}_A(c) \cap \text{r.ann}_A(b)$. Hence $\text{r.ann}_A(c + b) \subseteq \text{r.ann}_A(c)$ and $\text{r.ann}_A(c + b) \subseteq \text{r.ann}_A(b)$. From the choice of $\text{r.ann}_A(c)$ we have that $\text{r.ann}_A(c + b) = \text{r.ann}_A(c) = \text{r.ann}_A(b)$ for all $b \in B$. Therefore $(B \cap I)\text{r.ann}_A(c) = 0$. Since A is a prime ring and $B \cap I \neq 0$, $\text{r.ann}_A(c) = 0$. Therefore we again have that $c \in C_A(0)$. □

Theorem 5.4.23. [273] *Let A be a prime right Goldie ring. Then each essential right ideal I in A is generated by regular elements which belong to I.*

Proof. Let I be an essential right ideal of a prime Goldie ring A, and let T be an ideal which is generated by all regular elements which belong to I. It is obviously that $T \subseteq I$. We will show that $I \subseteq T$. Let $a \in I$. Then by lemma 5.4.22 $a + T$ contains a regular element $a + t \in C_A(0) \cap I \subseteq T$ with $t \in T$. Therefore $a \in T$ as well, i.e., $I \subseteq T$ and so $I = T$. □

Remark 5.4.24. Note that theorem 5.4.23 was proved by J.C. Robson [273] in a more general case, namely for semiprime right Goldie rings.

5.5 Reduced Rank and Artinian Classical Ring of Fractions

The famous Goldie theorem 1.11.3 states that a ring A has a classical right ring of fractions, which is a semisimple ring[4], if and only if A is a semiprime right Goldie ring. In 1966 L. Small studied the question of when a ring has a classical right ring of fractions which is right Artinian. He found necessary and sufficient conditions in the case of right Noetherian rings. Another proof of this theorem applying the technique of reduced rank was obtained by A.W. Chatters, A.W. Goldie, C.R. Hajarnavis, T.H. Lenagan in [45]. In [321] R.B. Warfield, Jr., applied this technique to prove the generalization of Small's theorem without the Noetherian hypothesis. In this section the proof of this theorem is given following to Warfield.

We first define the notion of the reduced rank for modules over a semiprime right Goldie ring, and then extend this concept for the case of arbitrary rings.

Definition 5.5.1. Let A be a semiprime right Goldie ring. Then by Goldie's theorem A has a classical right ring of fractions Q which is semisimple. Let M be a right A-module, then $M \otimes_A Q$ is a semisimple Q-module whose length and uniform

[4]Note that in this book a semisimple ring always means an Artinian semisimple ring.

(Goldie) dimension coincide. Then the **reduced rank** of M is defined as

$$\rho(M) = \text{length}_Q(M \otimes_A Q) = \text{u.dim}(M \otimes_A Q) \tag{5.5.2}$$

If $\text{u.dim}(M \otimes_A Q) = \infty$, then we write $\rho(M) = \infty$. Therefore $\rho(M) = n$ if and only if $M \otimes_A Q$ is a direct sum of n simple Q-modules.

If M is a finitely generated A-module over a semiprime right Goldie ring then $\rho(M)$ is obviously finite.

Proposition 5.5.3. *Let A be a semiprime right Goldie ring. Then for any right A-modules N and M such that $N \subseteq M$:*

1. $\rho(M) = 0$ *if and only if M is a singular module.*
2. $\rho(M) = \rho(N) + \rho(M/N)$.
3. $\rho(M) = \rho(M/\mathcal{Z}(M))$.

Proof.

1. Let Q be the classical right ring of fractions of a ring A. Consider the right A-module homomorphism

$$f : M \longrightarrow M \otimes_A Q$$

given by $f(m) = m \otimes 1$ for every $m \in M$. Then $\text{Ker}(f) = \mathcal{Z}(M)$. Indeed, let $x \in M$ and write $\mathcal{I} = \text{r.ann}_A(x)$. Consider the monomorphism $\alpha : A/\mathcal{I} \to M$ given by $\alpha(a+\mathcal{I}) = xa$. Since A is a right order in Q, $Q/\mathcal{I}Q \cong A/\mathcal{I} \otimes_A Q$ by corollary 5.4.12. So that we obtain a monomorphism $\beta : Q/\mathcal{I}Q \longrightarrow M \otimes_A Q$ defined by $\beta(q + \mathcal{I}Q) = x \otimes q$ for every $q \in Q$. Therefore $x \in \text{Ker} f$ if and only if $Q/\mathcal{I}Q = 0$, that is, $\mathcal{I}$ contains a regular element of A. Then by corollary 5.4.7 the ideal $\mathcal{I} = \text{r.ann}_A(x)$ is essential in A. This means that $\text{Ker} f = \mathcal{Z}(M)$. On the other hand, the Q-module $M \otimes_A Q$ is generated by $\text{Im} f$, therefore $M \otimes_A Q = 0$ if and only if $\text{Ker} f = M$. In this case $M = \mathcal{Z}(M)$. Thus $\rho(M) = \text{u.dim}(M \otimes_A Q) = 0$ if and only if M is a singular module.
2. Consider an exact sequence $0 \to N \longrightarrow M \longrightarrow M/N \to 0$ of right A-modules. Since Q is a flat left A-module, by proposition 5.4.11, there is an exact sequence

$$0 \to N \otimes_A Q \longrightarrow M \otimes_A Q \longrightarrow M/N \otimes_A Q \to 0. \tag{5.5.4}$$

Therefore

$$\text{length}(M \otimes_A Q) = \text{length}(N \otimes_A Q) + \text{length}(M/N \otimes_A Q),$$

and so $\rho(M) = \rho(N) + \rho(M/N)$.
3. This follows immediately from (2) and (1). □

Proposition 5.5.5. *Let A be a semiprime right Goldie ring, and M a right A-module. Then for any A-submodule $N \subseteq M$ and any family of right A-modules M_i, $i \in I$, one has:*

1. *If N is essential in M then $\rho(N) = \rho(M)$.*
2. $\rho(\bigoplus_{i\in I} M_i) = \sum_{i\in I} \rho(M_i)$.
3. *If M is uniform and nonsingular then $\rho(M) = 1$.*

Proof.

1. By proposition 5.5.3(2), $\rho(M) = \rho(N) + \rho(M/N)$. Since N is essential in M, M/N is a singular module by proposition 5.3.11. Hence $\rho(M/N) = 0$ by proposition 5.5.3(1).
2. This follows immediately from corollary 5.1.23.
3. Let M be a uniform nonsingular A-module, and $x \in M \otimes_A Q$ a non-zero element. Then there exist $m \in M$ and a regular element $s \in A$ such that $x = m \otimes s^{-1}$. Since mA is essential in M, the inclusion $mA \hookrightarrow M$ induces a monomorphism $mA \otimes_A Q \to M \otimes_A Q$. Since M/mA is a singular module, $M/mA \otimes_A Q = 0$ by proposition 5.5.3(1). Therefore we obtain an isomorphism $mA \otimes_A Q \cong M \otimes_A Q$. It follows that $m \otimes 1$ generates the Q-module $M \otimes_A Q$, so that $x = m \otimes s^{-1}$ also generates $M \otimes_A Q$. Thus, $M \otimes_A Q$ is simple and $\rho(M) = 1$. □

The following proposition gives an equivalent definition of the reduced rank of a module over a semiprime right Goldie ring.

Proposition 5.5.6. *Let A be a semiprime right Goldie ring with semisimple classical right ring of fractions Q. Then for a right A-module M*

$$\rho(M) = \text{u.dim}(M/\mathcal{Z}(M)) \tag{5.5.7}$$

Proof. To show this equality one can assume without loss of generality that M is nonsingular. If $\text{u.dim}(M) = \infty$, then M contains an infinite direct sum $\bigoplus_{i\in I} M_i$ of non-zero submodules. Hence

$$\rho(M) \geq \rho(\bigoplus_{i\in I} M_i) = \sum_{i\in I} \rho(M_i) = \infty$$

by proposition 5.5.3(2) and proposition 5.5.5(2). If $\text{u.dim}M = n < \infty$, then M contains an essential submodule $N = \bigoplus_{i=1}^{n} M_i$, that is, a direct sum of n uniform modules M_i. Therefore, by proposition 5.5.5(2),

$$\rho(M) = \rho(N) = \sum_{i=1}^{n} \rho(M_i) = n. \tag{5.5.8}$$

□

Remark 5.5.9. From the equality (5.5.7) it follows that

$$\text{u.dim}(M) = \rho(M) + \text{u.dim}(\mathcal{Z}(M)), \tag{5.5.10}$$

whence $\rho(M) \leq \text{u.dim}(M)$, which explains why $\rho(M)$ is called the reduced rank[5].

Definition 5.5.11. Let A be a ring with prime radical N such that A/N is a semiprime right Goldie ring and N is nilpotent. Let M be a right A-module. Define the **reduced A/N-rank** of M, denoted by $\rho_N(M)$, as follows. Consider a finite chain of submodules of M

$$M = M_0 \supseteq M_1 \supseteq \cdots \supseteq M_k = 0 \tag{5.5.12}$$

such that $M_i N \subseteq M_{i+1}$. (Note that such a chain always exists for the ring A. For example, one can take $M_i = MN^i$.) Since each M_i/M_{i+1} is an A/N-module, one can define

$$\rho_N(M) = \sum_{i=0}^{k} \rho(M_i/M_{i+1}), \tag{5.5.13}$$

where $\rho(M_i/M_{i+1})$ is the reduced rank of M_i/M_{i+1} as an A/N-module.

One shows that this definition does not depend on the choice of a chain like (5.5.12). The proof of this fact uses essentially the Schreier refinement theorem[6], which is a more general variant of the well known Jordan-Hölder theorem.

Recall that any finite chain of submodules of the form (5.5.12) is called a **submodule series** (or **normal series**) for a module M.

A submodule series

$$M = N_0 \supseteq N_1 \supseteq \cdots \supseteq N_t = 0 \tag{5.5.14}$$

is called a **refinement of series** (5.5.12) if all the terms of (5.5.12) occur in the series (5.5.14).

Two submodule series (5.5.12) and (5.5.14) are called **equivalent** (or isomorphic) if $k = t$ and there is a bijection between the factors of these series such that the corresponding factors are isomorphic, i.e., there is a permutation σ such that

$$N_{\sigma(i)}/N_{\sigma(i)+1} \simeq M_i/M_{i+1}$$

[5]Note that originally equality (5.5.10) was taken to be the definition of reduced rank by A.W. Goldie in paper [103]. In paper [101] A.W. Goldie introduced the notion of the rank of a module, which he later, in [102] named finite dimension.

[6]This theorem is named after Otto Schreier, who proved this theorem in 1928 and used it to give an improved proof of the Jordan-Hölder theorem.

for all $i = 1, 2, ..., k$.

The proof of the Schreier refinement theorem is given here using the "butterfly lemma" or "the Zassenhaus lemma"[7].

Proposition 5.5.15 (Zassenhaus lemma) [337]. *Let M, N be modules with submodules $M_1 \subseteq M$ and $N_1 \subseteq N$. Then*

$$(M_1 + (M \cap N))/(M_1 + (M \cap N_1)) \cong (N_1 + (M \cap N))/(N_1 + (N \cap M_1)) \quad (5.5.16)$$

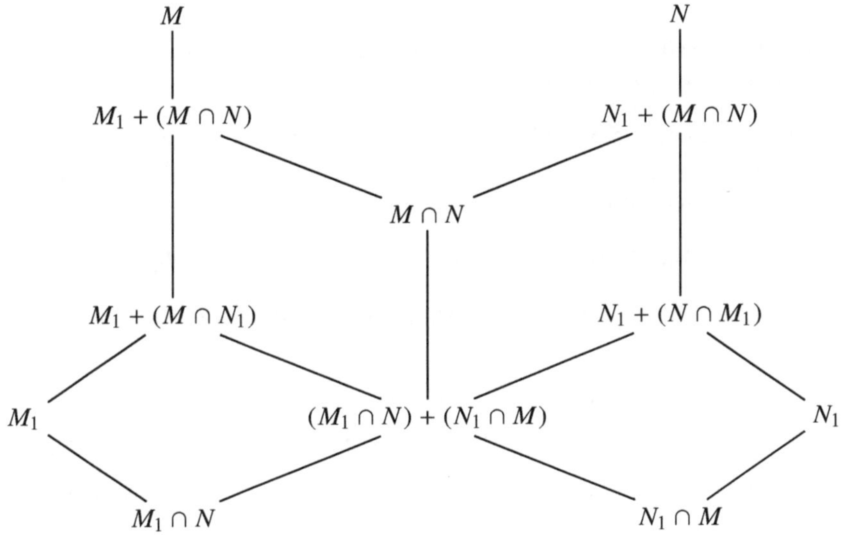

Proof. We show first that the left side of (5.5.16) is isomorphic to

$$(M \cap N)/((M_1 \cap N) + (N_1 \cap M))$$

Since $(M \cap N_1) \subseteq (M \cap N)$, obviously

$$M_1 + (M \cap N) = (M \cap N) + (M_1 + (M \cap N_1)).$$

By the modular law 1.1.2,

$$(M \cap N) \cap (M_1 + (M \cap N_1)) = (M \cap N \cap M_1) + (M \cap N_1) = (M_1 \cap N) + (M \cap N_1).$$

[7]This lemma is named after Hans Julieus Zassenhaus. It was proved in 1934 and used to give a more elegant proof of the Schreier theorem. H.J.Zassenhaus was a doctorate student under Emil Artin at the time. The name **"butterfly lemma"** originates with Serge Lang; it is based on the shape of the Hasse diagram of the various submodules involved.

By the first isomorphism theorem (See [146, theorem 1.3.3]),

$$(M_1+(M\cap N))/(M_1+(M\cap N_1)) = ((M\cap N)+(M_1+(M\cap N_1)))/(M_1+(M\cap N_1))$$
$$\cong (M\cap N)/((M\cap N)\cap(M_1+(M\cap N_1)))$$
$$= (M\cap N)/((M_1\cap N)+(M\cap N_1)).$$

By symmetry, there also results

$$(N_1+(N\cap M))/(N_1+(N\cap M_1)) \cong (M\cap N)/((M_1\cap N)+(M\cap N_1)).$$

□

Theorem 5.5.17 (Schreier refinement theorem.) [285]. *Any two submodule series of a module M have equivalent refinements.*

Proof. Let (5.5.12) and (5.5.14) be two submodule series for a module M. Define $M_{ij} = (M_i \cap N_j) + M_{i+1}$ and $N_{ji} = (M_i \cap N_j) + N_{j+1}$ for $i = 0,1,2,\dots,k-1$; $j = 0,1,2,\dots,t-1$. Then we get two submodule series for M:

$$M = M_{01} \supseteq M_{02} \supseteq \cdots \supseteq M_{0k} = M_1 = M_{11} \supseteq \cdots \supseteq M_{1,k}$$
$$= M_2 \supseteq \cdots \supseteq M_{t-1,k} = M_k = 0 \qquad (5.5.18)$$

and

$$M = N_{01} \supseteq N_{02} \supseteq \cdots \supseteq N_{0t} = N_1 = N_{11} \supseteq \cdots \supseteq N_{1,t}$$
$$= N_2 \supseteq \cdots \supseteq N_{k-1,t} = N_t = 0 \qquad (5.5.19)$$

which are refinements of two given chains (5.5.12) and (5.5.14). Applying the Zassenhaus lemma to the submodules $M_i, N_j, M_{i+1}, N_{j+1}$ one obtains

$$M_{i,j+1} = (M_i \cap N_{j+1}) + M_{i+1} \subseteq M_{ij} = (M_i \cap N_j) + M_{i+1}$$

and

$$N_{j,i+1} = (M_{i+1} \cap N_j) + N_{j+1} \subseteq (M_i \cap N_j) + N_{j+1}$$

From the same lemma it follows that

$$M_{ij}/M_{i,j+1} \cong N_{ji}/N_{j,i+1}.$$

Thus submodule series (5.5.18) and (5.5.19) are equivalent. □

Lemma 5.5.20. *Let A be a ring with nilpotent prime radical N such that A/N is a semiprime right Goldie ring. Then*

1. *The reduced rank $\rho_N(M)$ is independent on the choice of a submodule series.*
2. *The reduced rank ρ_N is additive on short exact sequences of right A-modules.*

Proof.

1. Since, by the Schreier refinement theorem, any two submodule series have equivalent refinements, it suffices to show that the sum in (5.5.13) is unchanged if we refine a given submodule series. But this is true by proposition 5.5.3(2).
2. The statement follows immediately from statement (1) taking a suitable refinement of the chain $0 \subseteq M_1 \subset M$ coming from an exact short sequence $0 \to M_1 \to M \to M/M_1 \to 0$. □

Corollary 5.5.21. *Let A be a ring with nilpotent prime radical N such that A/N is a semiprime right Goldie ring. If $L \subseteq M$, and $\rho_N(M)$ is finite, then $\rho_N(M) = \rho_N(L)$ if and only if $\rho_N(M/L) = 0$.*

Proof. This is a special case of lemma 5.5.20. □

Proposition 5.5.22. *Let A be a right Noetherian ring with prime radical N. Then $\rho_N(M)$ is defined for any right A-module M. Moreover, $\rho_N(M) < \infty$ for any finitely generated right A-module M.*

Proof. Let $N = Pr(A)$ be the prime radical of A. Then, by theorem 1.1.28, N is the largest nilpotent ideal of A. Therefore A/N is a semiprime ring. Since A is a right Noetherian ring, A/N is also right Noetherian. So A/N is a semiprime right Goldie ring with nilpotent prime radical. Therefore $\rho_N(M)$ is defined for any A-module M. Suppose that M is a finitely generated A-module. It suffices to consider the case when $MN = 0$. In this case M is a finitely generated A/N-module. Since $S = A/N$ is a semiprime right Goldie ring, it has a semisimple classical right ring of fractions Q. Therefore $M \otimes_S Q$ is a finitely generated Q-module and

$$\rho_N(M) = \text{length}_Q(M \otimes_S Q) < \infty$$

□

Let $\mathcal{I}$ be a two-sided ideal of a ring A. Denote by

$$C_A^r(\mathcal{I}) = \{x \in A \ : \ [x + \mathcal{I}] \text{ is right regular in } A/\mathcal{I}\} \tag{5.5.23}$$

the set of right regular elements in $A/\mathcal{I}$. Analogously denote by $C_A^l(\mathcal{I})$ the set of left regular elements in $A/\mathcal{I}$. Then $C_A(\mathcal{I}) = C_A^r(\mathcal{I}) \cap C_A^l(\mathcal{I})$ is the set of regular elements in $A/\mathcal{I}$. It is clear that $C_A(\mathcal{I})$ is a multiplicative set of elements.

Lemma 5.5.24. *Let M be a right A-module, where A is a ring with nilpotent prime radical N such that A/N is a semiprime right Goldie ring. Then $\rho_N(M) = 0$ if and only if for any element $m \in M$ there exists an element $r \in C_A(N)$ such that $mr = 0$.*

Proof. Suppose that $N^n = 0$ and consider the submodule series

$$0 = M_n \subseteq M_{n-1} \subseteq \cdots \subseteq M_1 \subseteq M_0 = M,$$

where $M_i = MN^i$. First assume that $\rho_N(M) = 0$. Then from lemma 5.5.20 it follows that each $\rho(M_i/M_{i+1}) = 0$. This means that M_i/M_{i+1} is a torsion

A/N-module. Therefore for any $m \in M$ there are regular elements $r_i \in C_A(N)$ such that $mr_1r_2 \cdots r_n \in M_n = 0$. And due to multiplicativity of $C_A(N)$ there is an element $r = r_1r_2 \cdots r_n \in C_A(N)$.

Conversely, assume that for any element $m \in M$ there exists an element $r \in C_A(N)$ such that $mr = 0$. This means that each M_i/M_{i+1} is a torsion A/N-module. Then from the definition it follows that $\rho(M_i/M_{i+1})=0$ for each i and so $\rho_N(M) = \sum_{i=0}^{k} \rho(M_i/M_{i+1}) = 0$. □

Lemma 5.5.25. (**Pseudo-Ore condition.**) *Let M be a right A-module, where A is a ring with nilpotent prime radical N such that A/N is a semiprime right Goldie ring. Suppose that $\rho_N(M) < \infty$. If $a \in A$ and r is a right regular element in A then there exists an element $x \in C_A(N)$ such that $ax \in rA$. In particular, $C_A^r(0) \subseteq C_A(N)$.*

Proof. If r is a right regular element in A then $rA \simeq A$ and so $\rho_N(rA) = \rho_N(A)$. Then, by corollary 5.5.21, $\rho_N(A/rA) = 0$. By lemma 5.5.24 this implies that for any element $\bar{a} \in A/rA$ there is an element $x \in C_A(N)$ such that $\bar{a}x = 0$. Let $\bar{a} = a + rA$, then $ax \in rA$. In particular, this implies that $(A/N)/r(A/N)$ is A/N-torsion, so $r+N$ is a regular element in A/N. □

Lemma 5.5.26. *Let A be a ring with prime radical N and Q a classical right ring of fractions of A. Then for any i*

$$(N^i/N^{i+1}) \otimes_A Q \simeq N^iQ/N^{i+1}Q. \tag{5.5.27}$$

Proof. By proposition 5.4.11, Q is a flat left A-module. Therefore by the flatness test, theorem 1.5.15, $\mathcal{I} \otimes_A Q \cong \mathcal{I}Q$ for any right ideal $\mathcal{I}$ of A. This gives in particular the isomorphisms (5.5.27). □

Theorem 5.5.28 (L. Small-R.B. Warfield, Jr. [321]). *Let A be a ring with prime radical N. Then A is a right order in a right Artinian ring if and only if*

1. *A/N is a semiprime right Goldie ring;*
2. *N is nilpotent;*
3. *$\rho_N(A_A) < \infty$;*
4. *$C_A(0) = C_A(N)$.*

Proof. Suppose that A is a ring with prime radical N and all conditions (1), (2), (3), (4) are fulfilled. Then, by lemma 5.5.25, for any $a \in A$ and any regular element $r \in A$ there exists a regular element $x \in C_A(N) = C_A(0)$ and an element $b \in A$ such that $ax = rb$, which implies that A is a right Ore ring. Therefore A has a classical right ring of fractions Q. We now show that Q is a right Artinian ring.

We will show that NQ is a nilpotent two-sided ideal in Q. By lemma 5.4.8, NQ is a right ideal in Q. Suppose that $a \in N$ and $r \in C_A(0)$. Then from the equality $ax = rb$ it follows that $b = r^{-1}ax \in N$, as $r^{-1} \in C_A(0) = C_A(N)$. So $r^{-1}a = bx^{-1} \in NQ$, whence $QN \subseteq NQ$, which implies that NQ is a two-sided ideal and $QNQ = NQ$. Therefore $(NQ)^i = NQNQ \cdots NQ = N^iQ$. Since N is nilpotent, NQ is nilpotent as well. Thus, NQ is a nilpotent two-sided ideal in Q. This implies that $NQ \cap A = N$ and $NQ \subseteq N_1$, where $N_1 = Pr(Q)$ is the prime radical of Q. Therefore Q/NQ is the right ring of fractions of A/N with respect to the subset $C_A(N) = \{r + N \ : r \in C_A(0)\}$. Since $C_A(N) = C_A(0)$, Q/NQ is a classical right ring of fractions of a semiprime right Goldie ring A/N. Therefore Q/NQ is a semisimple ring by theorem 1.11.3. Whence $NQ = N_1$ by theorem 1.1.24 and proposition 1.1.25. So N_1 is nilpotent, i.e., there is $k > 1$ such that $N_1^k = 0$.

Using lemma 5.5.26, one obtains

$$(N^i/N^{i+1}) \otimes_A Q \simeq N^iQ/N^{i+1}Q \simeq N_1^i/N_1^{i+1}. \tag{5.5.29}$$

Since $\rho_N(A_A) < \infty$, and N is nilpotent, $\rho(N^i/N^{i+1}) < \infty$ as well. From (5.5.29) it follows that

$$\text{length}(N_1^i/N_1^{i+1}) = \rho(N^i/N^{i+1}) < \infty.$$

Since Q/N_1 is semisimple and N_1 is nilpotent, from the filtration

$$0 = N_1^k \subseteq \cdots \subseteq N_1^2 \subseteq N_1 \subseteq Q$$

it follows that Q is a right Artinian ring.

Conversely, suppose that Q is a right Artinian classical right ring of fractions of A. Let J be the Jacobson radical of Q and $N_1 = Pr(Q)$ the prime radical of Q. By proposition 1.1.25, $N_1 = J$, and so it is nilpotent. Therefore $N_1 \cap A$ is a nilpotent two-sided ideal in A. Since Q is right Artinian and $N_1 = J$, Q/N_1 is a semisimple ring. We will show that Q/N_1 is the classical right ring of fractions of $A/(N_1 \cap A)$. It is clear that $A/(N_1 \cap A) \subseteq Q/N_1$. Since any regular element $r \in C_A(0)$ is invertible in Q, any element $r + N_1$ is also invertible in Q/N_1, and therefore it is regular in $A/(N_1 \cap A)$. Since any element of Q has the form ar^{-1}, where $r \in C_A(0)$, any element of Q/N_1 has the form $\bar{a}\bar{r}^{-1}$, where $\bar{a} = a+N_1$ and $\bar{r} = r+N_1$. Therefore Q/N_1 is the classical right ring of fractions of $A/(N_1 \cap A)$. Since Q/N_1 is a semisimple ring, $A/(N_1 \cap A)$ is a semiprime right Goldie ring. So that, in particular, $N_1 \cap A$ is a semiprime ideal in A. Thus $N_1 \cap A$ is a semiprime nilpotent ideal of A. Therefore $N_1 \cap A = N$ is the prime radical of A. In particular, N is nilpotent, and A/N is a semiprime right Goldie ring. So statements (1) and (3) are proved.

Since $C_A(N) \subseteq C_Q(J) \cap A = C_Q(0) \cap A \subseteq C_A(0)$, $C_A(0) = C_A(N)$.

From lemma 5.4.8 it follows that $N_1 = J = (J \cap A)Q = (N_1 \cap A)Q = NQ$ is a nilpotent two-sided ideal in Q. And in the same way as above one shows that $J^i = N^iQ$ for any $i \geq 0$.

By lemma 5.5.26

$$(N^i/N^{i+1}) \otimes_A Q \simeq N^iQ/N^{i+1}Q \simeq J^i/J^{i+1}.$$

Assume $J^k = 0$, then

$$\rho_N(A_A) = \sum_{i=0}^{k} \rho(N^i/N^{i+1}) = \sum_{i=0}^{k} \text{length}((N^i/N^{i+1}) \otimes_A Q) =$$

$$= \sum_{i=0}^{k} \text{length}_Q(J^i/J^{i+1}) = \text{length}_Q(Q_Q) < \infty$$

□

Theorem 5.5.30 (L. Small, T.D. Talintyre). *A right Noetherian ring A with prime radical N is a right order in a right Artinian ring if and only if $C_A(0) = C_A(N)$.*

Proof. Since A is a right Noetherian ring, the prime radical N of A is nilpotent by theorem 1.1.28, and A/N is a semiprime right Goldie ring. From proposition 5.5.22 it follows that $\rho_N(A_A) < \infty$. Therefore this theorem is a special case of theorem 5.5.28. □

Note that theorems 5.5.28 and 5.5.30 are true under certain weaker condition. This arises because of the next lemma.

Lemma 5.5.31. *If conditions* (1)-(3) *of theorem* 5.5.28 *are satisfied then $C_A(N) \subseteq C_A(0)$ implies $C_A(N) = C_A(0)$.*

Proof. By lemma 5.5.25, $C_A^r(0) \subseteq C_A(N)$. Suppose that $C_A(N) \subseteq C_A(0)$. Then we have the set of inclusions

$$C_A(0) \subseteq C_A^r(0) \subseteq C_A(N) \subseteq C_A(0)$$

which implies $C_A(0) = C_A(N)$. □

Thanks to this lemma theorems 5.5.28 and 5.5.30 can be rewritten in the following form:

Theorem 5.5.28*. (L. Small-R.B.Warfield, Jr. [321]). *Let A be a ring with prime radical N. Then A is a right order in a right Artinian ring if and only if*

1. *A/N is a semiprime right Goldie ring*;
2. *N is nilpotent*;
3. $\rho_N(A_A) < \infty$;
4. $C_A(N) \subseteq C_A(0)$.

Theorem 5.5.30* (L. Small). *A right Noetherian ring A with prime radical N is a right order in a right Artinian ring if and only if $C_A(N) \subseteq C_A(0)$.*

Corollary 5.5.32. *A Noetherian ring A with prime radical N has an Artinian classical ring of fractions if and only if $C_A(N) \subseteq C_A(0)$.*

5.6 Classical Krull Dimension

Originally the concept of Krull dimension, named after Wolfgang Krull (1899-1971), was introduced for commutative Noetherian rings using the prime ideals of rings.

First such a dimension was considered by E. Noether for algebraic varieties over fields defined by ideals in polynomial rings. Later, in 1928, W. Krull developed her ideas for arbitrary commutative Noetherian rings.

Definition 5.6.1. Let A be a commutative ring and let

$$P_0 \subsetneq P_1 \subsetneq P_2 \subsetneq \cdots \subsetneq P_n$$

be a chain of prime ideals of a ring A. We say that this chain has length n, if all prime ideals are different and $\neq A$. The (**Krull**) **dimension** of A is equal to m (written $\dim A = m$) if m is the supremum of the lengths of such chains of prime ideals in A. If there is no upper bound on the lengths of such chains of proper prime ideals in A, then A is said to be (Krull) **infinite-dimensional** and one writes $\dim A = \infty$.

By definition, a commutative ring A is (Krull) zero-dimensional if and only if each proper prime ideal of A is maximal. Examples of (Krull) zero-dimensional commutative rings are:

1. Fields.
2. Artinian rings. Really, let P be a prime ideal of an Artinian ring A. Then A/P is an Artinian domain, which is always a field. So every prime ideal of A is maximal, which implies that $\dim A = 0$.
3. Finite rings. This follows from the fact that every finite ring is Artinian.
4. Boolean rings [8]. Really, a ring A/P is a Boolean domain for any prime ideal P of a Boolean ring A . So any prime ideal P of A is maximal, which implies that $\dim A = 0$.

Another interesting examples of zero-dimensional commutative rings can be found in [4].

Examples of (Krull) one-dimensional commutative rings are:

1. The ring of integers $\mathbf{Z}$, since all chains of prime ideals in $\mathbf{Z}$ are $0 \subset p\mathbf{Z}$, where p is a prime.
2. A commutative discrete valuation domain A with maximal ideal M, since the only prime ideals of A are M and 0. Therefore $\dim(A) = 1$.
3. Any Dedekind domain which is not a field, since any prime ideal of a Dedekind domain is maximal.

A polynomial ring $F[x_1, x_2, \ldots, x_n, \ldots]$ over a field F is a (Krull) infinite-dimensional ring, since there is the infinite chain of prime ideals which are not maximal:

$$(x_1) \subset (x_1, x_2) \subset (x_1, x_2, x_3) \subset \cdots \subset (x_1, x_2, \ldots, x_n) \subset \cdots$$

---

[8]Recall that an associative ring A is Boolean if each its element is an idempotent.

There are a few different generalizations of (Krull) dimension given above to the case of noncommutative rings. The most important ones were given by R.Renntschler and P. Gabriel [270] and later by G. Krause [197] for noncommutative rings. The general definition of the Krull dimension in theoretical-module form will be considered in the next section.

Consider one of the direct generalizations of definition 5.6.1 for noncommutative rings taking only finite or infinite values.

Definition 5.6.2. (B. Gordon J.C. Robson [117]). Let A be an arbitrary ring. A **little classical Krull dimension**[9], denoted by $\text{k.dim}(A)$, is the supremum of the lengths of chains of prime ideals in A. If no such supremum exists then $\text{k.dim}\, A = \infty$.

Extending this definition G. Krause introduced the classical Krull dimension for noncommutative rings using ordinal values.

For convenience, consider -1 as the least ordinal (instead of 0).

Definition 5.6.3 (G. Krause [197, Definition 11]). The **classical Krull dimension** of a ring A, denoted by $\text{cl.K.dim}(A)$, is defined by transitive recursion as follows.

1. $\text{cl.K.dim}(A) = -1$ if and only if $A = 0$.
2. Define $\mathcal{R}_{-1}(A) = \emptyset$. For each ordinal $\alpha \geq 0$, if $\mathcal{R}_\beta(A)$ has been defined for each $\beta < \alpha$, define $\mathcal{R}_\alpha(A)$ to be the set of prime ideals P such that all prime ideals $Q \supset P$ are contained in $\bigcup\limits_{\beta} \mathcal{R}_\beta(A)$ (e.g. $\mathcal{R}_0(A)$ is the set of maximal ideals of A). If some $\mathcal{R}_\gamma(A)$ contains all prime ideals of A, then the **classical Krull dimension** $\text{cl.K.dim}(A)$ is defined to be the smallest such γ.

Remark 5.6.4. Let $\text{Spec}(A)$ be the set of all prime ideals of a ring A. Then the classical Krull dimension of A is defined to be the first ordinal γ such that $\mathcal{R}_\gamma(A) = \text{Spec}(A)$.

It is possible that there is no ordinal γ such that $\text{cl.K.dim}(A) = \gamma$. In this case one says that A fails to have Krull dimension, or that $\text{cl.K.dim}(A)$ does not exist.

Note that if P is a prime ideal of A then $P \in \mathcal{R}_\beta(A)$ means precisely that $\text{cl.K.dim}(A/P) \leq \beta$.

Example 5.6.5.

Since by definition $\text{cl.K.dim}(A) = 0$ if and only if each prime ideal of A is maximal, every skew field and every simple Artinian ring has classical Krull dimension which equal to 0.

Remark 5.6.6. In general the classical Krull dimension of a ring A need not be finite even for a commutative Noetherian ring A. The first example of a commutative Noetherian ring with infinite classical Krull dimension was given by Nagata (See e.g. [79, Example 9.6]).

---

[9] Sometimes it is called **f-Krull dimension** (see e.g. [327]).

In general a ring may not have the classical Krull dimension. The next theorem yields the criterium of existness of the classical Krull dimension of a ring.

Theorem 5.6.7 (Goodearl, Warfield [115, Proposition 12.1], Albu [1, Exercise 12A]). *If A is a ring then* cl.K.dim(A) *exists iff A satisfies the a.c.c. on prime ideals. In this case* cl.K.dim(A) *is the first γ such that $\mathcal{R}_\gamma = \mathcal{R}_{\gamma+1}$.*

Proof. Suppose A is a ring having classical Krull dimension. Then cl.K.dim$(A) = \alpha$ for some ordinal α and $\mathcal{R}_\alpha(A) = \text{Spec}(A)$. Assume that there is an infinite strictly ascending chain $P_1 \subset P_2 \subset \cdots$ of prime ideals of A. Since $\mathcal{R}_\alpha(A) = \text{Spec}(A)$, $P_1 \in \mathcal{R}_\alpha$ for any such a chain. Choose the least ordinal β such that $P_1 \in \mathcal{R}_\beta(A)$. Then $P_2 \in \mathcal{R}_\gamma(A)$ with $\gamma < \beta$. On the other hand for the chain $P_2 \subset P_3 \subset \cdots$ we have $P_2 \in \mathcal{R}_\alpha(A)$, which contradicts the minimal choice of β. Therefore there are no such infinite strictly ascending chain of prime ideals.

Conversely, suppose that a ring A satisfies the a.c.c. on prime ideals. Consider the chain of sets:

$$\mathcal{R}_0(A) \subseteq \mathcal{R}_1(A) \subseteq \mathcal{R}_2(A) \subseteq \cdots$$

Since the cardinality of these sets is bounded (e.g., by $2^{|A|}$), there is an ordinal α such that $\mathcal{R}_\alpha(A) = \mathcal{R}_{\alpha+1}(A)$. Assume that A does not have the classical Krull dimension. Then $\mathcal{R}_\alpha(A) \neq \text{Spec}(A)$. So there is a prime ideal $P \in \text{Spec}(A) \setminus \mathcal{R}_\alpha(A)$. Choose a prime ideal P which is maximal in $\text{Spec}(A) \setminus \mathcal{R}_\alpha(A)$. If there is a prime ideal Q such that $P \subset Q$ then $Q \in \mathcal{R}_{\alpha+1}(A) = \mathcal{R}_\alpha(A)$, which contradicts to maximal choice of P. Therefore $P \in \mathcal{R}_{\alpha+1}(A) = \mathcal{R}_\alpha(A)$. A contradiction. So A has the classical Krull dimension. □

From this theorem we immediately obtain the sufficient condition for a ring to have classical Krull dimension.

Corollary 5.6.8. *Any ring with a.c.c. on two-sided ideals has classical Krull dimension. In particular, any commutative Noetherian ring has classical Krull dimension.*

From theorem 5.6.8 we obtain the following corollary which yields the example of a ring with infinite classical Krull dimension.

Corollary 5.6.9. *The polynomial ring in infinite indeterminates $A[x_1, x_2, \ldots, x_n, \ldots]$ over any ring A with unit does not have classical Krull dimension.*

Note that in the case of finite classical Krull dimension it coincides with little classical Krull dimension. First this fact was proved by P.Gabriel in [91] for commutative Noetherian rings, and later by R.Gordon and J.C.Robson for rings without Noetherian assumption in [117]. In the general case this fact was proved by A. Woodward in [327].

Proposition 5.6.10. (A. Woodward [327, Lemma 5.2.2]). *A ring A has a finite classical Krull dimension if and only if it has finite little classical Krull dimension.*

In this case

$$\text{k.dim}(A) = \text{cl.K.dim}(A). \tag{5.6.11}$$

Proof. Suppose that a ring A has a finite classical Krull dimension and cl.K.dim$(A) = n$ for some integer $n > 0$. Assume that there is an ascending chain of prime ideals:

$$P_0 \subsetneq P_1 \subsetneq P_2 \subsetneq \cdots \subsetneq P_n \subsetneq P_{n+1}$$

of length $n+1$. Therefore P_n is not a maximal ideal and so $P_n \notin \mathcal{R}_0(A)$. Thus $\text{Spec}(A) \neq \mathcal{R}_0(A)$. Since $P_{n-1} \subsetneq P_n$ and $P_n \notin \mathcal{R}_0(A)$, we obtain, by definition, that $P_{n-1} \notin \mathcal{R}_1(A)$ and thus $\text{Spec}(A) \neq \mathcal{R}_1(A)$. Continuing this process we obtain that $\text{Spec}(A) \neq \mathcal{R}_n(A)$. A contradiction. So there are no a strictly ascending chain of prime ideals of length $> n$. Therefore

$$\text{k.dim}(A) \leq \text{cl.K.dim}(A). \tag{5.6.12}$$

Conversely, suppose that a ring A has a finite little classical Krull dimension and k.dim$(A) = n$ for some integer $n > 0$. In this case A has classical Krull dimension, by theorem 5.6.7. We show that cl.K.dim$(A) \leq n$. Assume that cl.K.dim$(A) > n$, then $\text{Spec}(A) \neq \mathcal{R}_n(A)$. Therefore there is a prime ideal $P_0 \in \text{Spec}(A) \setminus \mathcal{R}_n(A)$. Hence there is a prime ideal $P_1 \supsetneq P_0$ such that $P_1 \in \text{Spec}(A) \setminus \mathcal{R}_{n-1}(A)$. Continuing this process we obtain a strictly ascending chain of prime ideals of A:

$$P_0 \subsetneq P_1 \subsetneq P_2 \subsetneq \cdots \subsetneq P_n \subsetneq P_{n+1}$$

of length $n+1$ with $P_i \notin \mathcal{R}_{n-i}(A)$. This contradicts to the assumption that k.dim$(A) = n$. Therefore

$$\text{cl.K.dim}(A) \leq \text{k.dim}(A). \tag{5.6.13}$$

Hence two obtained inequalities (5.6.12) and (5.6.13) yield the required equality (5.6.11). □

Now we give some important properties concerning the classical Krull dimension of rings considering by G.Krause in [198].

Proposition 5.6.14. (G. Krause [198, Lemma 1.3]). *Let I be an ideal of a ring A and let $\alpha \geq 0$ be an ordinal.*

a. *If P is a prime ideal and $I \subseteq P$, then $P \in \mathcal{R}_\alpha(A)$ if and only if $P/I \in \mathcal{R}_\alpha(A/I)$.*
b. cl.K.dim$(A) = \alpha$ *implies* cl.K.dim$(A/I) \leq \alpha$.
c. *If A is a prime ring with* cl.K.dim$(A) = \alpha$ *and $P \neq 0$ is a prime ideal, then* cl.K.dim$(A/P) < \alpha$.

Proof.

a. We will proceed the proof by induction on α. Note that there is a one-to-one correspondence between prime ideals P containing I and prime ideals of A/I, given by $P \mapsto P/I$. Therefore if $\alpha = 0$ the statement is obvious, since both P and P/I are prime and maximal.
   Let $\alpha > 0$ and assume that statement holds for all ordinals $0 \leq \beta < \alpha$. From the definitions of sets $\mathcal{R}_\alpha(A)$, $P \in \mathcal{R}_\alpha(A)$ if and only if for every $Q \in \mathrm{Spec}(A)$ with $P \subsetneq Q$, $Q \in \mathcal{R}_\beta(A)$ for some $\beta < \alpha$. Then by the induction hypothesis we obtain that this holds if and only if $P/I \subsetneq Q/I \in \mathrm{Spec}(A/I)$ implies $Q/I \in \mathrm{Spec}(A/I)$ for some $\beta < \alpha$, which again, by definition, holds if and only if $P/I \in \mathrm{Spec}(A/I)$.
b. Let $\bar{P} \in \mathrm{Spec}(A/I)$. Then there is a prime ideal $P \in \mathrm{Spec}(A)$ such that $\bar{P} = P/I$ and $I \subseteq P$. Since $\mathrm{cl.K.dim}(A) = \alpha$, $\mathrm{Spec}(A) = \mathcal{R}_\alpha(A)$. Therefore $P \in \mathcal{R}_\alpha(A)$ and so, $P/I \in \mathcal{R}_\alpha(A/I)$ by statement a. Hence $\mathrm{Spec}(A/I) = \mathcal{R}_\alpha(A/I)$, which implies that $\mathrm{cl.K.dim}(A/I) \leq \alpha = \mathrm{cl.K.dim}(A)$.
c. Note that classical Krull dimension of a ring A/P exists by statement b. Since A is a prime ring, $0 \in \mathrm{Spec}(A) = \mathcal{R}_\alpha(A)$. Let $0 \neq P \in \mathrm{Spec}(A)$, then by definition $P \in \mathcal{R}_\beta(A)$ for some $\beta < \alpha$. If $\bar{Q} \in \mathrm{Spec}(A/P)$, then $\bar{Q} = Q/P$ for some $Q \in \mathrm{Spec}(A)$ such that $P \subseteq Q$. Therefore $Q \in \mathcal{R}_\gamma(A)$ for some $\gamma \leq \beta$. So, by statement a, $Q/P \in \mathcal{R}_\gamma(A/P)$. Thus,

$$\mathrm{Spec}(A/P) \subseteq \bigcup_{\gamma \leq \beta} \mathcal{R}_\gamma(A/P) = \mathcal{R}_\beta(A/P)$$

   which implies $\mathrm{cl.K.dim}(A/P) \leq \beta < \alpha$. □

5.7 Krull Dimension

R. Rentschler and P. Gabriel [270] and later G. Krause extended the concept of the Krull dimension given originally for commutative rings to certain modules over noncommutative rings. The properties of rings and modules with Krull dimension were studied in detail by R.Gordon and J.C. Robson [118] (See also [327]).

Definition 5.7.1. Let A be a ring, and M a right A-module. The **Krull dimension** of M (denoted by $\mathrm{K.dim}\, M$ if it exists) is defined by transfinite recursion as follows:

1. $\mathrm{K.dim}\, M = -1$ if and only if $M = 0$.
2. If α is an ordinal and $\mathrm{K.dim}\, M \neq \delta$ for any ordinal $\delta < \alpha$, then $\mathrm{K.dim}\, M = \alpha$ if for every countable descending chain $M_0 \supseteq M_1 \supseteq \ldots$ of submodules M_i of M one has that $\mathrm{K.dim}(M_{i-1}/M_i) < \alpha$ for all but finitely many indices i.
   It is possible that there is no ordinal α such that $\mathrm{K.dim}\, M = \alpha$. In this case one says that M fails to have Krull dimension, or that $\mathrm{K.dim}\, M$ is not exists.
   The **right Krull dimension** of a ring A, written by $\mathrm{r.K.dim}\, A$, is defined as the Krull dimension of the right regular module A_A (if it exists). Similarly, one can define the **left Krull dimension** of a ring A, written by $\mathrm{l.K.dim}\, A$.

It may be worthwhile to note again that in general a ring A need not have a Krull dimension in the sense above. And even if a ring A has a right Krull dimension, A need not have the same left Krull dimension; indeed, it might not have a left Krull dimension at all (See the examples below). In general left and right dimension can be different. Nevertheless there exist some classes of rings in which the existence of the right Krull dimension guarantees that it has the same left Krull dimension. Some examples of such rings are right fully bounded right Noetherian rings (**right FBN-rings**)[10] [167], and serial rings [329].

Examples 5.7.2.

1. An A-module M has Krull dimension 0 if and only if M is a non-zero Artinian module. Indeed, by definition, $\text{K.dim}\, M = 0$ if and only if $M \neq 0$ and for every descending chain
$$M_0 \supseteq M_1 \supseteq \dots$$
of submodules M_i of M one has $\text{K.dim}(M_{i-1}/M_i) = -1$, i.e., $M_{i-1}/M_i = 0$ for all but finitely many indices i. Therefore in this case M satisfies the descending chain condition on submodules, i.e., if M is Artinian, $\text{K.dim}(M) = 0$. In particular, A is a right Artinian ring if and only if $\text{r.K.dim}(A) = 0$.
2. Let O be a discrete valuation ring with unique maximal ideal $\mathcal{M} = \pi O$. Then there is a unique infinite descending chain of ideals of O:
$$O \supseteq \mathcal{M} \supseteq \mathcal{M}^2 \supseteq \cdots$$
So $\text{r.K.dim} O > 0$. Since every factor $\mathcal{M}^n/\mathcal{M}^{n+1}$ is isomorphic to a division ring $O/\mathcal{M}$, it follows that $\text{r.K.dim}(O) = 1$.
3. If A is a commutative principal ideal domain, which is not a field, then $\text{r.K.dim}(A_A) = 1$. In particular, if $\mathbf{Z}$ is a ring of integers, then $\text{K.dim}\, \mathbf{Z} = 1$.
4. If a module M contains an infinite direct sum of copies of a non-zero module N, then the Krull dimension of M is not defined. To prove this first consider a simplified version. Let M be a countably infinite sum of copies of a non-zero module N. Consider $M' = \bigoplus_{i=1}^{\infty} M_i$ where $M_i = M$ for all i. Let M_j, $(j = 1,2,\dots)$, be the submodule of all vectors $(m_1, m_2, m_3, \dots) \in M'$ which have the first j entries equal to zero. This gives a descending chain
$$M' = M_0 \supset M_1 \supset \cdots \supset M_n \supset \cdots$$
such that each quotient M_n/M_{n+1} is isomorphic to M. Now suppose that $\text{K.dim}(M) = \alpha$. Then by the existence of this chain $\text{K.dim}(M') > \alpha$ (or does not exist). But $M \simeq M'$ so $\text{K.dim}(M') = \alpha$. A contradiction.

---

[10] A ring is **right bounded** if every essential right ideal of it contains an ideal which is essential as a right ideal. A **right FBN-ring** A is a right Noetherian ring such that A/P is right bounded for every prime ideal P.

The more general statement above now follows by applying Krause's lemma 5.6.3 below.

5. Let
$$A = \begin{pmatrix} \mathbf{Q} & \mathbf{R} \\ 0 & \mathbf{R} \end{pmatrix},$$
where $\mathbf{Q}$ is the field of rational numbers and $\mathbf{R}$ is the field of real numbers. Then A is a right Artinian ring, and therefore $\text{r.K.dim}(A) = 0$. On the other hand ${}_AA$ contains an infinite direct sum of copies of a simple left module, and so $\text{l.K.dim}(A)$ is not defined.

Lemma 5.7.3. (G. Krause). *Let A be a ring. If M is an A-module and N is a submodule of M then*

$$\text{K.dim}\, M = \sup\{\text{K.dim}(M/N), \text{K.dim}\, N\} \tag{5.7.4}$$

if either side exists.

Proof. First assume that $\text{K.dim}\, M$ exists. Since the natural map of the lattice of submodules of N to the lattice of submodules of M is a lattice map that preserves proper inclusions,
$$\text{K.dim}\, N \leq \text{K.dim}\, M.$$
Similar arguments show that
$$\text{K.dim}(M/N) \leq \text{K.dim}\, M.$$
Therefore if $\text{K.dim}\, M$ exists then both $\text{K.dim}\, N$ and $\text{K.dim}(M/N)$ exist, and
$$\text{K.dim}\, M \geq \sup\{\text{K.dim}(M/N), \text{K.dim}\, N\}.$$

Now suppose that both $\text{K.dim}\, N$ and $\text{K.dim}(M/N)$ exist. Let $\alpha = \sup\{\text{K.dim}(M/N), \text{K.dim}\, N\}$. We will prove that $\text{K.dim}\, M$ exists and $\text{K.dim}\, M \leq \alpha$. We will proceed the proof by induction on α. Suppose that this statement is true for all modules with Krull dimension $< \alpha$.

The case $\alpha = -1$ is obvious.

Let $M_0 \supseteq M_1 \supseteq \dots$ be a descending chain of submodules of M. Then the submodules $(M_i + N)/N$ form a descending chain of submodules of M/N, and so, as $\text{K.dim}\, M/N$ exists, we have
$$\text{K.dim}((M_{i-1} + N)/(M_i + N)) < \text{K.dim}(M/N) \leq \alpha$$
for all but finitely many indices i.

Analogously, the submodules $M_i \cap N$ form a descending chain of submodules of N, and as $\text{K.dim}\, N$ exists we have
$$\text{K.dim}((M_{i-1} \cap N)/(M_i \cap N)) < \text{K.dim}(N) \leq \alpha$$
for all but finitely many indices i.

For any i, the kernel of the natural epimorphism

$$\varphi_i : M_{i-1}/M_i \longrightarrow (M_{i-1} + N)/(M_i + N)$$

is isomorphic to $(M_{i-1} \cap N)/(M_i \cap N)$. So that $\text{K.dim}\,(\text{Ker}(\varphi_i)) < \alpha$. Thus by induction hypothesis

$$\text{K.dim}(M_{i-1}/M_i) = \sup\{\text{K.dim}((M_{i-1} + N)/(M_i + N)), \text{K.dim}(\text{Ker}(\varphi_i))\} < \alpha$$

for all but finitely many indices i. This shows that $\text{K.dim}M$ exists and $\text{K.dim}M \le \alpha$. □

Corollary 5.7.5. *Let $M_1, M_2, \dots, M_n$ be A-modules with Krull dimension. Then*

$$\text{K.dim}(M_1 \oplus M_2 \oplus \dots \oplus M_n) = \sup\{\text{K.dim}(M_1), \dots, \text{K.dim}(M_n)\}.$$

Corollary 5.7.6. *Let A be a ring with right Krull dimension, and let M be a finitely generated right A-module. Then M has Krull dimension and*

$$\text{K.dim}\, M \le \text{r.K.dim}\, A.$$

Proof. From lemma 5.7.3 it follows that $\text{K.dim}(A^n) = \text{r.K.dim}\, A$. Then the epimorphism $A^n \to M \to 0$ shows that $\text{K.dim}\, M \le \text{r.K.dim}\, A$, by lemma 5.7.3 and corollary 5.7.5. □

Corollary 5.7.7. *Let A be a ring, then*

$$\text{r.K.dim}\, A = \sup\{\text{K.dim}\, M \ : \ M \text{ is a finitely generated right } A\text{-module}\}$$

if either side exists.

Proposition 5.7.8.

1. *Every homomorphic image of a ring A with Krull dimension has Krull dimension less than or equal to* $\text{K.dim}\, A$.
2. *The Krull dimension of a ring has the Morita equivalence property, i.e., Morita equivalent*[11] *rings have the same Krull dimension.*
3. *If A is a ring with Krull dimension and P is a finitely generated projective A-module, then* $\text{End}\, P$ *has Krull dimension less than or equal to* $\text{K.dim}\, A$.

Proof.

1. This follows immediately from lemma 5.7.3.

[11] Recall that rings are Morita equivalent if their categories of modules are equivalent. See e.g. [146, Section 10].

2. This follows from the fact that a category equivalence preserves lattices of submodules.
3. Let $S = \operatorname{End} P$. There is a monomorphism from the lattice of right ideals I of S into the lattice of A-submodules of P given by $I \mapsto IP$. Thus $\text{K.dim}\, S \leq \text{K.dim}\, P_A \leq \text{K.dim}\, A$. □

Not every module has Krull dimension. But there is a wide class of modules with Krull dimension.

Proposition 5.7.9. (P. Gabriel [91]). *Every Noetherian module has Krull dimension. In particular, any right (left) Noetherian ring has right (left) Krull dimension.*

Proof. Assume otherwise. Let M be a Noetherian module which has no Krull dimension. Using Noetherian property, one can assume by induction that every proper factor module of M has Krull dimension. Since any Artinian module has Krull dimension 0, we can assume that M is not Artinian. Let

$$\alpha = \sup\{\text{K.dim}(M/N) \ : \ N \text{ is a non-zero submodule of } M\}.$$

Let $M = M_0 \supseteq M_1 \supseteq \ldots$ be a descending chain of submodules M_i of M and $M_i \neq 0$ for all i. Then every proper factor module of M in this chain has Krull dimension $\text{K.dim}(M_i/M_{i+1}) \leq \alpha$ for each i. The definition of the Krull dimension then shows that $\text{K.dim}\, M \leq \alpha + 1$, a contradiction. □

Proposition 5.7.10. (G. Krause [197]). *A module with Krull dimension has finite uniform dimension. In particular, any ring with right (left) Krull dimension has finite uniform dimension.*

Proof. The result is trivial for a module with Krull dimension −1.

Suppose that the statement is false, i.e., there are modules with Krull dimension and infinite uniform dimension. Amongst all such modules choose one, M, with minimal Krull dimension, α say. Then $M \supseteq \bigoplus_{i=1}^{\infty} N_i$ for non-zero submodules N_i. Set $M_n = \bigoplus_{j=1}^{\infty} N_{j2^n}$ for each integer n. Consider an infinite descending chain $M_0 \supset M_1 \supset M_2 \supset \ldots$ of submodules of M. Each factor M_n/M_{n+1} of this chain contains an infinite direct sum of submodules and so has infinite uniform dimension. Since $\text{K.dim}(M_n/M_{n+1}) \leq \text{K.dim}\, M = \alpha$, by minimality of α we obtain that $\text{K.dim}\, M_n/M_{n+1} = \alpha$ for all $n > 0$. Whence, by the definition of the Krull dimension, $\text{K.dim}\, M > \alpha$. A contradiction. □

Proposition 5.7.11. (G. Michler [241], G. Krause [197]). *If M is a module with Krull dimension and*

$$\alpha = \sup\{\text{K.dim}(M/E) + 1 \ : \ E \text{ is an essential submodule of } M\} \qquad (5.7.12)$$

then $\text{K.dim}\, M \leq \alpha$.

Proof. Suppose K.dim $M > \alpha$. Then there is an infinite chain of submodules of M:

$$M_0 \supset M_1 \supset \dots$$

such that K.dim $M_i/M_{i+1} \geq \alpha$ for almost all i. By proposition 5.7.10 there is an integer n such that M_n and M_{n+1} have the same uniform dimension. Choose a submodule N of M which is maximal with respect to $N \cap M_n = 0$. Then $N \oplus M_n$ is an essential submodule of M by proposition 5.1.27. Since u.dim(M_n) = u.dim(M_{n+1}), $N \oplus M_{n+1} \subseteq_e N \oplus M_n$ by proposition 5.1.9(2). Therefore $N \oplus M_{n+1} \subseteq_e M$ by [146, Lemma 5.3.2]. However, $M_n/M_{n+1} \simeq (N \oplus M_n)/(N \oplus M_{n+1}) \subseteq M/(N \oplus M_{n+1})$, whence K.dim $M_n/M_{n+1} + 1 \leq \alpha$. A contradiction. □

Definition 5.7.13. A non-zero A-module M is called α-**critical** for some ordinal α if K.dim $M = \alpha$ and K.dim$(M/N) < \alpha$ for each non-zero submodule N. A non-zero module M is called **critical** if it is α-critical for some α. A non-zero right ideal of A is called **critical** if it is critical as a right A-module.

Remark 5.7.14. Note, that K.dim N = K.dim M for any non-zero submodule N of a critical module M, which follows immediately from lemma 5.7.3.

Lemma 5.7.15. *Any non-zero module with Krull dimension contains a critical submodule.*

Proof. Let A be a ring, and let M be a non-zero right A-module with Krull dimension. Among all submodules of M choose a submodule N whose Krull dimension is minimal. Suppose K.dim $N = \alpha$ for some ordinal $\alpha \geq 0$. If N is not critical it contains a submodule N_1 such that K.dim$(N/N_1) = \alpha$. By the minimality of α and lemma 5.7.3, K.dim$(N_1) = \alpha$. Applying the same argument to N_1 and so on we obtain a chain $N = N_0 \supseteq N_1 \supseteq N_2 \supseteq \dots$ of submodules of M with K.dim$(N_i/N_{i+1}) = \alpha$ for all $i \geq 1$. Since K.dim $N = \alpha$, this chain must terminate, which means that in this chain there is a critical submodule. □

Lemma 5.7.16. *Let A be a ring with right Krull dimension, and I a critical right ideal of A.*

1. *If $I^2 \neq 0$ then there exists $a \in I$ such that* $\text{r.ann}_A(a) \cap I = 0$.
2. *If I is a nil-ideal, then $I^2 = 0$.*

Proof.

1. Suppose that $I^2 \neq 0$. Then there exists $a \in I$ such that $aI \neq 0$. Define a non-zero map $\varphi : I \to I$ by $\varphi(x) = ax$ for all $x \in I$. Since I is a critical ideal and Im $\subseteq I$, taking into account remark 5.7.14 we obtain

$$\text{K.dim}(I/\text{Ker}(\varphi)) = \text{K.dim}(\text{Im}(\varphi)) = \text{K.dim}(I).$$

   Therefore $\text{Ker}(\varphi) = 0$, whence $\text{r.ann}_A(a) \cap I = 0$.
2. Suppose that I is a nil-ideal and $I^2 \neq 0$. Then, by statement (1), there exists $a \in I$ such that $\text{r.ann}_A(a) \cap I = 0$. So there exists n such that $a^n \neq 0$ and $a^{n+1} = 0$. Then $a^n \in \text{r.ann}_A(a) \cap I = 0$. A contradiction. □

Proposition 5.7.17. *If a semiprime ring A has right Krull dimension then A is a right Goldie ring.*

Proof. By proposition 5.7.10, A has finite uniform dimension. So, by theorem 5.4.10, it suffices to prove that A is a right nonsingular ring. Suppose $\mathcal{Z}(A_A) \neq 0$, then it has Krull dimension. So, by lemma 5.7.15, $\mathcal{Z}(A_A)$ contains a critical right ideal I which is non-zero by definition. Since A is semiprime, $I^2 \neq 0$. Then, by lemma 5.7.16(1), there exists $a \in I$ such that $\text{r.ann}_A(a) \cap I = 0$. Since $a \in I \subset \mathcal{Z}(A_A)$, and therefore $\text{r.ann}_A(x)$ is an essential ideal in A_A, we obtain a contradiction. □

Proposition 5.7.18. *If a semiprime ring A has right Krull dimension then*

$$\text{K.dim}(A) = \sup\{\text{K.dim}(A/E) + 1 \mid E \text{ is an essential right ideal of } A\}$$

Proof. Denote

$$\alpha = \sup\{\text{K.dim}(A/E) + 1 \mid E \text{ is an essential right ideal of } A\}.$$

Then the inequality $\text{K.dim}(A) \leq \alpha$ follows from proposition 5.7.11.

We prove now the inverse inequality, i.e., $\text{K.dim}(A) \leq \alpha$. Let E be an essential right ideal of A. By proposition 5.7.17, A is a right Goldie ring. Therefore there is a regular element $x \in E$, by corollary 5.4.7. Then $x^n A/x^{n+1}A \simeq A/xA$ for any $n \geq 1$. So that $\text{K.dim}(x^n A/x^{n+1}A) = \text{K.dim}(A/xA) \geq \text{K.dim}(A/E)$. Since x is a regular element, we have the infinite strictly descending chain of right ideals $A \supsetneq xA \supsetneq x^2 A \supsetneq \cdots$. Therefore $\text{K.dim}A \geq \text{K.dim}(A/E) + 1$. So that $\text{K.dim}(A) \geq \alpha$, which implies $\text{K.dim}(A) = \alpha$ that required. □

Corollary 5.7.19. *If a semiprime ring A has right Krull dimension then*

$$\text{K.dim}(A) = \text{K.dim}(E) \tag{5.7.20}$$

for any essential ideal E in A.

Proof. This follows immediately from proposition 5.7.18 and lemma 5.7.3. □

Proposition 5.7.21. *Let A be a ring with right Krull dimension. Then A has the ascending chain condition on prime ideals.*

Proof. Let $P_1 \subsetneq P_2$ be two prime ideals. Then P_2/P_1 is a non-zero ideal in a ring A/P_1 which is prime and have the Krull dimension. Since $\text{u.dim}(A/P_1) = \text{u.dim}(P_2/P_1) < \infty$, P_2/P_1 is an essential ideal of A/P_1, by proposition 5.1.9(2). Then from proposition 5.7.18 it follows that $\text{K.dim}(A/P_1) \geq \text{K.dim}(A/P_2) + 1 > \text{K.dim}(A/P_2)$. Therefore any strictly ascending chain $P_1 \subsetneq P_2 \subsetneq P_3 \subsetneq \cdots$ of prime ideals of A generates a strictly decreasing chain of ordinals $\text{K.dim}(A/P_1) > \text{K.dim}(A/P_2) > \cdots$. A contradiction. □

From the proof of this proposition we obtain the following useful fact.

Corollary 5.7.22. *Let A be a ring with right Krull dimension. If $P_1 \subsetneq P_2$ are prime ideals of A then* $\text{r.K.dim}(A/P_2) < \text{r.K.dim}(A/P_1)$.

Proposition 5.7.23. *Let e be a non-zero idempotent of a ring A, and let M be an A-module. Then*

1. *There is an injective homomorphism from the lattice of right eAe-submodules of M to the lattice of right A-submodules of M, and there is a surjective homomorphism from the lattice of right A-submodules of M to the lattice of right eAe-submodules of M.*
2. *If a right A-module M has Krull dimension, then a right eAe-module Me has Krull dimension.*
3. *If a ring A has Krull dimension, then the ring eAe has Krull dimension.*

Proof.

1. It is easy to verify that for any set $\{N_i\}_{i\in I}$ of submodules of M one has the equalities:
$$(\bigcap_{i\in I} N_i)e = \bigcap_{i\in I}(N_i e)$$
$$(\sum_{i\in I} N_i)e = \sum_{i\in I}(N_i e)$$
Define a homomorphism φ from the lattice of right A-submodules of M to the lattice of right eAe-submodules of M by the rule $\varphi(N) = Ne$ for any $N \subseteq M$. We now show that φ is surjective, i.e., every eAe-submodule of M has the form Ne for some A-module $N \subseteq M$. Let $N' \subseteq M$ be a right eAe-submodule of M where M is regarded as a right eAe-module using the inclusion $eAe \subset A$. Consider $N'eA$ as a right A-submodule of M. Then $(N'eA)e = N'(eAe) = N'$. So every right eAe-submodule of M is of the form Ne for some right A-submodule N of M. This shows that φ is surjective. On the other hand we can define a homomorphism from the lattice of right eAe-submodules of M to the lattice of right A-submodules of M given by the rule $\psi(Ne) = NeA$ for any right eAe-submodules of M. Then $\psi\varphi(Ne) = \psi(NeA) = (NeA)e = N(eAe) = N$. And so ψ is injective.
The lattice preserving properties were already stated above.
2. This follows from statement 1 and the fact that if $M_A = \sum_{i\in I} M_i$ then $Me_{eAe} = \sum_{i\in I} M_i e$.
3. This follows from statements 1 and 2. □

Recall that an FDI-ring is a ring with finite decomposition of identity into a sum of primitive pairwise orthogonal idempotents (See [146, page 56]).

Proposition 5.7.24. *Let A be an FDI-ring and $1 = e_1 + e_2 + \ldots + e_n$ a sum of non-zero pairwise orthogonal idempotents. Then*

1. *If M is a right A-module such that each right e_iAe_i-module Me_i has Krull dimension α_i for all $i = 1,2,\dots,n$, then M has Krull dimension and that* $\mathrm{K.dim}(M)$ *is equal to* $\max\{\alpha_i\}$.
2. *If for all i and j, the right e_iAe_i-module e_jAe_i has Krull dimension α_{ij}, then A_A has Krull dimension and it is equal to* $\max\{\alpha_{ij}\}$.
3. *If A is a right Noetherian ring and the right Krull dimension of the right Noetherian ring e_iAe_i is equal to γ_i for $i = 1,2,\dots,n$, then*
$$\mathrm{K.dim}(A_A) = \max\{\gamma_i\}.$$
4. *If A is a right Noetherian ring and the right Krull dimension of the Noetherian ring e_iAe_i does not exceed 1 for $i = 1,2,\dots,n$, then*
$$\mathrm{K.dim}(A_A) \leq 1.$$

Proof.

1. This follows from the definition of the Krull dimension and corollary 5.7.5.
2. Let $\beta_j = \max\limits_{1\le i\le n}\{\alpha_{ij}\}$. From statement 1 it follows that $\mathrm{K.dim}(e_jA_A) = \beta_j$. Since $A_A = \bigoplus\limits_{j=1}^{n} e_jA$ and $e_jA = \bigoplus\limits_{i=1}^{n} e_jAe_i$, the Krull dimension of A exists and equals to $\max\limits_{1\le j\le n}\{\beta_j\} = \max\limits_{1\le i,j\le n}\{\alpha_{ij}\}$.
3. Fix $i,j \in \{1,2,\dots,n\}$. Since A is a right Noetherian ring, e_jA is a right Noetherian A-module, and e_jAe_i is a right Noetherian e_iAe_i-module. By proposition 5.7.9 e_jAe_i has Krull dimension, say α_{ij}. Since e_jAe_i is a finitely generated e_iAe_i-module, and e_iAe_i has Krull dimension, from corollary 5.7.6 it follows that $\alpha_{ij} \le \gamma_i \le \mathrm{K.dim}(A_A)$. By statement 2, $\mathrm{K.dim}(A_A) = \max\limits_{1\le i,j\le n}\{\alpha_{ij}\}$. Whence $\mathrm{K.dim}(A_A) = \max\limits_{1\le i\le n}\{\gamma_i\}$.
4. This follows immediately from statement 3. □

The next our aim is to prove the following impotent result.

Theorem 5.7.25 (Gordon, Robson [118, Lemma 5.6], [217, Theorem 5]). *Each nil-ideal N of a ring A with right Krull dimension is nilpotent.*

For this end we first prove some statements which are of some independent interest.

Lemma 5.7.26. *If a ring A has right Krull dimension then the Jacobson radical of A does not contain non-zero idempotent*[12] *right ideals.*

Proof. Let $J(A)$ be the Jacobson radical of a ring A, I a right ideal of A, and $I = I^2 \subseteq J(A)$. Let M be a right A-module which has Krull dimension (note, that the

[12]Recall that an ideal is called idempotent if $I^2 = I$.

ideal I has Krull dimension). We show that $MI = 0$. Define the critical socle series of M as follows: $S_0(M) = 0$, $S_{\gamma+1}(M)/S_\gamma(M)$ is the sum of all critical submodules of $M/S_\gamma(M)$ for every ordinal γ, and $S_\alpha(M) = \bigcup\limits_{\lambda<\alpha} S_\lambda(M)$ for every limit ordinal α. We prove that $MI = 0$ for all M with Krull dimension α. We will proceed the proof by induction on α. The case $\alpha = -1$ is trivial. Suppose that $\alpha \geq 0$. Since $M = S_\lambda(M)$ for some ordinal λ and $I^2 = I$, it suffices to show that $NI = 0$ for each β-critical module N such that $\beta \leq \alpha$. If $\beta < \alpha$ then $NI = 0$ by induction hypothesis. So one can suppose that $\beta = \alpha$. Let D be any non-zero submodule of N. Then K.dim$(N/D) < \alpha$ and so $(N/D)I = 0$, that is, $NI \subseteq D$. Thus NI is contained in every non-zero submodule of N. Therefore NI is either simple or $NI = 0$. If NI is simple then $NI = NI^2 \subset (NI)J(A) = 0$ by Nakayama's lemma 1.1.20, a contradiction. Therefore $NI = 0$. Thus, by induction, $MI = 0$ for each M with Krull dimension, and $I = I^2 = 0$ as a particular case. □

Proposition 5.7.27 (T.H. Lenagan [218, Lemma]). *Let M be a module with an ascending chain of submodules $0 = X_0 \subseteq X_1 \subseteq X_2 \subseteq \cdots$ such that $M = \bigcup\limits_{i=1}^{\infty} X_i$. If there is an infinite descending chain of submodules*

$$M = M_0 \supseteq M_1 \supseteq \cdots$$

such that for all i

$$M_i \cap X_{i+1} \nsubseteq M_{i+1} + X_i \tag{5.7.28}$$

then M does not have Krull dimension.

Proof. Put $B = \sum\limits_{i=1}^{\infty} (X_i \cap M_i)$. Note that $B \subseteq X_n + M_{n+1}$, for all n. Indeed, if $i < n$, then $X_i \cap M_i \subseteq X_n \subseteq X_n + M_{n+1}$. If $i \geq n$ then $X_i \cap M_i \subseteq M_{n+1} \subseteq X_n + M_{n+1}$.

We now show that the module M/B has an infinite uniform dimension. Namely, we show that this module contains a submodule which is an infinite direct sum of non-zero submodules. Taking into account that $B \subseteq X_n + M_{n+1}$ for all n, and (5.7.28), we obtain that $X_{n+1} \cap M_n \nsubseteq B$ for all n. So that N_i/B is a non-zero submodule in M/B for all i, where

$$N_i = X_{i+1} \cap M_i + B.$$

From $X_i \subseteq X_{i+1}$ it follows that $B \subseteq \sum\limits_{i=1}^{\infty} (X_{i+1} \cap M_i) = C \subseteq M$. We claim that $C/B = \bigoplus\limits_{i=1}^{\infty} (N_i/B)$. It suffices to prove that

$$\Big(\sum_{i=1}^{n} (N_i/B)\Big) \cap (N_{n+1}/B) = 0$$

for each $n \geq 1$, or equivalently, that

$$\Big(\sum_{i=1}^{n} N_i\Big) \cap N_{n+1} \subseteq B \tag{5.7.29}$$

Let $x \in (\sum_{i=1}^{n} N_i) \cap N_{n+1}$. Then

$$x \in \sum_{i=1}^{n} N_i = \sum_{i=1}^{n}(X_{i+1} \cap M_i + B) = \sum_{i=1}^{n}(X_{i+1} \cap M_i) + \sum_{i=1}^{\infty}(X_i \cap M_i) =$$

$$= \sum_{i=1}^{n}(X_{i+1} \cap M_i) + \sum_{i=1}^{n}(X_i \cap M_i) + \sum_{i=n+1}^{\infty}(X_i \cap M_i) =$$

$$= \sum_{i=1}^{n}(X_{i+1} \cap M_i) + \sum_{i=n+1}^{\infty}(X_i \cap M_i) \subseteq X_{n+1} + \sum_{i=n+1}^{\infty}(X_i \cap M_i),$$

since $X_i \subseteq X_{i+1}$ for all i.

Therefore $x = y + z$, where $y \in X_{n+1}$ and $z \in \sum_{i=n+1}^{\infty}(X_i \cap M_i) \subseteq B$. On the other hand $x \in X_{n+2} \cap M_{n+1} + B$. If $x \in B$ then we have done, i.e. (5.7.29) holds. So we can assume that $x \in X_{n+2} \cap M_{n+1}$. Then $x - z \in M_{n+1} + \sum_{i=n+1}^{\infty}(X_i \cap M_i) \subseteq M_{n+1}$, since $M_i \subseteq M_{n+1}$ for all $i > n$. Thus, $y = x - z \in X_{n+1} \cap M_{n+1} \subseteq B$. Whence $x \in B$, i.e. (5.7.29) holds.

Thus the module M/B contains a submodule C/B which is an infinite direct sum of non-zero submodules, that is, $\text{u.dim}(M/B) = \infty$. Thus, M does not have Krull dimension, by lemma 5.7.3 and proposition 5.7.10. □

Lemma 5.7.30 (T.H. Lenagan [217, Corollary 4].) *Let N be a module with Krull dimension, and an ascending chain of submodules*

$$0 = B_0 \subseteq B_1 \subseteq B_2 \subseteq \cdots$$

in N. If $I = \bigcup_{i=1}^{\infty} B_i$, then there exist some n, j such that

$$I^n \subseteq I^{n+1} + B_j. \tag{5.7.31}$$

Proof. Suppose that $I^n \not\subseteq I^{n+1} + B_j$ for all n, j. We can put $A_0 = 0$. Suppose we have constructed the first n terms of a chain of submodules $0 = X_0 \subseteq X_1 \subseteq X_2 \subseteq \cdots$ in N. Since $I = \bigcup_{i=1}^{\infty} B_i$ and $I^n \not\subseteq I^{n+1} + B_n$, for all n we obtain that $I^n \cap X_j \not\subseteq I^{n+1} + B_n$ for some j such that $X_n \subseteq B_j$. Set $X_{n+1} = B_j$. So we have two chains of submodules $I \supseteq I^2 \supseteq \cdots \supseteq I^n \supseteq \cdots$ and $0 = X_0 \subseteq X_1 \subseteq X_2 \subseteq \cdots$ such that $X_{i+1} \cap I^n \not\subseteq I^{n+1}$ for all i. Then from proposition 5.7.27 it follows that the module N does not have Krull dimension. A contradiction. □

Proof of theorem 5.7.25. Let $N \neq 0$ be a nil-ideal of a ring A. Put $C_0 = 0$. Since A is a ring with right Krull dimension, by lemma 5.7.15 and lemma 5.7.16 there is a right ideal C in N such that $C^2 = 0$. Then AC is a two-sided ideal in A and $(NC)^2 = NC^2 = 0$. So N contains a two-sided ideal I such that $I^2 = 0$. By Zorn's lemma one can choose an ideal C_1 in N to be maximal with respect to the property $C_1^2 = 0 = C_0$. Since the ring A/C_1 has the Krull dimension one can repeat the same argument, i.e., there is an ideal I_2/C_1 in N/C_1 such that $(I_2/C_1)^2 = 0$. Applying Zorn's lemma we can choose such an ideal C_2 that $N \supseteq C_2 \supseteq C_1$ which is maximal to the property $(C_2/C_1)^2 = 0$. So $0 = C_0 \subseteq C_1 \subseteq N$ and $C_1^2 \subseteq C_0$. If there has already been chosen an ideal C_i we can choose by Zorn's lemma an ideal $C_{i+1} \subseteq N$ such that $C_i \subseteq C_{i+1}$ and which is maximal with respect to the property $(C_{i+1}/C_i)^2 = 0$. So there exists a chain of ideals $0 = C_0 \subseteq C_1 \subseteq C_2 \subseteq \cdots \subseteq C_n \subseteq \cdots$ in N such that C_{i+1} is maximal with respect to inclusion

$$C_{i+1}^2 \subseteq C_i \tag{5.7.32}$$

for all i. Since $C_0 = 0$ all ideals C_i are obviously nilpotent. Set $I = \bigcup_{i=1}^{\infty} C_i$. Consider the infinite descending chain of ideals $0 = I^0 \supset I \supset I^2 \supset \cdots$. Since N has Krull dimension, from lemma 5.7.30 it follows that there exist n, j such that $I^n \subseteq I^{n+1} + C_j$. Consider a ring A/C_j which has Krull dimension. Then $J = (I^n + C_j)/C_j \subseteq N/C_j$ is an idempotent ideal in A/C_j. Since N/C_j is a nil-ideal in A/C_j, it is contained in the Jacobson radical of A/C_j. Therefore $J = 0$, by lemma 5.7.26. Therefore $I^n \subseteq C_j$ for some n, j. If $n > 1$ then

$$(I^{n-1} + C_{j+1})^2 = I^{2(n-1)} + I^{n-1}C_{j+1} + C_{j+1}I^{n-1} + C_{j+1}^2 \subseteq C_j.$$

Since C_{j+1} is the maximal ideal with respect to inclusion (5.7.32), $I^{n-1} \subseteq C_{j+1}$. Analogously for all m such that $n - m > 0$ one can prove by induction on m that $I^{n-m} \subseteq C_{j+m}$. In particular for $m = n-1$ we obtain that $I \subseteq C_k$ where $k = j+n-1$. Therefore $C_k \subseteq \bigcup_{i=1}^{\infty} C_i = I \subseteq C_k$, which implies that $I = C_k$. Thus $I \subseteq N$ is nilpotent.

Suppose $N \neq I$. Then N/I is a non-zero nil ideal in A/I with Krull dimension. Then N/I contains a non-zero critical ideal whose square is zero. On the other hand, since $C_{k+m} \subseteq I = C_k \subseteq C_{k+m}$, $C_k = C_{k+m}$ for all $m \geq 0$. Therefore the ring A/I contains no non-zero ideals whose square is zero. This contradiction shows that $N = I$, which implies that N is nilpotent. □

Now we give some results which follows from theorem 5.7.25

Theorem 5.7.33 (R. Gordon, J.C. Robson [117, Lemma 5.6], T.H. Lenagan [215, Theorem 3.2]). *The prime radical of a ring A with right Krull dimension is nilpotent.*

Proof. Since the prime radical of any ring is a nil ideal by proposition 1.1.27, the statement follows immediately from theorem 5.7.25. □

Theorem 5.7.34 (T.H. Lenagan [217, Theorem 8]). *Let A be a ring with right Krull dimension, and N the prime radical of A. Then A has a classical right ring of fractions if and only if* $C_A(0) = C_A(N)$.

Proof. By theorem 5.7.33 the prime radical N is nilpotent. Since A/N is a semiprime ring and has right Krull dimension, it is a right Goldie ring, by proposition 5.7.17. Therefore $\rho(N^i/N^{i+1})$ is finite for every i, by propositions 5.5.6 and 5.7.10. So $\rho(A_A) < \infty$, and the statement follows from theorem 5.5.28. □

Proposition 5.7.35 (R. Gordon, J.C. Robson [117, Proposition 7.3, Theorem 7.4]). *Let A be a ring with right Krull dimension and let I be a proper ideal of A. Then A contains only a finite number of prime ideals in A minimal over I. The ideal I contains some finite product of these prime ideals.*

Proof. Let I be a proper ideal of a ring A with right Krull dimension. Since there is a one-to-one correspondence between prime ideals in A minimal over I and minimal prime ideals in A/I, we may assume that $I = 0$ and consider the case of minimal prime ideals. Let $N = Pr(A)$ be the prime radical of A which is the intersection of all prime ideals in A. Since N is a semiprime ideal, A/N is a semiprime ring having right Krull dimension. So that A/N is a semiprime right Goldie ring, by proposition 5.7.17. Therefore there are only a finite number of minimal prime ideals in A and their intersection is equal to 0, by proposition 5.4.21. Therefore there are only a finite number of minimal prime ideals in A and their intersection is N. Since N is the prime radical in A, N is nilpotent by theorem 5.7.33. Hence some finite product of minimal prime ideals of A is zero. □

As an immediate corollary of this proposition we obtain the following useful fact.

Proposition 5.7.36. *If A is a ring with right Krull dimension, then there are a finite number of minimal prime ideals $P_1, P_2, \dots, P_n$ in A such that $P_1 P_2 \cdots P_n = 0$.*

Since any right Noetherian ring has right Krull dimension we obtain the following well known fact.

Proposition 5.7.37. *If A is a right Noetherian ring, then there are a finite number of minimal prime ideals $P_1, P_2, \dots, P_n$ of A such that $P_1 P_2 \cdots P_n = 0$.*

Lemma 5.7.38. *Let A be a ring with right Krull dimension. Then there is a prime ideal P in A such that* $\text{r.K.dim}(A) = \text{r.K.dim}(A/P)$.

Proof. By proposition 5.7.36 there are n prime ideals $P_1, P_2, \dots, P_n$ of A such that $P_1 P_2 \cdots P_n = 0$. Consider the descending chain of ideals

$$A \supseteq P_1 \supseteq P_1 P_2 \supseteq \cdots \supseteq P_1 P_2 \cdots P_{n-1} \supseteq P_1 P_2 \cdots P_n = 0$$

For each $2 \le i \le n$ we have that $(P_1 P_2 \cdots P_{i-1})/(P_1 P_2 \cdots P_i)$ is a A/P_i-module and so, by corollary 5.7.6,

$$\text{K.dim}((P_1 P_2 \cdots P_{i-1})/(P_1 P_2 \cdots P_i)) \le \text{r.K.dim}(A/P_i)$$

Since $P_1P_2\cdots P_n = 0$,

$$\text{r.K.dim}(A) = \sup\{\text{r.K.dim}(A/P_1), \text{r.K.dim}(A/P_2), \dots, \text{r.K.dim}(A/P_n)\}$$

by lemma 5.7.3. Therefore r.K.dim(A) = r.K.dim(A/P_i) for some $1 \le i \le n$. □

5.8 Relationship between Classical Krull Dimension and Krull Dimension

In this section we consider some relationship between the concepts of the classical Krull dimension of rings considered in Section 5.6 and the general module-theoretic Krull dimension of rings considered in the previous section.

First note that a ring with classical Krull dimension need not have the (right) Krull dimension that shows the following example.

Example 5.8.1 ([107], [327, Example 5.3.1]).
Let F be a field with char$(F) = p$, where p is a prime number. Consider the Prüfer p-group

$$G = \mathbf{Z}_{p^\infty} = \{\exp(2\pi i m/p^n) \ : m \in \mathbf{Z}, n \in \mathbf{Z}^+\}$$

which is an infinite Abelian p-group having a presentation:

$$\mathbf{Z}_{p^\infty} = \langle x_1, x_2, x_3, \dots \ : \ x_1^p = 1, x_{i+1}^p = x_i \ \text{ for all } i \ge 1\rangle.$$

Let $A = F[G]$ be a group F-algebra. Consider the augmentation ideal N of A, i.e., $N = \text{Ker}(\varphi)$, where $\varphi : F[G] \to G$ defined by

$$\varphi(\sum_{g \in G}(f_g g)) = \sum_{g \in G} g.$$

It is well known that φ is surjective and N as a free module over F is generated by elements $g - 1_g$, where $g \neq 1_g \in G$. Then $(x_i - 1)^p = x_i - 1 \in N$. So $A \subseteq N^p$, which implies that $N = N^2 = \cdots = N^p$, i.e., N is an idempotent ideal. Moreover, N is a maximal ideal, since $A/N \simeq F$.

Show now that N is a nil-ideal. Let $0 \neq y \in N$. Then there is an integer $n \ge 1$ such that $y \in F[H]$, where $H = \langle x_n \rangle$ is a finite cyclic group of order p^n generated by x_n. In fact $y \in K$, where K is the augmentation ideal of $F[H]$. Since K as free F-module is generated by $x_n - 1$, $K = F[H](x_n - 1)$. Therefore K is nilpotent, since A is commutative, char$(F) = p$, and so $(x_n - 1)^{p^n} = x_n^{p^n} - 1 = 0$. Thus $K^{p^n} = 0$, and hence y is nilpotent. So N is a nil-ideal.

Let P be a prime ideal of A and $a \in N$. Then $a^n = 0 \in P$ for some $n \ge 1$. Therefore $a \in P$, i.e., $N \subseteq P$. This means that $P = N$ is a maximal ideal. Hence cl.K.dim$(A) = 0$.

On the other hand, if A had the Krull dimension, N would be nilpotent, by theorem 5.7.25, since N is a nil ideal. In this case $N = N^2 = 0$. This contradiction shows that A does not have the Krull dimension.

At the same time a ring having the right Krull dimension necessarily has the classical Krull dimension.

Proposition 5.8.2. *If a ring A has a right Krull dimension then A has a classical Krull dimension.*

The proof follows immediately from proposition 5.7.21 and theorem 5.6.7. □

The main relationship between the concepts of Krull dimension by R. Rentschler and P.Gabriel, considered in the previous section, and the classical Krull dimension introduced by Krause yields the following theorem which was obtained by A.W. Goldie and L.Small:

Theorem 5.8.3. (A.W. Goldie, L. Small [107]). *If a ring A has right Krull dimension then it has classical Krull dimension and*

$$\text{cl.K.dim}(A) \leq \text{r.K.dim}(A).$$

Proof. If $A = 0$ then $\text{r.K.dim}(A) = \text{cl.K.dim}(A) = -1$. Suppose that $A \neq 0$ and $\text{r.K.dim}(A) = \alpha$ for some ordinal α. If $\alpha = 0$ then A is a right Artinian ring. By [146, Proposition 3.5.5] any right Artinian prime ring is simple, and so the unique its maximal ideal (which is equal in this case the Jacobson radical) is 0. Therefore any prime ideal of a right Artinian ring is maximal, so that $\text{cl.K.dim}(A) = 0 = \text{r.K.dim}(A)$.

Suppose A is not a right Artinian ring and $\alpha = 0$. Let P be a prime ideal of A. Then A/P is a prime ring with right Krull dimension. Therefore A/P is a prime right Goldie ring, by proposition 5.7.17. Let Q be any prime ideal of A such that $P \subsetneq Q$. Then Q/P is an essential right ideal of A/P and hence, by proposition 5.7.18,

$$\text{r.K.dim}(A/Q) = \text{r.K.dim}((A/P)/(Q/P)) < \text{r.K.dim}(A/P) \leq \text{r.K.dim}(A) = \alpha.$$

Then, by induction hypothesis, the ring A/Q has classical Krull dimension and $\text{cl.K.dim}(A/Q) \leq \text{r.K.dim}(A/Q)$. Therefore $Q \in \mathcal{R}_\beta(A)$ for some $0 \leq \beta < \alpha$. So that $P \in \mathcal{R}_\alpha(A)$ and $\mathcal{R}_\beta(A) = \text{Spec}(A)$. Hence A has classical Krull dimension and $\text{cl.K.dim}(A) \leq \alpha = \text{r.K.dim}(A)$. □

Now consider some classes of rings for which this theorem has some stronger form, namely for these rings the right Krull dimension and the classical Krull dimension are in fact equal.

Definition 5.8.4. A ring A is called **right bounded** if every essential right ideal of it contains a two-sided ideal which is essential as a right ideal. A ring A is called **right fully bounded** if A/P is right bounded for each prime ideal P in A. A right fully bounded ring having right Krull dimension is called **right FBK-ring**.

There exist example of rings which are right FBK-rings but not right bounded rings. One of such example was presented by A. Woodward in [327].

Example 5.8.5 [327, Example 4.4.1].
Let p be a prime integer. Write

$$\mathbf{Z}(p^\infty) = \{a/p^n + \mathbf{Z} \ : \ a \in \mathbf{Z}, n \geq 0\} \tag{5.8.6}$$

Then $\mathbf{Z}(p^\infty)$ is a $\mathbf{Z}$-submodule in $\mathbf{Q}/\mathbf{Z}$ all which $\mathbf{Z}$-submodules are given as terms of the following ascending chain:

$$0 \subset (1/p + \mathbf{Z})\mathbf{Z} \subset (1/p^2 + \mathbf{Z})\mathbf{Z} \subset \cdots \subseteq \bigcup_{n\geq 1}(1/p^n + \mathbf{Z})\mathbf{Z} = \mathbf{Z}(p^\infty) \tag{5.8.7}$$

So that $\mathbf{Z}(p^\infty)$ is an Artinian but not Noetherian $\mathbf{Z}$-module, and every proper submodule of it is cyclic.

Consider the triangular matrix ring

$$A = \begin{pmatrix} \mathbf{Z} & \mathbf{Z}(p^\infty) \\ 0 & \mathbf{Z} \end{pmatrix} \tag{5.8.8}$$

Since $\mathbf{Z}(p^\infty)$ is an Artinian $\mathbf{Z}$-module,

$$\text{K.dim}(\mathbf{Z}(p^\infty)) = 0 \tag{5.8.9}$$

Consider a two-sided ideal of A:

$$N = \begin{pmatrix} 0 & \mathbf{Z}(p^\infty) \\ 0 & 0 \end{pmatrix} \tag{5.8.10}$$

Then $N^2 = 0$ and every right ideal in N has the form

$$I = \begin{pmatrix} 0 & K \\ 0 & 0 \end{pmatrix}, \tag{5.8.11}$$

where K is of the $\mathbf{Z}(p^\infty)$-submodule in the chain (5.8.7). Therefore $\text{K.dim}_A(N) = \text{K.dim}_\mathbf{Z}(\mathbf{Z}(p^\infty)) = 0$.

Since $A/N \simeq \mathbf{Z} \oplus \mathbf{Z}$, $\text{K.dim}_A(A/N) = \text{K.dim}_{A/N}(A/N) = 1$.

So that $\text{r.K.dim}(A) = \sup\{\text{K.dim}_A(A/N), \text{K.dim}_A(A/N)\} = 1$. Analogously, $\text{l.K.dim}(A) = 1$.

If P is a prime radical of A then $N \subseteq P$ since $0 = N^2 \subset P$. Therefore $A/P \simeq (A/N)/(P/N)$, so A/P is a commutative ring and hence A/P is bounded on both side. Thus A is both right and left FBK-ring.

But A is not either right or left bounded ring.

Theorem 5.8.12 (R. Gordon, J.C. Robson [117, Theorem 8.12]). *If A is a right FBK-ring then A has the classical Krull dimension and*

$$\text{r.K.dim}(A) = \text{cl.K.dim}(A).$$

Proof. By proposition 5.7.38 there is a prime ideal of A such that $\text{r.K.dim}(A) = \text{r.K.dim}(A/P)$. Suppose that $\text{r.K.dim}(A/P) = \text{cl.K.dim}(A/P)$. Then by theorem 5.8.3

$$\text{r.K.dim}(A/P) = \text{cl.K.dim}(A/P) \leq \text{cl.K.dim}(A) \leq \text{r.K.dim}(A) = \text{r.K.dim}(A/P)$$

which implies that $\text{cl.K.dim}(A) = \text{r.K.dim}(A)$.

So without loss of generality we can assume that A is a prime ring.

Suppose $\text{r.K.dim}(A) = \alpha$. If $\alpha = -1$, then $A = 0$, and $\text{cl.K.dim}(A) = \text{cl.K.dim}(A) = -1$ by definition. If $\alpha = 0$ then A is a prime Artinian ring and in this case any prime ideal is maximal, so that $\text{cl.K.dim}(A) = \text{cl.K.dim}(A) = 0$. Therefore we can assume that $\alpha \geq 1$. Let $\beta < \alpha$. By proposition 5.7.18 there is an essential right ideal E of A such that $\beta \leq \text{r.K.dim}(A/E) < \alpha$. Since A is a prime fully bounded ring there is a two-sided ideal I of A such that $I \subseteq E$. Then I is also an essential right ideal of A. Therefore using proposition 5.7.18 again one obtains that $\beta \leq \text{r.K.dim}(A/E) < \text{r.K.dim}(A/I) < \alpha$. By lemma 5.7.38 there is a prime ideal Q of A such that $\text{r.K.dim}(A/I) = \text{r.K.dim}(A/Q)$. So that there is a prime ideal Q in A such that $\beta \leq \text{r.K.dim}(A/Q) < \alpha$.

Now we will proceed the proof by induction on α. Assume that for each prime right FBK-ring B with $\text{r.K.dim}(B) < \alpha$ we have $\text{r.K.dim}(B) = \text{cl.K.dim}(B)$. Let $\text{r.K.dim}(A) = \alpha$. Then by theorem 5.8.3 A has classical Krull dimension and $\text{cl.K.dim}(A) \leq \text{r.K.dim}(A) = \alpha$. Suppose $\text{cl.K.dim}(A) < \alpha$ then by proving above there is a prime ideal Q of A such that

$$\text{cl.K.dim}(A) \leq \text{r.K.dim}(A/Q) < \alpha.$$

By induction hypothesis $\text{r.K.dim}(A/Q) = \text{cl.K.dim}(A/Q)$. Therefore

$$\text{cl.K.dim}(A) \leq \text{r.K.dim}(A/Q) = \text{cl.K.dim}(A/Q) \leq \text{cl.K.dim}(A)$$

which implies that $\text{cl.K.dim}(A) = \text{cl.K.dim}(A/Q)$. But this impossible, by proposition 5.6.14(c), since Q is a non-zero prime ideal of a prime ring A. Thus $\text{cl.K.dim}(A) = \alpha = \text{r.K.dim}(A)$, that required. □

Since any commutative ring is obviously fully bounded ring, from this theorem we obtain immediately the following corollary.

Theorem 5.8.13. *If a commutative ring A has the Krull dimension (in particular, commutative Noetherian ring) then it also has the classical Krull dimension and*

$$\text{K.dim}(A) = \text{cl.K.dim}(A).$$

Definition 5.8.14. A right fully bounded and right Noetherian ring is called a **right FBN-ring**.

Obviously, commutative Noetherian rings are FBN-rings. Examples of right FBN-rings are also Noetherian PI-rings and Artinian rings. Since any right

Noetherian ring has right Krull dimension by proposition 5.6.9, every right FBN-ring is a right FBK-ring. Therefore the following theorem is immediately follows from theorem 5.8.5.

Theorem 5.8.15 (G. Krause [198, Proposition 2.4]). *If A is a right FBN-ring then A has the classical Krull dimension and*

$$\text{r.K.dim}(A) = \text{cl.K.dim}(A).$$

At the end of this section we give some examples of rings with Krull dimension. In general, a ring can have right Krull dimension and do not have left Krull dimension, which shows the following examples.

Example 5.8.16.
Let

$$A = \begin{pmatrix} K & L \\ 0 & L \end{pmatrix},$$

where $K \subset L$ are fields and $\dim_K L = \infty$. This ring is right Noetherian and right Artinian, but it is neither left Noetherian nor left Artinian (See e.g. [146, Corollary 3.6.2]). Therefore it has right Krull dimension equal to 0. Since $\mathbf{Q}$ is an infinitely generated $\mathbf{Z}$-module, so the left ideal Ae_{22} has infinite Goldie dimension. By proposition 5.7.10, A has not left Krull dimension.

Moreover from the description of all right ideals of A it is not difficult to show that A is a right FNB-ring.

Example 5.8.17.
Let

$$A = \begin{pmatrix} \mathbf{Z} & \mathbf{Q} \\ 0 & \mathbf{Q} \end{pmatrix}.$$

This ring is right Noetherian, but not left Noetherian. It is neither left Artinian nor left Artinian (See e.g. [146, Example 3.6.1]). Therefore it has right Krull dimension. Since $\mathbf{Q}$ is an infinitely generated $\mathbf{Z}$-module, so the left ideal Ae_{22} has infinite Goldie dimension. By proposition 5.7.10, A has not left Krull dimension.

Like as in the previous example one can show A is a right FNB-ring.

The next class of examples are connected with polynomial rings and skew polynomial rings. The following important result we give for completeness here without proof.

Theorem 5.8.18. (R. Rentschler, P. Gabriel [270]). *If $A[x]$ is a polynomial ring over a right Noetherian ring A, then*

$$\text{r.K.dim}(A[x]) = \text{r.K.dim}(A) + 1. \tag{5.8.19}$$

Proposition 5.8.20. *If $A[x_1, x_2, \ldots, x_n]$ is a polynomial ring over a right Artinian ring A, then*

$$\text{r.K.dim}(A[x_1, x_2, \ldots, x_n]) = n.$$

Proof. Since $A[x_1,x_2,\dots,x_n] = A[x_1,x_2,\dots,x_{n-1}][x_n]$ and $A[x_1,x_2,\dots,x_{n-1}]$ is a right Noetherian ring, we can apply proposition 5.8.18 and then using induction on n we obtain that $\text{r.K.dim}(A[x_1,x_2,\dots,x_n]) = \text{r.K.dim} + n$. Now result follows from the fact that $\text{r.K.dim}(A) = 0$ for an Artinian ring A. □

Since any field is an Artinian ring, from this theorem it follows immediately the well known result which is very important in algebraic geometry.

Theorem 5.8.21. *If $A = F[x_1,x_2,\dots,x_n]$ is a polynomial ring over a field F then* $\text{K.dim}(A) = n$.

Proposition 5.8.22 (Gordon, Robson [117]). *The polynomial ring $A[x]$ over ring A has right Krull dimension if and only if A is right Noetherian.*

Proof. If A is a right Noetherian ring, then the statement follows proposition 5.8.18.

To prove the converse statement suppose that A is not a right Noetherian ring. Then there is an infinite strictly ascending chain of right ideals of A:

$$I_0 \subset I_1 \subset I_2 \subset \cdots$$

We set

$$X = I_0 + I_1 x + I_2 x^2 + \cdots + I_n x^n + \cdots$$

and

$$Y = I_1 + I_2 x + I_3 x^2 + \cdots + I_{n+1} x^n + \cdots$$

It is clear that X, Y are right ideals of $A[x]$ and

$$Y/X \simeq I_1/I_0 \oplus I_2/I_1 \oplus \cdots \oplus I_{n+1}/I_n \oplus \cdots$$

as A-modules and $A[x]$-modules. Therefore the $A[x]$-module Y/X has no Krull dimension since it has infinite uniform dimension. So it is true for the polynomial ring $A[x]$. □

Corollary 5.8.23. *The polynomial ring $A[x_1,x_2,\dots,x_n,\cdots]$ in infinite indeterminates over any ring A with identity does not have Krull dimension.*

Proof. Let $S = A[x_2,x_3,\dots]$, then $A[x_1,x_2,\dots,x_n,\cdots] = S[x_1]$. Since S is not Noetherian, result follows immediately from proposition 5.8.22. □

The following results give the possibility to calculate the Krull dimension for Laurent series rings, skew polynomial rings and skew Laurent polynomial rings over right Noetherian rings.

Proposition 5.8.24 (K.R. Goodearl, L. Small [112]). *Let A be a right Noetherian ring, and let $T = A((x))$ be a Laurent series ring. Then*

$$\text{r.K.dim}(T) = \text{r.K.dim}(A).$$

Proposition 5.8.25 (K.R. Goodearl, R.B. Warfield, [114]). *Let A be a right Noetherian ring, σ an automorphism and δ a derivation. Then*

1. $\text{K.dim}(A) \leq \text{K.dim}(A[x;\sigma,\delta]) \leq \text{K.dim}(A) + 1$. *In particular, if $\delta = 0$ then* $\text{K.dim}(A[x;\sigma] = \text{K.dim}(A) + 1$.
2. *If A is a right Artinian ring then* $\text{K.dim}(A[x;\sigma,\delta]) = 1$.

Proposition 5.8.26 (T.J. Hodges [152]). *Let A be a right Noetherian ring, σ an automorphism and δ a derivation. Then*

1. $\text{K.dim}(A) \leq \text{K.dim}(A[x,x^{-1};\sigma]) \leq \text{K.dim}(A) + 1$. *In particular, if $\sigma = 0$ then* $\text{K.dim}(A[x,x^{-1}] = \text{K.dim}(A) + 1$.
2. *If A is a right Artinian ring then* $\text{K.dim}(A[x,x^{-1};\sigma]) = 1$.

Recall that the **Weyl algebra** is the ring of differential operators with polynomial coefficients in one variable. More precisely, the Weyl algebra $A(D)$ is the noncommutative algebra over a domain D on two generators x, y with defining relation $xy - yx = 1$, i.e.,

$$A(D) = D\langle x, y\rangle/(xy - yx - 1).$$

If $D = k$ is a field of characteristic 0 we define the first Weyl algebra $A(k)$ over k, denoted by $A_1(k)$. The n-th Weyl algebra $A_n(k)$ is defined by $A_n(k) = A(A_{n-1})(k)$. This is the ring of differential operators with polynomial coefficients in n variables.

In the case if k is a field of characteristic 0, the n-th Weyl algebra $A_n(k)$ is a simple Noetherian domain. The Weyl algebra $A(k)$ is an example of a simple ring that is not a matrix ring over a division ring.

For the Weil algebra $A_n(k)$ there were obtained the following results:

Proposition 5.8.27. *Let k be a field, and let D be the quotient division ring of the Weil algebra $A_n(k)$ and $S = M_n(D)$. Then* $\text{K.dim}(S) = 2n$.

Proposition 5.8.28. *Let $A_n(k)$ be the n-th Weil algebra over a field k with* $\text{char}\, k > 0$ *then* $\text{K.dim}(A_n(k)) = 2n$ for each n.

Proposition 5.8.29 (Y. Nouazé, P. Gabriel [255], Rentschler, P.Gabriel [270]). *Let $A_n(k)$ be the n-th Weil algebra over a field k with* $\text{char}\, k = 0$ *then* $\text{K.dim}(A_n(k)) = n$ *for each positive integer n.*

Note also the following interesting result which was obtained independently by R.Gordon, J.C. Robson in [117] and by Gulliksen in [130].

Theorem 5.8.29. ([117], [130]). *There exist commutative Noetherian domains of arbitrary Krull dimension.*

5.9 Notes and References

Uniform right ideals and uniform modules were introduced by A.W. Goldie (see [101], [102]). Simple Artinian rings with uniform right ideals were studied by R. Hart (see [144]). The concept of finite-dimensional modules, or modules of finite rank, was also first introduced by A.W. Goldie (see [101], [102]).

Infinite Goldie dimension of hereditary rings and modules was studied by H.Q. Dinh, P.A. Guil Asensio, S.R. Lopez-Permouth in [65]. They found a bound on the Goldie dimension of hereditary modules in terms of the cardinality of the generator sets of their quasi-injective hull. In particular, they show that every right hereditary module with a countably generated quasi-injective hull is Noetherian, and that every right hereditary ring A with a finitely generated injective hull $E(A_A)$ has finite Goldie dimension and therefore is right Artinian.

Right and left singular ideals of a ring were first introduced by R.E. Johnson in [173]. Essentially finitely generated modules were introduced by Catefories in [38].

The equivalence $1 \iff 3$ in theorem 5.4.10 is due to A.W. Goldie (See [102, theorems 4.1, 4.4]).

The concept of reduced rank of a module over a noncommutative Noetherian ring has been introduced by A.W. Goldie in [103] and used in many applications to study Noetherian rings (see [46]).

Theorem 5.5.17 is a still stronger variant of the well-known Jordan-Hölder theorem. The Schreier refinement theorem was first proved by O.Schreier for normal series of subgroups in [285]. In this paper in a footnote O.Schreier remarked that his proof also works for groups with operators. O.Schreier used this theorem to give an improved proof of the fundamental Jordan-Hölder theorem, 39 year after the publication of Hölder's paper. "It is rare that such a widely used and basic theorem can be deepened after such a long time." (see [40]). Otto Schreier (1901–1929) was an Austrian mathematician who made major contributions in combinatorial group theory. The axiomatic proof of Zassenhaus's lemma and the Jordan-Hölder-Schreier refinement theorem were obtained by N.J.S. Hughes, and he applied this theorem to groups, rings, lattices and abstract algebraic systems (see [156]). Later generalizations of the Jordan-Hölder-Schreier theorem were obtained for infinite normal systems, and in particular for totally ordered normal and composition series (see [204]), and for normal categories (see [287]). Various generalizations of this theorem were obtained in the language of lattice theory and posets (see [22], [90], [99]).

Proposition 5.5.15 was proved by Hans Julius Zassenhaus (1912–1991). He published his proof in 1934 [337] to provide a new and simpler proof of the Jordan-Hölder-Schreier theorem. Zassenhaus was a doctorate student under Emil Artin's supervision at the time. Now this proposition is known as the "Zassenhaus lemma". The name "butterfly lemma" was given to it by Serge Lang based on the shape of the Hasse diagram of the various submodules involved.

The famous Goldie theorem which gives necessary and sufficient conditions for a ring being a right order in a semisimple ring was proved in 1960 in [102]. Theorem

5.5.30, which is a generalization of this theorem and provides a criterion for a Noetherian ring being a right order in a right Artinian ring, was proved by L.Small (see [291]). A.W. Chatters gave a new much simpler proof of Small's theorem for Noetherian rings. This proof with another applications of reduced rank was given by A.W. Chatters, A.W. Goldie, C.R. Hajarnavis, T.H. Lenagan in [46] (see also [48]). A variant of Small's theorem without the Noetherian hypothesis, theorem 5.5.28, was given by R.B. Warfield, Jr. using the ideas of A.W. Chatters and A.W. Goldie in [321]. Condition 4 of theorem 5.5.28 was shown to hold independently by T.D. Talintyre in [301]. The notion of reduced rank was introduced by A.W. Goldie in [103]. The different criteria for a non-Noetherian ring to be a right order in a right Artinian ring were obtained L.Small in [293] and Robson in [273].

The notion of the (classical) Krull dimension as a powerful tool for arbitrary commutative Noetherian rings was considered by W. Krull [203]. Originally, the Krull dimension was defined to be the supremum of the length of chains of prime ideals in a ring, being ∞ if no such supremum exists. The module-theoretic form of the Krull dimension in the general case for noncommutative rings was introduced by R. Rentschler and P.Gabriel in 1967 (see [270]) for finite ordinals and later, in 1970, extended to arbitrary ordinals by G.Krause [197]. This notion and their basic properties were studied in detail by R. Gordon, J.C. Robson (see [118]). The concept of critical modules was introduced by R.Hart in [145] (under the name "restricted modules") and A.W. Goldie in [105]. In this chapter we have considered the notion of the Krull dimension in the sense of Rentschler and Gabriel. Note that a ring with classical Krull dimension need not have Krull dimension (see [327], Example 5.3.1). However a ring with right Krull dimension necessarily has classical Krull dimension.

Proposition 5.7.10 is due to G.Krause [197]. Proposition 5.7.17 was proved independently by B. Lemonnier [213] and R. Gordon, J.C. Robson [117]. The proof of theorem 5.7.25 is due to T.H. Lenagan [217], R. Gordon, J.C. Robson [118] and A.Facchini [133]. Lemma 5.7.26 was proved by G.Krause and T.H. Lenagan in [199, lemma 7].

CHAPTER 6

Rings with Finiteness Conditions

An important role in the theory of rings and modules is played by various finiteness conditions. Many types of finiteness conditions on rings can be formulated in terms of d.c.c. (descending chain condition) or a.c.c. (ascending chain condition) on suitable classes of one-sided ideals. The d.c.c. (minimal condition) on right (left) ideals defines right (left) Artinian rings. Analogously, right (left) Noetherian rings are defined as rings which satisfy the maximal condition, or the a.c.c. on right (left) ideals. These rings were considered in [146, chapter 3].

Section 6.1 gives some examples of Noetherian rings in connection with the basic constructions of rings considered in chapter 2. Section 6.2 considers various other finiteness conditions for rings and modules, and relations between them. Dedekind-finite rings, stable finite rings and IBN rings are examples of rings which are considered in this section.

FDI-rings, i.e., rings with a finite decomposition of identity into a sum of pairwise orthogonal primitive idempotents, form the next class of rings with finiteness conditions. These rings are considered in Section 6.3. In Section 6.4 we prove a criterium for a semiprime FDI-ring to be decomposable into a direct product of prime rings.

6.1 Some Noetherian Rings

Artinian and Noetherian rings and modules were considered in [146, chapter 3]. This section gives some examples of Artinian and Noetherian rings in connection with the basic constructions of rings considered in the previous chapters.

Example 6.1.1 ([237], p.18).

Let K be a field and $A = K[y]$. Define an endomorphism $\sigma : A \longrightarrow A$ by $\sigma(f(y)) = f(0)$, in particular $\sigma(y) = 0$. Then the ring

$$T = A[x;\sigma] = \{\sum_{i=0}^{n} a_i x^i \ : \ a_i \in A\}$$

with multiplication defined by the distributive law and the rule $xa = \sigma(a)x$ for all $a \in A$, has the following properties:

1. The ring $T = A[x;\sigma]$ is not an integral domain since $xy = \sigma(y)x = 0$.
2. The ring $T = A[x;\sigma]$ is neither a prime nor a semiprime ring. Indeed, consider the non-zero ideal $J = yxT = \{yx \sum_{i=0}^{n} b_i(y)x^i \ : \ b_i(y) \in K[y]\}$. Then for any element $p(x,y) \in xTy$ we have
$$p(x,y) = x(\sum_{i=0}^{n} b_i(y)x^i)y = xb_0(y)y = xyb_0(y) = 0.$$
Therefore $xTy = 0$ which implies that $J^2 = (yxT)^2 = y(xTy)xT = 0$.
3. The ring $T = A[x;\sigma]$ is neither a right Noetherian nor a left Noetherian ring, since the sums $\sum_i y^i xT$ and $\sum_i Tyx^i$ are both direct sums. Really, any element of $y^i xT$ has the form
$$p(x,y) = y^i x \sum_{j=0}^{n} b_j(y)x^j = y^i \sum_{j=0}^{n} b_j(0)x^{j+1} = y^i \sum_{j=0}^{n} c_j x^{j+1}$$
where $c_j \in K$. Therefore $y^i xT \subseteq y^i K[x]$, and so $y^i xT \cap y^j xT = 0$ for $i \neq j$. Analogously, $Tyx^i \cap Tyx^j = 0$ for $i \neq j$.

In the case when an endomorphism σ satisfies special conditions one obtains the following result.

Theorem 6.1.2. *Let $T = A[x;\sigma,\delta]$ be a skew polynomial ring defined by a skew derivation (σ,δ) satisfying* (2.3.24). *If σ is an automorphism of A and A is a right (left) Noetherian ring then T is a right (left) Noetherian ring as well.*

Proof. The proof is a variation of a standard proof of the Hilbert basis theorem 1.1.13.

Assume that A is a right Noetherian ring and σ is an automorphism of A. Let J be a right ideal of $T = A[x;\sigma,\delta]$. Let S be a set of leading coefficients of all elements of J. We first show that S is a right ideal of A.

Obviously, $0 \in S$. Suppose $f(x), g(x) \in J$ with leading coefficients a, b respectively. If $a+b = 0$ then $a+b \in S$. Suppose $a+b \neq 0$. Assume $\deg(f(x)) = n$, $\deg(g(x)) = m$, and $m \leq n$. Then $f(x)+g(x)x^{n-m} \in J$ with leading coefficient $a+b$. Therefore $a+b \in S$. Suppose $c \in A$ and $c \neq 0$. Since $\sigma \in \mathrm{Aut}(A)$, $\sigma^{-n}(c) \in A$. Then $f(x)\sigma^{-n}(c) = (ax^n + \cdots)\sigma^{-n}(c) = acx^n + \cdots$, i.e., the leading coefficient of $f(x)\sigma^{-n}(c)$ is ac, and consequently $ac \in S$. Thus S is a right ideal of A.

Since A is a right Noetherian ring, S is a finitely generated ideal. Therefore there is a finite set of generators $\{b_1, b_2, \ldots, b_k\}$ for S. Denote by $f_i(x)$ a polynomial of J with the leading coefficient b_i and degree n_i. If $n = \max\{n_1, n_2, \ldots, n_k\}$ then each $f_i(x)$ can be replaced by $f_i(x)x^{n-n_i}$. So without loss of generality one can assume that all $f_1, f_2, \ldots, f_k$ have the same degree n.

Let $\mathcal{I}_n = \{p(x) \in T \ : \ \deg(p) < n\}$. Then $\mathcal{I}_n$ is a finitely generated right A-submodule of T generated by $1, x, \ldots, x^{n-1}$. Therefore $\mathcal{I}_n$ is a Noetherian

A-module, so that $\mathcal{I}_n \cap J$ is also a finitely generated A-module, generated by, say, $g_1, g_2, \ldots, g_t \in T$.

Set $X = f_1T + f_2T + \cdots + f_kT + g_1T + \cdots + g_tT$. We will now show that $X = J$. Obviously, $\mathcal{I}_n \cap J \subseteq X$. So X contains all elements of J with degree less than n.

Suppose with induction that X contains all elements of J of degree less than m. Let $f(x)$ be an arbitrary polynomial of J with degree $m \geq n$ and leading coefficient a. Then $a = b_1c_1 + b_2c_2 + \cdots + b_kc_k$ for some $c_1, \ldots, c_k \in A$. Consider the polynomial

$$p(x) = \sum_{i=1}^{k} f_i(x)\sigma^{-n}(c_i)x^{m-n}.$$

Clearly, $p(x) \in X$ and $\deg(p) = m$ and leading coefficient a. Therefore $q(x) = f(x) - p(x) \in J$ and $\deg(q) < m$. By assumption $q(x) \in X$, whence $f(x) \in X$. Therefore, by induction, $X = J$, which means that J is a finitely generated right ideal of T.

The left Noetherian case is similar, taking into account that $A[x;\sigma,\delta]^{op} \cong A^{op}[x;\sigma^{-1}, \delta\sigma^{-1}]$. □

If σ is not an automorphism, then right (left) Noetheriness of A does necessarily not imply right (left) Noetheriness of $A[x;\sigma,\delta]$, which shows the following example.

Example 6.1.3. (See [115, Example 1N])

Let K be a field and $A = K[y]$. Suppose that the endomorphism $\sigma : A \longrightarrow A$ is given by $\sigma(f(y)) = f(y^2)$.

Consider the ring

$$T = A[x;\sigma] = \{\sum_{i=0}^{n} a_i x^i \ : \ a_i \in A\}$$

with multiplication defined by the distributive law and the rule $xa = \sigma(a)x$ for all $a \in A$.

Since $xf(y) = \sigma(f(y))x = f(y^2)x$, $xy = y^2x$, and $xy^i = y^{2i}x$, $x^iy = y^{2^i}x^i$ for all positive integer i.

Consider the right ideal I in $T = A[x;\sigma]$ generated by the elements $\{yx, xyx, x^2yx, \ldots, x^nyx, \ldots\}$. Each element of T has the following form:

$$t(x,y) = \sum_{j=0}^{n} b_j(y)x^j, \text{ where } b_j(y) \in A.$$

Each element of yxT has the following form

$$yxt(x,y) = yx\sum_{j=0}^{n} b_j(y)x^j = y\sum_{j=0}^{n} b_j(y^2)x^{j+1},$$

and each element of $xyxT$ is of the form:

$$xyxt(x,y) = xy\sum_{j=0}^{n} b_j(y^2)x^{j+1} = y^2\sum_{j=0}^{n} b_j(y^4)x^{j+2}.$$

Analogously each element of $x^i yxT$ for any positive integer i has the following form

$$x^i yxt(x,y) = y^{2^i} \sum_{j=0}^{n} b_j(y^{2\cdot 2^i})x^{j+i+1}.$$

An arbitrary term of this polynomial has the form $y^u x^v$ where

$$u = 2^i + s \cdot 2^{i+1} = 2^i(2s+1).$$

Since there are no positive integers $i \neq k, s, r$ such that $2^i(2s+1) = 2^k(2r+1)$, we get that $x^i yxT \cap x^k yxT = 0$ for $i \neq k$. Therefore I is an infinitely generated right ideal in T, which shows that T is not a right Noetherian ring.

Theorem 6.1.4. *Let $T = A[x,x^{-1};\sigma]$ be a skew Laurent polynomial ring. If σ is an automorphism of A and A is a right (resp., left) Noetherian ring then T is a right (resp., left) Noetherian ring as well.*

Proof. Let $S = A[x;\sigma]$. As the ring A is right Noetherian and σ is an automorphism σ of A, then by theorem 6.1.2, S is also a right Noetherian ring. Let I be a right ideal of T, then $I \cap S$ is a right ideal of S, so that $I \cap S$ is a finitely generated right ideal in S. Now $(I \cap S)T$ is a finitely generated right ideal in T. Obviously, $(I \cap S)T \subseteq I$. We will now show that $I \subseteq (I \cap S)T$. Let $y \in I$. Then $y = \sum\limits_{i=-n}^{m} a_i x^i$, where $a_i \in A$. So $yx^n \in S$. Therefore $yx^n \in I \cap S$ and $y = yx^n \cdot x^{-n} \in (I \cap S)T$, which means that $I \subseteq (I \cap S)T$. Thus $I = (I \cap S)T$ is a finitely generated right ideal in T.

The left-sided case is proved analogously, taking into account that $A[x,x^{-1};\sigma]^{op} \cong A^{op}[x,x^{-1};\sigma^{-1}]$. □

Theorem 6.1.5. *Let A be a right Noetherian ring. Then the formal power series ring $A[[x]]$ is right Noetherian as well.*

Proof. The proof of this theorem is a modification of the Hilbert basis theorem taking into account that the leading coefficient of a power series is the non-zero coefficient of the least power of x occurring.

Let A be a right Noetherian ring, and let $\mathcal{I}$ be a right ideal of $T = A[[x]]$. Consider the set

$$\mathcal{I}_k = \{a_k \in A : \sum_{i=k}^{\infty} a_i x^i \in \mathcal{I}, a_k \neq 0\} \cup \{0\}.$$

It is easy to show that $\mathcal{I}_k$ is a right ideal of A. If $f(x) \in \mathcal{I}$ and $f(x) = \sum\limits_{i=k}^{\infty} a_i x^i \neq 0$ with $a_k \neq 0$, then $a_k \in \mathcal{I}_k$. Since $f(x)x = \sum\limits_{i=k}^{\infty} a_i x^{i+1} \in \mathcal{I}$, $a_k \in \mathcal{I}_{k+1}$. Thus, we have a non-decreasing sequence

$$\mathcal{I}_0 \subseteq \mathcal{I}_1 \subseteq \mathcal{I}_2 \subseteq \cdots \subseteq \mathcal{I}_i \subseteq \mathcal{I}_{i+1} \subseteq \cdots$$

of right ideals of A. Since A is a right Noetherian ring, this sequence must stabilize. Therefore there exists $t \in \mathbf{N}$ such that $\mathcal{I}_t = \mathcal{I}_{t+1} = \cdots = \mathcal{I}_{t+s} = \cdots$.

Since A is a right Noetherian ring, each $\mathcal{I}_k$ for $0 \leq k \leq t$ is a finitely generated ideal, i.e., there is a finite set of generators $b_{k1}, b_{k2}, \ldots, b_{ks_k} \in A$ such that $\mathcal{I}_k = \{b_{k1}, b_{k2}, \ldots, b_{ks_k}\}$. Denote by $f_{ij}(x)$ a power series of $\mathcal{I}$ with the leading coefficient b_{ij}.

We claim that the set of power series $f_{ij}(x)$ for $i = 0, 1, 2, \ldots, t$ and $j = 1, 2, \ldots, s_i$ generates the ideal $\mathcal{I}$.

Let $f(x)$ be an arbitrary power series of $\mathcal{I}$ of order[1] d with the leading coefficient a. Assume that $d \leq t$. Since $a \in \mathcal{I}_d$, there are elements $c_{dj} \in A$ for $1 \leq j \leq s_d$ such that $a = \sum\limits_{j=1}^{s_d} b_{dj}c_{dj}$. Then the power series

$$p(x) = f(x) - \sum_{j=1}^{s_d} f_{dj}(x)c_{dj}$$

has the order $\geq d + 1$. Consistently applying this process we obtain that any power series of order $\leq t$ can be written as a linear combination of a power series of order $> t$ and a linear combination of the power series $f_{ij}(x)$ for $i = 0, 1, 2, \ldots, t$ and $j = 1, 2, \ldots, s_i$. Therefore to finish the proof we must show that any power series of order $d > t$ can be written as a linear combination of the power series $f_{ij}(x)$ for $i = 0, 1, 2, \ldots, t$ and $j = 1, 2, \ldots, s_i$. We will now show a somewhat stronger statement, namely that any power series of order $d > t$ can be written as a linear combination of the power series $f_{tj}(x)$ for $j = 1, 2, \ldots, s_t$.

Assume that $d > t$. We will prove the statement by induction on d. Let $f(x)$ be a power series of order d with the leading coefficient a. Then $a \in \mathcal{I}_t$ and so there are elements $c_{tj} \in A$ for $1 \leq j \leq s_t$ such that $a = \sum\limits_{j=1}^{s_t} b_{tj}c_{tj}$. Then the power series

$$p(x) = f(x) - \sum_{j=1}^{s_t} x^{d-n_{tj}} f_{tj}(x)c_{tj}$$

has order $\geq d + 1$. Here $n_{tj} = o(f_{tj})$. Consistently applying this result we obtain that for any n the power series

$$p(x) = f(x) - \sum_{j=1}^{s_t} f_{tj}(x)g_j(x),$$

where $g_j(x) = \sum\limits_{v=t}^{n} c_{vj}x^{v-n_{tj}}$ for $j = 1, 2, \ldots, s_t$, has order $> n$. Therefore

$$f(x) = \sum_{j=1}^{s_t} f_{tj}(x)g_j(x)$$

[1]The order, $o(f)$, of a power series $f = \sum\limits_{i=0}^{\infty} a_i x^i$ is d if a_d is the first non-zero coefficient, i.e., the leading coefficient.

where $g_j = \sum_{\nu=t}^{\infty} c_{\nu j} x^{\nu - n_{tj}}$ for $j = 1,2,\dots,s_t$.

Thus, I is a finitely generated right ideal of $A[[x]]$. □

Corollary 6.1.6. *Let A be a right Noetherian ring. Then the formal power series ring $A[[x_1,x_2,\dots,x_n]]$ is right Noetherian as well.*

Proof. The proof is immediate from the previous theorem by induction on the number of variables n.

Theorem 6.1.7. *Let A be a right Noetherian ring, and $\sigma \in \mathrm{Aut}(A)$. Then the skew power series ring $A[[x;\sigma]]$ is right Noetherian as well.*

Proof. The proof of this theorem is a variation of theorem 6.1.2 and theorem 6.1.5, taking into account that $\sigma^{-n}(a) \in A$ for any $a \in A$ and $n \in \mathbf{N}$, since $\sigma \in \mathrm{Aut}(A)$. □

6.2 Dedekind-finite Rings and Stably Finite Rings

In the formulations of finiteness conditions for rings various special classes of ideals are considered. One such important class of ideals is formed by principal right (left) ideals. A famous theorem of H. Bass [15] (See also theorem 1.9.8) states that the d.c.c. on principal left ideals defines right perfect rings.

Other important special class of ideals is formed by right (left) annihilators. It is easy to show that the set of all right (left) annihilators of a ring A forms a lattice. There is a lattice anti-isomorphism between the lattice of right annihilators of A and the lattice of left annihilators of A. This implies the following simple fact, which was earlier proved in chapter H5 (proposition 5.3.23).

Proposition 6.2.1. *The following conditions are equivalent for a ring A:*

1. *A satisfies d.c.c. on right (left) annihilators.*
2. *A satisfies a.c.c. on left (right) annihilators.*

Other important "finiteness condition" considered by Goldie defines when a module A does not contain an infinite direct sum of submodules. In this case M is said to be finite-dimensional, and is this case $\mathrm{u.dim}\, M < \infty$. A ring A is right finite-dimensional if it contains no infinite direct sum of non-zero right ideals, i.e., the right regular module A_A is finite dimensional. In this case $\mathrm{u.dim} A_A < \infty$. A right Goldie ring is right finite-dimensional and satisfies the a.c.c. on right annihilators.

Connected to this condition for a module M there are also closely related "finiteness conditions" on complements of M, as was shown by Goldie. Recall that a submodule C of M is called a **complement** in M if there exists a submodule N of M such that C is maximal with respect to the property that $N \cap L = 0$. As follows from theorem 5.1.38, the a.c.c. on right complements in a ring A is equivalent to a ring A being right finite-dimensional.

Let $\mathcal{I}$ be a right direct summand of the right regular module A_A. Then $\mathcal{I} = eA$ for some idempotent e of A. Therefore $\mathcal{I}$ is a principal right ideal. On the other hand $eA = \text{r.ann}_A(A(1-e))$, so $\mathcal{I}$ is also a right annihilator.

Thus there are some classes of ideals with which one can connect different "finiteness conditions". From the observations above one obtains the following diagram of different classes of ideals (here an arrow means an inclusion):

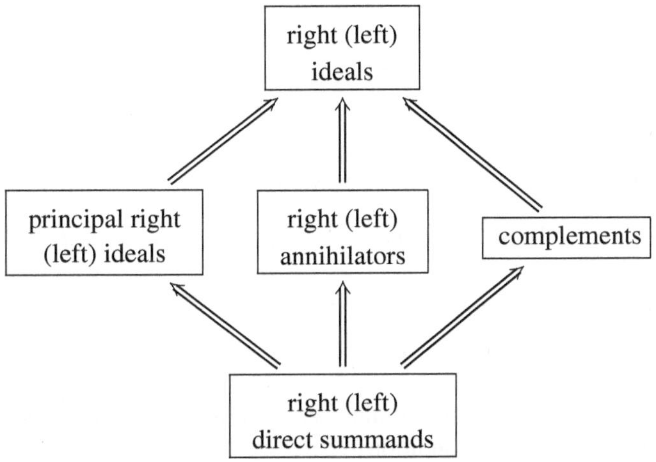

Diag. 6.2.2.

One important finiteness condition for a ring is connected with the cardinal number of a set of pairwise orthogonal non-zero idempotents. Following to A.A. Tuganbaev [308] one has the definition.

Definition 6.2.3 [308]. A ring A is called **orthogonally finite** if it does not contains an infinite set of pairwise orthogonal non-zero idempotents.

Using this definition there is the following theorem:

Theorem 6.2.4 [207, Proposition 6.59]. *For any ring A the following statements are equivalent*:

1. *A is orthogonally finite.*
2. *A satisfies the ascending chain condition on right direct summands of* A_A.
3. *A satisfies the descending chain condition on left direct summands* ${}_AA$.

Proof.

$2 \iff 3$. Suppose that e and f are idempotents of a ring A, and $eA \subsetneq fA$. Since $\text{l.ann}_\text{A}(\text{eA}) = \text{A}(1-\text{e})$ for any idempotent e of A, it follows that $A(1-e) \subset A(1-f)$. This inclusion is strict, otherwise taking right annihilators again one finds $eA = fA$ which is a contradiction with the assumption above. This remarks shows the equivalence of statements 2 and 3.

$3 \Longrightarrow 1$. Suppose that A has an infinite number of pairwise orthogonal non-zero idempotents $\{e_1, e_2, \ldots\}$. Let $f_i = e_1 + \cdots + e_i$ for all $i \geq 1$. It is easy to see that all the f_i are idempotents of A and

$$f_i f_{i+1} = f_{i+1} f_i = f_i \neq f_{i+1}.$$

This implies that $f_i A \subsetneq f_{i+1} A$ for all $i \geq 1$ which contradicts condition 3.

$1 \Longrightarrow 2$. Assume that there exists an infinite sequence of right direct summands of A:

$$A = A_0 \supsetneq A_1 \supsetneq \cdots$$

Then $A_i = B_i \oplus A_{i+1}$ for suitable right ideals B_i of A for $i \geq 1$. Write $1 = e_1 + f_1$ with $e_1 \in B_1$ and $f_1 \in A_1$, and write $f_1 = e_2 + f_2$ with $e_2 \in B_2$ and $f_2 \in A_2$, and so on. Then $B_i = Ae_i$, so $e_i \neq 0$ for each $i \geq 1$. Moreover,

$$1 = e_1 + f_1 = e_1 + e_2 + f_2 = \cdots = e_1 + e_2 + \cdots + e_i + f_i$$

is a decomposition of 1 with respect to the direct sum expression

$$A = B_1 \oplus B_2 \oplus \cdots \oplus B_i \oplus A_i.$$

It follows that $\{e_1, e_2, \ldots\}$ is an infinite set of mutually orthogonal non-zero idempotents, which contradicts condition 1. □

Using the theorem above and definition 6.2.3 one has the following examples of orthogonally finite rings:

1. Semisimple rings.
2. Noetherian rings.
3. Artinian rings.
4. Local rings and fields (they have at most one non-zero idempotent).
5. Right (left) finite-dimensional rings, i.e., rings which right (left) regular modules do not contain infinite sum of right (left) ideals.
6. Semiperfect rings. This follows by theorem 1.9.3.
7. Perfect rings. This follows from theorem 1.9.8.

We proceed to the next:

Definition 6.2.5. A ring A is called **Dedekind-finite**[2] if $ba = 1$ whenever $ab = 1$ for $a, b \in A$. Otherwise A is said to be **non-Dedekind-finite** (or **Dedekind-infinite**). In this case there are elements a and b such that $ab = 1$, but $ba \neq 1$.

The following easy fact gives simple examples of Dedekind-finite rings.

---

[2]Such a ring is also sometimes called **von Neumann-finite**, **directly finite**, **weakly 1-finite**, or **inverse symmetric**.

Proposition 6.2.6. *A ring A is Dedekind-finite if it satisfies any of the following conditions*:

1. *A is a unital subring*[3] *of a Dedekind-finite ring.*
2. *A is a direct product of Dedekind-finite rings.*
3. *A is an epimorphic image of a Dedekind-finite ring.*
4. *A has no right or left zero divisors.*

Proof.
1, 2, 3 are trivial.

4. Let $ab = 1$ for non-zero $a, b \in A$, then $a(1 - ba) = 0$. Since A has no right zero divisor, $1 - ba = 0$, i.e., $ba = 1$. The proof for left zero divisors is similar. □

From this proposition and definition 6.2.5 we immediately obtain the following examples of Dedekind-finite rings.

Examples 6.2.7.

1. Any commutative ring.
2. Any domain.
3. Any division ring.
4. Any ring which is a finite dimensional vector space over a division ring.
5. The ring of endomorphisms of a finitely dimensional vector space over a division ring.
6. The ring of matrices $M_n(D)$ over a division ring D, since $M_n(D) \simeq \text{End}_D(V)$ where $V = D^n$.
7. Any semisimple ring, since it is isomorphic a finite direct sum of matrix rings over division rings.
8. Any finite ring. Indeed, let A be a finite ring and $ab = 1$ for $a, b \in A$. Consider a map $f : A \longrightarrow A$ defined by $f(x) = ax$. Then $f(bx) = a(bx) = (ab)x = x$ for every $x \in A$. So f is a surjective map. Since A is finite, f is also injective. Since $f(ba) = a(ba) = (ab)a = 1 \cdot a = f(1)$, we obtain that $ba = 1$. So A is Dedekind-finite.

The following example shows the existence of Dedekind-infinite rings.

Example 6.2.8. (See [208, p.4].)

Let V be a K-vector space $Ke_1 \oplus Ke_2 \oplus \cdots$ with a countably infinite basis $\{e_i : i \geq 1\}$ over a field K, and let $A = \text{End}_K(V)$ be the K-algebra of all vector space endomorphisms of V. Let $a, b \in A$ be defined by

$$b(e_i) = e_{i+1} \text{ for all } i \geq 1$$

and

$$a(e_1) = 0, \quad a(e_i) = e_{i-1} \text{ for all } i \geq 2,$$

[3] Recall that A_1 is said to be a unital subring of A if the identity of A is also the identity of A_1.

then $ab = 1 \neq ba$, so a is right-invertible without being left-invertible. Therefore A gives an example of a Dedekind-infinite ring.

Remark 6.2.9. This example, together with examples 6.2.7(4,5,6) above, also illustrates why Dedekind-finite is thought of a finiteness condition. A stronger illustration that we are here dealing with a finiteness condition is Jacobson's theorem 6.2.15 below.

Definition 6.2.10. An A-module M is called **Dedekind-finite** (or **directly finite**) if $M = M_1 \oplus N$, where N is an A-module and $M \cong M_1$, implies that $N = 0$. In other words, a Dedekind-finite module is not isomorphic to any proper direct summand of itself.

From this definition we obtain immediately the following examples of Dedekind-finite modules.

Example 6.2.11. The following modules are Dedekind-finite:

1. Finite modules.
2. Modules with finite length.
3. Indecomposable modules.
4. Finite dimensional modules [4], i.e., modules which contain no infinite direct sum of non-zero submodules, or equivalently modules with finite uniform dimension (See proposition 5.1.16).

The proposition below gives a simple relation between Dedekind-finite modules and Dedekind-finite rings.

Proposition 6.2.12. *The following conditions are equivalent for an A-module M:*

1. *M is Dedekind-finite.*
2. *The ring* $\mathrm{End}_A(M)$ *is Dedekind-finite.*

Proof.
$1 \Longrightarrow 2$. Let $f, g \in \mathrm{End}_A(M)$ and $fg = 1$. Therefore f is an epimorphism, and g is a monomorphism. Consider an exact sequence

$$0 \longrightarrow \mathrm{Ker}(f) \longrightarrow M \xrightarrow{f} M \longrightarrow 0.$$

Since $fg = 1$, this sequence is split by [146, proposition 4.2.1]. So $M \simeq M \oplus \mathrm{Ker}(f)$. Since M is Dedekind-finite, $\mathrm{Ker}(f) = 0$, i.e., f is an isomorphism. So we have $gf = 1$, that is, $\mathrm{End}_A(M)$ is a Dedekind-finite ring.

$2 \Longrightarrow 1$. Suppose that M is not Dedekind-finite. Then there is a decomposition $M = M_1 \oplus N$, where $N \neq 0$ and $M \cong M_1$. Let $\varphi : M \longrightarrow M_1$ be an isomorphism of A-modules, $\iota : M_1 \longrightarrow M$ the inclusion map, and $g = \iota\varphi$. Define a map $f : M \longrightarrow M$

[4] In sense of A.W. Goldie, see definition 5.1.13.

such that $f|_{M_1} = \varphi^{-1}$ and $f|_N = 0$. Then $fg = 1 \in \mathrm{End}_A(M)$, but $gf \neq 1$ since $gf(N) = 0$. Thus, $\mathrm{End}_A(M)$ is not Dedekind-finite. $\square$

Since the second condition of proposition 6.2.12 is left-right symmetric we obtain immediately the following result.

Proposition 6.2.13. *A ring A is Dedekind-finite if and only if the right regular module A_A (or the left regular module ${}_AA$) is Dedekind-finite.*

From this proposition it follows immediately that a ring A is Dedekind-infinite if and only if A_A (or ${}_AA$) is Dedekind-infinite, which means that $A_A \cong A_A \oplus N$ for some right A-module N. So there exists an idempotent $1 \neq e \in A$ such that $A = eA \oplus (1-e)A$ and $A \cong eA$ as right A-modules. Thus one has the following useful statement:

Proposition 6.2.14. *A ring A is Dedekind-infinite if and only if there exists an idempotent $1 \neq e \in A$ such that $A = eA \oplus (1-e)A$ and $A \cong eA$ as right A-modules.*

With respect to example 6.2.8 notice the next important result due to N. Jacobson [161].

Proposition 6.2.15. (N. Jacobson [161]). *Any Dedekind-infinite ring A contains a countable infinite set of pairwise orthogonal non-zero idempotents.*

Proof. Let A be a ring which is Dedekind-infinite, i.e., there exist elements $a, b \in A$ such that $ab = 1$ but $ba \neq 1$. Note that also $ba \neq 0$ because $a \neq 0$ and $ab = 1$. Write $e = ba$. Then $e^2 = b(ab)a = ba = e$, so e is a non-trivial idempotent. For $i, j \geq 0$ let

$$e_{ij} = b^i(1-e)a^j.$$

Then $\{e_{ij}\}$ is a set of matrix units in the sense that $e_{ij}e_{kl} = \delta_{jk}e_{il}$. To see this note that $a^ib^i = 1$ for all i, and $a(1-e) = 0 = (1-e)b$. If $j \neq k$, then a^jb^k is either $a^{|j-k|}$ or $b^{|j-k|}$, so

$$e_{ij}e_{kl} = b^i(1-e)a^jb^k(1-e)a^l = 0$$

On the other hand, since $1-e$ is an idempotent,

$$e_{ij}e_{jl} = b^i(1-e)a^jb^j(1-e)a^l = b^i(1-e)a^l = e_{il}.$$

Note that each $e_{ij} \neq 0$. Indeed, if $b^i(1-e)a^j = 0$, then $0 = a^ib^i(1-e)a^jb^j = (1-e)$, a contradiction.

In particular, $\{e_{ii} : i \geq 1\}$ is a countable infinite set of pairwise orthogonal non-zero idempotents in A, and so A contains an infinite direct sum of non-zero right ideals $\bigoplus\limits_{i \geq 0} e_{ii}A$. $\square$

From the proof of the above proposition it follows, in particular, that there is a countable infinite set of pairwise orthogonal non-zero idempotents $\{e_{ii} : i \geq 1\}$ in A so that A contains an infinite direct sum of non-zero right ideals $\bigoplus_{i\geq 0} e_{ii}A$. Therefore any Dedekind-infinite ring is neither orthogonal finite rings nor right (left) finite-dimensional rings. So we obtain the following corollary.

Corollary 6.2.16.

1. *Any orthogonally finite ring is Dedekind finite.*
2. *Any right (left) finite-dimensional ring is Dedekind-finite.*

Corollary 6.2.17. *Let A be a ring with a.c.c. or d.c.c. on principal left (right) ideals generated by idempotents. Then A is Dedekind-finite.*

Proof. Suppose that A is a Dedekind-infinite ring, then one can construct a descending chain of principal left ideals:

$$(1 - e_{11})A \supset (1 - e_{11} - e_{22})A \supset \cdots \supset (1 - \sum_{i=1}^{n} e_{ii})A \supset \cdots$$

where e_{ii} are idempotents as constructed in proposition 6.2.15.

Similarly one can construct an ascending chain of left ideals. □

Corollary 6.2.18. *Let A be a ring such that $B = A/\mathrm{rad}(A)$ is a right (left) finite-dimensional ring then A is Dedekind-finite.*

Proof. From corollary 6.2.16 it follows that $B = A/I$ is Dedekind-finite. Assume that $ab = 1$ in A. Let $R = \mathrm{rad}\, A$ and let $\varphi : A \longrightarrow A/R$ be the natural epimorphism. Since A/R is Dedekind-finite, $\varphi(ba) = \varphi(b)\varphi(a) = \varphi(a)\varphi(b) = \varphi(ab) = \bar{1}$. Therefore $ba = 1 + r$, where $r \in \mathrm{rad}\, A$, is invertible in A, i.e., there exists $u \in U(A)$ such that $bau = 1$, whence $a = abau = au$. So $ba = 1$, as required. □

From this corollary and proposition 6.2.6(3) we immediately obtain the following fact.

Corollary 6.2.19. *A ring A is Dedekind-finite if and only if $B = A/\mathrm{rad}A$ is Dedekind-finite.*

The next statement can be easy obtained from the above corollaries and proposition 6.2.15 and gives further examples of Dedekind-finite rings.

Proposition 6.2.20. *A ring A is Dedekind-finite if it satisfies any of the following conditions:*

1. *A is right (left) Artinian.*
2. *A is right (left) Noetherian.*
3. *A satisfies a.c.c. on right (left) direct summands.*
4. *A satisfies a.c.c. on right (left) annihilators.*

5. *A is a local ring.*
6. *A is a semilocal ring.*
7. *A is a semiperfect ring.*
8. *A is a perfect ring.*

Proposition 6.2.21.

1. *Let N be a nilpotent ideal of a ring A. Then A is Dedekind-finite if and only if $B = A/N$ is Dedekind-finite.*
2. *Let A be a Dedekind-finite ring. Then for any idempotent $e \in A$ the ring eAe is also Dedekind-finite.*
3. *Let A, B be rings and X be an A-B-bimodule. Then the ring*

$$M = \begin{pmatrix} A & X \\ 0 & B \end{pmatrix}$$

is Dedekind-finite if and only if A and B are both Dedekind-finite.

Proof.

1. The "only if" part follows from proposition 6.2.6(3). Suppose that A/N is Dedekind-finite and $\varphi : A \longrightarrow A/N$ is the natural epimorphism. Suppose $ab = 1$ for $a, b \in A$. Since A/N is Dedekind-finite, $\varphi(ba) = \varphi(b)\varphi(a) = \varphi(a)\varphi(b) = \varphi(ab) = \bar{1}$. Therefore $ba = 1 + x$, where $x \in N$. Since N is nilpotent, $(1 + x)^n = 0$ for enough large n. So ba is invertible in A, i.e., there exists $u \in U(A)$ such that $bau = 1$, whence $a = abau = au$. Thus $ba = 1$, as required.
2. Suppose that $ab = e$ for $a, b \in eAe$. Put $f = 1 - e$. Since e, f are orthogonal idempotents, and $a, b \in eAe$, $af = 0 = fb$. So $(a + f)(b + f) = ab + f = e + f = 1$. Since A is Dedekind-finite, $(a + f)(b + f) = 1$ implies that $(b + f)(a + f) = 1$, whence $ba = 1 - f = e$, as desired.
3. The proof follows immediately from statement 2. □

Proposition 6.2.22. *The following statements are equivalent for a ring A and any $n \geq 1$:*

1. *$M_n(A)$ is a Dedekind-finite ring.*
2. *If $A^n \simeq A^n \oplus M$ for some A-module M then $M = 0$.*
3. *Any module epimorphism $\varphi : A^n \longrightarrow A^n$ is an isomorphism.*
   Here A^n denotes a direct sum of n copies of the right regular module A_A over a ring A.

Proof. The equivalence of conditions 1 and 2 follows immediately from proposition 6.2.12 for $M = A^n$. The equivalence of conditions 2 and 3 follows from the fact that the A-module A^n is projective. □

Definition 6.2.23. A ring A which satisfies the equivalent conditions of proposition 6.2.22 is called **stably finite**. If these conditions are satisfied for some fixed n, one said that A is **n-finite**. For $n = 1$ observe that the notion of 1-finite is the same that Dedekind-finite.

So proposition 6.2.22 can be written in the following equivalent form:

Proposition 6.2.24. *A ring A is stably finite if and only if the ring $M_n(A)$ is Dedekind-finite for all n, and if and only if the A-module A^n is Dedekind-finite for all n.*

The following rings are stably finite:

1. Skew fields, since for any skew field D the ring $M_n(D)$ for $n \geq 1$ is simple, and so D stably finite.
2. Right (left) Noetherian rings, since A is a right (left) Noetherian ring implies that $M_n(A)$ is right (left) Noetherian for any positive integer n, and any right (left) Noetherian ring is Dedekind-finite.
3. Right (left) Artinian rings, since $M_n(A)$ is an Artinian ring for any Artinian ring A.
4. Commutative rings. Let A be a commutative ring. It is obviously Dedekind-finite. Suppose $X, Y \in M_n(A)$ and $XY = I_n$. Then $\det(X)\det(Y) = 1$, i.e., $\det(Y) \in U(A)$. Then from the adjugate equation $Y\mathrm{adj}(Y) = \det(Y)I_n$ we obtain that $(\det(Y))^{-1}\mathrm{adj}(Y)$ is a right inverse of Y. Thus Y is both left inverse and right inverse, so it is invertible and $XY = YX = I_n$. Therefore $M_n(A)$ is Dedekind-finite for any n, i.e., A is stably finite.
5. Finite rings. This follows from the facts that for any finite ring A the ring $M_n(A)$ is also finite for any $n \geq 1$, and any finite ring is Dedekind-finite.

Proposition 6.2.25. (S. Montgomery [244, Lemma 2]). *Let A be a ring with Jacobson radical R, and I any ideal of A contained in R. Then for any $n > 0$, A is n-finite if and only if A/I is n-finite.*

Proof. It suffices to prove this theorem for $n = 1$, since $\mathrm{rad}(M_n(A)) = M_n(R) \supseteq M_n(I) = I_1$, an ideal of $M_n(A)$.

Recall [146, proposition 3.4.6], that if $x \in R$, $1 + x$ is two-sided invertible. Write the inverse as $1 + y$, then $1 = (1 + x)(1 + y) = 1 + x + y + xy$, so that $x + y + xy = 0$ which implies $y \in R$. If $x \in I \subseteq R$, $y \in I$.

Now assume that A/I is 1-finite, i.e., A/I is Dedekind finite. Let $a, b \in A$ and $ab = 1$. Write $\bar{a}, \bar{b} \in A/I$ for the images of a, b. Since A/I is Dedekind finite, $\bar{a}\bar{b} = \bar{b}\bar{a} = \bar{1}$. So $ba = 1 + x$ with $x \in I \subseteq R$. Let $1 + y$, $y \in I$, be the inverse of $1 + x$. Then $ba(1 + y) = 1$. So $aba(1 + y) = a$, which implies that $a(1 + y) = a$, and hence $ay = 0$. So $1 = ba(1 + y) = ba + bay = ba$, i.e., A is Dedekind finite.

Inversely, let A be Dedekind finite, $\bar{a}, \bar{b} \in A/I$ and $\bar{a}\bar{b} = \bar{1}$ in A/I. Let $a, b \in A$ be lifts of $\bar{a}, \bar{b}$. Then $ab = 1 + x$, $x \in I$. Let $1 + y$, $y \in I$, be the two-sided inverse of $1 + x$. Then $ab(1 + y) = 1$. So, as A is Dedekind finite, $b(1 + y)a = 1$. But $bya \in I$, so $\bar{b}\bar{a} = 1$. □

Note, that from this theorem it follows, in particular, that any perfect ring is stably finite. Indeed, if A is a perfect ring with Jacobson radical R, then A/R is a semisimple ring, and so it is stably finite. Therefore, A/I is also stably finite for each ideal $I \subseteq R$.

Corollary 6.2.26. *Let A be a ring. Then the following statements are equivalent:*

1. *A is n-finite.*
2. *$A[x]$ is n-finite.*
3. *$A[[x]]$ is n-finite.*

Since $A \subset A[x] \subset A[[x]]$ then $3 \Longrightarrow 2 \Longrightarrow 1$. So it suffices to prove that $1 \Longrightarrow 3$. Suppose that A is n-finite. Let $I = (x) \subseteq \mathrm{rad}(A[[x]])$. Then $A[[x]]/I \cong A$, so by proposition 6.2.25 $A[[x]]$ is also n-finite. □

If a ring A is stably finite then it is obviously Dedekind-finite. The inverse statement is not true.

Theorem 6.2.27 (J.C. Shepherdson [286]) *There exists a ring A with a unit element and no divisors of zero over which there exists 2×2-matrices B, C, X such that $BC = I$, $BX = 0$, $X \neq 0$, where I is the 2×2 unit matrix.*

To prove this theorem J.C.Shepherdson in [286] constructed a ring A having elements satisfying certain equations and inequalities. Namely, let A be the K-algebra over a field K generated by non-commuting elements $\{s, t, u, v, w, x, y, z\}$ subject to the relations:

$$sx + uz = 1; \qquad sy + uw = 0$$

$$tx + vz = 0; \qquad ty + vw = 1$$

Shepherdson then proved that this ring is a domain. The proof of this fact uses a normal form for monomials in $\{s, t, u, v, w, x, y, z\}$ of a special form.

Set

$$B = \begin{pmatrix} s & u \\ t & v \end{pmatrix}, \qquad C = \begin{pmatrix} x & y \\ z & w \end{pmatrix} \tag{6.2.28}$$

Then it is easy to check that

$$BC = I, \qquad CB \neq I, \tag{6.2.29}$$

where I is the 2×2 unit matrix over A. So setting $X = CB - I \neq 0$, one obtain that $BX = B(CB - I) = (BC)B - BI = 0$, as required. □

Corollary 6.2.30. *There exists a ring A with a unit element and no divisors of zero (and so it is Dedekind-finite), but so that $M_2(A)$ is not Dedekind-finite.*

Proof. Consider the matrices B, C as in theorem 6.2.27 given by (6.2.28) for which $BC = I$ and $CB \neq I$. Therefore the ring $M_2(A)$, where A as in theorem 6.2.27, is not Dedekind-finite. At the same time A is a domain, and so Dedekind-finite. □

Remark 6.2.31. If A is the ring constructed by J.C. Shepherdson, then also $M_n(A)$ contains matrices satisfying conditions (6.2.29), and so $M_n(A)$ is also Dedekind-infinite. Using similar methods, P.M. Cohn in [58] constructed a ring A such that for any $n \geq 1$ the ring $M_t(A)$ is Dedekind-finite for all $t \leq n$, but $M_{n+1}(A)$ is Dedekind-infinite.

Using the construction of tensor products Susan Montgomery in [244] constructed an other example of a Dedekind-infinite ring. She showed that if a field K is not algebraically closed, L is an algebraic extension of the field K, $L \neq K$, then there exists a K-algebra A which is a domain, hence a Dedekind-finite ring, so that $A \otimes_K L$ is Dedekind-infinite.

Recall that a ring A is called **von Neumann regular** if for every $a \in A$ there exists an $x \in A$ such that $a = axa$. Such a ring is sometimes called a regular ring. These rings were invented and named by von Neumann [248]. Simple examples of von Neumann regular rings are skew fields, Boolean rings and the ring $M_n(K)$ of $n \times n$ square matrices with entries from some field K. The study of von Neumann rings is the special topic of the well-known book of K.R. Goodearl [110].

One of the important characterizations of von Neumann rings was obtained by von Neumann [248].

Theorem 6.2.32. *For a ring A the following conditions are equivalent*:

1. *A is a von Neumann regular ring.*
2. *Every principal right (left) ideal in A is generated by an idempotent.*
3. *Every finitely generated right (left) ideal in A is generated by an idempotent.*

Proof.

$1 \Longrightarrow 2$. Let $a \in A$. Since A is a von Neumann ring, there exists $x \in A$ such that $axa = a$. Then the element $e = ax$ is an idempotent and $aA = axaA = axA = eA$.

$2 \Longrightarrow 3$. It suffices to prove that an ideal $I = aA + bA$ is principal for every $a, b \in A$. From statement 2 it follows that there exists an idempotent $e = e^2 \in A$ such that $aA = eA$. Consider the element $b - eb \in I$. Then $eA + (b - eb)A \subseteq I$ and $I \subseteq eA + bA - ebA$, so that $I = eA + (b - eb)A$. By statement 2 there exists an idempotent $f = f^2 \in A$ such that $(b - eb)A = fA$. Let $f = (b - be)x$ for some $x \in A$. Then $ef = e(b - eb)x = 0$. Consider the element $g = f - fe$ which is an idempotent, since $ef = 0$ and so $g^2 = (f - fe)(f - ef) = f - fe = g$. Since $eg = e(f - fe) = 0$ and $ge = (f - fe)e = 0$, e and g are mutually orthogonal idempotents. Moreover, $fg = f(f - fe) = g$ and $gf = (f - fe)f = f$. Therefore $gR = fgR \subseteq fR$ and $fR = gfR \subseteq gR$. So that $gR = fR$ and $I = eA + gA$. Since $eg = ge$, $I = (e + g)A$ where $(e + g)^2 = e + g$ is an idempotent.

$3 \Longrightarrow 2$. It is obvious.

$2 \Longrightarrow 1$. Let $a \in A$. Then there exists an idempotent $e \in A$ such $aA = eA$. If $e = ax$ for some $x \in A$ then $axa = ea = a$.

Since the definition of a ring being von Neumann is left-right symmetric we obtain that the corresponding statements 2 and 3 for left ideals also hold. □

From this theorem we immediately obtain the following corollary.

Theorem 6.2.33. *For a ring A the following conditions are equivalent*:

1. *A is a von Neumann regular ring.*
2. *Every principal right* (*left*) *ideal in A is a direct summand of the right* (*left*) *regular module of A.*
3. *Every finitely generated right* (*left*) *ideal in A is a direct summand of the right* (*left*) *regular module of A.*

Proof. The proof follows from theorem 6.2.32 and [146, proposition 2.1.1] taking into account that for any idempotent e of a ring A we have a direct decomposition of a right regular module A_A: $A = eA \oplus (1-e)A$. □

Corollary 6.2.34. *Let A be a von Neumann regular ring. Then*

1. *Every finitely generated right* (*left*) *ideal in A is projective*;
2. *A Is a semihereditary ring*;
3. *Every right* (*left*) *annihilator of any element in A is generated by an idempotent, and so it is projective and it is a direct summand of the right* (*left*) *regular module of A*;
4. *Every finitely generated submodule of a projective A-module is projective.*

Proof.

1. This follows immediately from theorem 6.2.33.
2. This follows from the previous statement.
3. Suppose that $a \in A$. Consider the exact sequence
$$0 \longrightarrow \mathrm{r.ann}(a) \longrightarrow A \longrightarrow aA \longrightarrow 0$$
which is split, since the right ideal aA is projective. Therefore $\mathrm{r.ann}(a)$ is a direct summand of A, and so $\mathrm{r.ann}(a) = eA$ for some idempotent $e \in A$.
4. This follows from theorem 4.4.9. □

Theorem 6.2.35 (M. Auslander, M. Harada). *A ring A is von Neumann regular if and only if all right* (*left*) *A-modules are flat.*

Proof. Suppose that A is a von Neumann regular ring. Let M be a right A-module. Consider an exact sequence of right A-modules
$$0 \longrightarrow N \longrightarrow F \longrightarrow M \longrightarrow 0$$
with F free. Any finitely generated left ideal I in A is principal and generated by an idempotent, i.e., $I = Ae$ for some $e^2 = e \in A$. Therefore $FI = Fe$ and $NI = Ne$. Let

$x \in N \cap Fe$, then $x \in N$ and $x = fe$ for some $f \in F$. So $x = fe = fe^2 = xe \in Ne$, i.e., $N \cap Fe \subseteq Ne$. On the other hand if $x \in Ne$ then $x = ne$ for some $n \in N$, and so $x \in N$ and $x = ne \in Fe$ since $N \subseteq F$. Thus $N \cap Fe = Ne$ and so M is a flat module by the flatness test 1.5.15.

Conversely, suppose that every A-module M is flat. Let $a \in A$ and consider an exact sequence

$$0 \longrightarrow aA \longrightarrow A \longrightarrow A/aA \longrightarrow 0.$$

By hypothesis the cyclic right A-module A/aA is flat. Therefore by the flatness test 1.5.15 $aA \cap AI = (aA)I$ for every finitely generated left ideal I in A. In particular, if $I = Aa$ we have $aA \cap Aa = aAa$. This implies that there exists an element $x \in A$ such that $axa = a$, i.e., A is a von Neumann regular ring. □

Corollary 6.2.36. *A ring A is von Neumann regular if and only if* $\text{w.gl.dim}\, A = 0$.

The example above 6.2.38 shows that not all von Neumann regular ring are Dedekind-finite. For this end we shall prove the following statement.

Proposition 6.2.37. *If M is a semisimple module over a ring A then $B = \text{End}_A(M)$ is a von Neumann regular ring.*

Proof. Let $f \in \text{End}_A(M) = B$. Since M is a semisimple module, every its submodule is a direct summand of M by [146, proposition 2.2.4]. Therefore there are submodules N and L of M such that $M = \text{Ker}(f) \oplus N = \text{Im}(f) \oplus L$. Let $g = f|_N$. Then $g(N) = \text{Im}(f)$. Therefore g is an isomorphism of N onto $\text{Im}(f)$. We define $t \in B$ such that $t(x+y) = g^{-1}(x) + g(y)$ for all $x \in \text{Im}(f)$ and $y \in L$. Then $ftf = f$, i.e., B is von Neumann regular. □

Example 6.2.38. Consider the ring $A = \text{End}_K(V)$ from the example 6.2.8, where $V = Ke_1 \oplus Ke_2 \oplus \cdots$ is an infinite vector space over a field K. Then V is obviously a semisimple module and so A is a von Neumann regular ring by the previous proposition. On the other hand this ring is Dedekind-infinite that was shown in example 6.2.8.

At the same time there is a special class of von Neumann regular rings all of which are Dedekind-finite.

Definition 6.2.39. A ring A is called **unit-regular** if for any $a \in A$ there exists a unit $u \in \mathcal{U}(A)$ such that $a = aua$.

Lemma 6.2.40. *A unit-regular ring is Dedekind-finite.*

Proof. Suppose $ab = 1$. Since there exists $u \in \mathcal{U}(A)$ such that $a = aua$, we obtain $1 = ab = auab = au$. An element u^{-1} exists. So multiplying on the right with u^{-1} gives $a = u^{-1} \in \mathcal{U}(A)$, and so $b = u \in \mathcal{U}(A)$. Therefore $ba = 1$, i.e., A is Dedekind-finite. □

Note that from this lemma it follows that the example 6.2.38 yields the example of von Neumann regular ring which is not a unit-regular ring.

Proposition 6.2.41. *Let M be an A-module and $B = \mathrm{End}_A(M)$ is a unit-regular ring. Then $M = N \oplus X = L \oplus Y$ with $N \simeq L$ implies $X \simeq Y$.*

Proof. Let $f \in B$ such that $f(X) = 0$ and f induces an isomorphism N onto L. Since B is a unit-regular ring there is an automorphism $u \in \mathrm{Aut}_A(M)$ such that $fuf = f$. Since

$$M = uf(M) + (1 - uf)M = uf(N) + (1 - uf)(M) = u(L) + (1 - uf)(M)$$

and $f(1 - fu) = 0$, $(1 - uf)(M) \subseteq \mathrm{Ker}(f) = X$. So that $M = u(L) + X$. Since $u(L) \cap X = uf(M) \cap \mathrm{Ker}(f) = 0$, $M = u(L) \oplus X$. Taking into account that u is an automorphism of M, we have that $M = u(M) = u(L \oplus Y) = u(L) + u(Y)$. Therefore $X \simeq u(Y) \simeq Y$. □

Some interesting characterization of unit-regular rings was obtained by V.P. Camillo and D. Khurana:

Theorem 6.2.42. (V.P. Camillo and D. Khurana [32]). *A ring A is unit-regular if and only if for every $a \in A$ there is a unit $u \in U(A)$ and an idempotent e such that $a = e + u$ and $aA \cap eA = 0$.*

Proof. Suppose that A is a unit-regular ring and $a \in A$. Then by theorem 6.2.33 and corollary 6.2.34 $L = aA$ and $N = \mathrm{r.ann}_A(a)$ are direct summands of A_A. Since $\mathrm{r.ann}_A(a)$ is a principal ideal, by corollary 6.2.34(3), $M = aA + \mathrm{r.ann}_A(a)$ and $K = aA \cap \mathrm{r.ann}_A(a)$ are finitely generated ideals and so, by theorem 6.2.33 and corollary 6.2.34, they are direct summands of A_A, which implies that they are projective as well.

Consider an exact sequence

$$0 \longrightarrow L \longrightarrow M \longrightarrow M/L \longrightarrow 0. \qquad (6.2.43)$$

Since L and M are principal and projective ideals, M/L is a finitely generated A-module which is flat by theorem 6.2.35. Therefore M/L is projective by theorem 4.6.21. Therefore the sequence (6.2.43) splits, i.e., there is a right ideal X of A such that $M = L \oplus X$. Since $X \subset M = L + N$ and $X \cap L = 0$, $X \subseteq N$.

Next consider an exact sequence

$$0 \longrightarrow M \longrightarrow A \longrightarrow A/M \longrightarrow 0 \qquad (6.2.44)$$

which splits, since M is projective and principle ideal, A is free and so that A/M is projective, by theorem 4.6.24. Thus $A_A = M \oplus Y$ for some right ideal Y of A.

Analogously there is a right ideal Z of A such that $L = K \oplus Z$. Therefore we have the following decomposition:

$$A_A = K \oplus X \oplus Y \oplus Z \qquad (6.2.45)$$

Since $X \subseteq N$, using the modular law (1.1.4) we have

$$N = N \cap M = N \cap (X \oplus L) = X \oplus (N \cap L) = X \oplus K. \tag{6.2.46}$$

On the other hand

$$N = \mathrm{r.ann}_A(a) \simeq A/aA = A/L \simeq X \oplus Y. \tag{6.2.47}$$

Therefore comparing (6.2.46) and (6.2.47) we get

$$X \oplus K \simeq X \oplus Y. \tag{6.2.48}$$

Since A is a unit-regular ring, applying proposition 6.2.41 to the decomposition (6.2.46) and taking into account (6.2.48) one obtains that $K \simeq Y$. Let $\varphi : K \longrightarrow Y$ be an isomorphism. Considering the decomposition (6.2.45) one can construct two endomorphisms $\psi, h \in \mathrm{End}_A(A)$ such that $\psi(k + x + y + z) = \varphi(k) + y$ and $h(k + x + y + z) = \varphi(k) + \varphi^{-1}(y)$ for all $k \in K$, $x \in X$, $y \in Y$ and $z \in Z$. We may regard ψ and h as elements of A. Then

$$\psi h\psi(k + x + y + z) = \psi h(\varphi(k) + y) = \psi(\varphi^{-1}(\varphi(k) + y) = \psi(k + \varphi^{-1}(y)) =$$

$$= \varphi(k) + \varphi\varphi^{-1}(y) = \varphi(k) + y = \psi(k + x + y + z).$$

So that $\psi h\psi = \psi$, which implies that $e = \psi h$ is an idempotent. As $a = \psi h + (a - \psi h)$ it remains only to show that $a - \psi h$ is a unit element. To this end we will show that $\mathrm{r.ann}_A(a - \psi h) = 0$. Suppose that $(a - \psi h)(k + x + y + z) = 0$. Taking into account that $ak = ax = 0$, we get

$$a(k + x + y + z) = a(y + z) = \psi h(k + x + y + z) = \psi(\varphi(k) + \varphi^{-1}(y)) =$$

$$= \varphi(k) + \varphi\varphi^{-1}(y) = \varphi(k) + y \in aA \cap Y = 0,$$

so that $\mathrm{r.ann}_A(a - \psi h) = 0$. Since A is a unit-regular ring, for the element $b = a - \psi h$ there is a unit u such that $bub = b$, which implies that $b(1 - ub) = 0$, and so $1 = ub$. By lemma 6.2.40, $bu = 1$, as well. Thus $b = a - \psi h$ is a unit. Since $eA = \psi hA \subseteq X \oplus Y$ and $aA \cap (X \oplus Y) = 0$, we get $eA \cap aA = 0$, that required.

Conversely, suppose that for any element $a \in A$ there are an idempotent $e \in A$ and a unit $u \in A$ such that $a = e + u$, and $eA \cap aA = 0$. Consider

$$x = au^{-1}e = (e + u)u^{-1}e = eu^{-1}e + e \in aA \cap eA = 0.$$

Therefore $0 = au^{-1}(a - u) = au^{-1}a - a$, i.e., $au^{-1}a = a$. So that A is a unit-regular ring. □

Remark 6.2.49. Note that a ring A is called **clean** if every element of A is a sum of a unit and an idempotent. Such rings were introduced by W.K. Nicholson in [249] studying exchange rings, which will be considered in this book in chapter H7.

Next we discuss the problem of rank for finitely generated free modules over a ring A, i.e., when this rank is well-defined and when it is not defined. Let A^n denote a direct sum of n copies of the right regular module A_A over a ring A.

Definition 6.2.50. A ring A is said to have **right invariant basis number** (abbreviated **IBN**) **property** if for all positive integers m,n, an isomorphism $A^n \simeq A^m$ as right A-modules implies $n = m$.

Equivalently, this means that there do not exist distinct positive integers m,n such that $A^n \simeq A^m$ as right A-modules.

This definition can be rephrase in terms of matrices. For this end we prove the following lemma.

Lemma 6.2.51. *$A^n \simeq A^m$ as right A-modules if and only if there are matrices $X \in M_{n\times m}(A)$, $Y \in M_{m\times n}(A)$ such that*

$$XY = I_n, \quad YX = I_m \tag{6.2.52}$$

where I_n, I_m are identity matrices.

Proof. Let $u_1,u_2,\dots,u_n$ be a basis of A^n, and $v_1,v_2,\dots,v_m$ be a basis of A^m. Consider an isomorphism $\varphi : A^n \longrightarrow A^m$. Then $\varphi(u_i) = \sum\limits_{j=1}^{m} x_{ij}v_j$, and $\varphi^{-1}(v_j) = \sum\limits_{k=1}^{n} y_{jk}u_k$, where $x_{ij}, y_{jk} \in A$. So that $u_i = \varphi^{-1}\varphi(u_i) = \sum\limits_{j=1}^{m}\sum\limits_{k=1}^{n} x_{ij}y_{jk}u_k$ which implies that $XY = I_n$, where $X = (x_{ij}) \in M_{n\times m}(A)$, $Y = (y_{jk}) \in M_{m\times n}(A)$, and I_n is the identity matrix.

In a similar way we obtain that $YX = I_m$.

Conversely, let there are two matrices $X = (x_{ij}) \in M_{n\times m}(A)$, $Y = (y_{jk}) \in M_{m\times n}(A)$ such that the equality (6.2.52) holds. Consider two maps $\varphi : A^n \longrightarrow A^m$ given by $\varphi(u_i) = \sum\limits_{j=1}^{m} x_{ij}v_j$, and $\psi : A^m \longrightarrow A^n$ given by $\psi(v_j) = \sum\limits_{k=1}^{n} y_{jk}u_k$. Then $\varphi\psi(v_j) = v_j$ for all $i = 1,2,\dots,n$, and $\psi\varphi(u_i) = u_i$ for all $j = 1,2,\dots,m$. Since these equalities hold for all elements of basis, $\varphi\psi(v) = v$ for every $v \in A^m$, and $\psi\varphi(u) = u$ for every $u \in A^n$, that is $\varphi\psi = 1$ and $\psi\varphi = 1$. So that φ and ψ are inverse homomorphisms, i.e., $A^n \simeq A^m$. □

Using this lemma one can obtain the proposition which gives the equivalent definition of IBN-rings.

Lemma 6.2.53. *A ring A has the right* IBN *property if and only if for all positive integers $m \neq n$ there are no matrices $X \in M_{n\times m}(A)$, $Y \in M_{m\times n}(A)$ such that*

$$XY = I_n, \quad YX = I_m \tag{6.2.54}$$

where I_n, I_m are identity matrices.

Remark 6.2.55. Since this lemma is right-left symmetric, a ring has the right IBN property if and only if it has the left IBN property. So we can speak simply about the **IBN property** for a ring A. A ring with IBN property is also called an **IBN-ring**.

Corollary 6.2.56.

1. *If A is an* IBN*-ring then so is $M_n(A)$ for all n.*
2. *If $A \longrightarrow B$ is a homomorphism of rings and A is not an* IBN*-ring then so is B.*

Proof.

1. Suppose that $B = M_n(A)$ is not an IBN-ring for some $n \geq 1$. Then by lemma 6.2.53 there are positive integers $k \neq m$ and matrices $X \in M_{k\times m}(B)$, $Y \in M_{m\times k}(B)$ such that $XY = I_k$ and $YX = I_m$. Then the matrix X can be considered as a matrix $X_1 \in M_{kn\times mn}(A)$, and the matrix Y can be considered as a matrix $Y_1 \in M_{mn\times kn}(A)$, and we obtain the equalities $X_1Y_1 = I_{kn}, Y_1X_1 = I_{mn}$, which show that A is not an IBN-ring. A contradiction.
2. Suppose that A is not an IBN-ring. Then by lemma 6.2.53 there are positive integers $n \neq m$ and matrices $X = (x_{ij}) \in M_{n\times m}(A)$, $Y = (y_{jk}) \in M_{m\times n}(A)$ such that the matrix equalities (6.2.54) hold. Applying the homomorphism $\varphi : A \longrightarrow B$ to this equalities we obtain the matrix equalities $X_1Y_1 = I_n$, $Y_1X_1 = I_m$, where $X_1 = (\varphi(x_{ij})) \in M_{n\times m}(B)$, $Y_1 = (\varphi(y_{jk})) \in M_{m\times n}(B)$. Therefore B is not an IBN-ring by lemma 6.2.53. □

Example 6.2.57. The following non-zero rings are IBN-rings:

1. Any field.
2. Any division ring.
3. Any non-zero commutative ring. This follows from the fact that A/M is a field for every maximal ideal M of a commutative ring A. Therefore A/M is an IBN-ring, and so A is also an IBN-ring by corollary 6.2.56(2).
4. Any finite ring.
5. Any (Artinian) semisimple ring. This follows from Krull-Schmidt theorem for semisimple modules (See e.g. [146, proposition 3.2.5].
6. Any right (left) Artinian ring. Indeed, a right Artinian ring A as the right regular module has a finite length $l(A) = k$. If $A^n \simeq A^m$ then $l(A^n) = nk$ and $l(A^m) = mk$. Therefore $nk = mk$ implies $n = m$. Another proof follows from the fact that if R is the Jacobson radical of A then A/R is a semisimple ring, and so is an IBN-ring. Therefore A is also an IBN-ring by corollary 6.2.56(2).
7. Any right (left) Noetherian ring. If N is the prime radical of a right Noetherian ring then A/N is a semiprime right Noetherian ring which is a subring in a semisimple ring, and so A/N is an IBN-ring. Therefore A is also an IBN-ring by corollary 6.2.56(2).
8. Any stably finite ring, in particular any Dedekind-finite ring. This follows from proposition 6.2.24.

Example 6.2.58. This is an example of a ring which does not have IBN.

Let V be a vector space over a field K. Suppose that V has a countable basis over K. Since the union of two countable sets is again a countable set, $V \oplus V \simeq V$. Consider the ring $A = \mathrm{End}_K(V)$. Then

$$A = \mathrm{End}_K(V) \simeq \mathrm{Hom}_V(V, V \oplus V) = \mathrm{Hom}_V(V,V) \oplus \mathrm{Hom}_V(V,V) = A \oplus A = A^2.$$

By induction one obtains $A^n \simeq A$, and so $A^n \simeq A^m$ for all positive integers n, m. Therefore A is a non-IBN-ring.

Example 6.2.59. This is an example of a ring which is an IBN-ring but is not stably finite.

Let A be the algebra over a commutative ring $K \neq 0$ generated by two elements x, y with the single relation $xy = 1$. Then $yx \neq 1$, and so A is Dedekind-infinite, in particular, A is not stably finite. On the other hand, K being a non-zero commutative ring has IBN property. So A has also IBN property, by corollary 6.2.56(2), because it admits a homomorphism $f : A \longrightarrow K$ defined by $f(x) = f(y) = 1$.

Summarizing we obtain the following diagram for rings considering in this section:

Unit-regular $\Longrightarrow$ Dedekind-finite $\Longrightarrow$ IBN-rings

At the end of this section note two interesting statements which give the examples of Dedekind-finite rings. These theorem are given here without proof.

Theorem 6.2.60 (Y.Utumi [311]). *Any right and left self-injective*[5] *ring is Dedekind-finite.*

Theorem 6.2.61 (Kaplansky [180]). *If K is a field of characteristic 0, then for any group G, the group algebra KG is Dedekind-finite.*

6.3 FDI-Rings

Recall that an idempotent $e \in A$ is called **primitive** if it cannot be written as a sum of two non-zero pairwise orthogonal idempotents of A. Consider the next important class of rings with finiteness condition.

Definition 6.3.1 [146, Chapter 2]. A ring A is called an **FDI-ring** if there exists a decomposition of the identity $1 \in A$ into a finite sum $1 = e_1 + e_2 + \cdots + e_n$ of pairwise orthogonal primitive idempotents e_i. In this case the right (left) regular A-module A_A $({}_AA)$ can be decomposed into a finite direct sum of indecomposable modules e_iA (Ae_i).

---

[5]Recall that a ring A is called **right self-injective** if the right regular module A_A is injective.

Note that the decomposition of $1 \in A$, given in the definition of an FDI-ring, may not be unique.

Examples 6.3.2. The following rings are FDI-rings:

1. Division rings.
2. Finite direct sums of FDI-rings.
3. Rings which are finite dimensional vector spaces over a division ring.
4. The ring of matrices $M_n(D)$ over a division ring D.
5. Semisimple rings.
6. Right (left) Artinian rings.
7. Right (left) Noetherian rings.
8. Semiperfect rings.
9. Right (left) finite-dimensional rings, i.e., rings which do not contain infinite direct sum of non-zero right (left) ideals (See definition 5.1.13).
10. Perfect rings.

Definition 6.3.3. A finite set of orthogonal idempotents $e_1, e_2, \ldots, e_m \in A$ is called **complete** if

$$e_1 + e_2 + \ldots + e_m = 1 \in A$$

A ring A is said to **have enough idempotents** if it has a (finite) complete set of orthogonal primitive idempotents.

Remark 6.3.4. Note that the concept of a ring to have enough idempotents coincides with concept of an FDI-ring.

Lemma 6.3.5. *Let A be an* FDI-*ring with finite complete set $S = \{e_1, e_2, \ldots, e_n\}$ of orthogonal primitive idempotents, and let $P \subseteq S$. Then fAf is also an* FDI-*ring for any idempotent $f = e_{i1} + e_{i2} + \cdots + e_{ik}$ where each $e_{ij} \in P$.*

Proof. Let A be an FDI-ring with a complete set of primitive pairwise orthogonal idempotents $S = \{e_1, e_2, \ldots, e_n\}$, i.e., there is a decomposition $1 = e_1 + e_2 + \cdots + e_n$ of the identity $1 \in A$ into a sum of pairwise orthogonal primitive idempotents. Let $f = e_{i1} + e_{i2} + \cdots + e_{ik}$, where each $e_{ij} \in P \subseteq S$, be a non-zero idempotent of A. We will show that if $fe_if \neq 0$ then fe_if is a primitive idempotent in fAf. To this end it suffices to show that a ring $B = (fe_if)(fAf)(fe_if)$ has the only idempotent fe_if. If $e_i \notin P$ then $fe_if = 0$. If $e_i \in P$ then $fe_if = fe_i = e_if = e_i$. Therefore $B = e_iAe_i$ and $fe_if = e_i$ is the only idempotent in B. So that $f = e_{i1} + e_{i2} + \cdots + e_{ik}$ is a decomposition of the identity $f \in fAf$ in a finite sum of primitive idempotents $fe_{ij}f = e_{ij}$, i.e., fAf is an FDI-ring. □

Recall the following important statement:

Proposition 6.3.6. (See [146, theorem 2.4.14, corollary 2.4.15].)

1. *Let A be an* FDI-*ring. Then the identity of A can be written as a sum of a finite number of orthogonal centrally primitive idempotents.*

2. *Any* FDI-*ring can be uniquely decomposed into a direct product of a finite number of indecomposable rings.*

Recall that an idempotent $e \in A$ is called **central** if $ea = ae$ for all $a \in A$. It is easy to verify that an idempotent $e \in A$ is central if and only if $Ae = eA = eAe$ if and only if $eA(1-e) = (1-e)Ae = 0$, and if and only if a two-sided Peirce decomposition of A has the following form:

$$A = \begin{pmatrix} eAe & 0 \\ 0 & (1-e)A(1-e) \end{pmatrix}.$$

A central idempotent $c \in A$ is called **centrally primitive** if it cannot be written as a sum of two non-zero pairwise orthogonal central idempotents of A.

Some generalizations of a notion of a central idempotent are left (or right) semicentral idempotents introduced by G.F. Birkenmeier et al. in [21].

Definition 6.3.7 [21]. An idempotent $e \in A$ is called **left** (resp. **right**) **semicentral** in A if $Ae = eAe$ (resp. $eA = eAe$).

Obviously, any central idempotent is left and right semicentral.

An idempotent $e \in A$ is left semicentral if and only if

$$(1-e)Ae = (1-e)eAe = 0,$$

and if and only if a two-sided Peirce decomposition of A has the following form:

$$A = \begin{pmatrix} eAe & eA(1-e) \\ 0 & (1-e)A(1-e) \end{pmatrix}.$$

An idempotent $e \in A$ is left semicentral if and only if for each $x \in A$:

$$xe = [e + (1-e)]xe = exe + (1-e)xe = exe.$$

So that we obtain the following lemma.

Lemma 6.3.8. (See [21, Lemma 1.1].) *The following conditions are equivalent for an idempotent* $e^2 = e \in A$:

1. $Ae = eAe$
2. $(1-e)Ae = 0$.
3. $xe = exe$ *for each* $x \in A$.
4. *A two-sided Peirce decomposition of* A *has the following form:*

$$A = \begin{pmatrix} eAe & eA(1-e) \\ 0 & (1-e)A(1-e) \end{pmatrix}.$$

Similar conditions hold for right semicentral idempotents which we formulate in the following lemma.

Lemma 6.3.9. (See [21, Lemma 1.1].) *The following conditions are equivalent for an idempotent* $e^2 = e \in A$:

1. $eA = eAe$
2. $eA(1-e) = 0$.
3. $ex = exe$ *for each* $x \in A$.
4. *A two-sided Peirce decomposition of* A *has the following form:*

$$A = \begin{pmatrix} eAe & 0 \\ (1-e)Ae & (1-e)A(1-e) \end{pmatrix}.$$

Lemma 6.3.10. (See [21, Lemma 2.13].) *Let* $c = c^2$ *be a non-zero left (right) semicentral idempotent of an* FDI-*ring* A. *Then* cAc *is also an* FDI-*ring. Moreover, if* $S = \{e_1, e_2, \dots, e_n\}$ *is a complete set of primitive idempotents of* A *then there is a subset* $P \subseteq S$ *such that* $\{cg_ic \mid g_i \in P\}$ *is a complete set of primitive idempotents of* cAc. *If* $c \neq 1$ *then the set* P *has less than* n *elements.*

Proof.

Let A be an FDI-ring with a complete set of primitive pairwise orthogonal idempotents $S = \{e_1, e_2, \dots, e_n\}$, i.e., there is a decomposition $1 = e_1 + e_2 + \cdots + e_n$ of the identity of A into a sum of orthogonal primitive idempotents. Let $c^2 = c \in A$ be a non-zero left semicentral idempotent of A, i.e., it satisfies the equivalent conditions of lemma 6.3.8.

We will show that if $ce_ic \neq 0$ then ce_ic is a primitive idempotent in cAc. To this end it suffices to show that a ring $B = (ce_ic)(cAc)(ce_ic)$ has the only idempotent ce_ic.

Let $u = u^2$ be a non-zero idempotent in B, i.e., $u = (ce_ic)(cac)(ce_ic)$. Since c is a left semicentral idempotent, $e_ic = ce_ic$ and $uc = cuc$, so that one obtain $uc = u = cu = cuc$, $e_iu = e_i(ce_ic)(cac)(ce_ic) = u$, $ue_i = e_iue_i$. Therefore $(ue_i)(ue_i) = ue_i$. If $ue_i = 0$ then $u = ue_ic = 0$. Since $u \neq 0$, ue_i is a non-zero idempotent in e_iAe_i. Since e_i is a primitive idempotent, $ue_i = e_i$. Then $u = ue_ic = e_ic = ce_ic$, i.e., ce_ic is the only idempotent in B.

So that all ce_ic are primitive idempotents in B and $e_ice_i = (ue_i)ce_i = ue_i = e_i$. Some of that elements may be equal to zero. In the set $S = \{e_1, e_2, \dots, e_n\}$ we choose a subset P such that $ce_ic \neq 0$ for all $e_i \in P$. Without loss of generality we may consider $P = \{e_1, e_2, \dots, e_m\}$ where $m \leq n$. Since c is a left semicentral idempotent, $e_ic = ce_ic$, and so $ce_ic = 0$ implies $e_ic = 0$. Therefore $c = (e_1 + e_2 + \cdots + e_n)c = (e_1 + e_2 + \cdots + e_m)c = e_1c + e_2c + \cdots + e_mc = ce_1c + ce_2c + \cdots + ce_mc$. Since $(ce_ic)(ce_jc) = ce_je_jc = 0$ for $i \neq j$, P is a complete set of primitive pairwise orthogonal idempotents for cAc. Suppose that $m = n$. Then taking into account that $e_i = e_ice_i$ and $e_ic = ce_ic$ we obtain $1 = e_1 + e_2 + \cdots + e_n = e_1ce_1 + e_2ce_2 + \cdots + e_nce_n = ce_1ce_1 + ce_2ce_2 + \cdots + ce_nce_n = c(e_1ce_1 + e_2ce_2 + \cdots + e_nce_n) = c$. This contradiction shows that $m < n$. □

From this lemma and proposition 6.3.6 we immediately obtain the following corollary.

Corollary 6.3.11.

1. *Let A, B be rings and X be an A-B-bimodule. Then the ring*

$$M = \begin{pmatrix} A & X \\ 0 & B \end{pmatrix}$$

is an FDI-*ring if and only if A and B are both* FDI-*rings.*

2. *Any* FDI-*ring can be uniquely decomposed into a direct product of a finite number of indecomposable* FDI-*rings.*

Remark 6.3.12. Note that there exist FDI-rings which are not Dedekind-finite. It follows from the Stepherdson example 6.2.27 that there exists a ring R with 1 and matrices $A, B \in M_2(R)$ such that $AB = I$ and $BA \neq I$. Therefore the ring $M_2(R)$, where R as in theorem 6.2.27, is not Dedekind-finite. At the same time it is an FDI-ring, since $1 = e_{11} + e_{22}$, and, moreover, $e_{11}M_2(R)e_{11} \simeq e_{22}M_2(R)e_{22} \simeq R$, where R is a domain.

Thus there are FDI-rings, which are Dedekind-infinite.

Remark 6.3.13. The authors does not know the example of a ring which is Dedekind-finite but not a FDI-ring.

Proposition 6.3.14. *If a ring A is orthogonally finite then it is an* FDI-*ring.*

Proof. Suppose that A is not FDI-ring. Then from definition 6.3.1 it follows that A contains an infinite set of pairwise orthogonal primitive idempotents, so that A is not an orthogonally finite ring. A contradiction. □

Remark 6.3.15. An example constructed by K.R.Goodearl in [110, Example 5.15] shows that there exist unit-regular rings (and therefore, by lemma 6.2.40, Dedekind-finite rings) which contain uncountable direct sums of non-zero pairwise isomorphic right ideals. Moreover, an example constructed by L.A. Skornyakov in [290] shows that there exist FDI-rings which contain an infinite set of pairwise orthogonal idempotents, and thus contain an infinite direct set of principal right (left) ideals generated by idempotents.

From the above one can obtain the following diagram of relationships between the main classes of rings with finiteness conditions. Arrows in the diagram below mean containments of classes of rings.

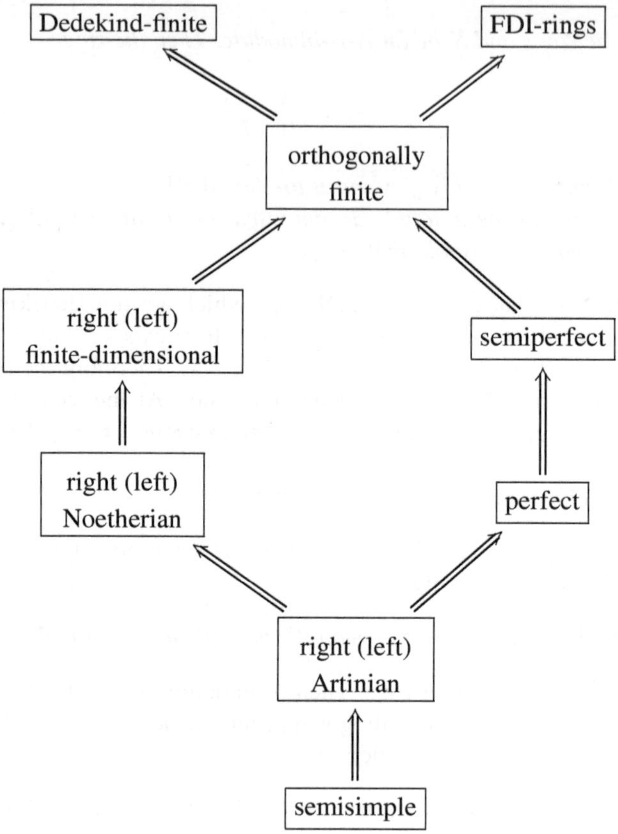

Diag. 6.3.16.

6.4 Semiprime FDI-Rings

In this section we prove a criterion for a semiprime FDI-ring to be decomposable into a finite direct product of prime rings. This theorem can be considered as a generalization of the following theorem proved by V.V. Kirichenko and M. Khibina [191]:

Theorem 6.4.1. (See [146, theorem 14.4.6].) *A semiprime semiperfect ring A is a finite direct product of prime rings if and only if all endomorphism rings of principal A-modules are prime.*

The notion of nilpotency is very important in the theory of rings and algebras. Recall that a right ideal I of a ring A is called nilpotent if $I^n = 0$ for some positive integer n. The smallest integer $n > 0$ such that $I^n = 0$ but $I^{n-1} \neq 0$ is called the **nilpotency index** (or **index of nilpotency**) of the ideal I and we will denote it $t(I)$.

Proposition 6.4.2 ([275, Lemma 2.7.13]). *Suppose e is an idempotent of a ring A and I is a two-sided ideal of A. Then I is nilpotent if and only if eIe and $(1-e)I(1-e)$ are nilpotent.*

Proof.
$\Longrightarrow$. Let I be a two-sided nilpotent ideal of A with nilpotency index $t(I) = n$. Suppose that $e^2 = e$ is an idempotent of A and $x_1, x_2, ..., x_n \in eIe$. Then $x_i = ey_ie$ for some $y_i \in I$, $i = 1,2,...,n$, and

$$x_1x_2\cdots x_n = ey_1eey_2e\cdots ey_ne = ey_1ey_2e\cdots y_ne = z_1z_2\cdots z_n = 0,$$

where $z_i = ey_i \in eI$.
$\Longleftarrow$. Conversely, suppose that $e^2 = e$ is an idempotent of A and ideals eIe and $(1-e)I(1-e)$ are nilpotent with $t(eIe) = n$ and $t((1-e)I(1-e)) = m$. Put $k = m+n$. Let $x_1, x_2, ..., x_n \in I$. Since $1 = e + (1-e)$,

$$u = x_1x_2\cdots x_k =$$

$$= [e+(1-e)]x_1[e+(1-e)]x_2[e+(1-e)]\cdots[e+(1-e)]x_k[e+(1-e)] =$$

$$= \sum e_{i_1}x_1e_{i_2}x_2\cdots e_{i_k}x_ke_{i_{k+1}}, \tag{6.4.3}$$

where each idempotent e_{i_s} is either e or $1-e$.

Consider a summand of this sum $y = e_{i_1}x_1e_{i_2}x_2\cdots e_{i_k}x_ke_{i_{k+1}}$. Since there are $k+1 > n+m$ idempotents in this product, either the number of idempotent e is more than n or the number of idempotent $1-e$ is more than m in this product. In any case each summand y in the sum (6.4.3) is equal to 0, and so $u = 0$. Therefore I is a nilpotent ideal.

From the proof of this proposition it immediately follows:

Corollary 6.4.4. *Suppose e is an idempotent of a ring A and I is a two-sided ideal of A with nilpotency index $t(I)$. Then*

$$t(I) \leq n + m \tag{6.4.5}$$

where $n = t(eIe)$ and $m = t[(1-e)I(1-e)]$.

Proposition 6.4.6. *Let A be a ring with finite complete set of pairwise orthogonal idempotents $S = \{e_1, e_2, \ldots, e_k\}$. Then a two-sided ideal I in A is nilpotent if and only if eIe is nilpotent for any idempotent $e \in S$. Moreover,*

$$t(I) \leq n_1 + n_2 + \cdots + n_k \tag{6.4.7}$$

where $n_i = t(e_iIe_i)$.

Proof. Apply the induction to proposition 6.4.2 and corollary 6.4.4. By proposition 6.4.2, the statement of the proposition is valid if $k = |S| = 2$. Assume that the statement is true if $n \leq k - 1$. Set $e = e_1 + e_2 + \cdots + e_{k-1}$, $f = 1 - e = e_k$. Then eIe and fIf are nilpotent, by proposition 6.4.2, and $t(I) \leq m_{k-1} + n_k$, where $m_{k-1} = t(eIe)$ and $n_k = t(fIf)$. Since eIe is a nilpotent two-sided ideal in eAe, by assumption, $t(eIe) \leq n_1 + n_2 + ... + n_{k-1}$, where $n_i = t(e_iIe_i)$. Therefore $t(I) \leq n_1 + n_2 + \cdots + n_k$. □

Applying this proposition to semiperfect rings and taking into account that eAe is a semiperfect ring for any local idempotent e of a semiperfect ring A, we obtain the following result which is some extension of [146, Theorem 11.4.1].

Proposition 6.4.8. *Suppose A is a semiperfect ring with a finite complete set of pairwise orthogonal local idempotents $S = \{e_1, e_2, \ldots, e_k\}$. Then a two-sided ideal I in A is nilpotent if and only if e_iIe_i is nilpotent for any $e_i \in S$. Moreover,*

$$t(I) \leq n_1 + n_2 + ... + n_k,$$

(5) *where $n_i = t(eiIei)$. In particular, if $e_iIe_i = 0$ for every local idempotent $e_i \in S$ then I is nilpotent.*

In particular, if $I = J(A)$ is the Jacobson radical of a semiperfect ring A we immediately obtain the following corollary.

Corollary 6.4.9. *Suppose A is a semiperfect ring with a finite complete set of pairwise orthogonal local idempotents $S = \{e_1, e_2, \ldots, e_k\}$. Then the Jacobson radical $J(A)$ of A is nilpotent if an ideal $e_iJ(A)e_i$ is nilpotent for every local idempotent $e_i \in S$. In particular, if $e_iJ(A)e_i = 0$ for every local idempotent $e_i \in S$, then $J(A)$ is nilpotent.*

Now taking into account lemma 6.3.5 one can obtain from proposition 6.4.6 the similar result for FDI-rings.

Proposition 6.4.10. *Let A be an* FDI*-ring with a finite complete set of pairwise orthogonal primitive idempotents $S = \{e_1, e_2, \ldots, e_k\}$. Then a two-sided ideal I in A is nilpotent if and only if eIe is nilpotent for any idempotent $e \in S$. Moreover,*

$$t(I) \leq n_1 + n_2 + ... + n_k, \tag{6.4.11}$$

where $n_i = t(e_iIe_i)$. In particular, if $e_iIe_i = 0$ for every primitive idempotent $e_i \in S$, then I is nilpotent.

Lemma 6.4.12. *Let A be a semiprime ring and $1 = g_1 + g_2$ be a decomposition of $1 \in A$ in a sum of two pairwise orthogonal idempotents, $A = \begin{pmatrix} A_1 & X \\ Y & A_2 \end{pmatrix}$, where $A_1 = g_1Ag_1$, $A_2 = g_2Ag_2$, $X = g_1Ag_2$ and $Y = g_2Ag_1$. Let $M = \begin{pmatrix} M_{11} & M_{12} \\ M_{21} & M_{22} \end{pmatrix}$ be an ideal in A and $M_{12} \neq 0$. Then $M_{12}M_{21} \neq 0$, $M_{21} \neq 0$, $M_{21}M_{12} \neq 0$. Symmetrically,* *if* $M_{21} \neq 0$ *then* $M_{12} \neq 0$, $M_{12}M_{21} \neq 0$, $M_{21}M_{12} \neq 0$. *In particular, if* A *is an indecomposable ring and* $Y \neq 0$, *then* $YX \neq 0$, $X \neq 0$, $XY \neq 0$.

Proof.

Suppose $M_{12} \neq 0$. We will show that $M_{12}M_{21} \neq 0$.

Assume $M_{12}M_{21} = 0$. Denote by $I(M_{12})$ the two-sided ideal in A generated by M_{12}. Since

$$\begin{pmatrix} 0 & M_{12} \\ 0 & 0 \end{pmatrix} \begin{pmatrix} A_1 & X \\ Y & A_2 \end{pmatrix} = \begin{pmatrix} M_{12}Y & M_{12} \\ 0 & 0 \end{pmatrix}$$

and

$$\begin{pmatrix} A_1 & X \\ Y & A_2 \end{pmatrix} \begin{pmatrix} M_{12}Y & M_{12} \\ 0 & 0 \end{pmatrix} = \begin{pmatrix} M_{12}Y & M_{12} \\ YM_{12}Y & YM_{12} \end{pmatrix},$$

we obtain that $I(M_{12}) = \begin{pmatrix} M_{12}Y & M_{12} \\ YM_{12}Y & YM_{12} \end{pmatrix}$.

Obviously, $M_{12}YM_{12} \subseteq M_{12}$ and $YM_{12}Y \subseteq M_{21}$. Therefore

$$(M_{12}Y)^3 = (M_{12}YM_{12})(YM_{12}Y) \subseteq M_{12}M_{21} = 0.$$

Consequently, $M_{12}Y$ is nilpotent. In addition

$$YM_{12}YM_{12}YM_{12} \subseteq YM_{12}M_{21}Y = 0.$$

Therefore YM_{12} is also nilpotent and, by proposition 6.4.2, $I(M_{12})$ is a nilpotent ideal. So A is not semiprime by definition. A contradiction. So $M_{12}M_{21} \neq 0$.

The rest of the statement is verified analogously. The lemma is proved. □

Lemma 6.4.13. *Let* A *be a semiprime ring and* $1 = g_1 + g_2$ *be a decomposition of* $1 \in A$ *in a sum of two pairwise orthogonal idempotents,* $A = \begin{pmatrix} A_1 & X \\ Y & A_2 \end{pmatrix}$, *where* $A_1 = g_1Ag_1$, $A_2 = g_2Ag_2$, $X = g_1Ag_2 \neq 0$ *and* $Y = g_2Ag_1 \neq 0$. *If* $I = \begin{pmatrix} I_1 & I_{12} \\ I_{21} & I_2 \end{pmatrix}$ *is a two-sided ideal in* A *with* $I_{12} \neq 0$ *then* $I_1 \neq 0$ *and* $I_2 \neq 0$.

Proof. If $I_{12} \neq 0$ then $I_{21} \neq 0$ as well by lemma 6.4.12. Suppose that $I_1 = 0$ then $I = \begin{pmatrix} 0 & I_{12} \\ I_{21} & I_2 \end{pmatrix}$. Since I is a two-sided ideal,

$$I^2 = \begin{pmatrix} I_{12}I_{21} & I_{12}I_2 \\ I_2I_{21} & I_{21}I_{12} + I_2^2 \end{pmatrix} \subseteq \begin{pmatrix} 0 & I_{12} \\ I_{21} & I_2 \end{pmatrix}.$$

Therefore $I_{12}I_{21} = 0$ which implies $I_{12} = I_{21} = 0$ by lemma 6.4.12. This contradiction shows that $I_1 \neq 0$. Analogously one can show that $I_2 \neq 0$. □

Definition 6.4.14. An indecomposable projective right A-module P of an FDI-ring A will be called **principal** if $P \simeq e_iA$ for some primitive idempotent e_i, $i = 1, \ldots, n$.

Theorem 6.4.15. *Let A be a semiprime FDI-ring with a finite complete set of pairwise orthogonal primitive idempotents $S = \{e_1, e_2, \dots, e_k\}$. Then A is a finite direct product of prime rings if and only if all rings e_iAe_i are prime for all $e_i \in S$, $i = 1, 2, \dots, n$.*

Proof.

Let A be a semiprime ring, and let $1 = e_1 + \dots + e_n$ be a decomposition of the identity of A into a sum of pairwise orthogonal primitive idempotents. First suppose that A is a finite direct product of prime rings. Say $A = A_1 \times A_2 \times \dots \times A_s$. Let e be a non-trivial primitive idempotent of A. Then $eAe \subset A_i$ for some (unique) i. Now $\varphi : I \longrightarrow eIe$ is a product preserving surjective map from ideals in A_i to ideals in $eAe \subset A_i$. Indeed let J be an ideal of eAe. Then A_iJA_i is an ideal in A_i and $eA_iJA_ie = J$. Now let $J, J' \subset eAe$ be two non-zero ideals such that $JJ' = 0$. Then A_iJA_i and $A_iJ'A_i$ are two ideals in A_i with product zero. So at least one of them is zero as A_i is prime. So at least one of J, J' is zero. This proves that eAe is prime.

We will now prove the converse statement.

Let A be a semiprime FDI-ring with decomposition $1 = e_1 + e_2 + \dots + e_n$ of $1 \in A$ into a sum of primitive pairwise orthogonal idempotents.

Write $e_iAe_j = A_{ij}$ and $P_i = e_iA$. It follows from lemma 6.4.12 that either $A_{ij} = A_{ji} = 0$ or $A_{ij} \neq 0$ and $A_{ji} \neq 0$. This yields the possibility to introduce a relation $\sim$ on the set $\{1, 2, \dots, n\}$, setting $i \sim j$ if and only if $A_{ij} \neq 0$. In this case we write $e_iA \sim e_jA$ and the principal projective modules P_i and P_j will be called **equivalent**. We now show that $\sim$ is an equivalence relation on $\{1, 2, \dots, n\}$. Indeed, $i \sim i$ is obvious. Let $i \sim j$, i.e., $A_{ij} \neq 0$. Then from lemma 6.4.12 it follows that $A_{ji} \neq 0$, i.e., $j \sim i$. Suppose now that $i \sim j$ and $j \sim k$. Then A_{ij}, A_{ji}, A_{jk}, and A_{kj} are all non-zero. Suppose that $i \nsim k$, i.e., $A_{ik} = 0$. By lemma 6.4.12 $A_{ki} = 0$ as well. Suppose $f = e_i + e_j + e_k$ and consider a ring fAf which has the following two-sided Peirce decomposition:

$$fAf = \begin{pmatrix} A_{ii} & A_{ij} & 0 \\ A_{ji} & A_{jj} & A_{jk} \\ 0 & A_{kj} & A_{kk} \end{pmatrix}.$$

This implies that $A_{ij}A_{jk} = 0$. Therefore $IJ = (A_{ji}A_{ij})(A_{jk}A_{kj}) = 0$, where $I = A_{ji}A_{ij}$ and $J = A_{jk}A_{kj}$ are non-zero ideals in A_j. Since A_j is a prime ring, $I = 0$ or $J = 0$. This contradiction shows that $i \sim k$. So that $\sim$ is an equivalence relation on $\{1, 2, \dots, n\}$.

Let $C_1, C_2, \dots, C_m$ be equivalence classes on $\{1, 2, \dots, n\}$ with regard to the relation $\sim$. Set $f_i = \sum\limits_{j \in C_i} e_j$, then the $f_1, f_2, \dots, f_m$ form a complete set of orthogonal idempotents in A.

We now show that A is isomorphic to a direct product of rings: $A \simeq \prod\limits_{i=1}^{m} f_iAf_i$. Indeed, let $k \in C_i$ and $s \in C_j$ for $i \neq j$. This means that $k \nsim s$ and so $A_{ks} = A_{sk} = 0$, i.e., $f_iAf_j = f_jAf_i = 0$ for all $i \neq j$. Hence $A \simeq \prod\limits_{i=1}^{m} f_iAf_i$. Moreover, if $B = f_iAf_i$

then B_B is equal to a direct sum of equivalent principal modules $P_k = e_k A$ where $k \in C_i$.

So without loss of generality, taking into account corollary 6.3.11(2), one can assume that A is an indecomposable semiprime FDI-ring with decomposition $1 = e_1 + e_2 + \cdots + e_n$ of $1 \in A$ into a sum of primitive pairwise orthogonal idempotents and all principal right A-modules $P_i = e_i A$ are equivalent. This means that $A_{jk} = e_j A e_k \neq 0$ for all $j, k = 1, 2, \ldots, n$.

Suppose that all the $e_i A e_i$ are prime rings. We will show that A is a prime ring. Let I, J be a non-zero two-sided ideal in A such that $IJ = 0$. Since I,J are non-zero ideals there are indices i, j, k, s such that $I_{ij} \neq 0$ and $J_{ks} \neq 0$. Let $g = e_i + e_j$ and $f = e_k + e_s$. Then gIg and fJf are two-sided ideals in semiprime rings gAg and fAf respectively. Therefore by lemma 6.4.13 $I_i \neq 0$ and $J_k \neq 0$. Consider $M = I_i A_{ik} J_k \subseteq IJ$ where $A_{ik} \neq 0$ by assumption. We will show that $M \neq 0$.

Suppose that $L = A_{ik} J_k = 0$. Then $A_{ki} A_{ik} J_k = 0$. Since $A_{ki} \neq 0$, $S_k = A_{ki} A_{ik} \neq 0$ by lemma 6.4.12. So that S_k is a non-zero two-sided ideal in A_k and $S_k J_k = 0$ in a prime ring A_k with $S_k \neq 0$ and $J_k \neq 0$. This contradiction shows that $L \neq 0$. Suppose now that $LA_{ki} = A_{ik} J_k A_{ki} = 0$. Then $A_{ki} A_{ik} J_k A_{ki} A_{ik} = S_k J_k S_k = 0$ in a prime ring A_k. So that $V_i = A_{ik} J_k A_{ki} \neq 0$ and V_i is a two-sided ideal in A_i. Since $M = I_i A_{ik} J_k = 0$ implies $MA_{ki} = I_i A_{ik} J_k A_{ki} = I_i V_i = 0$ in a prime ring A_i with $I_i \neq 0$ and $V_i \neq 0$, one obtains that $M = I_i A_{ik} J_k \neq 0$. Thus $IJ \neq 0$, i.e., A is a prime ring. □

6.5 Notes and References

Dedekind-finite modules and rings were studied by Goodearl in [109] and [110]. Such a concept of finiteness for sets was introduced by R. Dedekind in 1887 in his paper [64], where a set S is considered to have the right finiteness condition if from bijection $F : S \longrightarrow T$ of a set S onto a subset $T \subseteq S$ it follows that $S = T$. In other words he allowed only sets which are not in bijection with a proper subset of themselves. This made it possible to avoid a number of paradoxes in the theory of sets by G. Cantor. The term "Dedekind-finite" for rings was first introduced by F.C. Leary in [212], where he generalized the idea proposed by Dedekind for sets. Namely, he called a ring A Dedekind-finite if each monomorphism $f : A \longrightarrow A$ of the regular A-modules is an isomorphism.

Proposition 6.2.15 was proved by N.Jacobson in [161]. Theorem 6.2.27 was constructed by J.C. Shepherdson in [286]. The results of Section 6.4 come from [66] and [127].

CHAPTER 7

Modules with Change Property and Change Rings

This chapter is devoted to important problems connected with uniqueness of decompositions of modules into direct sums of indecomposable modules. The famous Krull-Remak-Schmidt theorem was already considered in [146], where in Section 10.4 this theorem was proved for the case of finite direct sums of modules with local endomorphism rings. Actually, G. Azumaya proved this theorem in [12] for infinite direct sums in the general case for Abelian categories with some additional properties. In this chapter a proof of this theorem is given for the case of infinite direct sums of modules with local endomorphism rings following Peter Crawley and Bjarni Jónsson. They proved this theorem using the exchange property, which was introduced in 1964 for general algebras [61]. From that time on this notion has become an important theoretical tool for studying rings and modules.

Some properties of modules having the exchange property are studied in Section 7.1. It is proved that the 2-exchange property is equivalent to the finite exchange property for arbitrary modules, and the 2-exchange property is equivalent to the exchange property for indecomposable modules. These results were obtained by P. Crawley and B. Jónsson for general algebras in [61] and R.B. Warfied, Jr. for Abelian categories in [321]. In this section we also prove the important result obtained by R.B.Warfied, Jr. in [317] which states that an indecomposable module has the exchange property if and only if its endomorphism ring is local.

A proof of the Azumaya theorem for infinite direct sums of modules is given in Section 7.2.

The problem of the cancellation property and some properties of modules having the cancellation property are considered in Section 7.3.

At the end of this chapter we consider some different classes of rings connected with exchange rings. Section 7.4 is devoted to the study of some properties and some structure theorems for exchange rings.

7.1 The Exchange Property

In this chapter we will consider some class of modules which possess a very important property named the exchange property.

Following P. Crawley and B.Jónsson [61] introduce the following definition:

Definition 7.1.1. Given a cardinal $\mathfrak{N}$, an A-module M is said to have the $\mathfrak{N}$**-exchange property** if for any A-module X and any two decompositions of X:

$$X = M_1 \oplus N = \bigoplus_{i \in I} X_i \tag{7.1.2}$$

with $M_1 \simeq M$ and $|I| \leq \mathfrak{N}$ it follows that there are submodules $Y_i \subseteq X_i$ such that

$$X = M_1 \oplus \Big(\bigoplus_{i \in I} Y_i\Big). \tag{7.1.3}$$

A module M has the **exchange property** if M has the $\mathfrak{N}$-exchange property for any cardinal $\mathfrak{N}$. If M has the $\mathfrak{N}$-exchange property for any finite cardinal $\mathfrak{N}$, M is said to have the **finite exchange property**. If M has the $\mathfrak{N}$-exchange property for $|\mathfrak{N}| = n$, M is said to have the n**-exchange property**.

Remark 7.1.4. Note that in 7.1.3 M_1 is a direct summand, so that $M_1 \cap Y_i = 0$ for all i. Without this condition the definition would be empty. One could take $Y_i = X_i$ and have

$$X = M_1 + \Big(\bigoplus_{i \in I} Y_i\Big).$$

Roughly speaking the exchange property says that one can so speak split-off pieces from the X_i that together make up M_1. The remaining pieces of the X_i are the Y_i.

To illustrate the point suppose for the moment that X is a distributive module (which is of course not always the case) which means that for all submodules $U, V, W \subset X$ one has, among other properties,

$$U \cap (V + W) = (U \cap V) + (U \cap W) \tag{7.1.5}$$

and let the index set I have two elements (or, more generally, be finite). Then the two decompositions

$$X = M_1 \oplus N, \quad X = X_1 \oplus X_2$$

give rise to

$$X = (M_1 \cap X_1) \oplus (M_1 \cap X_2) \oplus (N \cap X_1) \oplus (N \cap X_2) =$$

$$= M_1 \oplus (N \cap X_1) \oplus (N \cap X_2)$$

So in this case the exchange property holds with $Y_1 = N \cap X_1$, $Y_2 = N \cap X_2$.

Remark 7.1.6. It is hard to overestimate the importance of the exchange concept.

To get some idea of why 7.1.1 is called an exchange property consider the following very simple example of finite dimensional vector spaces.

Let $X = K^n$, the n-dimensional vector space over a field K with its standard basis

$$e_1 = (1,0,\dots,0), \; e_2 = (0,1,\dots,0), \; \dots, e_n = (0,0,\dots,1)$$

and let M be the vector space generated by the first $(n-1)$ standard basis vectors $e_1, e_2, \dots, e_{n-1}$ and $N = Ke_n$. Let v be any vector with last component $\neq 0$, $X_1 = M$, $X_2 = Kv$. Then

$$X = M \oplus N = X_1 \oplus X_2.$$

A decomposition

$$X = M \oplus Y_1 \oplus Y_2$$

is now given by taking $Y_1 = 0$, $Y_2 = Kv = X_2$ to give

$$X = M \oplus 0 \oplus Y_2$$

with natural basis $\{e_1, \dots, e_{n-1}, v\}$. That is the basis element e_n is replaced with v. Inversely v can be replaced by e_n. That is they can be **exchanged**.

In this sense the exchange notion is widely used in much more general contexts. For instance in the context of abstract dependence relations (see [60, page 20ff]) and the theory of matroids.

Let S be a set and $\mathcal{B}$ a family of subsets (called, and thought of as, bases). Then $\mathcal{B}$ is said to have the exchange property if for any two $B, B' \in \mathcal{B}$ and $b' \in B'$ there is a $b \in B$ such that $B' \cup \{b\} \setminus \{b'\}$ is again a basis. A set S with such a family of subsets $\mathcal{B}$ is called a **matroid**. For the theory of matroids see, e.g. [325]. For this particular bit, Section 2.1ff.

Example and Remark 7.1.7.

1. Let K be a division ring, and let V be a vector space over K which is a subspace of a vector space W. Suppose W has a decomposition into a direct sum of vector subspaces $W = \bigoplus_{i \in I} W_i$. Then V has the exchange property. This is easy to see by extending a basis for V to a basis of W using only vectors in vector subspaces W_i.
2. If M is a finitely generated A-module, then it has the exchange property if and only if it has the finite exchange property.

Remark 7.1.8. Note that in the definition of the $\mathfrak{N}$-exchange property because of the modular law (1.1.4) it follows that Y_i is a direct summand of X_i for all $i \in I$. Indeed as $U_1' \subset U_1$, one has by the modular law

$$U_1 = U_1 \cap (M \oplus U_2' \oplus U_1') = U_1 \cap (M \oplus U_2') \oplus U_1'.$$

The general case is obvious from this.

Lemma 7.1.9. (P. Crawley, B. Jónsson [61]). *If a module M has the $\mathfrak{N}$-exchange property and $X = M \oplus N \oplus Y = \left(\bigoplus_{i \in I} M_i\right) \oplus Y$ where $|I| \le \mathfrak{N}$, then there are submodules $M_i' \subseteq M_i$ such that $X = M \oplus \left(\bigoplus_{i \in I} M_i'\right) \oplus Y$.*

Proof.

Let $\pi : X \to M \oplus N$ be the natural projection with the kernel Y. Set $U_i = \pi(M_i)$. Since $\operatorname{Ker} \pi = Y$, the restriction $\pi|_{M_i} : M_i \to U_i$ is an isomorphism for each $i \in I$, and $M \oplus N = \bigoplus_{i \in I} U_i$. Write $\varphi_i = \pi|_{M_i}^{-1}$. Since M has the $\mathfrak{N}$-exchange property, there exists $U_i' \subseteq U_i$ for each $i \in I$ such that $M \oplus N = M \oplus \left(\bigoplus_{i \in I} U_i'\right)$. Using remark 7.1.8, U_i' is a direct summand of U_i for each $i \in I$. Now set $M_i' = \varphi_i(U_i')$. Then $\left(\bigoplus_{i \in I} M_i'\right) \oplus Y = \left(\bigoplus_{i \in I} U_i'\right) \oplus Y$. Therefore

$$X = M \oplus N \oplus Y = M \oplus \left(\bigoplus_{i \in I} U_i'\right) \oplus Y = M \oplus \left(\bigoplus_{i \in I} M_i'\right) \oplus Y.$$

□

Lemma 7.1.10 (P. Crawley, B. Jónsson, [61]). *Let $M = M_1 \oplus M_2$. Then M has the $\mathfrak{N}$-exchange property if and only if both M_1 and M_2 have the $\mathfrak{N}$-exchange property.*

Proof.

Suppose that both M_1 and M_2 have the $\mathfrak{N}$-exchange property and $X = (M_1' \oplus M_2') \oplus N = \bigoplus_{i \in I} U_i$ where $M_1' \simeq M_1$, $M_2' \simeq M_2$ and $|I| \le \mathfrak{N}$. Since M_1 has the $\mathfrak{N}$-exchange property, there exist submodules $U_i' \subseteq U_i$ such that $X = M_1' \oplus \left(\bigoplus_{i \in I} U_i'\right)$. Then using the $\mathfrak{N}$-exchange property for M_2 and lemma 7.1.9, there results that there exist submodules $U_i'' \subseteq U_i'$ for each $i \in I$ such that

$$X = M_1' \oplus M_2' \oplus \left(\bigoplus_{i \in I} U_i''\right).$$

Thus $M = M_1 \oplus M_2$ has the $\mathfrak{N}$-exchange property.

Conversely, suppose $M = M_1 \oplus M_2$ has the $\mathfrak{N}$-exchange property. Suppose that

$$X = M_1' \oplus N = \bigoplus_{i \in I} U_i$$

where $M_1' \simeq M_1$ and $|I| \le \mathfrak{N}$. One can assume without loss of generality that $X \cap M_2 = 0$ (otherwise we will consider a module $Y = X \oplus M_2$ instead of X). Then

$$X \oplus M_2 = M' \oplus N = M_2 \oplus \left(\bigoplus_{i \in I} U_i\right).$$

The exchange property for M says that $X \oplus M_2 = M_1' \oplus N \oplus M_2 = M' \oplus N$, where $M' = M_1' \oplus M_2 \simeq M$, can also be written as a direct sum of M' with certain submodules

$M_2' \subset M_2$, $U_i' \subset U_i$. This direct sum property then implies that the intersection of M' with each of these submodules is zero. So in particular $M' \cap M_2' = 0$. But $M_2' \subseteq M_2 \subset M'$. So this can happen only if $M_2' = 0$. Hence we find

$$X \oplus M_2 = M' \oplus N = M' \oplus \Big(\bigoplus_{i \in I} U_i'\Big) = M_1' \oplus M_2 \oplus \Big(\bigoplus_{i \in I} U_i'\Big). \tag{7.1.11}$$

Since $M_1' \oplus \Big(\bigoplus_{i \in I} U_i'\Big) \subseteq X$, it follows from (7.1.11), that $M_1' \oplus \Big(\bigoplus_{i \in I} U_i'\Big) = X$, which means that M_1 has the $\mathfrak{N}$-exchange property. Analogously one can show that M_2 has the $\mathfrak{N}$-exchange property. □

Remark 7.1.12. Note that 7.1.11 by itself does not guarantee that $M_1 \oplus \Big(\bigoplus_{i \in I} U_i'\Big) = X$ but only that they are isomorphic, see lemma 7.1.15 below.

Using induction on the number of summands in lemma 7.1.10 one obtains the following corollary.

Corollary 7.1.13. *Let $M = M_1 \oplus M_2 \oplus \ldots \oplus M_n$. Then M has the $\mathfrak{N}$-exchange property if and only if each M_i $(i = 1, 2, \ldots, n)$ has the $\mathfrak{N}$-exchange property.*

In the course of the proof of theorem 7.1.15 below we will meet a situation where for four modules X, U, V, W with U, V, W submodules of X there is the equality

$$X = U \oplus V = U \oplus W$$

and we will wish to conclude that

$$V \simeq W.$$

This looks like a cancellation property, a topic that will be discussed below in Section 7.3. A module U is said to have the **cancellation property** if $U \oplus V \simeq U \oplus W$ implies $V \simeq W$. It does not always hold. Here, however, we have a stronger hypothesis that $U \oplus V$ is equal to $U \oplus W$ instead of just isomorphic. And then the desired conclusion does follow as the following slightly more general lemma shows. The proof is easy but given anyway because it illustrates the sometimes delicate difference between isomorphic and identical.

Lemma 7.1.14. *Let U, V, W be three modules and suppose that there is a morphism $\varphi : U \oplus V \longrightarrow U \oplus W$ that induces an automorphism $\varphi|_U$ on U, i.e., $\varphi(U) = U$, more precisely $\varphi(U \oplus 0) = U \oplus 0$. Let*

$$V \xrightarrow{i_V} U \oplus V \xrightarrow{\varphi} U \oplus W \xrightarrow{\pi_W} W$$

be a sequence of modules, where i_V is the canonical injection $V \longrightarrow U \oplus V$, $v \mapsto (0, v)$ and π_W is the canonical projection $\pi_W : U \oplus W \longrightarrow W, (u, w) \mapsto w$. Let $\psi : V \longrightarrow W$ be the morphism that is the restriction of $\pi_W \varphi$ to V. Then φ is an isomorphism if and only if ψ is an isomorphism.

Proof.
The data given amount to a commutative diagram with exact rows

$$\begin{array}{ccccccccc} 0 & \longrightarrow & U & \xrightarrow{i_U} & U \oplus V & \xrightarrow{\pi_V} & V & \longrightarrow & 0 \\ & & \downarrow{\scriptstyle \varphi|_U} & & \downarrow{\scriptstyle \varphi} & & \downarrow{\scriptstyle \psi} & & \\ 0 & \longrightarrow & U & \xrightarrow{i_U} & U \oplus W & \xrightarrow{\pi_W} & W & \longrightarrow & 0 \end{array}$$

This is now a consequence of the well-known five lemma 1.2.5 that if two of the vertical arrows are isomorphisms, then so is the third. See also [146, pages 89,90].

For gluttons (of punishment) here are the 'diagram chasing' details in the more general case of a general commutative diagram with exact rows:

$$\begin{array}{ccccccccc} 0 & \longrightarrow & M_1 & \xrightarrow{i} & M_2 & \xrightarrow{j} & M_3 & \longrightarrow & 0 \\ & & \downarrow{\scriptstyle \varphi_1} & & \downarrow{\scriptstyle \varphi_2} & & \downarrow{\scriptstyle \varphi_3} & & \\ 0 & \longrightarrow & N_1 & \xrightarrow{i'} & N_2 & \xrightarrow{j'} & N_3 & \longrightarrow & 0 \end{array}$$

for the case that φ_1 and φ_2 are isomorphisms and it is to prove that φ_3 is also an isomorphism. This is the most relevant part for the considerations at hand.

Surjectivity of φ_3 is easiest. Let $n_3 \in N_3$. There is an $n_2 \in N_2$ such that $j'(n_2) = n_3$. Take $m_2 = \varphi_2^{-1}(n_2)$ and $m_3 = j(m_2)$. Then $\varphi_3(m_3) = \varphi_3 j(m_2) = j'\varphi_2(m_2) = j'\varphi_2(\varphi_2^{-1}(n_2)) = j'(n_2) = n_3$.

Next suppose that $m_3 \in M_3$ is mapped to zero by φ_3. Take $m_2 \in M_2$ such that $j(m_2) = m_3$. Then $j'\varphi_2(m_2) = \varphi_3 j(m_2) = \varphi_3(m_3) = 0$. So by exactness of the bottom row there is an $n_1 \in N_1$ such that $i'(n_1) = \varphi_2(m_2)$. Let $m_1 \in M_1$ be such that $m_1 = \varphi_1^{-1}(n_1)$. Now calculate $\varphi_2(m_2 - i(m_1))$. We have $\varphi_2(m_2) = i'(n_1)$ and $\varphi_2 i(m_1) = i'\varphi_1(m_1) = i'(n_1)$. Therefore $\varphi_2(m_2 - i(m_1)) = 0$. So, φ_2 being an isomorphism $m_2 - i(m_1) = 0$. Hence $0 = j(m_2 - i(m_1)) = j(m_2) - ji(m_1) = j(m_2) = m_3$ because $ji = 0$. So $m_3 = j(m_2) = 0$ as desired. □

Theorem 7.1.15 (P. Crawley, B. Jónsson, [61]). *If a module M has the 2-exchange property, then M has the finite exchange property.*

Proof.
The statement is proved by induction. Assume that a module M has the n-exchange property (which means that it has the k-exchange property for any $k \leq n$). We will prove that M has the $(n+1)$-exchange property, as well. Suppose that $M \simeq M_1$ and

$$X = M_1 \oplus N = U_1 \oplus U_2 \oplus \ldots \oplus U_n \oplus U_{n+1}$$

and set $Y = U_1 \oplus \ldots \oplus U_n$. Then

$$X = M_1 \oplus N = Y \oplus U_{n+1}. \tag{7.1.16}$$

Since M has the 2-exchange property, there exist submodules $Y' \subseteq Y$ and $U'_{n+1} \subseteq U_{n+1}$ such that

$$X = M_1 \oplus N = M_1 \oplus Y' \oplus U'_{n+1}. \tag{7.1.17}$$

One has

$$Y = Y \cap X = Y \cap (M_1 \oplus Y' \oplus U'_{n+1}).$$

Taking into account that $Y' \subseteq Y$ we can use the modular law in the form (1.1.4). So setting $Y'' = Y \cap (M_1 \oplus U'_{n+1})$ one has

$$Y = Y' \oplus (Y \cap (M_1 \oplus U'_{n+1})) = Y' \oplus Y'' \tag{7.1.18}$$

Symmetrically,

$$U_{n+1} = U_{n+1} \cap X = U_{n+1} \cap (M_1 \oplus Y' \oplus U'_{n+1}).$$

Taking into account that $U'_{n+1} \subseteq U_{n+1}$ we can again use the modular law in the form (1.1.4). So

$$U_{n+1} = U'_{n+1} \oplus (U_{n+1} \cap (M_1 \oplus Y')) = U'_{n+1} \oplus U''_{n+1}, \tag{7.1.19}$$

where $U''_{n+1} = U_{n+1} \cap (M_1 \oplus Y')$. Combining (7.1.16)–(7.1.19) one obtains

$$X = M_1 \oplus Y' \oplus U'_{n+1} = Y \oplus U_{n+1} =$$

$$= (Y' \oplus Y'') \oplus (U'_{n+1} \oplus U''_{n+1}) = (Y'' \oplus U''_{n+1}) \oplus (Y' \oplus U'_{n+1}). \tag{7.1.20}$$

By lemma 7.1.14 it follows from (7.1.20) that Y'' is isomorphic to a direct summand of M_1, and so has the n-exchange property. Now

$$Y = Y' \oplus Y'' = U_1 \oplus \ldots \oplus U_n,$$

and so there exist submodules $U'_i \subseteq U_i$ such that

$$Y = Y'' \oplus U'_1 \oplus \ldots \oplus U'_n. \tag{7.1.21}$$

Taking into account that $Y'' \subseteq M_1 \oplus U'_{n+1} \subseteq Y'' \oplus (Y' \oplus U_{n+1})$ and using the modular law identity (for the third time) in the form:

$$(M_1 \oplus U'_{n+1}) \cap (Y'' \oplus (Y' \oplus U_{n+1})) = Y'' \oplus ((M_1 \oplus U'_{n+1}) \cap (Y' \oplus U_{n+1})) = Y'' \oplus Y_1$$

where $Y_1 = (M_1 \oplus U'_{n+1}) \cap (Y' \oplus U_{n+1})$, there results that

$$M_1 \oplus U'_{n+1} = Y'' \oplus Y_1. \tag{7.1.22}$$

So from (7.1.17), (7.1.18), (7.1.21) and (7.1.22) it follows that

$$X = Y' \oplus Y'' \oplus Y_1 = Y \oplus Y_1 = Y'' \oplus U'_1 \oplus \ldots \oplus U'_n \oplus Y_1 =$$

$$= M_1 \oplus U'_1 \oplus \ldots \oplus U'_n \oplus U'_{n+1}.$$

Thus, M has the $(n+1)$-exchange property, as required. □

Lemma 7.1.23 (P. Crawley, B. Jónsson, [61]). *If an indecomposable module M has the 2-exchange property, then M has the exchange property.*

Proof.
Suppose $X = M_1 \oplus N = \bigoplus_{i \in I} X_i$ where $M_1 \simeq M$. Since any element of X is contained in the sum of finitely many summands X_i, there exists a finite subset $J \subset I$ such that

$$M_1 \cap \Big(\bigoplus_{i \in J} X_i\Big) \neq 0. \tag{7.1.24}$$

Set $Y = \bigoplus_{i \in I \setminus J} X_i$. Then $X = M_1 \oplus N = (\bigoplus_{i \in J} X_i) \oplus Y$. Since M has the finite exchange property by theorem 7.1.15, there exist submodules $Y' \subseteq Y$ and $X_i' \subseteq X_i$ for each $i \in J$ such that

$$X = M_1 \oplus \Big(\bigoplus_{i \in J} X_i'\Big) \oplus Y'. \tag{7.1.25}$$

By the modular law, analogously to the proof of theorem 7.1.15, one can find submodules $Y'' \subseteq Y$ and $X_i'' \subseteq X_i$ for each $i \in J$ such that $Y = Y' \oplus Y''$ and $X_i = X_i' \oplus X_i''$. Then it follows, also as before, that

$$M_1 \simeq \Big(\bigoplus_{i \in J} X_i''\Big) \oplus Y''. \tag{7.1.26}$$

Since M_1 is indecomposable, only one of these direct summands can be different from 0. Suppose $Y'' \neq 0$, then $X_i'' = 0$ for all $i \in J$, and so $X_i = X_i'$ for each $i \in J$. In this case equality (7.1.25) is impossible because (7.1.24) holds. Thus, $Y'' = 0$ and so $Y = Y'$. Setting $X_i' = X_i$ for all $i \in I \setminus J$, one obtains

$$X = M_1 \oplus \Big(\bigoplus_{i \in I} X_i'\Big),$$

i.e., M has the exchange property. □

Remark 7.1.27. Note that lemma 7.1.23 proves that the 2-exchange property implies the exchange property only for an indecomposable module. The fundamental question whether it is true for an arbitrary module is still open.

Lemma 7.1.28. *Let M, N, H be submodules of a module X. Suppose that $X = M \oplus N$ and $\pi : X = M \oplus N \to M$ is the canonical projection with* Ker$\pi = N$. *Then*

1. *$X = H \oplus N$ if and only if $\pi|_H : H \to M$ is an isomorphism.*
2. *If $M = M_1 \oplus M_2$ and K is a submodule of X mapped isomorphically onto M_1 by π, then $X = K \oplus M_2 \oplus N$.*
3. *If $X = H \oplus N$ and $M = M_1 \oplus M_2$, then there is a direct decomposition $H = H_1 \oplus H_2$ such that $X = H_1 \oplus M_2 \oplus N$.*

Proof.

1. The data supplied give rise to a commutative diagram

$$\begin{array}{ccccccccc}
0 & \longrightarrow & N & \longrightarrow & N \oplus H & \longrightarrow & H & \longrightarrow & 0 \\
 & & \| & & \downarrow{\scriptstyle i_N \oplus i_H} & & \downarrow{\scriptstyle \pi|_H = \pi i_H} & & \\
0 & \longrightarrow & N & \longrightarrow & N \oplus M & \xrightarrow{\pi} & M & \longrightarrow & 0
\end{array}$$

Here i_H is a given inclusion of H into $X = N \oplus M$ and i_N is the canonical inclusion $N \longrightarrow N \oplus M$, $n \mapsto (n,0)$. Thus $i_N \oplus i_H$ is the morphism $(n,h) \mapsto n + i_H(h)$.
Now if $i_N \oplus i_H$ is an isomorphism (and hence the identity as a direct sum of inclusions), then $\pi|_H$ is an isomorphism (as was proved before in the proof of lemma 7.1.14). Inversely if $\pi|_H$ is an isomorphism so is $i_N \oplus i_H$. So it has kernel zero showing that $N \cap H = 0$ and $N \oplus H$ is the internal direct sum of N and H. Also $i_N \oplus i_H$ is surjective so that the inclusion $N \oplus H = N + H \to N \oplus M$ is the identity.
2. This is proved by a similar diagram

$$\begin{array}{ccccccccc}
0 & \longrightarrow & M_2 \oplus N & \longrightarrow & M_2 \oplus N \oplus K & \longrightarrow & K & \longrightarrow & 0 \\
 & & \| & & \downarrow{\scriptstyle i_{(M_2 \oplus N)} \oplus i_K} & & \downarrow{\scriptstyle \pi|_K} & & \\
0 & \longrightarrow & M_2 \oplus N & \longrightarrow & M_2 \oplus N \oplus M_1 & \longrightarrow & M_1 & \longrightarrow & 0
\end{array}$$

By assumption $\pi|_K$ (where π is still the canonical projection $N \oplus M \to M$) is an isomorphism. Hence so is $i_{(M_2 \oplus N)} \oplus i_K$. As before it follows that this sum of inclusions is the identity.
3. By statement 1 of the lemma $\pi|_H : H \to M = M_1 \oplus M_2$ is an isomorphism. Take $H_1 = (\pi|_H)^{-1} M_1$, $H_2 = (\pi|_H)^{-1} M_2$. Then $H = H_1 \oplus H_2$ and $\pi|_H$ respects the two decompositions. In particular, $\pi_H|_{H_1} = \pi|_{H_1}$ is an isomorphism $H_1 \to M_1$. Now consider the commutative diagram

$$\begin{array}{ccccccccc}
0 & \longrightarrow & N & \longrightarrow & N \oplus H_1 \oplus M_2 & \longrightarrow & H_1 \oplus M_2 & \longrightarrow & 0 \\
 & & \| & & \downarrow{\scriptstyle i_N \oplus i_{H_1} \oplus i_{M_2}} & & \downarrow{\scriptstyle \pi|_{H_1} \oplus i_{M_2}} & & \\
0 & \longrightarrow & N & \longrightarrow & N \oplus M_1 \oplus M_2 & \xrightarrow{\pi} & M_1 \oplus M_2 & \longrightarrow & 0
\end{array}$$

where i_{M_2}, i_N are the canonical embeddings in direct sums and i_{H_1} is the restriction of $i_H : H \to X$ to $H_1 \subset H$.
By assumption $\pi|_{H_1}$ is an isomorphism, so the right vertical arrow is an isomorphism. Thus the sum of inclusions, that is the middle vertical arrow, is an isomorphism and hence the identity by the same argument as used twice above. □

Lemma 7.1.29. *Let $\varphi, \psi \in \text{End}_A(M)$ with $\varphi + \psi = 1$. Let X_1, X_2 be submodules of $M \oplus M$ defined as*

$$X_1 = \{(m, \varphi(m)) \ : \ m \in M\}$$

$$X_2 = \{(m, -\psi(m)) \ : \ m \in M\}.$$

Then $M \oplus M = X_1 \oplus X_2$.

Proof.
We first show that $X_1 \cap X_2 = 0$. Suppose $(m_1, m_2) \in X_1 \cap X_2$. Since $(m_1, m_2) \in X_1$, $m_2 = \varphi(m_1)$. On the other hand $(m_1, m_2) \in X_2$, and so $m_2 = -\psi(m_1)$. Therefore $\varphi(m_1) = -\psi(m_1)$, and since $\varphi + \psi = 1$, $m_1 = (\varphi + \psi)m_1 = 0$, and $m_2 = \varphi(m_1) = 0$.
Let $m \in M$. Then $(0, m) = (m, \varphi(m)) - (m, -\psi(m)) \in X_1 \oplus X_2$. Analogously, $(m, 0) = (\psi(m), \varphi\psi(m)) + (\varphi(m), -\varphi\psi(m)) = (\psi(m), \varphi\psi(m)) + (\varphi(m), -\psi\varphi(m)) \in X_1 \oplus X_2$, since $\varphi + \psi = 1$ and so $\varphi\psi = \psi\varphi$. Thus $M \oplus M \subseteq X_1 \oplus X_2$, hence $X_1 \oplus X_2 = M \oplus M$. □

Recall that a ring A is **local** if it has a unique maximal right ideal. For another various equivalent definitions for a ring to be local see proposition 1.9.1.
In Section 5.2 we considered an important class of indecomposable modules which are strongly indecomposable modules. They also play a fundamental role in the study of cancellation, exchange property and unique decomposition of modules. These modules were considered by R.B. Warfield to prove a strengthened form of Krull-Remak-Schmidt-Azumaya theorem in [317].

Recall that a module M over a ring A is called **strongly indecomposable** if its endomorphism ring $\text{End}_A(M)$ is local.

Obviously, each strongly indecomposable module is indecomposable. Any simple module M over a ring A is strongly indecomposable, since $\text{End}_A(M)$ is a division ring and so it is a local ring.
If $A = \begin{pmatrix} D & D \\ 0 & D \end{pmatrix}$ where D is a division field then the right A-module $M = D \oplus D$ is strongly indecomposable, since $\text{End}_A(M) \simeq D$. In a more general case if A is a basic semiperfect ring, and $1 = e_1 + \cdots + e_n$ is a decomposition of the identity into a sum of primitive idempotents, then the principal right module $P_i = e_i A$ is strongly indecomposable, since $\text{End}_A(P_i) = e_i A e_i$ is a local ring.
For example, if $A = \begin{pmatrix} \mathbf{Z}_p & \mathbf{Q} \\ 0 & \mathbf{Q} \end{pmatrix}$, where p is a prime ring, then $P_1 = \begin{pmatrix} \mathbf{Z}_p & \mathbf{Q} \end{pmatrix} \simeq \mathbf{Z}_p \oplus \mathbf{Q}$ and $P_2 = \begin{pmatrix} 0 & \mathbf{Q} \end{pmatrix} \simeq \mathbf{Q}$ are strongly indecomposable right A-modules.
Another examples of strongly indecomposable modules are indecomposable modules which are both Artinian and Noetherian, since for such modules there endomorphism rings are local, by [146, proposition 10.1.5].
At the same time not any indecomposable module is strongly indecomposable in general. Since for any ring A the right regular module A_A is isomorphic to $\text{End}_A(A)$, the right regular module A_A is not strongly indecomposable for any indecomposable

non-local ring A. The simplest such example is the right regular module $M = \mathbf{Z}_{\mathbf{Z}}$, since M is indecomposable, but $\mathrm{End}_{\mathbf{Z}}(\mathbf{Z}) \simeq \mathbf{Z}$ is not local.

If $A = \begin{pmatrix} \mathbf{Z} & \mathbf{Q} \\ 0 & \mathbf{Q} \end{pmatrix}$, then $P_1 = \begin{pmatrix} \mathbf{Z} & \mathbf{Q} \end{pmatrix} \simeq \mathbf{Z} \oplus \mathbf{Q}$ is an indecomposable right A-module but not strongly indecomposable, since $\mathrm{End}_A(P_1) \simeq \mathbf{Z}$ is not local.

As was noted in Remark 5.2.4 for injective modules the notions of indecomposability and strong indecomposability are the same.

Proposition 7.1.30 (R.B. Warfield [317]). *Any strongly indecomposable module has the exchange property.*

Proof.
Let M be a strongly indecomposable module over a ring A. By lemma 7.1.23 it follows that it remains to prove that M has the 2-exchange property.

Suppose that an A-module X has two decompositions $X = M \oplus N = X_1 \oplus X_2$. Let $\pi_k : X \to X_k$ be the canonical projections, and $i_k : X_k \to X_1 \oplus X_2 = X$ the canonical inclusions ($k = 1,2$). Further let $i_M : M \to M \oplus N = X$ be the canonical inclusion and $\pi_M : X = M \oplus N \to M$ the corresponding projection with kernel N. Then, of course, $\pi_M i_M = 1_M$.

Further $i_1\pi_1 + i_2\pi_2 = 1_X : X = X_1 \oplus X_2 \to X_1 \oplus X_2 = X$ and so

$$\pi_M(i_1\pi_1 + i_2\pi_2)i_M = \pi_M i_M = 1_M.$$

Therefore at least one of $\pi_M(i_1\pi_1)i_M$, $\pi_M(i_2\pi_2)i_M$ is an automorphism because $\mathrm{End}_A(M)$ is local. It can be assumed that this is $\pi_M i_1 \pi_1 i_M$. Let the inverse be φ. Then we have the commutative diagram:

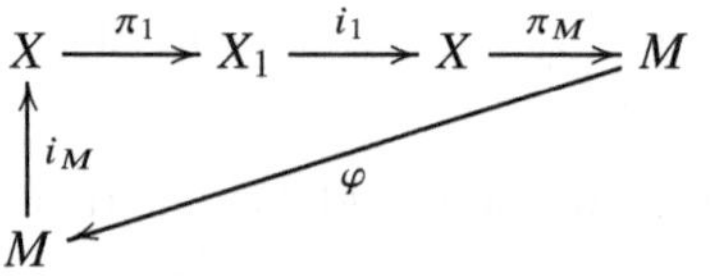

So $(\varphi\pi_M i_1)(\pi_1 i_M) = 1_M$. This makes $\pi_1 i_M$ a split monomorphism $M \to X_1$. So for some submodule $Y \subset X_1$ $X_1 = \pi_1 i_M(M) \oplus Y$ and $\pi_1 i_M(M) \simeq M$. Therefore by lemma 7.1.9 it follows that $X = M \oplus Y \oplus X_2$. This proves the 2-exchange property.
□

Theorem 7.1.31 (R.B. Warfield [317]). *An indecomposable module having the finite exchange property is strongly indecomposable.*

Proof.

Let M be an indecomposable module over a ring A which has the finite exchange property. Then $M \oplus 0$ is an exchangeable summand of $M \oplus M$. In order to prove that M is strongly indecomposable it suffices, by proposition 1.9.1, to show that if $\varphi \in \mathrm{End}_A(M)$ then φ or $1 - \varphi$ is an automorphism. Write $\psi = 1 - \varphi$. Consider the following two submodules in $M \oplus M$:

$$X_1 = \{(m, \varphi(m)) \ : \ m \in M\}$$

$$X_2 = \{(m, -\psi(m)) \ : \ m \in M\}.$$

By lemma 7.1.29, $M \oplus M = X_1 \oplus X_2$. There are two isomorphisms $f : M \to X_1$ given by $f(m) = (m, \varphi(m))$, and $g : M \to X_2$ given by $g(m) = (m, -\psi(m))$. Therefore X_1 and X_2 are indecomposable. Write $M_1 = M \oplus 0$ and $M_2 = 0 \oplus M$. Since M_1, M_2 have the exchange property, and $M_1 \oplus M_2 = X_1 \oplus X_2$, it follows that $M_1 \oplus M_2 = M_1 \oplus Y_1 \oplus Y_2$, where $Y_i \subseteq X_i$ for $i = 1, 2$. Hence there results that either $M_1 \oplus M_2 = M_1 \oplus X_1$ or $M_1 \oplus M_2 = M_1 \oplus X_2$ because X_1, X_2 are both indecomposable. If $M_1 \oplus M_2 = M_1 \oplus X_1$, then by lemma 7.1.28 the projection π_2 from $M_1 \oplus M_2$ onto M_2 restricts to an isomorphism from X_1 onto M_2. But then the composition

$$m \mapsto (m, \varphi(m)) \mapsto \pi_2(m, \varphi(m)) = \varphi(m)$$

is an automorphism of M, i.e., φ is invertible. Analogously, if $M_1 \oplus M_2 = M_1 \oplus X_2$ then ψ is invertible. Thus, $\mathrm{End}_A(M)$ is a local ring, as required. □

Since for injective modules the notions of indecomposability and strong indecomposability coincide, from proposition 7.1.30, theorem 7.1.31 and corollary 7.1.13 we obtain the following corollaries:

Corollary 7.1.32. *If a module M has finite Goldie dimension, then its injective hull $E(M)$ has the exchange property.*

Corollary 7.1.33. *If a module M has finite uniform dimension, then its injective hull $E(M)$ has the exchange property.*

Corollary 7.1.34. *The injective hull of a Noetherian module M has the exchange property. In particular, the injective hull of any finitely generated module over a Noetherian ring has the exchange property.*

7.2 The Azumaya Theorem

In ring theory one of the most important problems is the study of indecomposable modules and decompositions of modules into indecomposable ones. If a module can be decomposed into a direct sum of indecomposable modules there arises another important problem: When is such a decomposition unique? This section is devoted to the second problem, the problem of uniqueness of decompositions.

In [146, Section 10.4] the famous Krull-Remak-Schmidt theorem for uniqueness of finite direct sums of strongly indecomposable modules was proved (See theorem 1.9.5).

Taking into account theorem 5.2.2, and applying theorem 1.9.5 to injective modules, one obtains the following corollary.

Corollary 7.2.1. *Let M have two different decompositions as a finite direct sum of indecomposable injective submodules*

$$M = \bigoplus_{i=1}^{n} M_i = \bigoplus_{i=1}^{m} N_i. \tag{7.2.2}$$

Then $m = n$ and there is a permutation τ of the numbers $i = 1, 2, ..., n$ such that $M_i \simeq N_{\tau(i)}$ $(i = 1, ..., n)$.

From this corollary and corollaries 5.2.9 and 5.2.10, there immediately follows the next result.

Proposition 7.2.3 (R.B. Warfield, Jr.) *Let M be a Noetherian right A-module (in particular, let M be a finitely generated module over a right Noetherian ring). If the injective hull $E(M)$ has two different decompositions into a finite direct sum of indecomposable submodules $E(M) = \bigoplus_{i=1}^{n} M_i = \bigoplus_{i=1}^{m} N_i$, then $m = n$ and there is a permutation τ of the numbers $i = 1, 2, ..., n$ such that $M_i \simeq N_{\tau(i)}$ $(i = 1, ..., n)$.*

All these statements are only about *finite* direct sums of submodules. Actually, however, they also hold for *infinite* direct sums of modules, as has been proved by G. Azumaya in [12]. The main goal of this section is to prove the Azumaya theorem for an infinite direct sum of modules with local endomorphism rings.

Lemma 7.2.4. *Let an A-module M have two different decompositions into a direct sum of indecomposable submodules $M = \bigoplus_{i \in I} M_i = X \oplus Y$. Let X and M_i be strongly indecomposable modules for all $i \in I$. Then there is a number $j \in I$ such that X is isomorphic to M_j and $M = X \oplus \Big(\bigoplus_{\substack{i \in I \\ i \neq j}} M_i\Big)$.*

Proof.

Since X is strongly indecomposable, it has the exchange property, by proposition 7.1.30. Therefore, taking into account remark 7.1.8, $M = X \oplus (\bigoplus_{i \in I} M_i')$, where each M_i' is a direct summand of M_i. Since all the M_i are indecomposable, $M_i' = M_i$ or $M_i' = 0$. Since $X \neq 0$, there exist a number j such that $M_j' \neq 0$. Let $S = \{j \in I : M_j' = 0\}$. Then $M = X \oplus \Big(\bigoplus_{i \in I \setminus S} M_i\Big)$ and hence $X \simeq \bigoplus_{i \in S} M_i$. Since X is indecomposable, $S = \{j\}$, and so $X \simeq M_j$ and $M = X \oplus \Big(\bigoplus_{\substack{i \in I \\ i \neq j}} M_i\Big)$. □

Proposition 7.2.5. *Let M be a strongly indecomposable module. Then $M \oplus X = M_1 \oplus Y$ with $M \cong M_1$ implies $X \cong Y$ for any pair of A-modules X and Y.*

Proof.
Suppose M is a strongly indecomposable module and $W = M \oplus X = M_1 \oplus Y$ with $M \simeq M_1$. By proposition 7.1.30, M has the exchange property. Therefore either $W = M \oplus Y$ (hence $X \simeq Y$ by theorem 7.1.15) or $W = M \oplus Y_1 \oplus M_1$ with $Y_1 \subseteq Y$. In this case, $X \simeq M_1 \oplus Y_1$ and $Y \simeq M \oplus Y_1$, whence $X \simeq Y$, as required. □

Definition 7.2.6. Two decompositions $M = \bigoplus_{i \in I} M_i = \bigoplus_{j \in J} N_j$ of a module M into a direct sum of submodules are said to be **isomorphic** if the summands are isomorphic in pairs, i.e., there is a bijection $\tau : I \to J$ such that $M_i \simeq N_{\tau(i)}$ for all $i \in I$. The second decomposition is said to be a **refinement** of the first if there is an epimorphism $\varphi : J \to I$ such that $N_j \subseteq M_{\varphi(j)}$, from which it follows that the induced morphism $\bigoplus_{\varphi(j)=i} N_j \to M_i$ is an isomorphism.

Theorem 7.2.7. (**Krull-Remak-Schmidt-Azumaya Theorem.**) [12]. *Any two different decompositions of an A-module M into direct sums of strongly indecomposable submodules $M = \bigoplus_{i \in I} M_i = \bigoplus_{j \in J} N_j$ are isomorphic. Moreover, any indecomposable summand of M is isomorphic to one of M_i.*

Proof.
Take any strongly indecomposable module K and define

$$I(K) = \{i \in I \ : \ M_i \simeq K\}$$

$$J(K) = \{j \in J \ : \ N_j \simeq K\}.$$

To prove the equivalence of two decompositions it suffices to prove that for all K, the cardinalities, $\mathfrak{c}(I(K))$ and $\mathfrak{c}(J(K))$ are the same.

We have $\bigoplus_{i \in I} M_i = X = \bigoplus_{j \in J} N_j$. Take an $i \in I(K)$ (assuming $I(K)$ is not empty). Write the equality as

$$X = M_i \oplus Y = \bigoplus_{j \in J} N_j$$

where $Y = \bigoplus_{i' \neq i} M_i$. Being strongly indecomposable M_i has the exchange property, so there exist submodules $N_j' \subseteq N_j$ such that

$$X = M_i \oplus \bigoplus_{j \in J} N_j' = \bigoplus_{j \in J} N_j \tag{7.2.8}$$

By remark 7.1.8 the N_j' are direct summands of the N_j. So either $N_j' = 0$ or $N_j' = N_j$. Let $J(i) = \{j \in J \ : \ N_j' = 0\}$. Then

$$X = M_i \oplus (\bigoplus_{j \notin J(i)} N_j) = (\bigoplus_{j \in J(i)} N_j) \oplus (\bigoplus_{j \notin J(i)} N_j).$$

So by theorem 7.1.15

$$M_i \simeq \bigoplus_{j \in J(i)} N_j.$$

As M_i is indecomposable (and $\neq 0$), this is only possible if $J(i)$ consists of precisely one element $j(i)$. So

$$M_i \simeq N_{j(i)} \quad \text{and} \quad J(K) \neq \emptyset$$

and therefore

$$X = M_i \oplus \bigoplus_{j \neq j(i)} N_j. \tag{7.2.9}$$

(Note that this does not prove that this particular $j(i)$ is the only one for which (7.2.9) holds; $j(i)$ depends on the choice of the N_j' in (7.2.8). The exchange property says nothing about uniqueness of these N_j. And, indeed as a rule there are several choices; see below.)

Now define for each $j \in J(K)$

$$I(K, j) = \{i \in I(K) \ : \ X = M_i \oplus \bigoplus_{j' \neq j} N_{j'}\}.$$

The foregoing shows that

$$\bigcup_{j \in J(K)} I(K, j) = I(K).$$

The next thing to prove is that each of the $I(K, j)$ is finite. To this end take any $j \in J(K)$ and a non-zero element $\kappa \in N_j$. The element $\kappa \in N_j \subset \bigoplus_j N_j = \bigoplus_i M_i$ can be written

$$\kappa = \sum_i \kappa_i, \quad \kappa_i \in M_i$$

and only finitely many of the κ_i are non-zero. Let κ_i be zero. Then

$$N_j \xrightarrow{i} X = M_i \oplus \bigoplus_{j' \neq j} N_{j'} \xrightarrow{\pi} M_i \tag{7.2.10}$$

sends κ to zero. Here $i : N \to \bigoplus_j N_j = X$ is the canonical injection and $\pi : X = \bigoplus_j N_j \to N_j$ is the projection.

Now, using $X = M_i \oplus \bigoplus_{j' \neq j} N_{j'}$ (as in the definition of the $I(K, j)$ above), consider the diagram with exact rows:

$$\begin{array}{ccccccccc} & & & & X & & & & \\ & & & & \| & & & & \\ 0 & \longrightarrow & N' & \longrightarrow & N' \oplus N_j & \longrightarrow & N_j & \longrightarrow & 0 \\ & & \| & & \| & & \downarrow & & \\ 0 & \longrightarrow & N' & \longrightarrow & N' \oplus M_i & \longrightarrow & M_i & \longrightarrow & 0 \\ & & & & \| & & & & \\ & & & & X & & & & \end{array}$$

where $N' = \bigoplus_{j' \neq j} N_j$. Here the right vertical arrow is the restriction to N_j of the projection $X \to M_i$ (which is the morphism (7.2.10) just considered). If $\kappa_i = 0$ this morphism has non-zero kernel which cannot be by theorem 7.1.15. So $i \in I(K, j)$ can only hold if $\kappa_i \neq 0$ and there are only finitely many such i.

We now have

$$I(K) = \bigcup_j I(K, j)$$

and $I(K, j)$ is finite for all j. Now suppose that $J(K)$ is finite. Then $I(K) \leq \sum_{j \in J(K)} I(K, j)$ which is a finite sum of finite numbers. So $I(K)$ is also finite.

Using symmetry (or repeating the proof with M's replacing N's everywhere), we now have

$$I(K) = \emptyset \;\Leftrightarrow\; J(K) = \emptyset$$

$$I(K) \text{ is finite } \;\Leftrightarrow\; J(K) \text{ is finite}$$

and hence also

$$I(K) \text{ is infinite } \;\Leftrightarrow\; J(K) \text{ is infinite}$$

There are now two cases which are treated very differently.

Case 1. $J(K)$ is finite and nonempty. Then $I(K)$ is also finite and nonempty. In this case a stronger property holds. Namely, $\bigoplus_{i \in I} M_i \simeq \bigoplus_{j \in J} N_j$ implies $\mathfrak{c}(I(K)) = \mathfrak{c}(J(K))$ for every K for which $I(K)$ and $J(K)$ are finite. Indeed take an $i \in I(K)$. Then, as has been shown, there is a $j(i)$ such that $M_i \simeq N_j$. Then we have the situation

$$X = \Big(\bigoplus_{i' \neq i} M_{i'}\Big) \oplus M_i \simeq \Big(\bigoplus_{j \neq j(i)} N_j\Big) \oplus N_{j(i)}$$

with $M_i \simeq N_j$ and strongly indecomposable. So by proposition 7.2.5

$$\bigoplus_{i' \neq i} M_{i'} \simeq \bigoplus_{j \neq j(i)} N_j$$

and we have reduced the number of summands isomorphic to K by 1 on both sides. Induction finishes the proof.

Case 2. $J(K)$ is infinite. Then

$$I(K) = \bigcup_{j \in J(K)} I(K,j)$$

and $I(K,j)$ is finite imply $\mathfrak{c}(I(K)) \leq \aleph_0(\mathfrak{c}(J(K))) = \mathfrak{c}(J(K))$ because $\mathfrak{c}(J(K))$ is infinite. Here $\aleph_0$ is the countably infinite cardinal.

By symmetry also $\mathfrak{c}(J(K)) \leq \mathfrak{c}(I(K))$ and so the Cantor-Bernstein-Schröder theorem[1] from set theory says that $\mathfrak{c}(J(K)) = \mathfrak{c}(I(K))$. □

Remark 7.2.11. Taking into account theorem 7.2.7 one can see that the infinite version of corollary 7.2.1 for injective modules is also true:

Corollary 7.2.12. *Let M be an A-module which is a direct sum of indecomposable injective modules. Then any two different decompositions of M into direct sums of indecomposable injective submodules are isomorphic.*

Using the main results of P. Crawley and B. Jónsson in [61], R.B. Warfield in [317] proved the strengthened form of theorem 7.2.7. The Azumaya theorem says that any two direct decompositions of a module M into strongly indecomposable summands are isomorphic. The Cawley-Jónsson-Warfield theorem states that if there is one direct sum decomposition of a module M with countably generated and strongly indecomposable summands then for each summand N of M a similar decomposition exists.

Theorem 7.2.13. (Cawley-Jónsson-Warfield [317, Theorem 1]). *If an A-module M is a direct sum of indecomposable submodules, $M = \bigoplus_{i \in I} M_i$, where each M_i is countably generated and is strongly indecomposable (i.e., has local endomorphism ring), then any other direct sum decomposition of M refines to a decomposition isomorphic to this one, and (in particular) any summand of M is again a direct sum of modules, each isomorphic to one of the original summands M_i.*

From this theorem one may obtain the famous theorem of I. Kaplansky:

Theorem 7.2.14 (Kaplansky's Theorem). [179]. *Any projective module over a local ring is free.*

Proof.

The proof follows from theorem 7.2.13 taking into account that for any ring A there is an obvious isomorphism $A \cong \mathrm{End}(A_A)$. □

---

[1]The Cantor-Bernstein-Schröder theorem states that, if there exist injective functions $f : A \to B$ and $g : B \to A$ between the sets A and B, then there exists a bijective function $h : A \to B$. In terms of cardinality of the two sets, this means that if $\mathfrak{c}(A) \leq \mathfrak{c}(B)$ and $\mathfrak{c}(B) \leq \mathfrak{c}(A)$, then $\mathfrak{c}(A) = \mathfrak{c}(B)$.

Applying theorem 7.2.13 to local rings R.B. Warfield proved in [317] the following theorem:

Theorem 7.2.15 (R.B. Warfield [317]). *If M is a module over a local ring A and M is a summand of a module N where $N = \bigoplus_{i\in I} C_i$, and each C_i is of the form A/I_i, where I_i is a two-sided ideal, then M is also a direct sum of cyclic modules, each isomorphic to one of the C_i.*

Another application of theorem 7.2.7 to semiperfect rings gives a full description of projective modules.

Theorem 7.2.16. *Let A be a semiperfect ring and $1 = e_1 + e_2 + \ldots + e_n$ be a decomposition of the identity of A into a sum of pairwise orthogonal local idempotents. Then for any projective right A-module P there exist sets $I_1, I_2, \ldots, I_n$ (unique with up to cardinality and possibly empty) such that*

$$P \simeq P_1^{I_1} \oplus P_2^{I_2} \oplus \ldots \oplus P_n^{I_n}$$

where the $P_i = e_i A$ are the principal right A-modules.

Proof.
Since A is a semiperfect ring, the right regular module A_A has the following form:

$$A_A \simeq P_1^{k_1} \oplus P_2^{k_2} \oplus \ldots \oplus P_n^{k_n}$$

Let P be a projective right A-module. Then one can construct an exact sequence

$$0 \longrightarrow X \longrightarrow F \longrightarrow P \longrightarrow 0$$

where F is a free right A-module. Since P is projective, this sequence is split and so

$$P \oplus X \simeq F \simeq A^I \simeq P_1^{k_1 \times I} \oplus P_2^{k_2 \times I} \oplus \ldots \oplus P_n^{k_n \times I}$$

Since each P_i is strongly indecomposable, one can apply theorem 7.2.13. □

7.3 Cancellation Property

In general, for modules M, X, Y over a ring A, $M \oplus X \cong M \oplus Y$ does not imply that $X \cong Y$. Sometimes it is true, sometimes not. It is depend on the ring A and the modules considered. For example, if M is a finitely generated Abelian group, then for any Abelian groups X and Y, $M \oplus X \cong M \oplus Y$ implies that $X \cong Y$, as was proved independently by P.M. Cohn [58] and E.A. Walker [316].

On the other hand, for any two non-isomorphic modules X and Y one can consider the module $M = Y \oplus X \oplus Y \oplus X \oplus \cdots$. Then $X \oplus M \cong Y \oplus M$, since both these modules are isomorphic to M. But, by assumption, $X \ncong Y$.

There are several variations of the concept of cancellation. We start with the property of cancellation which is called an **internal cancellation** and is defined as follows.

Definition 7.3.1. An A-module M is said to satisfy the **internal cancellation property** (or M is **internally cancellable**) if whenever $M = X \oplus Y = X_1 \oplus Y_1$ and $X \cong X_1$, for A-modules X, Y, X_1, Y_1, then $Y \cong Y_1$.

The following statement follows trivially from the definition.

Lemma 7.3.2. *If an A-module M satisfies internal cancellation property, then M is not isomorphic to a proper summand of itself, i.e., M is Dedekind-finite.*

Proof.
Suppose $M = M \oplus 0 \simeq M \oplus X$ and M satisfies internal cancellation property, then $X = 0$. □

Definition 7.3.3. An A-module M is called a **cancellable summand** of $M \oplus X$ if whenever $M \oplus X = M_1 \oplus Y$ with $M \cong M_1$ then $X \cong Y$. An A-module M is said to have the **cancellation property** (or M is **cancellable**) if $M \oplus X \cong M \oplus Y$ implies $X \cong Y$ for any pair of modules X and Y.

Proposition 7.3.4. *The class of modules having the cancellation property is closed under finite direct sums and direct summands.*

Proof.
Suppose M and N have the cancellation property and $M \oplus N \oplus X \simeq M \oplus N \oplus Y$. Then since M has the cancellation property, $N \oplus X \simeq N \oplus Y$. Since N has the cancellation property, $X \simeq Y$. Hence $M \oplus N$ has the cancellation property.

Conversely, suppose $M \oplus N$ has the cancellation property, and $M \oplus X \simeq M \oplus Y$, then $M \oplus N \oplus X \simeq M \oplus N \oplus Y$, and so $X \simeq Y$, i.e., M has the cancellation property.
□

Note that proposition 7.2.5 plays an important role in proving the Azumaya theorem. Taking into account the notion of cancellation property this proposition can be written in the following form:

Proposition 7.3.5 (R.B. Warfield [317, Proposition 2]). *Every strongly indecomposable module has the cancellation property.*

The following examples show that not all modules have the cancellation property.

Example 7.3.6 (E. Lee Lady [205]).
Let A be a commutative Dedekind domain having an infinite number of prime ideals, and let Q be its quotient field. Let $\mathcal{P}_1$ and $\mathcal{P}_2$ be disjoint infinite sets of prime ideals not containing a certain non-unit non-zero element $b \in A$. Let a be another non-zero and non-unit element in A such that $a \in P$ for all $P \in \mathcal{P}_1 \cup \mathcal{P}_2$. Let X be a submodule of $Q \oplus Q$ generated by all $(P_1^{-1}, 0)$ for $P_1 \in \mathcal{P}_1$ and $(0, P_2^{-1})$ for

$P_2 \in \mathcal{P}_2$ together with $(1,1)/b$. Let Y be a submodule of $Q \oplus Q$ generated by the same set of elements except $(1,1)/b$ replaced with $(1,a)/b$. Let M be a submodule of Q generated by all P_2^{-1} for $P_2 \in \mathcal{P}_2$. Then X and Y are not isomorphic but $M \oplus X \simeq M \oplus Y$.

Another example was mentioned in the introductory paragraph of this section. Still another examples can be found in [276, page 248] and [209, Examples 3.2].

Note, that any indecomposable module is obviously internally cancellable, but it may not be cancellable as is shown by the following example.

Example 7.3.7 (T.Y. Lam [210]).
Let $A = \mathbf{R}[x,y,z]$ with the relation $x^2 + y^2 + z^2 = 1$, where $\mathbf{R}$ denotes the real numbers. Let P be the kernel of the epimorphism $A^3 \to A$ defined by $e_1 \mapsto x$, $e_2 \mapsto y$, and $e_3 \mapsto z$. Then $P \oplus A \cong A^3 \cong A^2 \oplus A$, but $P \ncong A^2$. Thus, A is not cancellable in the category of finite generated projective A-modules.

Proposition 7.3.8. *Any cancellable A-module M is internally cancellable.*

Proof.
Let $M = X \oplus Y = X_1 \oplus Y_1$ with $X \cong X_1$. Since M is a cancellable module and X is a direct summand of M, X is cancellable, by proposition 7.3.4. Thus from $X \oplus Y = X_1 \oplus Y_1 \cong X \oplus Y_1$ it follows that $Y \cong Y_1$. □

So that we have the following hierarchy of module-theoretic properties:

$$\text{Cancellation} \implies \text{Internal cancellation} \implies \text{Dedekind} - \text{finite}$$

The next theorem shows that there is a converse to this proposition when a module has the 2-exchange property (or equivalently, the finite exchange property, by theorem 7.1.15).

Theorem 7.3.9. (H.-P. Yu [336]). *Let M be a module with 2-exchange property. Then M is cancellable if and only if it is internally cancellable.*

Proof.
Assume that M is internally cancellable. Consider a module $N = M \oplus K = M_1 \oplus K_1$, where $M \cong M_1$. Since M is assumed to have 2-exchange property, one can write $N = M_1 \oplus X \oplus Y$ for suitable submodules $X \subseteq M$ and $Y \subset K$. Set $M = X_1 \oplus X$ and $K = Y_1 \oplus Y$. Then

$$M_1 \cong N/(X \oplus Y) = (M \oplus K)/(X \oplus Y) \cong X_1 \oplus Y_1.$$

Since $M_1 \cong M = X_1 \oplus X$ is internally cancellable, $X \cong Y_1$. Therefore, $K = Y_1 \oplus Y \cong X \oplus Y \cong N/M_1 \cong K_1$, which means that M has the cancellation property. □

Definition 7.3.10. A ring A is said to have **right stable range one** provided that whenever $a,b \in A$ and $aA + bA = A$, there exists an element $y \in A$ such that $a + by$

is invertible in A. In this case one simply writes $\mathrm{sr}_r(A) = 1$. Analogously one can define the notion of a ring which has a **left stable range one**. In this case one writes $\mathrm{sr}_l(A) = 1$.

For a discussion of stable range (and stable rank) in general see [237, definition 6.7.2, section 11.3].

The following theorem proved by L.N.Vaserstein [313] shows that these notions are left-right symmetric. So one can talk about a ring having stable range one.

Theorem 7.3.11 (L.N. Vaserstein [313, Theorem 2]) *Let A be a ring. Then* $\mathrm{sr}_r(A) = 1$ *if and only if* $\mathrm{sr}_l(A) = 1$.

Proof.
Suppose $\mathrm{sr}_r(A) = 1$. We have to prove that $\mathrm{sr}_l(A) = 1$. Let $Ac + Ad = A$ for some $c, d \in A$. So there are elements $a \in A$ and $t \in A$ such that $ac + td = 1$. Then for these elements a and $b = td$ one has $aA + bA = A$. Since $\mathrm{sr}_r(A) = 1$, there exists $x \in A$ such that $u = a + bx \in U(A)$. Let $uv = 1$. Set $w = a + x(1 - ca)$. Then

$$w(1 - cx) = (a + x(1 - ca))(1 - cx) = a + x(1 - ca) - acx - xc(1 - ac)x =$$

$$= a + x - xca - acx - xc(u - a) = a + x - acx - xcu =$$

$$= a + bx - xcu = u - xcu = (1 - xc)u.$$

Set $y = (1 - cx)v$. Then $wy = w(1 - cx)v = (1 - xc)uv = 1 - xc$. Therefore

$$w(c + yb) = [a + x(1 - ca)](c + yb) = ac + x(1 - ca)c + (1 - xc)b =$$

$$ac + xc - xcac + (1 - xc)b = ac + xc(1 - ac) + (1 - xc)b = 1.$$

Thus, the element $c + yb = c + ytd \in U(A)$, which means that $\mathrm{sr}_l(A) = 1$. □

Theorem 7.3.12. (H. Bass [15]). *A semilocal ring has stable range 1.*

Proof.
Note that any element $x \in A$ is invertible in A if and only if $\bar{x} = x + R$ is invertible in A/R, where R is the Jacobson radical of A. Therefore one can replace A by A/R. Since A is semilocal, A/R is Artinian. So one can assume that A is a semisimple ring, see theorem 1.1.24. Let $a, b \in A$ and $aA + bA = A$. Consider the right ideal $\mathcal{I} = aA \cap bA$ in A. Since A is semisimple, there is a right ideal S in A such that $bA = \mathcal{I} \oplus S = A$. Since $\mathcal{I}$ is also a right ideal in aA, there results $A = aA \oplus S$. Consider the exact sequence:

$$0 \to T \to A \xrightarrow{f} aA \to 0,$$

where f is an epimorphism defined by $f(x) = ax$ for any $x \in A$ and $T = \mathrm{Ker}\, f$. This sequence splits and so there exists an epimorphism $g : A \to T$ such that $(f, g) : A \to aA \oplus T$ is an isomorphism. Since $A = aA \oplus S$ and the Krull-Schmidt-Remak theorem holds for semisimple rings, there is an isomorphism $\varphi : T \to S$. Then $(1,\varphi)\circ(f,g) : A = aA \oplus T \to aA \oplus S$ is an automorphism of A_A. Therefore the image of the identity in A is an invertible element in A. But $(1,\varphi)\circ(f,g) = a + \varphi(g(1)) = a + y$, where $y = g(\varphi(1)) \in S \subset bA$. Thus, $a + y$ is an invertible element in A and $y \in bA$. □

Theorem 7.3.13 (I. Kaplansky [182]) *If A is a ring having right stable ring one then A is Dedekind-finite.*

Proof.
Suppose $ba = 1$ for $a, b \in A$. Let $x \in A$ be an arbitrary element of A. Then $xb \in A$ and $x = a(bx) - (1 - ab)x \in aA + (1 - ab)A$. Therefore $aA + (1 - ab)A = A$. Since $\mathrm{sr}_r(A) = 1$, there exists an element $y \in A$ such that $u = a + (1 - ab)y \in U(A)$. Then $bu = ba + (b - bab) = ba = 1$. So that $b \in U(A)$, since $u \in U(A)$, i.e., $ab = 1$ and A is Dedekind-finite. □

Theorem 7.3.14 (E.G. Evans [81]). *If the endomorphism ring* $\mathrm{End}_A(M)$ *of an A-module M has stable range one, then M has the cancellation property.*

Proof.
Suppose that $M \oplus X \cong M \oplus Y$ and that the endomorphism ring $E = \mathrm{End}_A(M)$ has stable range one. Note that if X_1, X_2, Y_1, Y_2 are A-modules, then any homomorphism $f \in \mathrm{Hom}_A(X_1 \oplus X_2, Y_1 \oplus Y_2)$ can be written in the matrix form: $f = \begin{pmatrix} f_{11} & f_{12} \\ f_{21} & f_{22} \end{pmatrix}$, where $f_{ij} \in \mathrm{Hom}_A(X_i, Y_j)$.

Since $M \oplus X \cong M \oplus Y$, there exists an isomorphism φ from $M \oplus Y$ to $M \oplus X$, and an isomorphism ψ from $M \oplus X$ to $M \oplus Y$ such that $\psi\varphi$ is the identity in $M \oplus Y$. Let $\varphi = \begin{pmatrix} \varphi_{11} & \varphi_{12} \\ \varphi_{21} & \varphi_{22} \end{pmatrix}$ and $\psi = \begin{pmatrix} \psi_{11} & \psi_{12} \\ \psi_{21} & \psi_{22} \end{pmatrix}$. Then

$$\psi\varphi = \begin{pmatrix} \psi_{11}\varphi_{11} + \psi_{12}\varphi_{21} & \psi_{11}\varphi_{12} + \psi_{12}\varphi_{22} \\ \psi_{21}\varphi_{11} + \psi_{22}\varphi_{21} & \psi_{21}\varphi_{12} + \psi_{22}\varphi_{22} \end{pmatrix} = \begin{pmatrix} 1_M & 0 \\ 0 & 1_Y \end{pmatrix}. \tag{7.3.15}$$

Hence

$$\psi_{11}\varphi_{11} + \psi_{12}\varphi_{21} = 1_M \tag{7.3.16}$$

$$\psi_{21}\varphi_{12} + \psi_{22}\varphi_{22} = 1_Y \tag{7.3.17}$$

From (7.3.16) it follows that $E\varphi_{11} + E\psi_{12}\varphi_{21} = E$. Since E has stable rang 1, it follows from theorem 7.3.11 that there exists an endomorphism $y \in E$ such that $\varphi_{11} + y\psi_{12}\varphi_{21} = u \in \mathrm{Aut}_A(M)$. Now consider the homomorphism:

$$\alpha = \begin{pmatrix} 1_M & y\psi_{12} \\ \psi_{21} & \psi_{22} \end{pmatrix} : M \oplus Y \longrightarrow M \oplus X.$$

Then

$$\alpha\varphi = \begin{pmatrix} u & \varphi_{12} + y\psi_{12}\varphi_{22} \\ 0 & 1_Y \end{pmatrix} \in \mathrm{Aut}_A(M \oplus Y). \tag{7.3.18}$$

Here the 0 and 1_Y in (7.3.18) come from (7.3.15). Since φ is an isomorphism, α is an isomorphism as well.

Consider the isomorphism

$$\beta = \begin{pmatrix} 1_M & 0 \\ -\psi_{21} & 1_X \end{pmatrix} \alpha \begin{pmatrix} 1_M & -y\psi_{12} \\ 0 & 1_Y \end{pmatrix} : M \oplus Y \longrightarrow M \oplus X.$$

Since

$$\beta = \begin{pmatrix} 1_M & 0 \\ 0 & \psi_{22} - \psi_{21} y \psi_{12} \end{pmatrix},$$

$\psi_{22} - \psi_{21} y \psi_{12} : Y \longrightarrow X$ is an isomorphism. So $X \cong Y$. □

Corollary 7.3.19. *If for an A-module M the endomorphism ring* $\mathrm{End}_A(M)$ *is semilocal then M has the cancellation property.*

The proof follows immediately from theorem 7.3.11 and theorem 7.3.14.

Lemma 7.3.20. (C.S. Hsü [153, Lemma 2]). *Let A be a Dedekind domain*[2]. *Then the right regular module* A_A *is cancellable.*

Proof.

Consider an A-module $M = X \oplus Y = X_1 \oplus Y_1$, where $X \cong X_1 \cong A$. Considering the natural projection of M onto X. Then the kernel of the restriction of this projection to Y_1 is $Y \cap Y_1$. So one obtains an exact sequence

$$0 \to Y \cap Y_1 \longrightarrow Y_1 \longrightarrow I \to 0,$$

where $I \subseteq X$. Since A is a Dedekind domain, I is projective. Therefore this sequence splits and one obtains $Y_1 = (Y \cap Y_1) \oplus I_1$, where $I_1 \cong I$. Similarly, $Y = (Y \cap Y_1) \oplus I_2$, where $I_1 \cong J \subseteq A$. Since both $X \oplus I_1$ and $X_1 \oplus I_2$ are complements of $Y_1 \cap Y$ in M, $X \oplus I_1 \cong X_1 \oplus I_2$. Then applying the Steinitz-Chevalley theory of torsion-free modules over Dedekind domains [297] (see also [177] and [62]), one obtains that $I_1 \cong I_2$, and hence $Y \cong Y_1$. □

Theorem 7.3.21. (C.S. Hsü [153, Theorem 1]). *Any finitely generated module over a Dedekind domain is cancellable.*

Proof.

By [146, theorem 8.5.5] any finitely generated module M over a Dedekind domain A is isomorphic to a direct sum of a finite number of ideals of A and a finite direct sum of Artinian uniserial modules of finite length of the form A/P^n, where P is a prime ideal of A. Let $M = I$, where I is an ideal of A. Since A is a hereditary ring, I is a f.g. projective A-module. Therefore I is cancellable, by proposition 7.3.4 and lemma 7.3.20. If $M = A/P^n$, then $\mathrm{End}_A(M)$ is a local ring by [146, proposition 10.1.5], since M is an Artinian uniserial module of finite length. So M is strongly

[2]Recall that a Dedekind domain is a commutative hereditary integral domain.

indecomposable and hence it is cancellable by proposition 7.3.5. Therefore M is cancellable by proposition 7.3.4. □

Remark 7.3.22. Note that this theorem first was proved independently by P.M. Cohn in [58] and E.A. Walker in [316] for modules over principal ideal domains. A nice source for cancellation and such is [209].

7.4 Exchange Rings

In this section we consider some various classes of rings which are closely connected with the exchange property. The most important class of them consists of exchange rings which were first introduced by R.B. Warfield in [319]. First we will consider rings considered by W.K. Nicholson [249] which are characterized by interesting idempotent property. Later we will show that these rings are precisely exchange rings.

Definition 7.4.1. A ring A is called **left suitable** if for any element $x \in A$ there exists an idempotent $e^2 = e \in Ax$ such that $1-e \in A(1-x)$. Analogously, a ring A is called **right suitable** if for any element $y \in A$ there exists an idempotent $f^2 = f \in yA$ such that $1 - f \in (1 - y)A$.

Remark 7.4.2. Note that conditions in these definitions can be written in more symmetric form. A ring A is called **left** (resp. **right**) **suitable** if for any elements $x, y \in A$ such that $x+y = 1$ there exist pairwise orthogonal idempotents $e^2 = e \in Ax$, $f^2 = f \in Ay$ (resp. $e^2 = e \in xA$, $f^2 = f \in yA$) such that $e + f = 0$.

Although the definition above deals separately with right and left suitable rings the theorem above shows that these notions are the same.

Theorem 7.4.3. *A ring A is right suitable if and only if A is left suitable.*

The proof of this theorem that is given below follows W.K. Nicholson [250].

Suppose that A is a right suitable ring. Then, by remark 7.4.2 for any $a, b \in A$ such that $a + b = 1$ there exists pairwise orthogonal idempotents $e^2 = e \in aA$ and $f^2 = f \in bA$ such that $e + f = 1$. Let $e = ax$, $f = by$, where $x, y \in A$. Then $e^2 = e = ax = axe$, and $xe = xe^2 = xeaxe$. Analogously $yf = yfbyf$. Write $xe = u$, $yf = v$. Then

$$uau = u, \qquad vbv = v$$

$$ubv = 0, \qquad uav = 0 \tag{7.4.4}$$

Now let

$$r = 1 - vb + ub, \qquad s = 1 - ua + va \tag{7.4.5}$$

Then taking into account (7.4.4) $rv = (1 - vb + ub)v = v - vbv + ubv = v - v = 0$ and analogously $su = 0$. Since $au + bv = axe + byf = e + f = 1$, it follows from (7.4.5) that

$$ar = a(1 - vb) + aub = a(1 - vb) + (1 - bv)b = (a + b)(1 - vb) = 1 - vb.$$

Then one has

$$rar = (1 - vb + ub)(1 - vb) = 1 - vb + ub - vb + vbvb + ubvb$$

$$= 1 - vb + ub - vb + vb + 0 = 1 - vb + ub = r$$

and so $rara = ra$, i.e., the element $e_1 = ra \in Aa$ is an idempotent in A. Analogously one can obtain that the element $f_1 = sb \in Ab$ is an idempotent in A. It remains only to show that $e_1 + f_1 = 1$. Since $a + b = 1$, $ab = ba$. Therefore $e_1 + f_1 = ra + sb = (1 - vb + ub)a + (1 - ua + va)b = a - vba + uba + b - uab + vab = a + b = 1$. Therefore A is a left suitable ring. □

Remark 7.4.6. Taking into account this theorem we will say simply about suitable rings.

Proposition 7.4.7. (W.K. Nicholson [249, Proposition 1.10]). *If A is a suitable ring then so is eAe for every idempotent $e^2 = e \in A$.*

Proof.
Let $x \in eAe$ and choose an idempotent $f^2 = f \in Ax$ such that $1 - f \in A(1 - x)$. Then $x = eae$, $f = bx$ and $1 - f = c(1 - x)$. So $fe = bxe = beae^2 = beae = f$, $(ef)^2 = (ef)(ef) = ef = efe \in eAe$. Setting $u = ef$ we obtain that $u^2 = u \in eAe$ and $e - u = e - e(1 - f)e = ec(1 - x)e = ece(e - x) \in eAe(e - x)$, i.e., eAe is left suitable. □

Proposition 7.4.8 (W.K. Nicholson [249, Corollary 1.3]). *A ring is suitable if and only if idempotents can be lifted modulo every left (right) ideal.*

Proof.
Suppose that A is a suitable ring. Let I be a left ideal in A, $\pi : A \to A/I$ the projection and $\bar{e}$ an idempotent of A/I. Let $x \in A$ be a lift of $\bar{e}$, i.e., $\pi(x) = \bar{e}$. Then $x - x^2 \in I$ and by a left suitability of A there exists an idempotent $e \in Ax$ such that $(1 - e) \in A(1 - x)$. Then

$$e - x = e(1 - x) - (1 - e)x \in A(x - x^2)$$

and so $\pi(e) = \pi(x)$, so that $\pi(e) = \bar{e}$ and e is an idempotent lift of $\bar{e}$, as desired.

Conversely, suppose that idempotents can be lifted modulo every left ideal I in a ring A. For $x \in A$ set $I = Ax$. Then $x - x^2 \in Ax$, and so there exists an idempotent $e^2 = e$ such that $e - x \in A(x - x^2)$. So $e - x = b(x - x^2)$ for some $b \in A$. Hence $1 - e = (1 - bx)(1 - x) \in A(1 - x)$ as required. □

Lemma 7.4.9. *A ring A is suitable if and only if for each* $x \in A$ *there exists an idempotent* $e^2 = e \in A$ *such that* $e - x \in A(x - x^2)$.

Proof.
Suppose that A is suitable and $x \in A$, then there exists an idempotent $e^2 = e \in Ax$ such that $1 - e \in A(1 - x)$. Therefore $e - x = e(1 - x) - (1 - e)x \in A(x(1 - x)) = A(x - x^2)$.

Conversely, suppose there exists an idempotent $e^2 = e \in A$ such that $e - x \in A(x - x^2)$. Let $e - x = a(x - x^2)$. Then $f = 1 - e = (1 - ax)(1 - x) \in A(1 - x)$ and $f^2 = f$, i.e., A is a suitable ring. □

The following class of rings introduced by W.K. Nicholson in [249] is connected with suitable rings and unit-regular rings considered in Section 6.2.

Definition 7.4.10. A ring A is called **clean** if for each element $x \in A$ can be written as $x = e + u$, where $e \in A$ is an idempotent and $u \in U(A)$ is a unit element.

From theorem 6.2.42 it follows that any unit-regular ring is clean.

Proposition 7.4.11 (W.K. Nicholson [249, Proposition 1.8(1)]). *Every clean ring is suitable.*

Proof.
Let $x \in A$, then there exist an idempotent $e \in A$ and a unit $u \in U(A)$ such that $x = e + u$. Then $f = 1 - e \in A$ and $h = u^{-1}fu$ are also idempotents. Consider

$$u(x - h) = u(e + u - u^{-1}fu) = u(e + u - u^{-1}(1 - e)u) = ue + u^2 - u + eu.$$

On the other side

$$x^2 - x = (e + u)^2 - (e + u) = eu + ue + u^2 - u.$$

Therefore $x - h = u^{-1}(x^2 - x) \in A(x^2 - x)$, which implies that A is suitable by lemma 7.4.9. □

Thus taking into account theorem 6.2.42 and proposition 7.4.11 we have the following hierarchy of rings:

Unit-regular rings $\Longrightarrow$ Clean rings $\Longrightarrow$ Suitable rings

Theorem 7.4.12. (W.K. Nicholson [249, Theorem 2.1]). *For a ring A the following conditions are equivalent for any right A-module M:*

1. $\mathrm{End}_A(M)$ *is a suitable ring*;
2. *M has the finite exchange property*;

Proof.
$1 \Longrightarrow 2$. Suppose that $\mathrm{End}_A(M)$ is a suitable ring and there are two direct decompositions $X = M \oplus N = X_1 \oplus X_2$ of left A-modules. Let $B = \mathrm{End}_A(X)$ and

let $\pi \in B$ be such that $M = \pi(X)$, $\mathrm{Ker}(\pi) = N$ and $\pi^2 = \pi$. Let $\tau_1, \tau_2 \in B$ such that $X_i = \tau_i(X)$ and $\mathrm{Ker}(\tau_1) = X_2$, $\mathrm{Ker}(\tau_2) = X_1$. Then τ_1, τ_2 are pairwise orthogonal idempotents of B, i.e., $\tau_1 + \tau_2 = 1$ and $\tau_i^2 = \tau_i$. Therefore $\pi = \pi(\tau_1 + \tau_2)\pi$. Since $\mathrm{End}_A(M)$ is a suitable ring, $\pi B \pi \simeq \mathrm{End}_A(M)$ is suitable as well, by proposition 7.4.7. Therefore from remark 7.4.2 it follows that there exist pairwise orthogonal idempotents $v_i \in (\pi B \pi)\pi\tau_i\pi = (\pi B \pi)\tau_i\pi$ such that $v_1 + v_2 = \pi$. Let $v_i = \alpha_i\tau_i\pi$, where $\alpha_i \in \pi B \pi$. Note that $\alpha_i\pi = \alpha_i = \pi\alpha_i$. We can assume that $v_i\alpha_i = \alpha_i$. Now we set $u_i = \tau_i\alpha_i\tau_i$ $(i = 1, 2)$. Since $u_i = \tau_i\alpha_i\tau_i = \tau_i v_i\alpha_i\tau_i = \tau_i\alpha_i\tau_i\pi\alpha_i\tau_i = \tau_i\alpha_i(\tau_i)^2\alpha_i\tau_i = u_i^2$, we obtain that u_1, u_2 are pairwise orthogonal idempotents, and $u_i(X) \subseteq X_i$ for $i = 1, 2$. Therefore $X_i = u_i(X) \oplus X_i'$ where $X_i' \subset X_i \cap \mathrm{Ker}(u_i)$ and so

$$X = (u_1(X_1) \oplus u_2(X_2)) \oplus (X_1' \oplus X_2').$$

We will now prove that $X = M \oplus (X_1' \oplus X_2')$.

Since $u_i\pi = \tau_i v_i$, we have $\alpha_i u_i \pi = \alpha_i\tau_i v_i = \alpha_i\tau_i\alpha_i\tau_i\pi = \alpha_i\tau_i\pi\alpha_i\tau_i\pi = (\alpha_i\tau_i\pi)^2 = v_i^2 = v_i$. Let $x \in M \cap (X_1' \oplus X_2')$. Then $u_1(x) = 0 = u_2(x)$ and so $x = \pi(x) = v_1(x) + v_2(x) = \alpha_1 u_1\pi(x) + \alpha_2 u_2\pi(x) = \alpha_1 u_1(x) + \alpha_2 u_2(x) = 0$. Therefore $M \cap (X_1' \oplus X_2') = 0$. It remains to show that if $x \in X$ then $x \in M \oplus (X_1' \oplus X_2')$. Let $x \in X$ and write $x = x_1 + x_2 + z$, where $x_i \in u_i(X)$ $(i = 1, 2)$ and $z \in X_1' \oplus X_2'$. Then

$$u_j\alpha_i u_i = u_j(\pi\alpha_i)(\tau_i u_i) = (\tau_j v_j)(v_i\alpha_i)(\tau_i u_i) = \delta_{ij}u_i,$$

where $\delta_{ij} = 1$ for $i = j$ and $\delta_{ij} = 0$ for $i \neq j$. Therefore $x_i - \alpha_i(x_i) \in X_i'$ for $i = 1, 2$. So

$$x - \alpha_1(x_1) - \alpha_2(x_2) = (x_1 - \alpha_1(x_1)) + (x_2 - \alpha_2(x_2)) + z \in X_1' \oplus X_2'.$$

Since $\alpha_i(x_i) \in M$ for $i = 1, 2$, we have $x \in M \oplus (X_1' \oplus X_2')$. Thus $X = M \oplus (X_1' \oplus X_2')$, i.e., M has the 2-exchange property, and so it has the finite exchange property, by theorem 7.1.15.

$2 \Longrightarrow 1$. Assume that a module M has the finite exchange property, in particular it has the 2-exchange property. Consider a module $X = M \oplus M$ and write $X_1 = \{(x, 0) : x \in M\}$, $X_2 = \{(0, x) : x \in M\}$, and $D = \{(x, x) : x \in M\}$. Suppose $\alpha, \beta \in \mathrm{End}_A(M) = B$ are such that $\alpha + \beta = 1$. Let $M_1 = \{(\alpha(x), -\beta(x)) : x \in M\}$. Then $M_1 \simeq M$ and $X = M_1 \oplus D = X_1' \oplus X_2'$. Since M has the 2-exchange property, there exist submodules $X_i' \subseteq X_i$ such that $X = M_1 \oplus X_1' \oplus X_2'$. Let $x \in X$, then we have the unique decomposition:

$$(x, x) = (\alpha(y), -\beta(y)) + (x_1, 0) + (0, x_2), \qquad (*)$$

where $(x_1, 0) \in X_1'$ and $(0, x_2) \in X_2'$. Equating components in (*) we obtain $x = x_1 + \alpha(y)$ and $x = x_2 - \beta(y)$, which yields

$$x_2 - x_1 = \alpha(y) + \beta(y) = (\alpha + \beta)(y) = y. \qquad (**)$$

We define now two endomorphisms α_1, β_1 of M such that $x_2 = \alpha_1(x)$ and $x_1 = \beta_1(x)$. Then consider the following decompositions:

$$(\alpha\alpha_1(x), \alpha_1\alpha(x)) = (\alpha\alpha_1(x), -\beta\alpha_1(x)) + (0,0) + (0, \alpha_1(x))$$

$$(\beta\beta_1(x), \beta_1\beta(x)) = (-\alpha\beta_1(x), \beta\beta_1(x)) + (\beta_1(x), 0) + (0,0)$$

which imply $\alpha\alpha_1\alpha = \alpha_1$ and $\beta\beta_1\beta = \beta$. So $e = \alpha\alpha_1$ and $f = \beta\beta_1$ are idempotents such that $e \in \alpha B$ and $f \in \beta B$. Moreover, from (*) and (**) we obtain that $x = \alpha(y) + x_1 = \alpha(y) + \beta_1(x) = \alpha(\alpha_1(x) - \beta_1(x)) + \beta_1(x) = (\alpha\alpha_1 - \alpha\beta_1 + \beta_1)(x) = (\alpha\alpha_1 - \alpha\beta_1 + (\alpha + \beta)\beta_1)(x) = (\alpha\alpha_1 + \beta\beta_1)(x)$, which implies $\alpha\alpha_1 + \beta\beta_1 = 1_M$, i.e., $e + f = 1_M$. Thus $B = \mathrm{End}_A(M)$ is a suitable ring. □

Since $A \simeq \mathrm{End}_A(A)$ for any ring A, from theorem 7.4.12 one immediately obtains

Corollary 7.4.13 (R.B. Warfield, [319]). *If A is a ring then the right regular module A_A has the finite exchange property if and only if ${}_AA$ has the finite exchange property.*

Following R.B. Warfield [319], we now introduce the following definition:

Definition 7.4.14. A ring A is called **exchange** if the right regular module A_A has the finite exchange property.

Note that although the definition given above considers only the right regular module the notion of an exchange ring is left-right symmetric in view of corollary 7.4.13.

Since $A \simeq \mathrm{EndA}_A(A)$, theorem 7.4.12 and corollary 7.4.13 implies:

Theorem 7.4.15. (R.B.Warfield [319, Theorem 3], W.K. Nicholson [249, Corollary 2.3]).

1. *A ring A exchange if and only if A is a suitable ring.*
2. *An A-module M has the finite exchange property if and only if the ring $\mathrm{End}_A(M)$ is an exchange ring.*

As a consequence from this theorem and proposition 7.1.30 one obtains:

Corollary 7.4.16. *Local rings and semiperfect rings are exchange rings.*

Theorem 7.4.17. (R.B. Warfield [319, Theorem 1]). *If A is an exchange ring, then any projective right A-module is a direct sum of right ideals generated by idempotents.*

Proof.
Let M be a projective right A-module. Then it is isomorphic to a direct summand of a free module. Therefore there is an A-module N such that $M \oplus N$ is a free A-module, that is, $M \oplus N \cong A^I$. Now applying the Crawley-Jónsson theorem 7.2.13 one obtains that M is a direct sum of submodules, each of which is isomorphic to a

summand of A_A. The result now follows taking into account that any direct summand of A_A is precisely a right ideal generated by an idempotent. □

Remark 7.4.18. Since any local ring is an exchange ring, this theorem can be considered as a generalization of a well known theorem of I.Kaplansky (See corollary 7.2.14).

Since any exchange ring is a suitable ring, from lemma 7.4.9 it follows immediately:

Proposition 7.4.19. *Every homomorphic image of an exchange ring is exchange.*

The next proposition gives one more equivalent definition of an exchange ring.

Proposition 7.4.20 (W.K. Nicholson [249, Corollary 2.4]). *A ring A with Jacobson radical R is exchange if and only if A/R is exchange and idempotents can be lifted modulo R.*

Proof.

If A is an exchange ring then conditions of the statement follows from theorem 7.4.15(1), propositions 7.4.8 and 7.4.19.

Assume that $B = A/R$ is an exchange ring and idempotents can be lifted modulo R. Let $x \in A$. Write $\bar{x} = x + R$. Since B is a suitable ring, there exists idempotent $\bar{e} \in B\bar{x}$ such that $\bar{1} - \bar{e} = \bar{a}(\bar{1} - \bar{x})$ for some $\bar{a} \in B$. We may assume that $e \in Ax$. Choose $f^2 = f \in A$ such that $\bar{f} = \bar{e}$. Then $u = 1 - f + e \in U(A)$, since $\bar{u} = \bar{1} - \bar{f} + \bar{e} = \bar{1}$. So $h = u^{-1}fu = u^{-1}fe \in Ax$ is an idempotent in A and $\bar{h} = \bar{f} = \bar{e}$. Therefore $1 - h - a(1 - x) \in R$. Since R is the Jacobson radical, $v = h + a(1 - x) \in U(A)$. Then $1 = v^{-1}h + v^{-1}a(1 - x) = th + s(1 - x)$, where $t = v^{-1}$ and $s = v^{-1}a$. Consider an element $g = h + (1 - h)th \in Ax$, which is an idempotent, since $g^2 = (h + (1 - h)th)(h + (1 - h)th) = h + (1 - h)th = g$ because $(1 - h)h = h(1 - h) = 0$. Then $1 - g = (1 - h) - (1 - h)th = (1 - h)(1 - th) = (1 - h)s(1 - x) \in A(1 - x)$, which shows that A is a suitable ring, and so A is an exchange ring. □

Proposition 7.4.21. *A ring A is exchange if and only if eAe is an exchange ring for every idempotent $e^2 = e \in A$.*

Proof.

If A is an exchange ring then it is a suitable ring, and so is eAe for every idempotent $e \in A$ by proposition 7.4.7.

Conversely, if $1 = e + f$, where e and $f = 1 - e$ are idempotents, then modules eA and fA has finite exchange properties by theorem 7.4.12. Therefore the right regular module $A_A = eA \oplus fA$ also has finite exchange property by lemma 7.1.10. So A is an exchange ring. □

The first element-wise characterization of an exchange ring was given by G.S. Monk in [243]:

Theorem 7.4.22. (G.S. Monk [243, Theorem 1]). *An A-module M has the finite exchange property if and only if, given* $a \in \mathrm{End}_A M$ *there exist elements* $b, c \in \mathrm{End}_A M$ *such that*

$$bab = b \text{ and } c(1-a)(1-ba) = 1-ba. \qquad (7.4.23)$$

In particular, a ring A is an exchange ring if and only if for each $a \in A$ *there exist elements* $b, c \in A$ *satisfying* (7.4.23).

Proof.
Write $B = \mathrm{End}_A M$ and let $a \in \mathrm{End}_A M$. If such elements $b, c \in \mathrm{End}_A M$ satisfying (7.4.23) exist let $h = ba$. Then $h^2 = h \in Ba$ and $(1-h)c(1-a)(1-h) = 1-h$. Set $e = 1-(1-h)c(1-a)$, then $e^2 = e = eh \in Ba$ and $1-e = (1-h)c(1-a) \in B(1-a)$. Hence B is a suitable ring, and so it an exchange ring.

Conversely, assume that B is an exchange ring. So it is a suitable ring. Let $a \in B$. Then there exists an idempotent $e^2 = e \in Ba$ such that $1-e \in B(1-a)$. If $e = ba$ and $1-e = c(1-a)$ we obtain that $b, c \in B$ satisfy (7.4.23). □

Remark 7.4.24. The proof presented above is given following W.K. Nicholson [249, corollary 2.5], and it is different from original proof in [243].

Collecting the equivalent definitions given above of an exchange ring one obtains the following theorem:

Theorem 7.4.25. *Let A be a ring with Jacobson radical R. Then the following conditions are equivalent:*

1. *A is an exchange ring;*
2. *A is a suitable ring;*
3. A/R *is an exchange ring and idempotents can be lifted modulo R;*
4. *All idempotents of A can be lifted modulo every right ideal of A;*
5. *All idempotents of A can be lifted modulo every left ideal of A;*
6. *For any* $x \in A$*, there is an* $e = e^2 \in xA$ *with* $1-e \in (1-x)A$;
7. *For any* $a \in A$*, there is an* $f = f^2 \in Aa$ *with* $1-f \in A(1-a)$.
8. eAe *is an exchange ring for every idempotent* $e^2 = e \in A$.
9. For each $a \in A$ there exist elements $b, c \in A$ such that $bab = b$ and $c(1-a)(1-ba) = 1-ba$.

A relationship between exchange rings and rings having stable rang one was studied by V.P. Camillo and H.-P. Yu in [34].

Lemma 7.4.26 (V.P. Camillo and H.-P. Yu [34, Lemma 2]). *Let A be an exchange ring. Then the following conditions are equivalent:*

1. *A has stable range one.*
2. *For any* $a \in A$, $e^2 = e \in A$, *if* $ax + e = 1$ *for some* $x \in A$*, then there exists* $y \in A$ *such that* $a + ey = u \in U(A)$.

Proof.

$1 \Longrightarrow 2$. This follows immediately from the definition 7.3.10.

$2 \Longrightarrow 1$. Let $aA + bA = A$. Then there exists an $y \in A$ such that $1 = ay + b$, so $1-b = ay$. Since A is en exchange ring, by theorem 7.4.25 there exists an idempotent $e^2 = e \in bA$ such that $(1 - e) \in (1 - b)A$. Therefore there exist elements $c, x \in A$ such that $e = bc$ and $1 - e = (1 - b)x$. Then $1 - e = ayx$, i.e., $e + a(yx) = 1$. So that by assumption there exists $z \in A$ such that $a + ez = u \in U(A)$. Then $u = a + ez = a + b(cz)$, i.e., A has stable range one. □

Recall that an element $x \in A$ is von Neumann regular if there exists $x \in A$ such that $axa = a$. If $aua = a$ for $u \in U(A)$ then $a \in A$ is called unit-regular.

Theorem 7.4.27. (V.P. Camillo and H.-P. Yu [34, Theorem 3]). *An exchange ring A has stable range one if and only if every von Neumann regular element of A is unit-regular.*

Proof.

Suppose that A is an exchange ring having stable range one. Assume that $a \in A$ is a von Neumann regular element, i.e., there exists $x \in A$ such that $axa = a$. Then elements ax and $e = 1 - ax$ are idempotents and $ax + (1 - ax) = 1$. Therefore by lemma 7.4.26 there exists an element $y \in A$ such that $a + (1 - ax)y = u \in U(A)$. Multiplying the both sides of the last equality by ax on the left we obtain that $a = axu$ which implies $ax = au^{-1}$. Therefore $a = axa = au^{-1}a$, i.e., a is a unit-regular element.

Conversely, assume that A is an exchange ring in which every von Neumann regular element is unit-regular. Let $a \in A$, $e^2 = e \in A$ and $ax + e = 1$ for some element $x \in A$. Assume that $axa \neq a$. Set $f = ax$ and $b = fa - a$, then $f^2 = (ax)(ax) = (1 - e)(1 - e) = 1 - e = ax = f$, $fb = f(fa - a) = fa - fa = 0$ and $bx = (axa - a)x = (ax - 1)ax = -e(1 - e) = 0$. Letting $c = a + b$ we have $cx = ax + bx = ax = f$, $cxc = axc = fc = fa + fb = fa + 0 = a + b = c$. Since $fb = 0$, $f^2 = f$, $e = 1 - f$ is an idempotent and $b \in (1 - f)A = eA$, $b = ed$. Then $1 = ax + e = (c - b)x + e = cx - bx + e = cx + e$, and $cxc = c$. Since $c + ey = a + b + ey = a + ed + ey = a + e(d + y)$, $c + ey = u \in U(A)$ implies $a + e(d + y) \in U(A)$.

Therefore we can assume that $ax + e = 1$, where $e^2 = e$ and $axa = a$. Note that $axa + ea = a$, so $axa = a$ if and only if $ea = 0$. Since by assumption any von Neumann regular element is unit-regular, there exists a unit $u \in U(A)$ such that $aua = a$. Then we have $1 - e = ax = (aua)x = (au)(ax) = au(1 - e)$ and

$$(au-e)(au-e) = auau - eau - aue + e = au - aue + e = au(1-e) + e(1-e) + e = 1.$$

Therefore $au - e = v \in U(A)$ and $a - eu^{-1} = vu^{-1} \in U(A)$, i.e., A has stable range one by lemma 7.4.26. □

Some other results on connections of exchange rings with rings having stable range one can been obtained by H.-P.Yu in [336].

The class of exchange rings is quite wide. As was shown above it includes local rings, semiperfect rings, unit-regular rings. Besides these rings there exists a number of different important classes of rings which are exchange.

The following theorem gives simple relationship of exchange rings and semiperfect rings.

Proposition 7.4.28 (V. Camillo and H.-P. Yu [33, Corollary 2, Proposition 3]). *Let A be an exchange ring then the following statements are equivalent*:

1. *A is an orthogonally finite ring, i.e., contains no infinite set of orthogonal idempotents*;
2. *A is an FDI-ring.*
3. *A is a semiperfect ring.*

Proof.

$1 \Longrightarrow 2$. This follows from proposition 6.3.7.

$2 \Longrightarrow 3$ Let $1 = e_1 + e_2 + \cdots + e_n$, where e_i are primitive pairwise orthogonal idempotents. Since A is an exchange ring, each $e_i A e_i$ is also an exchange ring, by theorem 7.4.25. Therefore $e_i A$ is an indecomposable module having finite exchange property. So $e_i A$ is a strongly indecomposable module, i.e., each $e_i A e_i$ is a local ring, by theorem 7.1.32. Then A is a semiperfect ring, by theorem 1.9.3.

$3 \Longrightarrow 1$. This is trivial. □

The next two theorems are given for completeness sake without proof. The first one gives an element-wise characterization of semiperfect rings and the second one gives relations of FDI-rings with perfect rings.

Theorem 7.4.29. (V. Camillo and H.-P. Yu [33, Corollary 2, Proposition 3]). *A ring A is semiperfect if and only if A is clean and orthogonally finite.*

Theorem 7.4.30 (M. Harada [139], K. Yamagata [334]). *If A is an FDI-ring then the following conditions are equivalent*:

1. *Each projective right A-module has the exchange property.*
2. *The countably generated free right A-module* $A^{\mathbf{N}}$ *has the finite exchange property.*
3. *A is a perfect ring.*

A short proof of this proposition can be found in [33, Corollary 4].

7.5 Notes and References

The notion of "exchange" and "finite exchange" was introduced by Peter Crawley and Bjarni Jónsson for general algebras in in 1964 in their classical paper [61]. The main result of this paper yields sufficient conditions for a group (with or without operators) to have the isomorphic refinement property, which is a generalization of the famous Krull-Remak-Schmidt theorem for groups. For operator groups this result said that if an operator group G has a direct decomposition such that the admissible center of each factor satisfies the minimal and local maximum conditions, then any two direct decompositions of G have centrally isomorphic refinements. Actually their results are obtained for more general cases, namely for general algebras. To obtain these results they introduced and used the exchange property. The main results of Section 7.1 were obtained by Peter Crawley and Bjarni Jónsson for general algebras and R.B.Warfied, Jr. for Abelian categories (see [321]). Propositions 7.1.30 and 7.1.31, which say that an indecomposable module has the exchange property if and only if its endomorphism ring is local, are due to R.B.Warfied, Jr. (see [317]). Properties of direct sum decompositions of injective objects in reasonable Abelian categories were studied by R.B. Warfied, Jr. (see [321]). In this paper he proved, in particular, the exchange property for injective objects in a reasonable Abelian category. By now the notion of the exchange property is an important theoretical tool for studying rings and modules. The main problem which is connected with exchange property is still open at this time and in a general form may be formulated as follows: when does the finite exchange property imply the full exchange property? Some results concerning this problem were obtained by Pace P. Nilson (see [251], [252]).

The definition of a left stable range was introduced by H.Bass in [16] for general linear groups in algebraic K-theory. Stable range of rings was considered by L.N. Vaserstein in [312], [313]. Rings having stable range one are studied by K.R. Goodearl and P.Menal [111], V.P. Camillo and H.-P. Yu [34], [336]. A brief survey on stable range and its application can be found in [206].

The proof of uniqueness of decomposition (the variant of the Krull-Remak-Schmidt theorem for an infinite number of summands) was obtained by Azumaya in the general case for Abelian categories with some additional condition (see [12]). In Section 7.2 we gave a proof of this theorem for the category of modules. The strengthened form of the Krull-Remak-Schmidt-Azumaya theorem for infinite sums was obtained by Peter Crawley and Bjarni Jónsson (see [61]) for general algebras and R.B. Warfield, Jr. for modules (see [317]).

The problem of the cancellation property and some properties of modules having the cancellation property has been studied by many authors. In particular, E.G. Evans in [81] showed that for a right A-module M, if $\mathrm{End}_A M$ has stable range 1, then M has the cancellation property. P.M. Cohn, in [58], showed that the integer ring $\mathbf{Z}$ has the cancellation property. H.B. Zhang and W.T. Tong in [338] proved that Dedekind domains have the cancellation property. They also showed that if A is a Prüfer domain, then $A \oplus X \simeq A \oplus Y$ implies $X \simeq Y$ for any pair of finitely generated A-modules. Proposition 7.2.3 was obtained by R.B. Warfield, Jr. in [317].

Suitable rings were introduced by W.K.Nicholson and their properties were studied in [249], [250]. Exchange rings were studied by many authors (see e.g. [319], [249], [309], [252]), [243]. A more complete review on modules with exchange property and exchange rings was given by A.A.Tuganbaev in [308].

The nice and complete survey on stable range, cancellation and exchange property can be found in [209].

CHAPTER 8

Hereditary and Semihereditary Rings. Piecewise Domains

The class of semisimple rings has been studied most extensively. Their structure is completely defined by the Wedderburn-Artin theorem. Semisimple rings are also very simple from the point of view of the homological properties of modules over them. These are rings whose global homological dimension is equal to zero.

Recall that a ring A is **right (left) hereditary** if each of its right (left) ideals is projective. Hereditary rings immediately follow semisimple ones in the homological ranking . By theorem 4.4.8, r.gl.dim $A \leqslant 1$ if and only if A is a right hereditary ring. If a ring A is both right and left hereditary A is said to be a **hereditary ring**. The structure of hereditary rings is not so well studied as in the case of semisimple rings. This chapter is devoted to the study of the structure and main properties of hereditary rings. Also semihereditary rings, which are close to hereditary rings, are considered in this chapter. A ring A is **right (left) semihereditary** if every right (left) finitely generated ideal is projective. If a ring A is both right and left semihereditary, A is said to be a **semihereditary ring**. Note that the only difference with "hereditary" is that here only finitely generated ideals are required to be projective. Obviously, each hereditary ring is semihereditary.

In Section 4.5 it was shown that for a large class of rings (coherent, semiperfect, right serial, right Noetherian) being right semihereditary implies being left semihereditary. In Section 8.1 this result is proved for orthogonally finite rings.

The main results of chapter 4 and 5 are applied to study the properties of right hereditary and right semihereditary rings in Section 8.2. In particular there is proved a Goldie theorem which gives equivalent conditions for a domain A to be right Ore. Here, there is also proved the important Small theorem which states that a right Noetherian right hereditary ring is a right order in a right Artinian ring.

The structure and properties of some classes of right hereditary (semihereditary) prime rings are considered in Section 8.3.

The next step up from hereditary and semihereditary rings are piecewise domains. These rings were first introduced and studied by R.Gordon and L.W.Small in 1972.

Section 8.4 considers properties of piecewise domains and their relationships with hereditary and semihereditary rings. It is proved that a piecewise domain is a nonsingular ring. This section also gives a proof of the theorem which states that a right perfect piecewise domain is semiprimary. This theorem first was proved by M.Teply in 1991.

The notion of a triangular ring was first introduced by S.U. Chase in 1961 for semiprimary rings. In 1966 L.Small extended this notion to Noetherian rings and proved that a right Noetherian right hereditary ring is triangular. M. Harada in 1964 introduced the notion of generalized triangular rings and proved that any hereditary semiprimary ring is isomorphic to a generalized triangular ring with simple Artinian blocks along the main diagonal. In 1980 Yu.A. Drozd extended the notion of a triangular ring to FDI-rings and described the structure of right hereditary (semihereditary) FDI-rings.

Section 8.5 introduces the notion of primely triangular rings which includes all notions of a triangular ring mentioned above. The main result of this section gives the structure of piecewise domains in terms of primely triangular rings. This theorem was proved by R.Gordon and L.W. Small in 1972 and it states that any piecewise domains is a primely triangular ring. From this statement there easily follows the theorem obtained by L.W. Small in 1966 about the structure of right Noetherian right hereditary rings.

Section 8.6 gives the criterion for a triangular FDI-ring to be right hereditary or right semihereditary, which was obtained by Yu.A.Drozd in 1980.

In Section 8.7 the results of Section 8.6 are applied to various concrete classes of rings. In particular, we give a criterion for a right Noetherian primely triangular ring to be right hereditary. From this result there follows the famous decomposition theorem of Chatters which states that a Noetherian hereditary ring is a direct sum of rings each of which is either an Artinian hereditary ring or a prime Noetherian hereditary ring.

Section 8.8 is devoted to the study of hereditary species and tensor algebras as introduced by Yu.A. Drozd.

Diagram 8.1 shows the relationships between the main classes of rings considered in this chapter. An arrow signifies containment of the class of rings at the source of the arrow into its target which is the class of rings to which the arrow points.

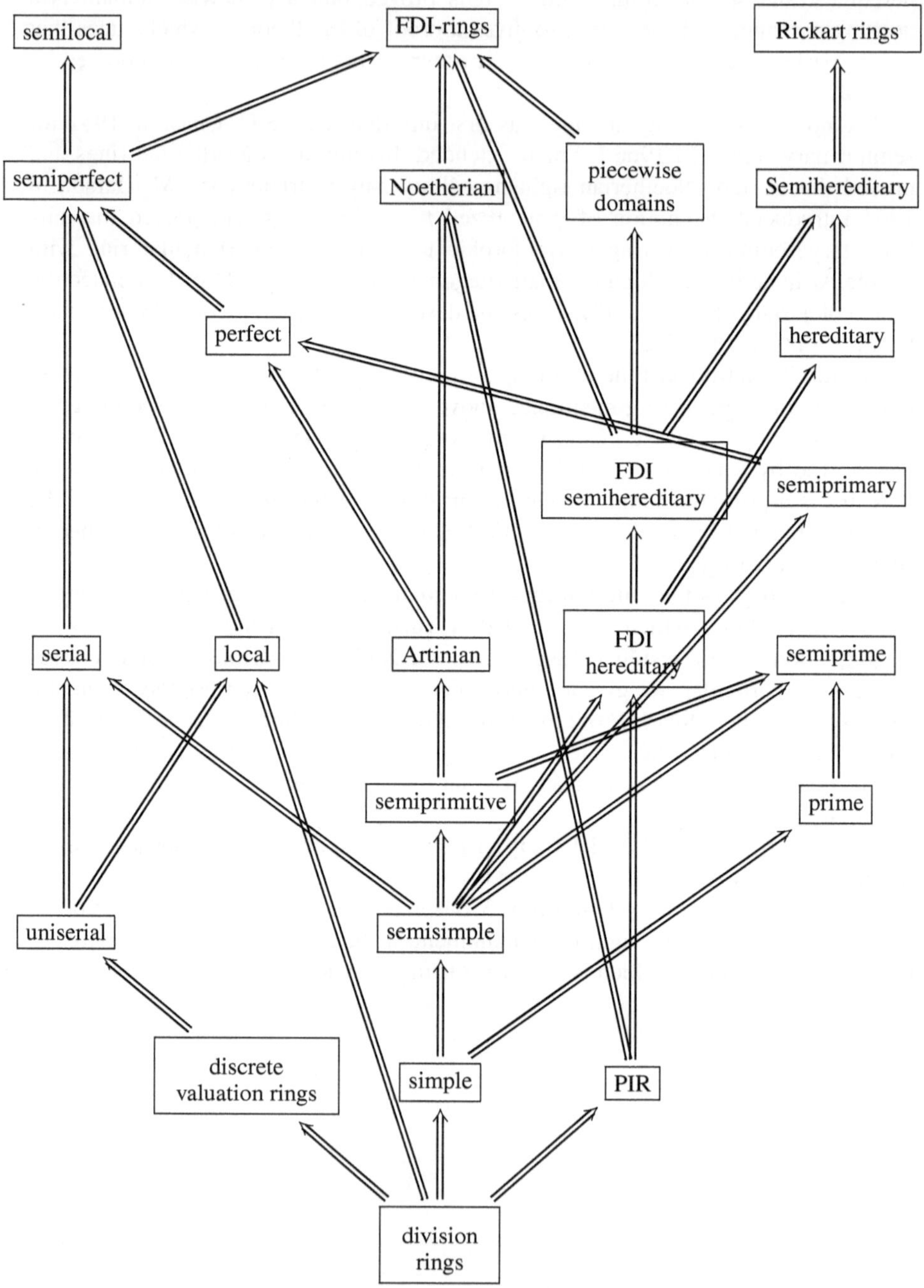
semilocal
FDI-rings
Rickart rings
semiperfect
Noetherian
piecewise domains
Semihereditary
perfect
hereditary
FDI semihereditary
semiprimary
serial
local
Artinian
FDI hereditary
semiprime
semiprimitive
prime
uniserial
semisimple
discrete valuation rings
simple
PIR
division rings

Diag. 8.1.

8.1 Rickart Rings and Small's Theorems

Definition 8.1.1. A ring A is called a **right (left) Rickart ring** if every principal right (left) ideal is projective. A ring which is both a right and left Rickart ring is called a **Rickart ring**.

Note that in general a right Rickart ring need not be a left Rickart ring.

Example 8.1.2. Any right semihereditary ring is a right Rickart ring.

Thus there is the following chain of inclusions:

$$\text{hereditary rings } \subsetneqq \text{ semihereditary rings } \subsetneqq \text{ Rickart rings}$$

Note that this sequence of classes of rings is not strict. There are both semihereditary rings which are not hereditary and Rickart rings which are not hereditary.

For example, a ring $A = \begin{pmatrix} \mathbf{Z} & \mathbf{Q} \\ 0 & \mathbf{Z} \end{pmatrix}$, where $\mathbf{Z}$ is the ring of integers and $\mathbf{Q}$ is the field of rational numbers, is a semihereditary ring but it is not neither right nor left hereditary.

Proposition 8.1.3. *Any right Rickart ring is right nonsingular.*

Proof.

For any $a \in A$ consider the homomorphism $\varphi : A \longrightarrow A$ which is given by $f(x) = ax$ for any $x \in A$. Then there is an exact sequence

$$0 \longrightarrow \text{r.ann}_A(a) \longrightarrow A \longrightarrow aA \rightarrow 0$$

with a projective principal right ideal aA. Therefore this sequence splits, and so $\text{r.ann}_A(a)$ is a direct summand of A, and so it cannot be essential in A. Hence $\mathcal{Z}(A_A) = 0$. □

Corollary 8.1.4. *Any right semihereditary ring is right nonsingular. In particular, any right hereditary ring is right nonsingular.*

A right Rickart ring was defined above as a ring which any principal right ideal is projective. The following statement gives an equivalent definition for a Rickart ring.

Proposition 8.1.5. *A ring A is right Rickart if and only if the right annihilator of any element in A is of the form eA for some idempotent $e \in A$.*

Proof.

1. Suppose that a ring A is right Rickart and $a \in A$. Then the right ideal aA is projective and so the exact sequence

$$0 \rightarrow \text{r.ann}(a) \rightarrow A \xrightarrow{f} aA \rightarrow 0 \tag{8.1.6}$$

is split, which implies that $\text{r.ann}_A(a) = eA$ for some idempotent $e \in A$.

2. Conversely, suppose that the right annihilator of any element in A is of the form eA for some idempotent $e \in A$. Let $a \in A$. Consider the exact sequence (8.1.6), where $f(x) = ax$ for $x \in A$. Then, by assumption $\text{r.ann}_A(a) = eA$ for some idempotent $e \in A$. Therefore this sequence is split, which implies that aA is a projective ideal. □

Lemma 8.1.7. *Let A be a right Rickart ring. Then any non-zero left annihilator in A contains a non-zero idempotent.*

Proof.
Let K be a non-zero left annihilator in A, i.e., $K = \text{l.ann}_A(S)$ for some subset S in A. Consider $L = \text{r.ann}_A(K)$. Since K is a left ideal in A, $K = \text{l.ann}_A(\text{r.ann}_A(K)) = \text{l.ann}_A(L)$ by lemma 5.3.19. Let $a \in K$. Since A is a right Rickart ring, $\text{r.ann}_A(a) = eA$ for some idempotent $e \in A$, by proposition 8.1.5. Since $L \subset \text{r.ann}_A(a) = eA$, $(1-e)L \subset (1-e)eA = 0$, i.e., $(1-e)L = 0$, so $f^2 = f = 1-e \in K$, as required. □

Theorem 8.1.8. *Let A be a right Noetherian right Rickart ring. Then A is a right order in a right Artinian ring.*

Proof.
Since A is a right Noetherian ring, it suffices to prove, by corollary 5.5.32, that $C_A(N) \subseteq C_A(0)$, where $N = Pr(A)$ is the prime radical of A, $C_A(0)$ is the set of all regular elements of A, and $C_A(N)$ is the set of all elements which are regular in A/N. Let $x \in C_A(N)$. Then $\text{r.ann}_A(x) \subseteq N$. Since A is a right Rickart ring, $\text{r.ann}_A(x) = eA$ for some idempotent $e \in A$, by proposition 8.1.5. So $e \in N$. Since N is a nil-ideal, $e = 0$. Therefore $\text{r.ann}_A(x) = 0$. We now show that $\text{l.ann}_A(x) = 0$ as well. Since $x \in C(N)$, $\text{l.ann}_A(x) \subseteq N$. Suppose $\text{l.ann}_A(x) \neq 0$. Then, by lemma 8.1.7, there is a non-zero idempotent $f^2 = f \in \text{l.ann}_A(x)$. So $f \in N$, which implies that $f = 0$, as N is a nil-ideal. This contradiction shows that $\text{l.ann}_A(x) = 0$. Thus $x \in C_A(0)$, i.e., $C_A(N) \subseteq C_A(0)$ as required. □

Theorem 8.1.9 (L. Small [292]). *A right Noetherian right hereditary ring is a right order in a right Artinian ring.*

Proof.
Since any right hereditary ring is right Rickart, this statement is a particular case of theorem 8.1.8. □

Definition 8.1.10. A ring A is called a **right Baer ring** if any right annihilator in A is of the form eA for some idempotent $e \in A$. In a similar way one can define a **left Baer ring**.

It follows from proposition 8.1.5 that a right Baer ring is a right Rickart ring.

Proposition 8.1.11. *A ring is right Baer if and only if it is left Baer.*

Proof.

Suppose that A is a right Baer ring. Then by assumption $\text{r.ann}_A(\text{l.ann}_A S) = eA$ for some $e^2 = e \in A$ and any subset S of A. Therefore

$$\text{l.ann}_A S = \text{l.ann}_A(\text{r.ann}_A(\text{l.ann}_A S)) = \text{l.ann}_A(eA) = A(1-e)$$

and $f = 1 - e$ is an idempotent in A. The inverse statement can be obtained analogously. □

Thanks to proposition 8.1.11 one can say about **Baer rings**.

Theorem 8.1.12. (L.Small [294, Theorem 1]). *Suppose that A is an orthogonally finite ring*[1]. *Then the following conditions are equivalent*:

1. *A is a Baer ring.*
2. *A is a right Rickart ring.*
3. *A is a left Rickart ring.*
   If any of these conditions holds then A satisfies a.c.c. and d.c.c. on left (right) annihilators.

Proof.

Since $1 \Longrightarrow 2$ and $1 \Longrightarrow 3$ follows from proposition 8.1.5, taking into account the symmetry of the theorem it suffices to prove that $2 \Longrightarrow 1$. Let A be a right Rickart ring. Let $S \subset A$ be any subset of A and $L = \text{l.ann}_A(S)$. By lemma 8.1.7, L contains an idempotent $e^2 = e$. Choose an idempotent e such that $N = \text{l.ann}_A(e) = A(1-e)$ is minimal with respect to inclusion. This is possible due to theorem 6.2.4, since A is orthogonally finite and $N(1-e)$ is a direct summand of A for every idempotent e.

Now we first show that $L \cap N = 0$. Suppose that $K = L \cap N \neq 0$, then $K = L \cap N = \text{l.ann}_A(S) \cap \text{l.ann}_A(e) = \text{l.ann}_A(S \cup e)$. By lemma 8.1.7, K contains an idempotent $f^2 = f$. It is clear that $fe = ef = 0$. Let $e_1 = e + (1-e)f$, then $e_1^2 = e_1$ and $e_1 e = e$. So e_1 is a non-zero idempotent and $\text{l.ann}_A(e_1) \subsetneq \text{l.ann}_A(e)$. Since $fe = 0$ and $fe_1 = f \neq 0$, this inclusion is proper, which contradicts the choice of e. Therefore $L \cap N = 0$.

Let $x \in L$, then $x(1-e) \in L \cap A(1-e) = 0$, i.e., $x = xe$, which implies that $L = Ae$. So A is a left Baer ring, and by proposition 8.1.11, A is a Baer ring.

Now again take into account theorem 6.2.4 and proposition 8.1.5, and there results the last statement of the theorem. □

Recall that two rings A and B are said to be **Morita equivalent** if their categories of modules are equivalent. (See e.g. [146, Section 10.7].) It follows from [146, Corollary 10.7.3] that for any natural number n rings A and $M_n(A)$ are Morita equivalent. If A and B is Morita equivalent rings then there exists an idempotent $e \in M_n(A)$ such that $B \simeq eM_n(A)e$ by [146, Corollary 10.7.4].

---

[1]Recall that a ring is orthogonally finite if it does not contain infinite set of pairwise orthogonal idempotents.

Proposition 8.1.13. (L. Small [294, Proposition]). *A ring A is right (left) semihereditary if and only if $M_n(A)$ is a right (left) Rickart ring for any $n \geq 1$.*

Proof.

Suppose that a ring A is right semihereditary. Since any right semihereditary ring is right Rickart, from Morita equivalence it follows that $M_n(A)$ is a right Rickart ring for any $n \geq 1$.

Conversely, suppose that $S = M_n(A)$ is a right Rickart ring for any $n \geq 1$. Let $I = x_1A + x_2A + \cdots + x_nA$ be a finitely generated ideal in A. Write

$$\mathbf{X} = \begin{pmatrix} x_1 & x_2 & \dots & x_n \\ 0 & 0 & \dots & 0 \\ \vdots & \vdots & \ddots & \vdots \\ 0 & 0 & \dots & 0 \end{pmatrix}.$$

Then the right ideal in $M_n(A)$ generated by $\mathbf{X}$ has the form

$$\mathcal{J} = \begin{pmatrix} I & I & \dots & I \\ 0 & 0 & \dots & 0 \\ \vdots & \vdots & \ddots & \vdots \\ 0 & 0 & \dots & 0 \end{pmatrix}$$

and is projective as a right S-module. On the other hand, S is a free A-module generated by the n^2 matrix units. Therefore $\mathcal{J}$ is a projective right A-module as well. Since $\mathcal{J}_A \cong nI_A$, it follows that I_A is also projective, i.e., A is a right semihereditary ring. □

Theorem 8.1.14. (L.Small [294, Theorem 3]). *Suppose $M_n(A)$ is orthogonally finite for any $n > 0$. Then A is left semihereditary if and only if it is right semihereditary.*

Proof.

Suppose that A is right right semihereditary. Then, by proposition 8.1.13, the ring $S = M_n(A)$ is right Rickart for any $n > 0$. Since S is orthogonally finite, it is left Rickart as well by proposition 8.1.12. Then from proposition 8.1.13 it follows that A is left semihereditary. □

Something different version of this theorem was given by T.Y.Lam.

Theorem 8.1.14***.** (T.Y. Lam [207, Theorem 7.64]). *Suppose A is right or left finite-dimensional ring. Then A is left semihereditary if and only if it is right semihereditary.*

Proof.

Suppose that A is right finite-dimensional ring, i.e., u.dim$A_A < \infty$. Let $S = M_n(A)$. By proposition 5.1.25,

$$\text{u.dim}S_S = n \cdot \text{u.dim}A_A < \infty.$$

Therefore, by theorems 6.2.4, S is orthogonal finite. Assume that A is right semihereditary. Then, by proposition 8.1.13, the ring S is right Rickart for any $n > 0$, and so S is left Rickart by proposition 8.1.12. Then again from proposition 8.1.13 it follows that A is left semihereditary. □

Taking into account that right (left) Noetherian rings and semiperfect rings are orthogonally finite, we obtain the following corollary:

Corollary 8.1.15. *A right (left) Noetherian ring is right semihereditary if and only if it is left semihereditary. A semiperfect ring is right semihereditary if and only if it is left semihereditary.*

8.2 Dimensions of Hereditary and Semihereditary Rings

This section gives some applications of the results obtained in chapters H5 and H6 to the case of hereditary and semihereditary rings.

Recall that module M is called **essentially finitely generated** if there exists a finitely generated submodule $N \subseteq M$ which is essential in M (see definition 5.2.13).

It is clear that any finitely generated module is essentially finitely generated setting $N = M$. On the other hand not any essentially finitely generated module is finitely generated. An easiest example of such module is a **Z**-module **Q** which is not finitely generated but it contains a finitely generated **Z**-module **Z** which is essential in **Q**.

In some particular cases these notions are the same. One of such examples shows the next theorem.

Proposition 8.2.1. *Let A be a right nonsingular ring, and M a projective right A-module. Then M is finitely generated if and only if it is essentially finitely generated.*

Proof.

Since any finitely generated module is obviously essentially finitely generated, it remains only to prove the inverse statement.

Let A be a right nonsingular ring. Suppose that M is an essentially finitely generated projective right A-module. Then there is a finitely generated submodule N which is essential in M. Let $f : M \to A$ be a homomorphism of right A-modules and $f(N) = 0$. Then f induces the homomorphism $g : M/N \to A$ defined by $g(m + N) = f(m)$ for any $m \in M$. Since A is a nonsingular ring, $\mathcal{Z}(A) = 0$. Since N is essential in M, M/N is singular, by proposition 5.3.11, i.e., $\mathcal{Z}(M/N) = M/N$. From proposition 5.3.8(2) it follows that $g(\mathcal{Z}(M/N)) \subset \mathcal{Z}(A)$. Therefore $g(M/N) = g(\mathcal{Z}(M/N)) \subset \mathcal{Z}(A) = 0$, hence $g(M/N) = 0$. So $f = 0$ as well.

Since M is a projective right A-module, by theorem 1.5.5 there exists a system of elements $m_i \in M$ and system of homomorphisms $\varphi_i : M \to A$, $i \in I$, such that any element $m \in M$ can be written in the form:

$$m = \sum_{i \in I} m_i \cdot \varphi_i(m),$$

where only a finite number of elements $\varphi_i(m) \in A$ are not equal to 0. Let $N = x_1A + x_2A + \ldots + x_nA$. Then for each x_j there is a finite subset $I_j \in I$ such that $\varphi_i(x_j) = 0$ for all $i \notin I_j$. If $J = \bigcup_{j=1}^{n} I_j$ then $\varphi_i(x_j) = 0$ for all $i \notin J$ and all $j = 1,2,\ldots,n$. Therefore $\varphi_i(N) = 0$ and by what has been proved above this implies that $\varphi_i = 0$ for all i. Thus

$$m = \sum_{i \in J} m_i \cdot \varphi_i(m),$$

where J is a finite set, i.e., M is a finitely generated A-module. □

Corollary 8.2.2. *Let A be a right nonsingular ring, and M a projective right A-module. If M has finite uniform dimension then it is finitely generated.*

Proof. This follows immediately from proposition 8.2.1 and lemma 5.2.14. □

Since, by corollary 8.1.4, any right semihereditary ring is right nonsingular, from proposition 8.2.1 and corollary 8.2.2 there immediately follow the next results:

Corollary 8.2.3. *Let A be a right semihereditary ring (right hereditary ring), and let M be a projective right A-module. Then M is finitely generated if and only if M is essentially finitely generated.*

Corollary 8.2.4. *Let A be a right semihereditary ring (right hereditary ring), and let M be a projective right A-module. Then M is finitely generated if and only if M has a finite uniform dimension.*

In [146, Section 3.6] it a criterion was obtained for a ring to be right Noetherian which is connected with the notion of a minor. Now there will be discussed another criterion for a right hereditary ring to be right Noetherian. This one is connected with the notion of uniform dimension.

Theorem 8.2.5. *Let A be a right hereditary ring. Then A is right Noetherian if and only if* $\text{u.dim}\, A_A < \infty$.

Proof.

Let A be a right hereditary ring. Then a right ideal in it is projective and, by corollary 8.1.2, A is a right nonsingular ring.

Suppose that A is right Noetherian, then $\text{u.dim}\, A_A < \infty$ by corollary 5.1.10.

Conversely, suppose that $\text{u.dim}\, A_A < \infty$. Then from corollary 8.2.2 it follows that any right ideal of it is finitely generated, i.e., A is right Noetherian. □

Recall that a ring A satisfying the right Ore condition (see Section 1.11) is called a **right Ore ring**. If, in addition, the ring is a domain, then it is called a **right Ore domain**.

Theorem 8.2.6 (A.W.Goldie). *The following conditions are equivalent for any domain A:*

1. *A is a right Ore domain.*
2. $\text{u.dim}(A_A) = 1$.
3. $\text{u.dim}(A_A) < \infty$.

Proof.
$1 \Longrightarrow 2$. Let A be a right Ore domain. Then it satisfies the right Ore condition, i.e., $aA \cap bA \neq 0$ for any two elements $a, b \in A$, which implies that any non-zero right ideal is essential in A. So, by proposition 5.1.3, the right regular module A_A is uniform, i.e., $\text{u.dim}(A_A) = 1$.

$2 \Longrightarrow 1$. By proposition 5.1.3, any right ideal of A is essential in A. Therefore for any two elements $a, b \in A$ the intersection $aA \cap bA \neq 0$, i.e., A is a right Ore ring.

$2 \Longrightarrow 3$. Obvious.

$3 \Longrightarrow 1$. Suppose there exist two non-zero elements $a, b \in A$ such that $aA \cap bA = 0$. It follows that the elements $a^i b$ for $i \geq 0$ are linearly independent over A. Indeed, assume that they are linearly dependent. Then $\sum\limits_{i \geq 0} a^i b x_i = 0$, where the $x_i \in A$ are almost all zero. Hence $bx_0 + a(bx_1 + abx_2 + \ldots) = 0$. Since $aA \cap bA = 0$ and A is a domain, $x_0 = 0$ and so $bx_1 + abx_2 + \ldots = 0$. Then $bx_1 + a(bx_2 + abx_3 + \ldots) = 0$, whence $x_1 = 0$. Continuing this process (which is finite) there results that all $x_i = 0$. Therefore A contains the infinite direct sum $\bigoplus\limits_{i \geq 0} a^i bA$, and so $\text{u.dim}(A_A) = \infty$. A contradiction. □

In [146, chapter 8] there was shown that for the commutative case any hereditary domain is a Dedekind domain, and so it is a Noetherian ring. In the noncommutative case this is not true in general. But it is true for Ore domains. (Here note that any commutative domain is an Ore domain.) That is the following result.

Corollary 8.2.7. *Any right hereditary Ore domain is right Noetherian.*

The proof is immediately follows from theorems 8.2.5 and 8.2.6.

Corollary 8.2.8. *Any right Noetherian domain is a right Ore domain.*

Proof.
Since, by corollary 5.1.10, for any right Noetherian ring $\text{u.dim}(A_A) < \infty$, the proof follows immediately from theorem 8.2.6. □

Proposition 8.2.9. *Let A be a hereditary Noetherian ring. If I is an essential right ideal in A then the right A-module A/I has finite length.*

Proof.
Let J be a right ideal in A and $I \subseteq J$. Consider the dual left A-modules J^* and I^*. Note that $J^* \subseteq I^*$. Since A is right hereditary, A is right nonsingular by corollary 8.1.4. Therefore all right ideals in A are nonsingular. Consider the exact sequence

$$0 \to I \longrightarrow J \longrightarrow J/I \longrightarrow 0 \qquad (8.2.10)$$

Since I, J are nonsingular and $I \subseteq_e J$, it follows from proposition 5.3.12, that J/I is singular. Applying the duality functor $* = \mathrm{Hom}_A(-, A)$ to (8.2.10) yields an exact sequence

$$0 \to (J/I)^* \to J^* \to I^*. \qquad (8.2.11)$$

Since A is nonsingular, and J/I is singular, it follows from proposition 5.3.15(2) that $(J/I)^* = \mathrm{Hom}_A(J/I, A) = 0$, which implies that $J^* \longrightarrow I^*$ is a monomorphism.
Now consider a descending chain of right ideals in A:

$$K_1 \supseteq K_2 \supseteq \ldots \supseteq I \qquad (8.2.12)$$

which, according to the above, yields an ascending chain of left ideals

$$K_1^* \subseteq K_2^* \subseteq \ldots \subseteq I^*. \qquad (8.2.13)$$

Since A is a Noetherian hereditary ring, I^* is a finitely generated left A-module, by proposition 4.5.9. Therefore the ascending chain of left ideals (8.2.13) terminates, and so the descending chain of right ideals (8.2.12) terminates, as well. This proves that A/I is an Artinian right A-module. Since it is also a Noetherian right A-ring, A/I has a finite length, by [146, proposition 3.2.2]. □

Proposition 8.2.14. *For any hereditary Noetherian ring A there is the inequality*:

$$\mathrm{K.dim}\, A \leq 1.$$

Proof.
The proof follows from proposition 5.7.11 taking into account that A/E is an Artinian right A-module for any essential right ideal E by proposition 8.2.9, so that $\mathrm{K.dim}(A/E) \leq 0$. □

Proposition 8.2.15. *A Noetherian semihereditary ring is hereditary.*

Proof. It follows from corollary 4.2.15 that

$$\mathrm{w.dim}A = \mathrm{r.gl.dim}A = \mathrm{l.gl.dim}A.$$

The proposition now follows from theorem 4.6.2, which states that $\mathrm{w.dim}A \leq 1$ for semihereditary rings. □

Recall that if A is a right hereditary ring then for any non-zero idempotent $e \in A$ the ring $S = eAe$ is also right hereditary, by proposition 4.4.6(4). An analogous statement is true also for right semihereditary rings.

Theorem 8.2.16 (P.L. Sandomierski [280]). *Let A be a right semihereditary ring. Then for any non-zero idempotent $e \in A$ the ring $S = eAe$ is also right semihereditary.*

Proof.
Let I be a finitely generated right ideal in S. Then $\bar{I} = IA$ is a finitely generated right ideal in A. By assumption, $\bar{I}$ is projective. Then there is a free A-module F which is a direct sum of copies of eA and $\bar{I} \simeq F/K$. So there is an exact sequence

$$0 \to K \to F \to \bar{I} \to 0 \tag{8.2.17}$$

Since Ae is a projective left A-module, the induced sequence of right S-modules

$$0 \to K \otimes Ae \to F \otimes Ae \to \bar{I} \otimes Ae \to 0 \tag{8.2.18}$$

is also exact. Since $\bar{I}$ is A-projective, the sequence (8.2.17) is split, and therefore so is (8.2.18). Hence $\bar{I} \otimes Ae$ is a direct summand of $F \otimes Ae$. Since $F \otimes Ae$ is a free S-module, $\bar{I} \otimes Ae$ is S-projective. Taking into account that $\bar{I} \otimes Ae \simeq IAAe = IeAe = I$, it follows that I is S-projective. Thus, indeed, S is a right semihereditary ring. □

Corollary 8.2.19. *Let P be a finitely generated projective right A-module and $S =$ $\mathrm{End}_A(P)$. If A is a right semihereditary ring then so is S.*

Proof.
Let P be a finitely generated projective right A-module, then it is a direct summand of a free right A-module with a finite basis, consisting, say, of n elements. Then $\mathrm{End}_A(F) \simeq M_n(A)$. If A is a semihereditary ring, then, by Morita equivalence, $M_n(A)$ is also a right semihereditary ring.

Since P is a direct summand of F, $P = eF$ for some idempotent $e \in M_n(A)$. Therefore $S \simeq eM_n(A)e$, which is semihereditary, by theorem 8.2.16. □

8.3 Right Hereditary and Right Semihereditary Prime Rings

This section studies the structure of some classes of hereditary and semihereditary prime rings.

Prime rings are a standard noncommutative generalization of integral domains. Recall that a ring A is **prime** if the product of any two non-zero two-sided ideals of A is not equal to zero.

One of the most important examples of prime rings are orders[2] in simple Artinian rings. By theorem 1.11.4, a ring A is a right order in a simple Artinian ring Q if and only if A is a prime right Goldie ring.

In Section 3.7 it was considered an interesting class of valuation rings that are Dubrovin valuation rings of simple Artinian rings[3].

Recall that a subring A of a simple Artinian ring S is called a **Dubrovin valuation ring** if there is an ideal M of A such that

1. A/M is a simple Artinian ring;
2. For each $s \in S \setminus A$ there are $a_1, a_2 \in A$ such that $sa_1 \in A \setminus M$ and $a_2 s \in A \setminus M$.

By proposition 3.7.21, any Dubrovin valuation ring of a simple Artinian ring Q is an order in Q. So that Dubrovin valuation rings are prime Goldie rings.

The next important class of orders are Bézout orders[4].

Recall that a ring A is said to be a **right (left) Bézout ring** if every finitely generated right (left) ideal of A is principal. A ring which is both right and left Bézout is called a **Bézout ring**. A **Bézout domain** is a (commutative) integral domain in which every finitely generated ideal is principal.

Any principal ideal ring is obviously Bézout. In a certain sense, a Bézout ring is a non-Noetherian analogue of a principal ideal ring.

The main examples of commutative Bézout domains which are neither principal ideal domains nor Noetherian rings are as follows.

1. The ring of all functions in a single complex variable holomorphic in a domain of the complex plane **C**.
2. The ring of holomorphic functions given on the entire complex plane **C**.
3. The ring of all algebraic integers.

It is clear that every Bézout order in a simple Artinian ring Q is a semihereditary order in Q.

The following proposition shows that this result has an inverse in the following form.

Proposition 8.3.1. [233, Lemma 5.7] *Let A be a semihereditary order in a simple Artinian ring Q. Assume that $A/J(A)$ is a simple Artinian ring, where $J(A)$ is the Jacobson radical of A. Then A is a Bézout order in Q.*

---

[2]For definition of orders see Section 1.11 and Section 5.4.

[3]For definition and properties of Dubrovin valuation rings see Section 3.7 and in particular Definition 3.3.9.

[4]See Definition 3.7.23.

Proof.

Note that as A is an order in a simple Artinian ring it is a prime Goldie ring by theorem 1.11.4. So we can apply all results for prime Goldie rings proved in chapter 5.

Let $\mathcal{I}$ be a finitely generated right ideal of a ring A. Since A is semihereditary, $\mathcal{I}$ is a projective ideal. We need to show that $\mathcal{I}$ is a principal right ideal.

If $\mathcal{I}$ is essential then it is generated by regular elements which belong to it by theorem 5.4.23.

If $\mathcal{I}$ is not essential then there are a finite number of uniform right ideals $\mathcal{I}_1, \mathcal{I}_2, \ldots, \mathcal{I}_n$ such that $K = \mathcal{I} \oplus \mathcal{I}_1 \oplus \mathcal{I}_2 \oplus \cdots \oplus \mathcal{I}_n$ is an essential ideal in A by lemma 5.1.15. Since any uniform right ideal is indecomposable, there exist $a_1, a_2, \ldots, a_n \in A$ such that $K = \mathcal{I} \oplus \bigoplus_{i=1}^{n} a_i A$ is essential. If we would prove that this ideal is principal, i.e., $K = cA$ for some element $c \in A$ then $c = x + \sum_{i=1}^{n} a_i x_i$ for some $x \in \mathcal{I}$ and $x_i \in \mathcal{I}_i$. Whence one obtains that $\mathcal{I} = xA$.

Thus in the both case we can assume that $\mathcal{I}$ is a finitely generated essential ideal which is generated by their regular elements. Moreover, by induction we can assume that $\mathcal{I} = rA + sA$ where $r, s \in C_A(0)$.

Consider an exact sequence

$$0 \longrightarrow \mathrm{Ker}(\pi) \longrightarrow A \oplus A \xrightarrow{\pi} \mathcal{I} = rA + sA \longrightarrow 0$$

where $\pi(a,b) = ra - sb$. Since A is a semihereditary ring, $\mathcal{I}$ is a projective ideal, and so this sequence is split. Therefore

$$A \oplus A = X \oplus Y, \tag{8.3.2}$$

where $X \simeq \mathcal{I}$ and $Y = \mathrm{Ker}(\pi) = \{(a,b) \mid ra = sb\}$.

Let $J(A)$ be the Jacobson radical of A. Write $\bar{A} = A/J(A)$ and $\bar{M} = M/MJ(A)$ for A-module M. We will also use the denotations which were introduced in Section 3.7. Namely, we write $d(\bar{A})$ for the number of all mutually non-isomorphic simple right $\bar{A}$-modules of a semisimple ring $\bar{A}$. We write $d_{\bar{A}}(\bar{M})$ the number of all direct summands in the decomposition of a module $\bar{M}$ into a direct sum of simple right $\bar{A}$-modules. As was mentioned in Section 3.7 to prove that an $\bar{A}$-module $\bar{M}$ is cyclic it suffices to show that $d_{\bar{A}}(\bar{M}) < d\bar{A}$.

From (8.3.2) it follows that $d_{\bar{A}}(\bar{X}) < d\bar{A}$ or $d_{\bar{A}}(\bar{Y}) < d\bar{A}$.

If $d_{\bar{A}}(\bar{X}) < d\bar{A}$ then $\bar{X}$ is a cyclic $\bar{A}$-module. Then $X \simeq \mathcal{I}$ is also a cyclic module by Nakayama's lemma 1.1.20, since X is a finitely generated A-module.

Suppose that $d_{\bar{A}}(Y) < d\bar{A}$. Then $Y = \mathrm{Ker}(\pi)$ is a cyclic A-module with generator $c = (x,y) \in \mathrm{Ker}(\pi)$. Then $rx = sy$, where $r, s \in C_A(0)$. On the other hand, since A is an Ore ring, for any pair elements (s,r) such that $s \in A$, $r \in C_A(0)$ there is a pair elements (a,b) such that $a \in C_A(0)$, $b \in A$ and $sa = rb$, that is $(a,b) \in Y$. This means that $\mathrm{r.ann}_A(c) = 0$. Therefore $Y \simeq A$ and

$$d_{\bar{A}}(\bar{X}) = 2d(\bar{A}) - d(\bar{A}) = d(\bar{A}),$$

i.e., again one obtains that $X \simeq I$ is cyclic as required. □

As was proved in Section 3.7 any Dubrovin valuation ring is a Bézout order, and so a semihereditary order by theorem 3.7.24. It turns out that this theorem has an inverse in the form of the following theorem which yields the equivalent definitions of Dubrovin valuation rings.

Theorem 8.3.3. *Let A be a subring of a simple Artinian ring Q. Then the following conditions are equivalent:*

1. *A is a Dubrovin valuation ring of Q.*
2. *A is a semihereditary order in Q with Jacobson radical $J(A)$ such that $A/J(A)$ is a simple Artinian ring.*
3. *A is a Bézout order in Q with Jacobson radical $J(A)$ such that $A/J(A)$ is a simple Artinian ring.*
4. *A is an n-chain ring in Q for some n with Jacobson radical $J(A)$ such that $\bar{A} = A/J(A)$ is a simple Artinian ring and $d\bar{A} \geq n$.*

Proof.
$1 \Longrightarrow 2$ and $1 \Longrightarrow 3$. These were proved in Theorem 3.7.24.
$3 \Longrightarrow 2$. This is trivial.
$2 \Longrightarrow 3$. This is proved in Proposition 8.3.1.
$3 \Longrightarrow 4$. This was proved in Theorem 3.7.30.
$4 \Longrightarrow 1$ was proved in Theorem 3.7.29. □

Another nice class of prime rings are hereditary Noetherian prime rings (HNP rings, for short). These rings and modules over them were subject of the study of many mathematics (see e.g. [79], [80], [224], [225], [226], [222], [237]).

A HNP ring A is a Goldie prime ring and therefore A is an order in a simple Artinian ring. Note also a well known fact that a commutative hereditary Noetherian ring is a finite direct product fields and Dedekind domains which follows from the following theorem of I. Kaplansky.

Theorem 8.3.4. ([181, Theorem 168]). *Let A be a commutative Noetherian ring such that A_M is an integral domain for every maximal ideal M. Then A is a direct sum of integral domains.*

Since any hereditary (commutative) integral domain is a Dedekind domain by [146, Theorem 8.3.4], and any prime ring is obviously indecomposable, it follows from theorem 8.3.5 that a commutative HNP ring, which is not Artinian, is a Dedekind domain. So noncommutative HNP rings can be considered as a natural generalization of Dedekind domains.

One of the most interesting examples of HNP rings are classical hereditary orders. When one consider orders in simple Artinian rings which are finite dimensional algebras over their centers it is preferable to consider a special class of orders, so said classical orders, which have some additional properties.

Definition 8.3.5. (See e.g. [237, Definition 3.5]). Let Q be a central simple algebra over a field C, i.e., C is the center of Q and Q is a simple Artinian ring which is finitely dimensional over C.

Suppose that K is a commutative integral domain, not necessary Noetherian, whose field of fractions is C. A subring A in Q is called **classical order** in Q over K if $K \subseteq A$, $AC = Q$ and A is a finite generated module over K.

An example of a classical order over a commutative integral domain K with field of fractions C is a subring $A = M_n(K)$ in a simple Artinian ring $Q = M_n(C)$.

Definition 8.3.6. A hereditary ring A is called a **classical hereditary order** if A is a classical order over a Dedekind domain being a center of A.

These rings are proved to be HNP rings (see e.g. [237]). They have been studied extensively in integral representation theory (see e.g. [62], [159], [269]).

The other important example of HNP rings, which are not classical hereditary orders, are first Weil algebras.

Let F be a field of characteristic 0. Let $A_1 = F[x, y]$ with $\alpha x = x\alpha$, $\alpha y = y\alpha$ for any $\alpha \in F$, and with defining relation $xy - yx = 1$. The algebra A_1 is called the **first Weyl algebra**. This is a simple ring and a hereditary Noetherian domain, but A is not a principal ideal domain.

Since char$F = 0$, the center of A is F, and hence A is not a classical hereditary order.

In this section right Noetherian and right hereditary semiperfect prime rings are considered. Obviously, these rings can be described up to Morita equivalence, that is, one can assume that A is a basic ring.

Recall that if A is a semiperfect ring then it decomposed as right A-module into a direct sum of principal right modules $A = P_1^{n_1} \oplus \cdots \oplus P_s^{n_s}$. If $B = P_1 \oplus \cdots \oplus P_s$ then the category of right A-modules and right B-modules are Morita equivalent. The semiperfect ring A is said to be **basic** if $n_1 = n_2 = \cdots = n_s = 1$, i.e., there are no isomorphic modules in the decomposition of the ring A into a direct sum of principal right A-modules.

Example 8.3.7.

Denote by $H_t(\mathcal{O})$ the ring of $t \times t$ matrices of the form:

$$H_t(\mathcal{O}) = \begin{pmatrix} \mathcal{O} & \mathcal{O} & \dots & \mathcal{O} \\ \mathcal{M} & \mathcal{O} & \dots & \mathcal{O} \\ \vdots & \vdots & \ddots & \vdots \\ \mathcal{M} & \mathcal{M} & \dots & \mathcal{O} \end{pmatrix}, \tag{8.3.8}$$

where $\mathcal{O}$ is a discrete valuation ring and $\mathcal{M} = \pi\mathcal{O} = \mathcal{O}\pi$ is its unique maximal ideal. Then the Jacobson radical of $H_t(\mathcal{O})$ is

$$R = \begin{pmatrix} \mathcal{M} & \mathcal{O} & \dots & \mathcal{O} \\ \mathcal{M} & \mathcal{M} & \dots & \mathcal{O} \\ \vdots & \vdots & \ddots & \vdots \\ \mathcal{M} & \mathcal{M} & \dots & \mathcal{M} \end{pmatrix},$$

which the unique maximal ideal in $H_t(\mathcal{O})$ and $H_t(\mathcal{O})/R$ is isomorphic to a finite direct sum of division ring $D = \mathcal{O}/\mathcal{M}$. As it was shown in [146, Section 12.3] the ring $H_t(\mathcal{O})$ is a Noetherian serial prime hereditary ring and moreover A is an order in a simple Artinian ring $M_n(D)$. Therefore $H_t(\mathcal{O})$ is also a Dubrovin valuation ring, by theorem 8.3.2.

G.O. Michler [240] proved the following statement which gives the structure of Noetherian semiperfect prime hereditary rings.

Theorem 8.3.9 (G.O. Michler [240]). (See e.g. [146, theorem 12.3.4].) *A Noetherian semiperfect prime basic hereditary ring A is either a division ring or it is isomorphic to a ring of the form $H_t(\mathcal{O})$, where $\mathcal{O}$ is a discrete valuation ring.*

This theorem implies the next results.

Proposition 8.3.10. (See [146, proposition 12.3.6].) *The quiver (left quiver) of a semiperfect prime Noetherian hereditary ring A is either a point or a simple cycle.*

Theorem 8.3.11. (See [146, corollary 12.3.7].) *Any prime Noetherian hereditary semiperfect basic ring is isomorphic to either a division ring, or a ring of the form $H_t(\mathcal{O})$, where $\mathcal{O}$ is a discrete valuation ring. Clearly, the above rings are serial hereditary Noetherian prime rings.*

Proposition 8.3.12. *A right Noetherian semihereditary local ring A is a principal right ideal domain.*

Proof.
Since A is a semihereditary local ring, from [146, proposition 10.7.9] and [146, proposition 10.7.8] it follows that A is a domain. Let $R = \text{rad}A$ be the Jacobson radical of A. If $R = 0$ then A is a division ring. Suppose $R \neq 0$. Then R is a finitely generated projective right A-module, as A is a right Noetherian semihereditary ring. By theorem 7.2.14 each projective module over a local ring is free, so that R is a free right A-module, i.e., $R \simeq A^n$. As was shown above A is a domain, so $R \simeq A_A$, i.e., R is a principal right ideal. Since A is a local ring, for any of its right ideals $I \subset R$. Moreover, I is a finitely generated projective right A-module, since A is right Noetherian and semihereditary. The same type of arguments which were used for R show that I is a principal right ideal. Therefore A is a principal right ideal domain. □

Corollary 8.3.13. *Any right ideal $\mathcal{I}$ of a right Noetherian semihereditary local ring A, viewed as a right A-module, has a unique maximal submodule which is $\mathcal{I}R$, where R is the Jacobson radical of A.*

Proof.
Let A be a right Noetherian semihereditary local ring with Jacobson radical R. Since A is a local ring, an ideal $\mathcal{I} \subset A$ as a right A-module contains a maximal submodule $\mathcal{I}R$. Suppose that $\mathcal{I}$ also contains a maximal right submodule $N \neq \mathcal{I}R$. Then $N+\mathcal{I}R = \mathcal{I}$. Since the ideal $\mathcal{I}$ is principal right ideal of A thanks to proposition 8.3.9, $N = \mathcal{I}$ by Nakayama's lemma. A contradiction. □

Proposition 8.3.14. *A right Noetherian semihereditary local ring A is a right uniserial ring.*

Proof.
Let A be a right Noetherian semihereditary local ring with Jacobson radical R. Suppose that A is not right serial. Then there exist two right ideals $\mathcal{I}_1$ such that $\mathcal{I}_2$ of A and $\mathcal{I}_1 + \mathcal{I}_2$ strongly contains $\mathcal{I}_1$ and $\mathcal{I}_2$. Then by corollary 8.3.13 $(\mathcal{I}_1 + \mathcal{I}_2)$ has the unique maximal submodule of $(\mathcal{I}_1 + \mathcal{I}_2)R$. Since it is maximal, it contains $\mathcal{I}_1$ and $\mathcal{I}_2$. So $(\mathcal{I}_1 + \mathcal{I}_2)R \supseteq \mathcal{I}_1 + \mathcal{I}_2$, a contradiction. □

Proposition 8.3.15. *A right Noetherian semiprime semihereditary local ring A is prime.*

Proof.
The proof follows immediately from propositions 8.3.12 and 8.3.14. □

Proposition 8.3.16. *Let A be a right Noetherian right hereditary (semihereditary) semiprime semiperfect ring A. Then for any non-zero idempotent $e^2 = e \in A$ the ring eAe is also a right Noetherian right hereditary (semihereditary) semiprime semiperfect ring.*

Proof.
By theorem 1.1.23 eAe is right Noetherian and by [146, proposition 9.2.13] eAe is semiprime. By proposition 4.4.6(4) (resp. theorem 8.2.16) eAe is right hereditary (resp. semihereditary). By [146, corollary 10.3.11] eAe is semiperfect. □

Theorem 8.3.17. *Every right Noetherian right hereditary (semihereditary) semiprime semiperfect ring A is a finite direct product of prime rings.*

Proof.
By the previous proposition for any primitive idempotent $e \in A$ a ring eAe is a right Noetherian right hereditary (semihereditary) semiprime semiperfect ring. Since A is a semiperfect ring, any primitive idempotent is local, and so eAe is a local ring. Therefore by proposition 8.3.15 eAe is a prime ring. Then from theorem 6.4.1 it follows that A is a direct product of prime rings. □

The next proposition shows that a right Noetherian semihereditary semiperfect prime ring A is a right serial ring.[5]

Proposition 8.3.18. *A right Noetherian semihereditary semiperfect prime ring A is a right serial ring.*

Proof.
Clearly one can assume that A is a basic ring. Let $A_A = P_1 \oplus P_2 \oplus \ldots \oplus P_n$ be a decomposition of the ring A into a direct sum of principal right A-modules and $1 = e_1 + e_2 + \cdots + e_n$ the corresponding decomposition of the identity of A into a sum of pairwise orthogonal local idempotents. Then for any $i = 1, \ldots, n$, $P_i = e_i A$. Assume that P_i is not uniserial. Then there exists a submodule $M \subseteq P_i$ which is a direct sum of submodules. Since A is right Noetherian and right semihereditary, M is a finitely generated projective right A-module. Since any finitely generated projective right module over a semiperfect ring has a unique up to isomorphism decomposition into a direct sum of principal right modules (see proposition 1.9.5), one can assume that $M = X \oplus Y \simeq P_j \oplus P_k$. Since A is a right semihereditary ring, there is a monomorphism $\varphi : P_j \to P_i$. Consider the element $\varphi(e_j) = e_i a_0$, where $a_0 \in A$. Since e_j is an idempotent and φ is a homomorphism of right A-modules, there results $\varphi(e_j^2) = \varphi(e_j)e_j = e_i a_0 e_j$. Therefore any element of $\operatorname{Im}\varphi$ has the form $\varphi(e_i a) = e_i a_0 e_j a$, where $a \in A$. So $X = \operatorname{Im}\varphi = e_i a_0 e_j A$. Analogously, $Y = e_i a_1 e_k A$. Therefore $M = e_i a_0 e_j A \oplus e_i a_1 e_k A$. Using the two-sided Peirce decomposition of the ring A there results that

$$e_i a_0 e_j A = (e_i a_0 e_j A e_1, e_i a_0 e_j A e_2, \ldots, e_i a_0 e_j A e_n)$$

and

$$e_i a_1 e_k A = (e_i a_1 e_k A e_1, e_i a_1 e_k A e_2, \ldots, e_i a_1 e_k A e_n).$$

By proposition 8.3.12, $e_i A e_i$ is a right principal ideal domain. Since $e_i a_0 e_j A e_i$ and $e_i a_0 e_k A e_i$ are right ideals in $e_i A e_i$, the direct sum $e_i a_0 e_j A e_i \oplus e_i a_0 e_k A e_i$ is also a right ideal and so it must be a principal right ideal. Therefore $e_i a_0 e_j A e_i = 0$ or $e_i a_1 e_k A e_i = 0$, which contradicts the primeness of A. □

Corollary 8.3.19. *Let A be a right Noetherian semihereditary semiperfect prime ring with Jacobson radical R. Then $P_i R$ is a principal right A-module for any principal right A-module P_i and $i = 1, \ldots, n$.*

Corollary 8.3.20. *Any right Noetherian semihereditary semiperfect semiprime ring is right serial.*

---

[5]Note that right serial rings were studied in [146, chapter 5].

8.4 Piecewise Domains. Right Hereditary Perfect Rings

Recall that a ring A is an FDI-ring if the identity of A can be decomposed into a finite sum of primitive pairwise orthogonal idempotents (see Section 6.3, in particular definition 6.3.1, and also [146, page 56]).

Definition 8.4.1. An indecomposable projective right A-module P of an FDI-ring A is called **principal** if $P \simeq e_i A$ for some primitive idempotent e_i, $i = 1, \ldots, n$.

Recall that a finite set of orthogonal idempotents $e_1, e_2, \ldots, e_m \in A$ is **complete** if

$$e_1 + e_2 + \ldots + e_m = 1 \in A$$

A ring A have enough idempotents if it has a (finite) complete set of orthogonal primitive idempotents. So that a ring A is an FDI-ring if and only if it has enough idempotents.

Definition 8.4.2. (R. Gordon, L. Small [118], see also [146, Section 10.7, p.259]). A ring A is called a **piecewise domain** (or simply **PWD**) with respect to a complete set of pairwise orthogonal idempotents $\{e_1, e_2, \ldots, e_n\}$ if $xy = 0$ implies $x = 0$ or $y = 0$ whenever $x \in e_i A e_j$ and $y \in e_j A e_k$ for $1 \leq i, j, k \leq n$.

Piecewise domains were first introduced and studied by R. Gordon and L. Small in [118].

Remark 8.4.3. Note that [146, proposition 10.7.8] gives an equivalent definition. A ring A is a PWD with respect to a complete set of pairwise orthogonal idempotents $\{e_1, e_2, \ldots, e_n\}$ if every non-zero homomorphism $e_i A \to e_j A$ is a monomorphism for $1 \leq i, j \leq n$.

Remark 8.4.4. Note also that the definition of PWD presupposes that the ring A has a (finite) complete set $\{e_1, e_2, \ldots, e_n\}$ of orthogonal primitive idempotents, since all $e_i A e_i$ for $i = 1, 2, \ldots, n$ are domains. So that any piecewise domain A is an FDI-ring.

Remark 8.4.5. Note that a piecewise domain can be considered as a generalization of an l-hereditary ring. Recall that a ring A is **l-hereditary** if it is Artinian and satisfies the following condition: given any pair of indecomposable projective left A-modules P and Q and given any A-homomorphism $\varphi : P \to Q$, either $\varphi = 0$ or φ is a monomorphism. There is the following equivalent condition: any local submodule[6] of a projective module is projective again. Thus the l-hereditary condition coincides with the "local" condition for projective modules. The notion of l-hereditary algebras was introduced by R.Bautista to study representations of incidence algebras of posets [17].

Important examples of piecewise domains are hereditary and semihereditary FDI-rings which shows the following proposition.

[6]Recall that a **local module** is a module with a unique maximal submodule.

Proposition 8.4.6. *Any right hereditary (semihereditary)* FDI*-ring is a piecewise domain.*

Proof. Let A be a right semihereditary FDI-ring and let $1 = e_1 + e_2 + \cdots + e_n$ be a decomposition of the identity $1 \in A$ into a sum of mutually orthogonal primitive idempotents. Write $P_i = e_i A$ for $i = 1, 2, \ldots, n$ and consider a non-zero homomorphism $\varphi_{ij} : P_i \longrightarrow P_j$ for $i, j \in \{1, 2, \ldots, n\}$. Since P_i, P_j are projective cyclic A-modules and A is a right semihereditary ring, $\mathrm{Im}(\varphi) \subseteq P_j$ is a projective A-module. Therefore an exact sequence $0 \longrightarrow \mathrm{Ker}(\varphi) \longrightarrow P_i \longrightarrow \mathrm{Im}(\varphi) \longrightarrow 0$ splits, i.e., $P_i \simeq \mathrm{Ker}(\varphi) \oplus \mathrm{Im}(\varphi)$. Since e_i is a primitive idempotent, P_i is indecomposable. So that $\mathrm{Ker}(\varphi) = 0$, i.e., φ is a monomorphism. Taking into account remark 8.4.3 this means that A is a piecewise domain. □

The next examples show that there exist a very wide class of rings which are piecewise domains but not hereditary rings.

Example 8.4.7. Let

$$T_2(\mathbf{Z}) = \begin{pmatrix} \mathbf{Z} & \mathbf{Z} \\ 0 & \mathbf{Z} \end{pmatrix},$$

where $\mathbf{Z}$ is the ring of integers. Then $A = T_2(\mathbf{Z})$ is a Noetherian ring. Since all non-zero $e_i A e_j$ are $\mathbf{Z}$ which are domains, A is a piecewise domain. We will show that A is not hereditary.

Suppose that A is right hereditary and apply proposition 4.7.3. Consider an ideal $(m) \subset \mathbf{Z}$ and a right $\mathbf{Z}$-module $\mathbf{Z}/(m)$. Suppose that $\mathbf{Z}/(m)$ is projective, and consider an exact sequence $0 \longrightarrow (m) \longrightarrow \mathbf{Z} \longrightarrow \mathbf{Z}/(m) \longrightarrow 0$. Then this sequence splits and so $\mathbf{Z} = (m) \oplus X$ with $X \simeq \mathbf{Z}/(m)$ which is impossible. Therefore $\mathbf{Z}/(m)$ is not projective, so that $T_2(\mathbf{Z})$ is not right hereditary by theorem 4.7.3. Analogously $T_2(\mathbf{Z})$ is not left hereditary.

Example 8.4.8.

Let $\mathcal{P} = \{p_1, p_2, \ldots, p_n\}$ be a partially ordered connected set with partial order $\leq$, and $A = M_n(\mathbf{Z})$ a ring of all square matrices of order n, where $\mathbf{Z}$ is the ring of integers. Denote by e_{ij} the matrix units of A $(i, j = 1, \ldots, n)$. Let $B_n(\mathbf{Z}, \mathcal{P})$ be a subring of A which consists of the matrices of the following form: $\mathbf{B} = (b_{ij}) \in A$, where $b_{ij} \in e_{ii} A e_{jj}$, and $b_{ij} = 0$ if $p_i \not\leq p_j$.

The ring $B_n(\mathbf{Z}, \mathcal{P})$ is a piecewise domain, but it is not a hereditary ring for $n \geq 2$, since there exists the idempotents $e = e_{ii} + e_{jj}$ (for $i \neq j$) such that the ring $eB_n(\mathbf{Z}, \mathcal{P})e \simeq T_2(\mathbf{Z})$ is not hereditary.

Example 8.4.9. (Small's example.) Let

$$A = \begin{pmatrix} \mathbf{Z} & \mathbf{Q} \\ 0 & \mathbf{Q} \end{pmatrix},$$

where $\mathbf{Z}$ is the ring of integers and $\mathbf{Q}$ is the field of rational numbers. Then A is a semihereditary FDI-ring, and so it is a piecewise domain by proposition 8.4.6. The ring A is right hereditary by theorem 4.7.3. However it is not left hereditary since $\mathbf{Q}$ is not projective as a left $\mathbf{Z}$-module.

Example 8.4.10. Let

$$A = \begin{pmatrix} \mathbf{Z} & \mathbf{Q} \\ 0 & \mathbf{Z} \end{pmatrix},$$

where $\mathbf{Z}$ is the ring of integers and $\mathbf{Q}$ is the field of rational numbers. Then A is a two-sided Noetherian ring. It is a piecewise domain since all non-zero e_iAe_j are domains. But A is neither a right nor left hereditary ring, which can be shown similarly as in Example 8.4.7.

Remark 8.4.11. In the paper [118] R. Gordon and L.W. Small posed the following question: "Can a PWD A possess a complete set $\{f_i\}_{i=1}^m$ of primitive orthogonal idempotents for which it is not true that $xy = 0$ implies $x = 0$ or $y = 0$ for some $x \in f_iAf_k$ and $y \in f_kAf_j$?" Moreover, they stated, "To avoid ambiguity, we sometimes say that A is a PWD with respect to $\{e_i\}_{i=1}^n$."

Let A be a semiperfect ring and $1 = e_1 + \ldots + e_n = f_1 + \ldots + f_m$ be two decompositions of $1 \in A$ into a sum of pairwise orthogonal local idempotents. Then, by [146, lemma 10.4.12], $m = n$ and there exists a permutation τ of numbers from 1 to n and an invertible element $a \in A$ such that $f_{\tau(i)} = ae_ia^{-1}$ for $i = 1,2,\ldots,n$.

Consider $f_{\tau(i)}A$ and e_iA for $i = 1,\ldots,n$. Since $f_{\tau(i)}A = ae_ia^{-1}A = ae_iA$, $f_{\tau(i)}A$ is isomorphic to e_iA.

Suppose that a semiperfect ring A is a piecewise domain with respect to $\{e_1,e_2,\ldots,e_n\}$, i.e., for any e_i and e_k each non-zero homomorphism $\varphi \in \mathrm{Hom}_A(e_iA, e_kA)$ is a monomorphism. Let $\psi : f_{\tau(i)}A \to f_{\tau(k)}A$ be a non-zero homomorphism. Then $f_{\tau(i)}A \simeq P_i = e_iA$, $f_{\tau(k)}A \simeq P_k = e_kA$ and ψ induces a non-zero homomorphism $P_i \to P_k$. Consequently, every non-zero homomorphism $f_{\tau(i)}A \to f_{\tau(k)}A$ is a monomorphism and a semiperfect ring A is a piecewise domain with respect to any decomposition of $1 \in A$ into a sum of mutually orthogonal local idempotents. Thereby it provides a positive answer to the "Question" of Gordon and Small for PWD semiperfect rings.

Proposition 8.4.12. *Any piecewise domain is nonsingular.*

Proof.
Suppose A is a piecewise domain with respect to a system of idempotents $\{e_1,e_2,\ldots,e_n\}$. Let $a \in \mathcal{Z}(A_A)$. Then $\mathrm{r.ann}_A(a)$ is essential in A. Hence for all i, there exists a non-zero element $b_i \in \mathrm{r.ann}_A(a) \cap e_iA$. Since A is a piecewise domain and $e_ib_i = b_i$, we obtain that $ae_i = 0$ for all i. Therefore $a = 0$ and so $\mathcal{Z}(A_A) = 0$. Analogously one can prove that $\mathcal{Z}({}_AA) = 0$. So A is nonsingular. □

Taking into account this proposition the next results follows immediately from corollary 8.2.2 and proposition 8.2.1.

Corollary 8.4.13. *Let A be a piecewise domain, and M a projective right A-module. Then M is finitely generated if and only if M is essentially finitely generated.*

Corollary 8.4.14. *Let A be a piecewise domain, and M a projective right A-module. Then M is finitely generated if and only if M has finite uniform dimension.*

Recall that a ring A with Jacobson radical R is called **semiprimary** if A/R is semisimple and R is nilpotent. It is obvious, that any semiprimary ring is perfect. The converse statement is not true in general. But it does hold in the following important case.

Theorem 8.4.15. *A right perfect piecewise domain is a semiprimary ring.*

Proof.
Let A be a right perfect piecewise domain. Since any one-sided perfect ring is semiperfect, the prime radical of A is nilpotent, by [147, corollary 4.9.3]. By [147, proposition 4.7.5] the prime radical of a one-sided perfect ring coincides with the Jacobson radical R of A. So R is nilpotent. Thus, A/R is semisimple and R is nilpotent, i.e., A is semiprimary. □

Since any perfect ring is a FDI-ring and any right hereditary FDI-ring is a piecewise domain by proposition 8.4.6, the next result immediately follows from theorem 8.4.15. It was proved by M. Teply in [303].

Theorem 8.4.16 (M. Teply). *A right perfect right hereditary ring is semiprimary.*

Any simple Artinian ring is obviously a perfect prime ring. It turns out that the inverse statement is also true.

Proposition 8.4.17. *A right perfect prime ring is a simple Artinian ring.*

Proof.
By [147, theorem 4.8.2] A is a left socular ring, i.e., every non-zero left A-module has a non-zero socle. Since any right perfect ring is semiperfect, it follows from [147, theorem 4.11.2] that A is a semisimple ring. Since A is prime, A is a simple Artinian ring. □

8.5 Primely Triangular Rings. Structure of Piecewise Domains

The main result of this section gives the structure of piecewise domains. First, here are the basic definitions.

Definition 8.5.1. Let A be a ring. A decomposition of the identity $1 = e_1 + e_2 + \cdots + e_n \in A$ into a sum of pairwise orthogonal idempotents is called **triangular** if $e_iAe_j = 0$ for all $i > j$. Such a decomposition is called **prime** if e_iAe_i is a prime ring for all $i = 1, \ldots, n$.

A ring A is called **triangular** if there exists a triangular decomposition of the identity of A. In this case the two-sided Peirce decomposition of a ring A can be represented in the following triangular form:

$$A = \begin{pmatrix} A_1 & A_{12} & \dots & A_{1,n-1} & A_{1n} \\ 0 & A_2 & \dots & A_{2,n-1} & A_{2n} \\ \vdots & \vdots & \ddots & \vdots & \vdots \\ 0 & 0 & \dots & A_{n-1} & A_{n-1,n} \\ 0 & 0 & \dots & 0 & A_n \end{pmatrix} \tag{8.5.2}$$

A ring A is said to be a **primely triangular ring** if it is triangular of the form (8.5.2) and e_iAe_i is a prime ring for all $i = 1,2,\dots,n$.

Remark 8.5.3. Note that the notion of "a triangular ring" was first introduced by S.U.Chase [42] in 1961 for semiprimary rings. According to S.U.Chase a semiprimary ring A with Jacobson radical R is triangular if there exists a complete set $e_1, e_2, \dots, e_k$ of mutually orthogonal primitive idempotents of A such that $e_iRe_j = 0$ for all $i > j$. This notion was used by L.W. Small in [292] for arbitrary right Noetherian rings. M. Harada [139] in 1966 introduced the notion of "generalized triangular matrix rings" for rings with a triangular decomposition of the identity where the $R_i = e_iAe_i$ are arbitrary rings. He studied the properties of such rings when all R_i are semiprimary rings. It is obvious that in this case the generalized triangular matrix rings are also semiprimary. Yu.A.Drozd [68] in 1980 used the term "triangular ring" for rings with a triangular prime decomposition of the identity.

Thus semiprime triangular rings as defined by S.U.Chase, and right Noetherian triangular rings as defined by Small are triangular rings in the sense of the definition given above. Triangular rings as defined by Yu.A. Drozd are primely triangular rings in the sense of the definition given above, which are semiprimary generalized triangular matrix rings as defined by M. Harada.

Theorem 6.4.10 and [146, proposition 9.2.13] immediately yield the next result:

Corollary 8.5.4. *Any semiprime primely triangular* FDI-*ring is isomorphic to a finite direct product of prime rings.*

Let A be a piecewise domain with respect to a complete set of orthogonal primitive idempotents $\{e_i\}_{i=1}^n$. Since A is a PWD, every non-zero homomorphism $f : e_iA \to e_jA$ is a monomorphism. Since $\mathrm{Hom}_A(e_iA, e_jA) \simeq e_jAe_i$, it is possible to introduce a relation $\sim$ on the set $\{1,2,\dots,n\}$, setting $i \sim j$ if and only if $e_iAe_j \neq 0$ and $e_jAe_i \neq 0$. In this case we write $e_iA \sim e_jA$ and the modules $P = e_iA$ and $Q = e_jA$ will be called **equivalent**. It follows directly from definition 8.4.2 that $\sim$ is an equivalence relation on $\{1,2,\dots,n\}$. Let $C_1, C_2, \dots, C_m$ be the equivalence classes on $\{1,2,\dots,n\}$ with regard to the relation $\sim$. Set $f_i = \sum\limits_{j \in C_i} e_j$, then the $f_1, f_2, \dots, f_m$ form a complete set of orthogonal idempotents in A.

On the set $\{1,2,\ldots,m\}$ one can introduce a relation $\preceq$, setting $i \preceq j$ if and only if $f_iAf_j \neq 0$. Then $\preceq$ is a partial ordering. Obviously, $\preceq$ is reflexive, since $f_iAf_i \neq 0$ for each $i = 1,2,\ldots,m$. Let $i \neq j$ and $i \preceq j$, i.e., $f_iAf_j \neq 0$. Then there exist $p \in C_i$ and $q \in C_j$ such that $e_pAe_q \neq 0$. If $s \sim p$ and $t \sim q$, then $0 \neq e_sAe_pAe_qAe_t \subseteq e_sAe_t$ because A is a PWD. So if $f_iAf_j \neq 0$, then $e_pAe_q \neq 0$ for each $p \in C_i$ and each $q \in C_j$. Suppose $f_jAf_i \neq 0$ as well. Then also $e_qAe_p \neq 0$ for each $p \in C_i$ and each $q \in C_j$. So $C_i = C_j$, a contradiction. This contradiction shows that $\preceq$ is antisymmetric. Since A is a PWD, $\preceq$ is transitive. Thus $\preceq$ is a partial ordering on $\{1,2,\ldots,m\}$ which can be extended to a total ordering on $\{1,2,\ldots,m\}$. Denote $Q_i = f_iA$, then using this total ordering we obtain a decomposition $A_A = \bigoplus_{i=1}^{m} Q_i$, where $f_iAf_j \simeq \mathrm{Hom}_A(Q_j,Q_i) = 0$ for $i > j$.

Lemma 8.5.5. *If A is a piecewise domain with respect to a complete set of orthogonal idempotents $\{e_i\}_{i=1}^n$, then the following conditions are equivalent:*

1. *A is prime.*
2. *$A_A = \bigoplus_{i=1}^{n} P_i$, where the $P_i = e_iA$ are equivalent principal modules, for $i = 1,2,\ldots,n$.*
3. *All the principal right A-modules e_iA are equivalent.*

Proof.

$1 \Longrightarrow 2$. Decompose as before the regular A-module A_A into a direct sum $A_A = \bigoplus_{i=1}^{m} Q_i$, where $Q_i = f_iA$ and $f_iAf_j \simeq \mathrm{Hom}_A(Q_j,Q_i) = 0$ for $i > j$, i.e., the corresponding two-sided Peirce decomposition of A form an upper triangular matrix. We now show that $m = 1$. Assume that $m > 1$ and $1 = f_1 + \cdots + f_m$ is the corresponding decomposition of the identity of A. Consider the ideal

$$\mathcal{I} = \begin{pmatrix} 0 & f_1Af_2 & \ldots & f_1Af_m \\ 0 & 0 & \cdots & f_2Af_m \\ \vdots & \vdots & \ddots & \vdots \\ 0 & 0 & \cdots & 0 \end{pmatrix} = \sum_{i<j} f_iAf_j,$$

which is obviously nilpotent. Since A is a prime ring, $\mathcal{I}$ must be zero. Suppose $\mathcal{I} = 0$, and $m > 1$. Then the $X_i = \mathrm{diag}(0 \ldots f_iAf_i \ldots 0)$ and $X_j = \mathrm{diag}(0 \ldots f_jAf_j \ldots 0)$ are ideals in A and $X_iX_j = 0$ for $i \neq j$, which contradicts the primeness of A. Therefore $m = 1$.

$2 \Longrightarrow 3$. Let P be a principal right A-module, i.e., $P \simeq e_iA$ for some $i = 1,\ldots,n$. So $P \sim P_i$.

$3 \Longrightarrow 1$. Let $\mathcal{I}$ and $\mathcal{J}$ be non-zero ideals in A. There exist principal modules $P = e_iA$ and $Q = e_jA$ such that $e_i\mathcal{J} \neq 0$, i.e., $P\mathcal{J} \neq 0$ and $\mathcal{I}e_j \neq 0$, i.e., $\mathrm{Hom}_A(e_jA,\mathcal{I}) \neq 0$. Since $P \sim Q$, then $\mathrm{Hom}_A(P,\mathcal{I}) \neq 0$. So, there is a non-zero monomorphism $f : P \to \mathcal{I}$, whence $\mathcal{I}\mathcal{J} \supset f(P)\mathcal{J} = f(P\mathcal{J}) \neq 0$. Therefore A is prime. □

Theorem 8.5.6 (R.Gordon-L.Small [118]). *A piecewise domain is a primely triangular ring.*

Proof.

Let A be a piecewise domain with respect to a complete set of pairwise orthogonal idempotents $\{e_i\}_{i=1}^n$. Then as before one finds a decomposition of the A-module A_A into a direct sum:

$$A_A = \bigoplus_{i=1}^{m} Q_i,$$

where $Q_i = f_iA$ and $\mathrm{Hom}_A(Q_j, Q_i) = 0$ for $i > j$. This is a triangular decomposition, and so it remains only to show that $f_iAf_i = A_i$ is prime for all $i = 1, \ldots, m$.

Since $f_i = \sum\limits_{p \in C_i} e_p$, where C_i is an equivalence class of $\{1, 2, \ldots, n\}$, $f_if_j = \delta_{ij}f_i$, where δ_{ij} is the Kronecker delta. Hence f_i is an identity in $A_i = f_iAf_i$, and by the definition of a piecewise domain it follows that a ring $A_i = f_iAf_i$ is a piecewise domain for all i. Consider the ring A_i for a given number i. Write $A_i = B$. Let $Q_i = \bigoplus\limits_{p \in C_i} e_pA$, where $e_jAe_k \neq 0$ and $e_kAe_j \neq 0$ for all $j, k \in C_i$. Then, setting $\overline{P}_k = e_kAf_i = P_kf_i$, we obtain a decomposition of the right regular B-module B_B into the direct sum of equivalent principal modules: $B_B = \bigoplus\limits_{s \in C_i} \overline{P}_s$, since $\mathrm{Hom}_B(\overline{P}_j, \overline{P}_k) \simeq \mathrm{Hom}_A(P_j, P_k) \simeq e_kAe_j \neq 0$, for $k, j \in C_i$. Then from lemma 8.5.5 it follows that $B = A_i$ is prime, for all $i = 1, 2, \ldots, m$. Therefore there is a triangular prime decomposition of the identity of A such that the A_i are prime for all i, i.e., A is a primely triangular ring as required. □

From this theorem, and taking into account proposition 8.4.6, there follow the corollaries obtained by Yu.A. Drozd [68] and L.W. Small [292]. These can be formulated as follows:

Corollary 8.5.7 (Yu.A. Drozd [68]). *A right hereditary (semihereditary) ring A is a primely triangular ring.*

Corollary 8.5.8 (L.W. Small). *If A is a right Noetherian right hereditary (semihereditary) ring then A is a primely triangular ring.*

Assume that $1 = e_1 + e_2 + \cdots + e_n$ is a decomposition of the identity of a piecewise domain A. If e is a sum of different idempotents from the set $\{e_1, e_2, \ldots, e_n\}$, then the ring eAe is again a piecewise domain. Therefore theorem 8.5.6 and theorem 6.4.15 yield the next fact.

Proposition 8.5.9 (R. Gordon and L. Small [118]). *Every semiprime piecewise domain is isomorphic to a finite direct product of prime piecewise domains.*

Recall that a ring A is called **weakly prime** if the product of any two non-zero ideals not contained in the Jacobson radical of A is non-zero. Such rings were studied in [147, Section 6.9]. Any weakly prime ring is indecomposable and any prime ring is weakly prime. Theorem 6.9.2 in [147] states that if $1 = e_1 + e_2 + \ldots + e_n$ is a decomposition of the identity of a semiperfect ring A into the sum of mutually orthogonal local idempotents and $A_{ij} = e_i A e_j$ $(i, j = 1, \ldots, n)$, then the ring A is weakly prime if and only if $A_{ij} \neq 0$ for all i, j. This theorem and theorem 8.5.6 imply the following corollary.

Corollary 8.5.10. *A weakly prime right semihereditary semiperfect ring is prime.*

There is well known theorem of Wedderburn-Mal'cev for separable subalgebras of a finite dimensional algebra A over a field K:

Theorem 8.5.11. (Wedderburn-Mal'cev theorem [70, Theorem 6.2.1], [**?**, Theorem 1]) *Let A be a finite dimensional algebra over a field K with Jacobson radical R. Suppose that $\overline{A} = A/R$ is a separable algebra. Then there is a subalgebra $B \subseteq A$ such that $A \simeq \overline{A}$ and $A = B \oplus R$ (as a direct sum of vector spaces). Moreover, if C is another subalgebra of A such that $C \simeq \overline{A}$ and $A = C \oplus R$ then B and C are unipotently conjugate, i.e., there always exist an element $r \in R$ such that $C = (1+r)B(1+r)^{-1}$.*

The following theorem can be considered as some analog of the Wedderburn-Mal'cev theorem for right semihereditary rings.

Proposition 8.5.12 (Yu.A.Drozd [68]). *Let A be a primely triangular* FDI*-ring with the prime radical N. Then N is nilpotent and A contains a subring B such that $A = B \oplus N$. Moreover, B is a direct product of prime rings, and if C is another subring of A such that $A = C \oplus N$, then B and C are unipotently conjugate, i.e., $C = u^{-1}Bu$, where $u \in 1 + N$. If A is right semihereditary (hereditary), so is B.*

Proof.
Let $1 = e_1 + \cdots + e_n$ be a triangular prime decomposition of the identity of A whose two-sided Peirce has the triangular form (8.5.2). Consider the ideal $\mathcal{I} = \sum_{i<j} e_i A e_j$, whose two-sided Peirce decomposition has the following triangular form:

$$\mathcal{I} = \begin{pmatrix} 0 & A_{12} & \ldots & A_{1,n-1} & A_{1n} \\ 0 & 0 & \ldots & A_{2,n-1} & A_{2n} \\ \vdots & \vdots & \ddots & \vdots & \vdots \\ 0 & 0 & \ldots & 0 & A_{n-1,n} \\ 0 & 0 & \ldots & 0 & 0 \end{pmatrix}. \tag{8.5.13}$$

Then obviously $\mathcal{I}^n = 0$ and $A/\mathcal{I} \simeq \prod_{i=1}^{n} A_i$ is a semiprime ring. So $\mathcal{I} = N$ and the subring $\bigoplus_{i=1}^{n} A_i$ can be taken for B. If the ring A is hereditary (resp., semihereditary), then B is hereditary, by proposition 8.3.6(4) (resp., semihereditary, by theorem 8.2.16). So it remains to check that B is unique up to unipotent conjugacy.

Let $C \subset A$ be another subring for which $A = C \oplus N$. Denote by f_i the image of e_i under the natural projection of A onto C. Then $C = \sum_{i=1}^{n} f_i A f_i$ and $f_i \sim e_i (\operatorname{mod} N)$. In particular, $f_1 = e_1 + x$, where $x \in N$. Since $Ne_1 = 0$, from the equality $f_1^2 = f_1$ it follows that $e_1 x = x$ and $x^2 = 0$. Thus $e_1(1+x) = f_1 = (1+x)f_1$, i.e., $f_1 = (1+x)^{-1}e_1(1+x)$. Replacing C by $(1+x)C(1+x)^{-1}$ one can assume that $f_1 = e_1$. Then $f_2 = e_2 + y$, where $y \in N$ and $e_1 y = 0$. Therefore also $ye_2 = 0$ as well, since $Ne_2 = e_1 N e_2$. Hence it can be similarly concluded that $f_2 = (1+y)^{-1}e_2(1+y)$, where $(1+y)^{-1}e_1(1+y) = e_1$. Continuing this process one finally finds a $u \in 1+N$ such that $f_i = u^{-1}e_i u$ for all i. Hence $C = \sum_{i=1}^{n} f_i A f_i = u^{-1}(\sum_{i=1}^{n} f_i A f_i)u = u^{-1}Bu$. □

Corollary 8.5.14. *If $1 = e_1 + \cdots + e_n = f_1 + \cdots + f_m$ are two triangular prime decompositions of the identity of an* FDI*-ring A, then they are unipotently conjugate, i.e., $m = n$ and the indices can be chosen so that $f_i = u^{-1}e_i u$ for all i, where $u \in 1+N$ (N being the prime radical of A).*

Proposition 8.5.15. (R. Gordon-L. Small [118]). *Each prime piecewise domain B has the following form:*

$$B = \begin{pmatrix} O_1 & \dots & \dots & O_{1t} \\ \vdots & O_2 & & \vdots \\ \vdots & & \ddots & \vdots \\ O_{t1} & \dots & \dots & O_t \end{pmatrix}, \tag{8.5.16}$$

where each O_j is a domain and each O_{jk} is isomorphic as a right O_k-module to a non-zero right ideal in O_k and as a left O_j-module to a non-zero left ideal in O_j.

Proof.

Let $1 = e_1 + \ldots + e_t$ be a decomposition of the identity of a prime piecewise domain B into a sum of pairwise orthogonal primitive idempotents. Let $O_i = e_i B e_i$ for $i = 1, \ldots, t$ and $O_{ij} = e_i B e_j$ for $i \neq j$ $(i, j = 1, \ldots, t)$. By the definition of the piecewise domain it follows that all the rings O_i are domains. Let $x \in O_{ij} = e_i B e_j$ be a non-zero element. Then $xe_i Be_j$ is a non-zero right ideal of $e_j B e_j = O_j$. Therefore $O_{ij} = e_i B e_j$ is isomorphic as a right O_j-module to a non-zero right ideal in O_j. The left case is proved analogously. □

Theorem 8.5.17. (R. Gordon-L. Small [118]). *If A is a piecewise domain, then there is a decomposition $1 = f_1 + \ldots + f_r$ of A into a sum of pairwise orthogonal idempotents such that the two-sided Peirce decomposition of a ring A can be* *represented in the following form*

$$A = \begin{pmatrix} B_1 & B_{12} & \dots & \dots & B_{1r} \\ 0 & B_2 & \ddots & & \vdots \\ \vdots & \ddots & \ddots & \ddots & \vdots \\ \vdots & & \ddots & \ddots & B_{r-1,r} \\ 0 & \dots & \dots & 0 & B_r \end{pmatrix}, \tag{8.5.18}$$

where each $B_i = f_iAf_i$ is a prime piecewise domain and each $B_{ij} = f_iAf_j$ is a B_i-B_j-bimodule for $i, j = 1, 2, \dots, r$. Each B_i satisfies the conditions of proposition 8.5.15. *Moreover, the integer r is uniquely determined by A.*

Proof.
The proof follows from theorem 8.5.6 and proposition 8.5.12. □

Corollary 8.5.19. *The prime radical $Pr(A) = N$ of a piecewise domain A is of the form:*

$$N = Pr(A) = \begin{pmatrix} 0 & B_{12} & B_{13} & \dots & \dots & B_{1r} \\ 0 & 0 & B_{23} & \ddots & & B_{2r} \\ \vdots & \ddots & \ddots & \ddots & \ddots & \vdots \\ \vdots & & \ddots & \ddots & B_{r-2,r-1} & B_{r-2,r} \\ \vdots & & & \ddots & 0 & B_{r-1,r} \\ 0 & \dots & \dots & \dots & 0 & 0 \end{pmatrix}. \tag{8.5.20}$$

Corollary 8.5.21. *The prime radical of a piecewise domain is nilpotent.*

Proof. The proof is obvious. □

Recall some useful definitions.

A ring A is called a **finitely decomposable ring** (or shortly, **FD-ring**), if it can be expressed as a direct product of a finite number of indecomposable rings.

Let $Pr(A)$ be the prime radical of a ring A. The quotient ring $\overline{A} = A/Pr(A)$ is called the **diagonal** of A. A ring A is called a **ring with finitely decomposable diagonal** (or simply, an **FDD-ring**) if $\overline{A}$ is an FD-ring.

From theorem 8.5.17, corollary 8.5.19, and corollary 8.5.21 there immediately follow the next two corollaries.

Corollary 8.5.22. *The diagonal $\overline{A} = A/Pr(A)$ of a piecewise domain A is a finite direct product $\overline{A} = B_1 \times B_2 \times \ldots \times B_r$ of prime piecewise domains B_i, $i = 1,2,\ldots,r$. Therefore $\overline{A}$ is an FD-ring, and A is an FDD-ring.*

Corollary 8.5.23. *If A is a piecewise domain then it decomposes as a direct sum of Abelian groups $A = A_0 \oplus N$ where $N = Pr(A)$ and the ring $A_0 \simeq \overline{A} = A/N$.*

Note that this not necessarily a decomposition as a direct sum of rings, i.e., the multiplication is not necessarily componentwise.

Since, by corollary 8.5.22, any piecewise domain A is an FDD-ring, one can construct the prime quiver $PQ(A)$ of A (see [146, Section 11.7]). Since the prime radical of a piecewise domain A is nilpotent, and so T-nilpotent, there immediately follows from [146, theorem 11.7.3] the proposition.

Proposition 8.5.24. *Let A be a piecewise domain. Then the prime quiver $PQ(A)$ is connected if and only if the ring A is indecomposable.*

Recall that a quiver without multiple arrows and multiple loops is called **simply laced**. The next statement immediately follows from theorem 8.5.17, corollary 8.5.19 and proposition 8.5.24.

Proposition 8.5.25. *Let A be a piecewise domain. Then the prime quiver $PQ(A)$ is an acyclic simply laced quiver.*

Lemma 8.5.26. *Let A be a primely triangular FDI-ring with prime radical N. Then $C_A(0) = C_A(N)$, where $C_A(0)$ is the set of regular elements of A, and $C_A(N)$ is the set of elements of A whose images are regular elements in A/N.*

Proof.
Let $1 = e_1 + e_2 + \cdots + e_n$ be a triangular prime decomposition of the identity of A. Then the two-sided Peirce decomposition of A has the form (8.5.2) and the prime radical N of A has the form (8.5.20). This lemma will be proved by induction on the number n. Suppose $n = 2$. Let

$$r = \begin{pmatrix} r_1 & x \\ 0 & r_2 \end{pmatrix} \tag{8.5.27}$$

be a non-zero regular element of A. It follows that r_i is a non-zero regular elements of $A_i = e_iAe_i$, $i = 1,2$. Indeed, suppose r_1 is not a non-zero regular element of A_1. Then there exists an element $0 \neq a_1 \in A_1$ such that $r_1a_1 = 0$ and therefore $ra = 0$, where $a = \begin{pmatrix} a_1 & 0 \\ 0 & 0 \end{pmatrix} \neq 0$. Suppose that r_2 is not a non-zero regular element of A_2. Then there is $0 \neq a_2 \in A_2$ such that $a_2r_2 = 0$ and therefore $ar = 0$, where $a = \begin{pmatrix} 0 & 0 \\ 0 & a_2 \end{pmatrix} \neq 0$. Therefore any regular element $r \in A$ has the form

$$r = \begin{pmatrix} r_1 & 0 \\ 0 & r_2 \end{pmatrix} + \begin{pmatrix} 0 & x \\ 0 & 0 \end{pmatrix} = r' + x', \tag{8.5.28}$$

where r_i is a non-zero regular element of A_i $(i = 1,2)$, $x' \in N$. Conversely, any element of form (8.5.28) is a non-zero regular elements of A, so $C_A(0) = C_A(N)$.

Suppose $n > 2$. Let $f = 1 - e_1$. Then

$$A = \begin{pmatrix} A_1 & X \\ 0 & B \end{pmatrix},$$

where $A_1 = e_1Ae_1$, $B = fAf$. By induction hypothesis any regular element of B has the form $s + y$, where $s = \mathrm{diag}(r_2,r_3,\dots,r_n)$, and all r_i are non-zero regular elements of A_i and $y \in Pr(B)$. Let $r = \begin{pmatrix} r_1 & x \\ 0 & s \end{pmatrix}$ be a non-zero regular element of A. Similarly r_1 is a non-zero regular element of A_1. Otherwise there is a non-zero element $a_1 \in A_1$ such that $r_1a_1 = 0$, and in this case $ra = 0$, where $a = \begin{pmatrix} a_1 & 0 \\ 0 & 0 \end{pmatrix} \neq 0$. Thus any regular element of A has the form

$$r = \mathrm{diag}(r_1,r_2,\dots,r_n) + x, \tag{8.5.29}$$

where r_i is a non-zero regular element of A_i $(i = 1,2,\dots,n)$, and $x \in N$. So $C_A(0) = C_A(N)$. □

From proposition 8.5.12 and this lemma there follows the following result.

Proposition 8.5.30. *Let A be a primely triangular FDI-ring with the prime radical N. Then*

1. *N is nilpotent.*
2. *A/N is a finite direct sum of prime rings.*
3. *$C_A(0) = C_A(N)$.*

Theorem 8.5.31. *Let A be a primely triangular* FDI-*ring with the prime radical N, such that $\rho_N(A_A) < \infty$ and each $A_i = e_iAe_i$ is a right Goldie ring, where $1 = e_1 + e_2 + \dots + e_n$ is a triangular prime decomposition of the identity of A. Then A has an Artinian classical ring of fractions $\tilde{A}$.*

Proof.
The proof follows immediately from corollary 8.5.30 and theorem 5.5.28. □

8.6 Right Hereditary Triangular Rings

In this section we give a criterion for an FDI-ring with a triangular decomposition of the identity to be right hereditary or right semihereditary. Let A be an FDI-ring and $1 = e_1 + \dots + e_n$ some fixed triangular decomposition of the identity of A. Set $A_{ij} = e_iAe_j$, $A_i = A_{ii}$ and $Q_i = e_iA$. Since $A_i \simeq \mathrm{Hom}_A(Q_i,Q_i)$, it is natural to consider Q_i as a left A_i-module.

Lemma 8.6.1. *The following conditions are equivalent for a module P over an FDI-ring A:*

1. *The right A-module P is projective.*
2. $P \simeq \bigoplus_{i=1}^{n}(P_i \otimes_{A_i} Q_i)$, *where P_i is a projective right A_i-module.*
3. *The right A_i-module $\overline{P}_i = Pe_i / \sum_{k<i} PA_{ki}$ is projective and the homomorphisms*

$$\xi_{ij} : \overline{P}_i \otimes_{A_i} A_{ij} \to Pe_j / \sum_{k<i} PA_{kj} \tag{8.6.2}$$

induced by the multiplication in A are monomorphisms for each $i, j = 1, 2, \ldots, n$.

Proof.

$2 \Longleftrightarrow 3$. If P_i is a projective A_i-module, then $P_i \otimes_{A_i} Q_i \simeq \bigoplus_{j=i}^{n}(P_i \otimes_{A_i} A_{ij})$, and the action of the operators from A is determined by the ring multiplication $A_{ik} \otimes_{A_k} A_{kj} \to A_{ij}$. This immediately implies the equivalence of conditions (2) and (3).

$2 \Longrightarrow 1$ follows from proposition 1.5.2(1) and [147, proposition 2.2.3].

$1 \Longrightarrow 2$. Set $N = \sum_{i<j} A_{ij}$ and $\overline{A} = A/N \simeq \prod_{i=1}^{n} A_i$. If P is projective, so is the $\overline{A}$-module $\overline{P} = P/PN \simeq \bigoplus_{i=1}^{n} \overline{P}_i$, i.e., all the A_i-modules $\overline{P}_i$ are projective. Write $Q = \bigoplus_{i=1}^{n}(\overline{P}_i \otimes_{A_i} Q_i)$. This is a projective A-module satisfying $Qe_k = \bigoplus_{i=1}^{n}(\overline{P}_i \otimes_{A_i} A_{ik})$, whence $Q/QN \simeq \overline{P} = P/PN$. Since N is a nilpotent ideal, this implies $Q \simeq P$ by [146, corollary 10.5.2]. □

Corollary 8.6.3. *Let A be an FDI-ring. Then every principal A-module is isomorphic to a module of the form $P \otimes_{A_i} Q_i$, where P is a principal A_i module.*

The following theorem gives a criterion for triangular rings to be right hereditary.

Theorem 8.6.4 (Yu.A. Drozd [68]). *Let $1 = e_1 + \cdots + e_n$ be a triangular decomposition of the identity of an FDI-ring A and let $\overline{A}_{ij} = A_{ij} / \sum_{i<k<j} A_{ik}A_{kj}$. For any right ideal $\mathcal{I} \subset A_i$, denote by $\mu^{\mathcal{I}}_{ikj}$ $(i < k < j)$ the homomorphism*

$$(\overline{A}_{ik}/\mathcal{I}\overline{A}_{ik}) \otimes_{A_k} A_{kj} \to A_{ij}/(\mathcal{I}A_{ij} + \sum_{i<s<k} A_{is}A_{sj}), \tag{8.6.5}$$

induced by the multiplication in A. The ring A is right hereditary if and only if the following conditions are satisfied:

1. *All the A_i are right hereditary FDI-rings;*
2. *The A_{ij} are flat left A_i-modules;*
3. *For any right ideal $\mathcal{I} \subset A_i$, the right A_j-modules $\overline{A}_{ij}/\mathcal{I}\overline{A}_{ij}$ are projective;*
4. *All the $\mu^{\mathcal{I}}_{ikj}$ are monomorphisms.*

Proof.
The proof will be given by induction on n.

If $n = 2$ then we have exactly theorem 4.7.3, while condition 4 is vacuous in this case.

In the general case, set $e = e_1$ and $f = e_2 + \cdots + e_n$, $B = fAf$ and $V = eAf$. Then $1 = e + f$ is a triangular decomposition of the identity of A and e is an identity for B. In view of the case $n = 2$, A is right hereditary if and only if A_1 and B are right hereditary, V is a flat A_1-module, and for any right ideal $\mathcal{I} \subset A_1$ the right B-module $V/\mathcal{I}V$ is projective. But $V = \bigoplus\limits_{i=2}^{n} A_{1i}$, so the flatness of V is equivalent to the flatness of all A_{1i}. By the induction hypothesis, one can assume that the hereditary property for B is equivalent to the fulfillment of conditions 1-4 for $i > 1$. Finally, the projectivity of $V/\mathcal{I}V$, by lemma 8.6.1, is equivalent to the fulfillment of conditions 3 and 4 for $i = 1$, which completes the proof of the theorem. □

Literally in the same way, a criterion for triangular rings to be right semihereditary can be proved. It is given by the following theorem:

Theorem 8.6.6 (Yu.A.Drozd). *In the notation of theorem* 8.6.4, *an FDI-ring A is right semihereditary if and only if the following conditions are satisfied:*

a. *All the A_i are right semihereditary FDI-rings;*
b. *The A_{ij} are flat left A_i-modules;*
c. *For any finitely generated right ideal $\mathcal{I} \subset A_i$, each finitely generated right A_j-submodule in $\overline{A}_{ij}/\mathcal{I}\overline{A}_{ij}$ is projective;*
d. *For any finitely generated right ideal $\mathcal{I} \subset A$ all the $\mu^{\mathcal{I}}_{ikj}$ are monomorphisms.*

8.7 Noetherian Hereditary Primely Triangular Rings

In this section theorems 8.6.4 and 8.6.6 will be applied for different concrete classes of FDI-rings. The notations of Section 8.6 will be used.

Theorem 8.7.1. *A primely triangular ring A is perfect and hereditary if and only if all the A_i are simple Artinian rings and all the homomorphisms*

$$\mu^0_{ikj} : \overline{A}_{ik} \otimes_{A_k} A_{kj} \to A_{ij}/ \sum_{i<s<k} A_{is}A_{sj} \tag{8.7.2}$$

are monomorphic.

Proof.

Let A be a primely triangular perfect and hereditary ring. Then $A_i = e_i A e_i$ is a perfect prime ring for any idempotent e_i, and therefore it is simple Artinian, by proposition 8.4.17. So equality (8.7.2) immediately follows from (8.6.5).

Conversely, since each A_i is a simple Artinian ring, conditions 1-3 of theorem 8.6.4 are satisfied automatically. Furthermore, every right ideal in A_i is a direct summand of A_i, so it is enough to check condition 4 for $\mathcal{I} = 0$. But in this case all conditions of theorem 8.6.4 are satisfied and so A is right hereditary. Since in this case $R = \text{rad}A$ is equal to the prime radical, A/R is semisimple and R nilpotent by proposition 8.5.11. So A is a perfect ring. Then from corollary 4.4.13 it follows that it is left hereditary as well. □

Remark 8.7.3. Since any semiprimary ring is automatically perfect, this theorem can be considered as insignificant generalization of the criterion which was obtained by M.Harada in [139] for semiprimary generalized triangular matrix rings.

Theorem 8.7.4. *A right Noetherian primely triangular ring A is right hereditary if and only if the following conditions satisfy:*

a. *All the A_i are right hereditary rings.*
b. $A_{ij} \simeq \tilde{A}_i \otimes_{A_i} A_{ij}$, *where $\tilde{A}_i$ is the right classical ring of quotients of A_i (it exists and is a simple Artinian ring, by Goldie's theorem).*
c. *The $\overline{A}_{ij}$ are projective right A_j-modules.*
d. *All the μ^0_{ikj} are monomorphisms.*

Proof. Let A be a right Noetherian right hereditary primely triangular ring. Then A_i is a right Noetherian right hereditary prime ring for any i. Conditions c, d follow immediately from theorem 8.6.4 for $\mathcal{I} = 0$. So it remains only to verify condition b.

Let $a \in A_i$ be a regular element, and $\mathcal{I} = aA_i$ a right ideal in A_i. Then, by theorem 8.6.4, A_{ij} is a flat left A_i-module and $\mathcal{I} \otimes_{A_i} A_{ij} \to A_{ij}$ is a monomorphism. By the flatness test 1.5.15, $\mathcal{I} \otimes_{A_i} A_{ij} \simeq \mathcal{I} A_{ij} = aA_{ij}$. Therefore multiplication by a on the left is a monomorphism of A_{ij} into itself.

Consider the exact sequence

$$0 \to aA_{ij} \to A_{ij} \to A_{ij}/aA_{ij} \to 0,$$

which splits because A_{ij}/aA_{ij} is a projective right A_i-module by lemma 8.6.1. So $A_{ij} \simeq aA_{ij} \oplus (A_{ij}/aA_{ij})$. Hence, since $A_{ij} \simeq aA_{ij}$ as a right A_j-module, and A is a right Noetherian ring, we obtain that $aA_{ij} = A_{ij}$, i.e., multiplication by a on the left is an automorphism of A_{ij}. But this means precisely that $\tilde{A}_i \otimes_{A_i} A_{ij} \simeq A_{ij}$.

Conversely, suppose all conditions of the theorem are satisfied. Then, since $\tilde{A}_i$ is a flat left A_i-module, conditions 1 and 2 of theorem 8.6.4 are also satisfied. Moreover, $\mathcal{I} A_{ij} = (\mathcal{I} \tilde{A}_i) A_{ij}$ and $\mathcal{I} \tilde{A}_i$ is a direct summand of $\tilde{A}_i$, so it suffices to check conditions 3 and 4 of theorem 8.6.4 for $\mathcal{I} = 0$. But in this case they coincide with conditions c and d of the theorem. □

Theorem 8.7.5. *A right Noetherian and hereditary ring is isomorphic to a ring of triangular matrices of the form* $\begin{pmatrix} A & V \\ 0 & B \end{pmatrix}$*, where A is an Artinian hereditary ring and B is a finite direct product of right Noetherian hereditary prime rings.*

Proof.

In view of corollary 8.5.7, it is enough to check that if A_1 and A_2 are prime Noetherian rings and M is a non-zero A_1-A_2-bimodule, which is Noetherian as a right A_2-module, so that the ring of triangular matrices

$$T = \begin{pmatrix} A_1 & M \\ 0 & A_2 \end{pmatrix}$$

is two-sided hereditary, then A_1 is Artinian.

By theorem 8.7.4, one can consider M as a left module over the simple Artinian ring $\tilde{A}_1$ (the right classical ring of quotients of A_1). If T is right hereditary, then M, and thus $\tilde{A}_1$, too, are projective left A_1-modules. Then by the Kaplansky theorem 1.5.5 there exist homomorphisms of left A_1-modules $f_\alpha : \tilde{A}_1 \to A_1$ and elements $x_\alpha \in \tilde{A}_1$ such that $x = \sum\limits_\alpha f_\alpha(x)x_\alpha$ for each $x \in \tilde{A}_1$ and only finitely many of the $f_\alpha(x)$ are not zero. It is easy to see that each f_α, considered as a mapping $\tilde{A}_1 \to \tilde{A}_1$, is a homomorphism of left $\tilde{A}_1$-modules, hence it is generated by multiplication on the right by the element $y_\alpha = f_\alpha(1)$. Here $\tilde{A}_1 y_\alpha \subset A_1$, and only finitely many y_α are non-zero. Then $\{x_\alpha : y_\alpha \neq 0\}$ is a finite system of generators of the left A_1-module $\tilde{A}_1$. By the Ore condition, there is a non-zero divisor $a \in A_1$ such that $x_\alpha a \in A_1$ for all α such that $y_\alpha \neq 0$, whence $\tilde{A}_1 a \subset A_1$; in particular $a^{-1} = a^{-2}a \in A_1$, and $A_1 = \tilde{A}_1$ is a simple Artinian ring. □

Taking into account that for a Noetherian ring the property to be right hereditary implies left hereditary, as a corollary of this theorem we obtain the well-known result of Chatters:

Theorem 8.7.6 (A.W. Chatters). *A Noetherian hereditary ring can be decomposed into a finite direct product of rings each of which is either an Artinian hereditary ring or a prime Noetherian hereditary ring.*

8.8 Right Hereditary Species and Tensor Algebras

The construction of species as introduced by P.Gabriel was considered in [147, Section 2.7]. This construction will be generalized in this section. These species will be used for describing a wide class of hereditary rings.

A **species** is a finite collection $S = (A_i, V_{ij})$, $(i,j = 1,\ldots,n)$, where the A_i are prime rings and the V_{ij} are A_i-A_j-bimodules. Set $B = \prod_{i=1}^{n} A_i$ and $V = \bigoplus_{i,j=1}^{n} V_{ij}$. Then V is a bimodule over the ring B and one can consider the tensor algebra $\mathfrak{T}_B(V) = \bigoplus_{i=0}^{\infty} T_i$, where $T_0 = B$ and $T_{i+1} = T_i \otimes_B V$, which is a tensor algebra of the B-bimodule V (see [147, section 2.2]). $\mathfrak{T}_B(V)$ is called the **tensor algebra** of the species S and denoted by $\mathfrak{T}(S)$. The **quiver** $\Gamma(S)$ of the species S is defined as a directed graph whose vertices are indexed by the numbers $i = 1,\ldots,n$, and there is an arrow from the vertex i to the vertex j if and only if $V_{ij} \neq 0$. A species is called **acyclic** if the quiver of it has no oriented cycles, i.e., the indices can be chosen so that $V_{ij} = 0$ for $i \leqslant j$. In this case, $\mathfrak{T}(S)$ is clearly a triangular ring, and $\mathcal{J} = \bigoplus_{i=1}^{\infty} T_i$ is the fundamental ideal of $\mathfrak{T}(S)$ which is obviously the prime radical of it.

On the other side, any primely triangular ring A can be assigned an acyclic species $S = S(A)$ by setting $A_i = e_i A e_i$ and $V_{ij} = A_{ij} / \sum_{i<k<j} A_{ik} A_{kj}$, where $A_{ij} = e_i A e_j$ for some triangular prime decomposition of the identity $1 = e_1 + \cdots + e_n$. By corollary 8.5.14, this species does not depend on the choice of such a decomposition. This species will be called the **species of a primely triangular ring** A and the quiver of it will be called the **quiver of a primely triangular ring** A.

Now let N be the prime radical of a primely triangular ring A, and let $\overline{A}$ be a subring in A such that $A = \overline{A} \oplus N$. Existence is guaranteed by proposition 8.5.12. A ring A is called **split** if there is an $\overline{A}$-submodule $\overline{N}$ in N such that $N = \overline{N} \oplus N^2$. By proposition 8.5.12, the splitting does not depend on the choice of a subring $\overline{A}$ because any two subrings with this property are unipotently conjugate. It is easy to see that A is split if and only if A_{ij} contains an A_i-A_j-bimodule $\overline{A}_{ij}$ such that $A_{ij} = \overline{A}_{ij} \oplus \sum_{i<k<j} A_{ik} A_{kj}$. In this case $\overline{N} = \sum_{i,j} \overline{A}_{ij}$. The tensor algebra of an acyclic species serves as an example of a split primely triangular ring. The following result shows that this example is, in a certain sense, universal.

Proposition 8.8.1. *Every split primely triangular ring A is isomorphic to a quotient ring $\mathfrak{T}(S)/\mathcal{I}$, where $S = S(A)$ is the species of the ring A and $\mathcal{I}$ is an ideal contained in $\mathcal{J}^2$.*

Proof.

Choose $\overline{A} \subset A$ so that $A = \overline{A} \oplus N$ and an $\overline{A}$-subbimodule $\overline{N} \subset N$ so that $N = \overline{N} \oplus N^2$. Then, if $S(A) = (A_i, V_{ij})$, set $B = \prod_{i=1}^{n} A_i$, and $V = \bigoplus_{i,j=1}^{n} V_{ij}$. It is easy to see that $B \simeq \overline{A}$ and $V \simeq \overline{N}$ as a B-bimodule. These isomorphisms are uniquely extended to a ring homomorphism $\varphi : \mathfrak{T}(S) = \mathfrak{T}_B(V) \to A$. Since N is nilpotent and $\overline{N}$ generates N, it follows that φ is an epimorphism and, since the restrictions of φ to $T_0 = B$ and $T_1 = V$ are monomorphisms, $\mathrm{Ker}\varphi = \mathcal{I} \subset \mathcal{J}^2$, as desired. □

Note one useful case where the splitting of a primely triangular ring can be seen from its quiver. A **circuit** in a graph is defined as an (non-oriented) path of length greater than 1 with the same beginning and end.

Proposition 8.8.2. *If there are no circuits in the quiver $\Gamma(A) = \Gamma(S(A))$, then the primely triangular ring A is split.*

Proof.
Obviously, if the quiver $\Gamma(A)$ has no paths beginning at i and ending at j, then $A_{ij} = 0$, and if i and j are not joined with an arrow, then $A_{ij} \subset N^2$. Suppose there is an arrow beginning at i and ending at j in the quiver $\Gamma(A)$. Since it has no circuits, for each point k either $A_{ik} = 0$ or $A_{kj} = 0$, whence $A_{ij} \cap N^2 = 0$, and as $\overline{N}$ one can take $\sum A_{ij}$, where the sum is taken over all arrows of the graph $\Gamma(A)$. □

A species is called **hereditary** (resp. **semihereditary**) if the tensor algebra of it is hereditary (resp. semihereditary). From theorem 8.6.4 one easily derives a criterion for an acyclic species to be hereditary.

Proposition 8.8.3. *An acyclic species $S = (A_i, V_{ij})$ is hereditary if and only if all the A_i are hereditary rings, the V_{ij} are flat left A_i-modules, and for any right ideal $\mathcal{I} \subset A_i$ the quotient modules $V_{ij}/\mathcal{I}V_{ij}$ are projective as right A_j-modules.*

Proof.
Suppose $V_{ij} = 0$ for $i \leq j$. Denote by e_i the identity of the ring A_i. Then $1 = e_1 + \cdots + e_n$ is a triangular prime decomposition of the identity of the ring $A = \mathfrak{T}(S)$, and

$$A_{ij} = e_i A e_j = \bigoplus_{(k_1,\ldots,k_s)} V_{ik_1} \otimes_{A_{k_1}} \cdots \otimes_{A_{k_s}} V_{k_s j},$$

where $A_i = e_i A e_i$. In particular, $\overline{A}_{ij} = V_{ij}$, so the conditions of the proposition are equivalent to conditions (a)-(c) of theorem 8.6.4 when applied to the ring $\mathfrak{T}(S)$. Moreover, the homomorphisms

$$\mu^0_{ikj} : V_{ik} \otimes_{A_k} A_{kj} \to A_{ij} / \sum_{i<t<j} A_{it} A_{tj}$$

are split bimodule monomorphisms, and therefore the homomorphisms $\mu^{\mathcal{I}}_{ikj}$ are also monomorphisms because $\mu^{\mathcal{I}}_{ikj} = \varepsilon \otimes \mu^0_{ikj}$, where ε is the identity homomorphism of the A_i-module $A_i/\mathcal{I}$. Thus, condition (d) of theorem 8.6.4 is always satisfied, and the proof of the theorem is complete. □

Similarly, theorem 8.6.6 implies a criterion for a species to be semihereditary.

Proposition 8.8.4. *An acyclic species $S = (A_i, V_{ij})$ is semihereditary if and only if all the A_i are semihereditary rings, the V_{ij} are flat left A_i-modules, and for any finitely generated right ideal $\mathcal{I} \subset A_i$ each finitely generated A_j-submodule of $V_{ij}/\mathcal{I}V_{ij}$ is projective.*

Propositions 8.8.3 and 8.8.4 immediately imply the next assertion:

Corollary 8.8.5. *The species of a hereditary (semihereditary) ring is always hereditary (semihereditary).*

Here is a summary of the previous observations.

Theorem 8.8.6. *Split hereditary (semihereditary) rings are, precisely, the tensor algebras of hereditary (semihereditary) acyclic species.*

Proof.
Obviously, the only thing requiring verification is that if a split primely triangular ring A is semihereditary, then the epimorphism $\varphi : T(S) \to A$, where $S = S(A)$, constructed in the proof of proposition 8.8.1, is a monomorphism (i.e., condition (d) of theorem 8.6.4 and 8.6.6 for $\mathcal{I} = 0$). □

Note that from the above proof there result the following corollaries.

Corollary 8.8.7. *A split primely triangular ring is hereditary (resp. semihereditary) if and only if the conditions* (a)-(c) *of theorem* 8.6.4 (*resp. theorem* 8.6.6) *are satisfied and all the homomorphisms μ_{ikj}^0 are monomorphisms.*

Corollary 8.8.8. *Split semihereditary rings A and B with prime radicals N and M, respectively, are isomorphic if and only if $A/N^2 \simeq B/M^2$.*

This follows from the fact that $S(A) = S(A/N^2)$.

From proposition 8.8.2 there also follows the result:

Corollary 8.8.9. *If the quiver of a semihereditary ring A contains no circuits, then A is isomorphic to the tensor algebra of its species.*

For finite dimensional algebras over a field k, the notion of species coincides with the notion of k-species introduced by P.Gabriel. If a field k is perfect then, since each semisimple k-algebra is then separable, each primely triangular algebra is split, and theorem 8.8.6 becomes a well-known result.

Corollary 8.8.10. *Finite dimensional hereditary algebras over a perfect field k are precisely the tensor algebras of acyclic k-species.*

Finally note another useful corollary.

Corollary 8.8.11. *Let A be a hereditary (semihereditary) ring with prime radical N, and let B be the graded ring associated to the N-adic filtration of A. Then B is also hereditary (semihereditary). If A is split, then $A \simeq B$.*

This follows from the fact that B is always split, $S(A) = S(B)$, and the homomorphisms μ^0_{ikj} for A and B are monomorphic simultaneously.

8.9 Notes and References

The various criteria for a non-Noetherian ring to be a right order in a right Artinian ring were obtained L.Small in [293] and Robson in [273]. Theorem 8.1.9, which states that a right Noetherian right hereditary ring is a right order in a right Artinian ring, was proved by L.Small in [291]. In this paper L. Small also noted that this theorem remains true for right Noetherian rings all of whose principal right ideals are projective, i.e., for right Noetherian right Rickart rings.

Some properties of hereditary rings were given by S. Eilenberg, J.P. Jans, H. Nagao and T. Nakayama in [75] and [163].

The notion of a triangular ring was first introduced by S.U. Chase [42], where he studied the properties of semiprimary triangular rings.

The definition of a generalized triangular matrix ring was introduced by M. Harada in [139], where he studied the main properties of such rings and proved that a hereditary semiprimary ring is isomorphic to a generalized triangular matrix ring with simple Artinian blocks along the main diagonal.

L.W. Small in [292] extended the idea of a triangular ring to right Noetherian rings and showed that any right Noetherian and right hereditary ring is triangular.

The structure theorem for semiperfect hereditary Noetherian rings was proved by G. Michler in [240].

The properties and structure of hereditary Noetherian prime rings were studied by D. Eisenbud and J.C. Robson in [80].

The decomposition theorem 8.7.6 for Noetherian hereditary rings was proved by A.W. Chatters [43].

The structure of triangular hereditary rings was studied by Yu.A. Drozd in [68]. Sections 8.6–8.8 are based on this paper.

Piecewise domains were introduced and studied by R.Gordon and L. Small in [118]. They gave a general structure theory for piecewise domains. The central result of this theory is that piecewise domains are (primely) triangular. In [118] it was also shown that every nil one-sided ideal in a piecewise domain is nilpotent, and a class of prime piecewise domains was characterized.

In [21] G.F. Birkenmeier, H.E. Heatherly and Jin Yong Kim studied generalized triangular matrix representations for K-algebras over a commutative ring K. They introduced the concept of a set of left (right) triangulating idempotents which determine a generalized triangular matrix representation for a K-algebra of the form (8.5.2). In this paper the authors also introduced a new class of K-algebras that are quasi-Baer rings which are some generalization of Baer rings and piecewise domains.

A ring *A* is called **quasi-Baer** if the right annihilator of every right ideal is generated by an idempotent as a right ideal. Examples of quasi-Baer rings are Baer rings, prime rings and piecewise domains. They also introduced a new class of piecewise prime generalization of piecewise domains that are quasi-Baer FDI-rings with a complete set of left triangulating idempotents. They called these rings as piecewise prime rings, (or **PWP rings**). For these rings they showed that if *A* is a PWP ring then it is a PWP ring with respect to any complete set of left triangulating idempotents. Therefore it provides the positive answer to the "Question" of R.Gordon and L. Small for PWP rings.

CHAPTER 9

Serial Nonsingular Rings. Jacobson's Conjecture

Serial rings and modules were considered in [146, Sections 12, 13]. This chapter is devoted to the further study of serial rings. See Sections 5.3 and 5.4 for initial material. In particular, the structure theorems for various different classes of serial nonsingular rings are presented. As will be illustrated in this section "serial", as intuitively expected, is a very strong property.

Serial right Noetherian piecewise domains are considered in Section 9.1. In particular, it is proved that a serial right Noetherian ring is a piecewise domain if and only if it is right hereditary. Section 9.2 is devoted to the study of serial nonsingular rings. In particular, there is the main result of R.B. Warfield, Jr. who proved that for a serial ring being semihereditary is equivalent to being nonsingular. In this section we also show that any serial nonsingular ring has a classical ring of quotients which is an Artinian ring. The structure of serial nonsingular rings is also studied in this section. The description of the prime quiver of a serial ring with Noetherian diagonal is considered in Section 9.3. There we also describe the structure of serial nonsingular rings with Noetherian diagonal.

Section 9.4 is devoted to the Krull intersection theorem. This theorem is very important and well known for Noetherian commutative rings. This section gives a proof of a version of this theorem for noncommutative rings.

Problems connected with the Jacobson conjecture are considered in Section 9.5. The Jacobson conjecture states that for any Noetherian ring with Jacobson radical R the intersection $\bigcap_{n\geq 0} R^n = 0$. It is well known that this conjecture is true for a commutative Noetherian ring. For noncommutative Noetherian rings this problem is still open in general. In this section it is shown that the Jacobson conjecture holds for at least certain classes of noncommutative Noetherian rings. In particular, it is proved that for Noetherian SPSD-rings and Noetherian serial rings the Jacobson conjecture holds.

9.1 Structure of Serial Right Noetherian Piecewise Domains

Let O be a discrete valuation ring with the unique maximal ideal $\mathcal{M}$ and division ring of fractions D. Recall that $H(O,m,n)$ is the formal triangular matrix ring of the form

$$H(O,m,n) = \begin{pmatrix} H_m(O) & X \\ 0 & T_n(D) \end{pmatrix}, \tag{9.1.1}$$

where

$$H_m(O) = \begin{pmatrix} O & O & \dots & O \\ \mathcal{M} & O & \dots & O \\ \vdots & \vdots & \ddots & \vdots \\ \mathcal{M} & \mathcal{M} & \dots & O \end{pmatrix}, \tag{9.1.2}$$

further, $T_n(D)$ is the ring of upper triangular $n \times n$ matrices over D, and $X = M_{m \times n}(D)$ is a set of all rectangular $m \times n$ matrices over D.

This ring, $H(O,m,n)$, as was shown in [146, Lemma 13.3.2, Theorem 13.5.2], is a right Noetherian right hereditary serial ring.

A full description of serial right Noetherian rings is given by theorem 1.10.5 and for Noetherian serial rings by theorem 1.10.8. The description of serial semiprime and right Noetherian rings is given by theorem 1.10.7.

From theorem 1.10.7 and corollary 8.5.22 there immediately results the following statement.

Proposition 9.1.3. [128, Proposition 3.1]. *Let A be a serial right Noetherian piecewise domain with prime radical $Pr(A)$. Then the diagonal $\overline{A} = A/Pr(A)$ is Morita equivalent*[1] *to a finite direct product of division rings and rings of the form $H_n(O)$, where O is a discrete valuation ring.*

Proposition 9.1.4. *Let A be a serial ring. Then the following conditions are equivalent:*

1. *A is an Artinian ring.*
2. *The Jacobson radical R of A is nilpotent.*

Proof.
1 $\Rightarrow$ 2 follows from the Hopkins theorem 1.1.21.
2 $\Rightarrow$ 1 follows from [146, proposition 12.1.1] □

Proposition 9.1.5. [128, Proposition 3.2]. *An Artinian serial piecewise domain A is Morita equivalent to a direct product of rings of the form $T_n(D)$, where D is a division ring. So A is a hereditary ring.*

Proof. The ring A can be assumed to be indecomposable. In view of theorems 1.12.1 and 1.12.2, one can also assume that the quiver of A is a chain or a simple cycle.

---

[1]See e.g. [146, Section 10.2], and Section 1.10.

Case 1. Suppose that the quiver $Q(A)$ of A is a chain:

$$\overset{1}{\bullet} \longrightarrow \overset{2}{\bullet} \longrightarrow \cdots \longrightarrow \overset{n-1}{\bullet} \longrightarrow \overset{n}{\bullet} \tag{9.1.6}$$

Then by results considering in [146, Section 12.3, p.306–308] A is a serial ring of the first type and it is Morita equivalent to a ring isomorphic to $T_n(D)/\mathcal{I}$, where $\mathcal{I}$ is a two-sided ideal in $T_n(D)$. Let $A_A = P_1 \oplus \ldots \oplus P_n$, where $P_i = e_{ii}A$, $i = 1,\ldots,n$. (Recall that the e_{ij} are the matrix units of the matrix ring $M_n(D)$.) Obviously, any non-zero two-sided ideal $\mathcal{I} \subset T_n(D)$ contains $e_{11}T_n(D)e_{nn}$. So, if $A \neq T_n(D)$ then any homomorphism $\varphi : P_n \to P_1$ is zero, i.e., $e_{11}Ae_{nn} = 0$.

The matrix units $e_{12}, e_{23}, \ldots, e_{n-1n}$ belong to A because there is an arrow $i \to i+1$ for each $i = 1,\ldots,n-1$ in the quiver $Q(A)$. Therefore by [146, proposition 10.7.8] the product $e_{12}e_{23}\ldots e_{n-1n} = e_{1n}$ is non-zero, since A is a piecewise domain, and $e_{1n} \in A$. Thus $e_{11}Ae_{nn} \neq 0$. A contradiction. So $A = T_n(D)$.

Case 2. Suppose that the quiver $Q(A)$ is a cycle:

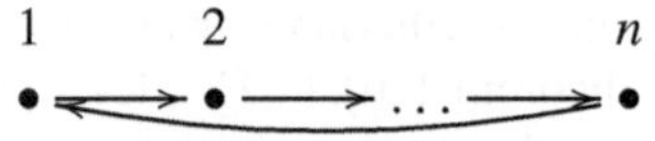

(9.1.7)

Since A is a right Artinian ring, the prime radical of A coincides with the Jacobson radical by proposition 1.1.25. So, in this case the prime quiver $PQ(A)$ can be obtained from the quiver $Q(A)$ by changing all arrows going from one vertex to another vertex to one arrow.

In our case $Q(A)$ is a simple cycle (9.1.7). Thus $Q(A) = PQ(A)$. By proposition 8.5.25, the prime quiver $PQ(A)$ of a piecewise domain is an acyclic simply laced quiver. A contradiction. □

Theorem 9.1.8. [128, Theorem 3.5]. *Let A be a serial right Noetherian ring. Then the following conditions are equivalent*:

1. *A is a right hereditary ring*;
2. *A is a piecewise domain.*

Proof.

$1 \Rightarrow 2$ follows from proposition 8.4.6.

$2 \Rightarrow 1$. By theorem 1.10.5, any serial right Noetherian ring A is Morita equivalent to a direct product of a finite number of rings of the following types:

a. An Artinian serial ring;

b. A ring isomorphic to a ring of the form $H_n(\mathcal{O})$;

c. A ring isomorphic to a quotient ring of $H(\mathcal{O}, m, n)$, where $\mathcal{O}$ is a discrete valuation ring.

Each case is considered separately.

Case (a). In this case A is hereditary, by proposition 9.1.5.

Case (b). $H_n(\mathcal{O})$ is hereditary.

Case (c). One can assume that a ring A is an indecomposable basic ring[2]. Then by propositions 8.5.24 and 8.5.25 the prime quiver $PQ(A)$ is a connected acyclic simply laced quiver. Since A is a serial ring, it follows from theorem 1.12.2 that $PQ(A)$ is a chain. Due to theorem 8.5.6 A is a primely triangular ring. So, there exists a decomposition of $1 \in A$ into a sum of mutually orthogonal idempotents $1 = f_1 + \ldots + f_r$ such that $f_iAf_j = 0$ for $i > j$, all rings $A_{ii} = A_i = f_iAf_i$ are prime, and $f_iAf_{i+1} \neq 0$ for $i = 1, \ldots, r-1$. Consequently, by theorem 8.3.5, A_i is either a division ring or a ring $H_m(O)$ and

$$A = \begin{pmatrix} A_1 & A_{12} & & & * \\ & A_2 & A_{23} & & \\ & & \ddots & \ddots & \\ & 0 & & A_{r-1} & A_{r-1,r} \\ & & & & A_r \end{pmatrix}$$

Since A is a serial piecewise domain, which is an indecomposable basic ring isomorphic to the quotient ring of $H(O,m,n)$, $A_1 = H_m(O)$, $A_2 = D$, ..., $A_r = D$. Consequently, the matrix units $e_{1,m+1}, e_{2,m+1}, \ldots, e_{m,m+1}, e_{m+1,m+2}, \ldots, e_{m+r-2,m+r-1}$ belong to A. Therefore, all Peirce components f_iAf_j for $i \leq j$ are non-zero by [146, proposition 10.7.8], and $A = H(O,m,r-1)$. Thus, a piecewise domain in case (c) is right hereditary. The theorem is proved. □

Corollary 9.1.9. [128, Corollary 3.1]. *The following conditions are equivalent for a Noetherian serial ring A*:

1. *A is a hereditary ring*;
2. *A is a piecewise domain.*

The next examples show that the conditions of theorem 9.1.8 are not equivalent in the case if the property of being serial is changed to the property of being an SPSD-ring[3] in the case of neither an Artinian ring nor a Noetherian ring.

Example 9.1.10.
Let a ring A have the following form:

$$A = \begin{pmatrix} D & D & D & D \\ 0 & D & 0 & D \\ 0 & 0 & D & D \\ 0 & 0 & 0 & D \end{pmatrix}, \tag{9.1.11}$$

[2]For the definition of a semiperfect basic ring see Section 8.3.

[3]Recall that **SPSD-ring** stands for a semiperfect semidistributive ring.

where D is a division ring. Note that as an upper 2×2 triangular ring with entries $\begin{pmatrix} D & D \\ 0 & D \end{pmatrix}$ this is indeed a ring.

Obviously, A is an Artinian ring. Since for any primitive orthogonal idempotents $e, f \in A$ the ring $(e+f)A(e+f)$ is either $\begin{pmatrix} D & D \\ 0 & D \end{pmatrix}$ or $\begin{pmatrix} D & 0 \\ 0 & D \end{pmatrix}$, A is semidistributive, by theorem 1.10.9.

Write $P_i = e_{ii}A$, $i = 1, \ldots, 4$. Let $\varphi : P_i \to P_j$ be a non-zero homomorphism. Then $\varphi(e_{ii}a) = \varphi(e_{ii})a = e_{jj}a_0e_{ii}a$, where $a_0, a \in A$ and $e_{jj}a_0e_{ii}$ is a non-zero element from $e_{jj}Ae_{ii} = D$. Thus $d_0 = e_{jj}a_0e_{ii}$ defines a monomorphism. Therefore the ring A is a piecewise domain.

But A is neither right hereditary nor left hereditary by Example 4.3.7.

The quiver $Q(A)$ has the following form:

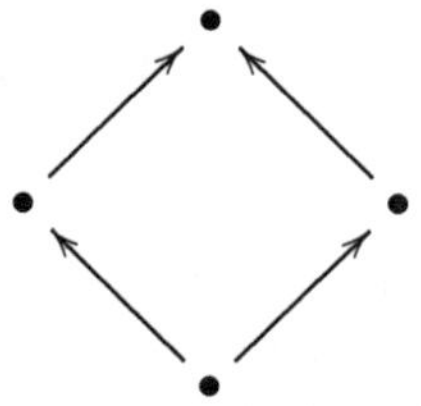

(9.1.12)

and is called a **rhombus** (admittedly a somewhat loose terminology).

Example 9.1.13.

Let O be a discrete valuation ring and $A = T_n(O)$, where, of course,

$$T_n(O) = \begin{pmatrix} O & O & \ldots & O & O \\ 0 & O & \ldots & O & O \\ \vdots & \vdots & \ddots & \vdots & \vdots \\ 0 & 0 & \ldots & O & O \\ 0 & 0 & \ldots & 0 & O \end{pmatrix} \qquad (9.1.14)$$

is the ring of all upper $n \times n$-matrices with elements from O. Then $T_n(O)$ is a Noetherian SPSD-ring and a piecewise domain, but it is not hereditary for $n \geq 2$, since its Jacobson radical:

$$J(T_n(O)) = \begin{pmatrix} \pi O & O & \ldots & O & O \\ 0 & \pi O & \ldots & O & O \\ \vdots & \vdots & \ddots & \vdots & \vdots \\ 0 & 0 & \ldots & \pi O & O \\ 0 & 0 & \ldots & 0 & \pi O \end{pmatrix},$$

where πO is the unique maximal ideal of O, is not projective for $n \geq 2$.

Example 9.1.15.

Let a ring A have the following form:

$$A = \begin{pmatrix} O & \pi O \\ \pi O & O \end{pmatrix}, \tag{9.1.16}$$

where O is a discrete valuation ring with unique maximal ideal $\mathcal{M} = \pi O = O\pi$. It is easy to show that A is a Noetherian semiperfect semidistributive prime PWD, but it is not a hereditary ring, since its Jacobson radical

$$A = \begin{pmatrix} \pi O & \pi O \\ \pi O & \pi O \end{pmatrix}$$

is not projective.

9.2 Structure of Serial Nonsingular Rings

This section is devoted to the study of serial nonsingular rings. The results given here generalize the results for serial right hereditary rings presented in [146, Section 12.3].

The following theorem gives a full description of serial right hereditary rings.

Theorem 9.2.1. (See [146, theorem 13.5.2].) *A serial right hereditary ring A is Morita equivalent to a direct product of rings isomorphic to rings $T_n(D)$ of upper triangular matrices over a division ring D, rings of the form $H_m(O)$ and rings of the form $H(O,m,n)$, where O is a discrete valuation ring. Conversely, all these rings are serial right hereditary.*

From this theorem we immediately get the structure of serial hereditary rings.

Theorem 9.2.2. *Every hereditary serial ring is Morita equivalent to a finite product of rings of the following types: $T_n(D)$ and $H_m(O)$, where D is a division ring and O is a discrete valuation ring.*

Note that any right serial ring is a semiperfect ring. By proposition 1.9.6 every finitely generated projective module over a semiperfect ring can be uniquely decomposed into a finite direct sum of indecomposable projective modules. This result has a generalization for finitely generated modules over semiperfect rings. It was proved by L.H.Rowen in [275] and given in the next proposition.

Proposition 9.2.3 (L.H.Rowen [275]). *Every finitely generated module M over a semiperfect ring A can be decomposed into a finite direct sum of indecomposable submodules.*

Proof. Since A is a semiperfect ring, every finitely generated A-module M has a projective cover $P \to M \to 0$, where P is a finitely generated projective A-module.

By proposition 1.9.6, P can be uniquely decomposed into a direct sum of principal right A-modules. Let the number of indecomposable summands of P be equal to n. Now induction on n is used to show that M is a finite direct sum of $t \leq n$ indecomposable submodules. If $n = 1$, i.e., P is indecomposable, then M is indecomposable, as well, by [146, Proposition 10.4.7]. Suppose that $n > 1$ and $M = M_1 \oplus M_2$. Let $P_i \to M_i \to 0$ be a projective cover of M_i for $i = 1, 2$. Then there is a projective cover $P \simeq P_1 \oplus P_2$ of M. So one can proceed inductively on M_1 and M_2. □

The following important result, obtained by R.B.Warfield, Jr. in [317], states that for a right serial ring being right semihereditary or right nonsingular is equivalent.

Theorem 9.2.4 (R.B. Warfield, Jr. [317]). *Let A be a right serial ring. Then A is right nonsingular if and only if A is right semihereditary.*

Proof. Since any right semihereditary ring is right nonsingular, by corollary 8.1.4, it remains only to prove the inverse statement. Let A be a right serial right nonsingular ring. Suppose M is a finitely generated submodule in a projective A-module B. Note that B is a direct summand of some free A-module A^I, and so it is nonsingular by proposition 5.3.13(2). Therefore M is a nonsingular A-module, by the same proposition.

Since any f.g. module over a semiperfect ring decomposes into a direct sum of indecomposable modules by proposition 9.2.3, one can assume that M is an indecomposable module. Since A is semiperfect, there is a projective cover of the module M:

$$0 \to \mathrm{Ker}\varphi \to P \to M \to 0,$$

where P is an indecomposable projective A-module. Then $P = eA$ for some local idempotent $e \in A$. Since A is a right serial ring, P is a uniserial A-module, and so any submodule of P is essential. Therefore $M \cong P/S$, where S is an essential submodule of P. Since A_A is nonsingular, $P = eA \subset A_A$ is also nonsingular. By proposition 5.3.12, if $S \neq 0$ then P/S is singular. Therefore, S must be zero. So $M \simeq P$ is projective. □

Theorem 9.2.5. *Let A be a right serial ring. Then the following statements are equivalent:*

1. *A is right semihereditary.*
2. *A is left semihereditary.*
3. *A is semihereditary.*
4. *A is right nonsingular.*
5. *A is nonsingular.*
6. *A is a piecewise domain.*
7. *A is a right Rickart ring.*
8. *A is a left Rickart ring.*
9. *A is a Rickart ring.*

Proof.

The equivalence of conditions 1, 2 and 3 follows from corollary 4.6.12.

Therefore the equivalence of conditions 1,2,3,4,5 now follows from theorem 9.2.4.

Since any set of pairwise orthogonal idempotents of a right serial ring A is finite, the equivalence of conditions 7, 8 and 9 follows from theorem 8.1.12.

Since any right serial ring is an FDI-ring, condition 1 implies condition 6. Therefore now the equivalence of conditions 1,2,3,4,5,6 follows from proposition 8.4.12.

Since any semihereditary ring is a Rickart ring, it follows from proposition 8.1.3 that all conditions of the theorem are equivalent. □

Remark 9.2.6. Thanks to this theorem for right serial rings one can simply speak about semihereditary rings or Rickart rings without specifying left or right.

Theorem 9.2.5 is a generalization of the following theorem proved by A. Facchini and G. Pununski in [83].

Proposition 9.2.7. ([83, Lemma 2.2].) *Let A be serial ring, and $A_{ij} = e_i A e_j$, where $1 = e_1 + e_2 + \cdots + e_n$ is a decomposition of the identity of A into a sum of pairwise orthogonal primitive idempotents. Then the following conditions are equivalent:*

1. *A is a nonsingular ring;*
2. *For any $i, j, k = 1, 2, \ldots, n$ if $0 \neq x \in A_{ij}$ and $0 \neq y \in A_{jk}$ then $xy \neq 0$, i.e., A is a piecewise domain.*

Note that by proposition 8.4.6 any semiperfect right semihereditary ring is a piecewise domain.

Let A be a serial semihereditary ring with prime radical $Pr(A) = \mathcal{I}$, and $\bar{A} = A/\mathcal{I}$, which is obviously a semiprime ring. Then by [147, theorem 5.5.2] there exists a decomposition $\bar{A} = \bar{A}_1 \times \bar{A}_2 \times \ldots \times \bar{A}_r$, where the $\bar{A}_i$ are prime serial rings for $i = 1, \ldots, r$. Let $\bar{1} = \bar{f}_1 + \ldots + \bar{f}_r$ be the corresponding decomposition of the identity of $\bar{A}$ into a sum of pairwise orthogonal idempotents. Since $\mathcal{I}$ is a nil-ideal by proposition 1.1.19, the idempotents $\bar{f}_1, \ldots, \bar{f}_r$ can be lifted modulo $\mathcal{I}$ preserving orthogonality by [146, Lemma 10.3.3]. In this case $1 = f_1 + \ldots + f_r$, where $f_i f_j = \delta_{ij} f_j$ $(i, j = 1, \ldots, r)$ and $\bar{f}_i = f_i + \mathcal{I}$. Write $f_i A f_j = A_{ij}$ for $i, j = 1, \ldots, r$. Then $f_i \mathcal{I} f_j = A_{ij}$ for $i \neq j$ and $\mathcal{I}_i = f_i \mathcal{I} f_i$ is the prime radical of the ring A_{ii}. Then the two-sided Peirce decomposition of $\mathcal{I}$ has the following form (see [146, Section 11.4]):

$$\mathcal{I} = \bigoplus_{i,j=1}^{r} f_i \mathcal{I} f_j, \tag{9.2.8}$$

where (as before) $f_i \mathcal{I} f_i = \mathcal{I}_i$ and $f_i \mathcal{I} f_j = A_{ij}$ for $i \neq j$; $(i, j = 1, 2, \ldots, r)$.

Let A be a serial ring, and $A_{ij} = e_i A e_j$, where $1 = e_1 + e_2 + \cdots + e_n$ is a decomposition of the identity of A into a sum of pairwise orthogonal primitive idempotents. Then we can introduce a relation $\prec$ on the set $\{1, 2, \ldots, n\}$, setting $i \prec j$

if and only if $A_{ij} \neq 0$ or $A_{ji} \neq 0$. The next lemma shows that this relation is an equivalence relation for the case of serial nonsingular rings, i.e., for PWD's. Note that this relation is some weaker than the other equivalence relation $\sim$ for PWD's considered in Section 8.5.

Lemma 9.2.9. ([310, Proposition 2.2], [265, Corollary 5.8].) *The relation $\prec$ is an equivalence relation for every serial nonsingular ring.*

Proof.
Obviously, this relation is reflexive and symmetric. We will show that it is also transitive. Suppose that $i \prec j$ and $j \prec k$. So that $A_{ij} \neq 0$ or $A_{ji} \neq 0$, and $A_{jk} \neq 0$ or $A_{kj} \neq 0$. If $A_{ij} \neq 0$ and $A_{jk} \neq 0$ then $A_{ik} \neq 0$ as well, since A is a PWD. Analogously, if $A_{kj} \neq 0$ and $A_{ji} \neq 0$ then $A_{ki} \neq 0$. So that it remains to consider only two cases

1. $A_{ij} \neq 0$ and $A_{kj} \neq 0$;
2. $A_{ji} \neq 0$ and $A_{jk} \neq 0$.

Consider the first case. Let $0 \neq x \in A_{ij}$ and $0 \neq y \in A_{kj}$. Then $x \in Ae_j$ and $y \in Ae_j$. Since A is a serial ring, Ae_j is a uniserial module. Therefore either $Ax \subseteq Ay$ or $Ay \subseteq Ax$. So that either $x \in Ay$ or $y \in Ax$. Suppose that $x \in Ay$, then $x = ay$ for some $a \in A$. Therefore $e_i x = x = ay = e_i ay = e_i a e_k b e_j = (e_i a e_k)(e_k b e_j) = a_0 y$ where $a_0 = e_i a e_k \in A_{ik}$. Since A is a PWD and $x \neq 0$, $y \neq 0$, we get that $a_0 \neq 0$, as well. Therefore $A_{ik} \neq 0$ and so $i \prec k$. The second case is similar. □

Proposition 9.2.10. ([310, Proposition 2.2], [265, Proposition 5.9].) *Let A be a serial nonsingular ring, and $A_{ij} = e_i A e_j$, where $1 = e_1 + e_2 + \cdots + e_n$ is a decomposition of the identity of A into a sum of pairwise orthogonal primitive idempotents. Then the following conditions are equivalent*:

1. *A is indecomposable.*
2. *$A_{ij} \neq 0$ or $A_{ji} \neq 0$.*

Proof.
$1 \Longrightarrow 2$. Suppose $A_{ij} = A_{ji} = 0$ for some i,j. Consider two subsets $I, J \subset \{1,2,\ldots,n\}$ such that $k \in I$ iff $i \prec k$; and $s \in J$ iff $i \nprec s$. Clearly, $I \bigcap J = \emptyset$ and $I \bigcup J = \{1,2,\ldots,n\}$. Moreover, $J \neq \emptyset$ since $j \in J$, and $I \neq \emptyset$ since A is indecomposable. Let $e = \sum\limits_{k \in I} e_k$ and $f = \sum\limits_{s \in J} e_s$ and consider two right A-modules $P = eA$ and $Q = fA$. Suppose $\mathrm{Hom}(e_k A, e_s A) = e_s A e_k \neq 0$ or $\mathrm{Hom}(e_s A, e_k A) = e_s A e_k \neq 0$ for some $k \in I$ and $s \in J$. Then it means that $k \prec s$, that is impossible since $I \bigcap J = \emptyset$. Therefore $\mathrm{Hom}(P,Q) = \mathrm{Hom}(Q,P) = 0$, i.e., A is decomposable. A contradiction. So that $A_{ij} \neq 0$ or $A_{ji} \neq 0$.

$2 \Longrightarrow 1$. Suppose A is decomposable and let $A = eA \oplus fA$, where $e = \sum\limits_{k \in I} e_k$ and $f = \sum\limits_{s \in J} e_s$ for some subsets $I, J \subset \{1,2,\ldots,n\}$ such that $I \bigcap J = \emptyset$ and $I \bigcup J = \{1,2,\ldots,n\}$. Then $\mathrm{Hom}(e_i A, e_j A) = e_j A e_i = 0$ and $\mathrm{Hom}(e_j A, e_i A) = e_i A e_j = 0$ for all $i \in I$ and $j \in J$. A contradiction. □

Proposition 9.2.11. ([122, Proposition 3].) *If A is an indecomposable serial nonsingular ring, then there is a prime decomposition of the identity $1 = f_1 + \ldots + f_r$ into a sum of pairwise orthogonal idempotents such that $A_{ii} = f_iAf_i$ are prime rings, $A_{ij} = f_iAf_j = 0$ and $A_{ji} = f_jAf_i \neq 0$ for all $i > j$ and $i, j = 1, 2, \ldots, r$.*

Proof.
By theorem 9.2.5 A is a piecewise domain. Therefore by theorem 8.5.17 there is a decomposition of the identity $1 = f_1 + \ldots + f_r$ of A into a sum of pairwise orthogonal idempotents such that the two-sided Peirce decomposition of a ring A has the following triangular form:

$$A = \begin{pmatrix} A_{11} & A_{12} & \ldots & A_{1,r-1} & A_{1r} \\ 0 & A_{22} & \ldots & A_{2,r-1} & A_{2r} \\ \vdots & \vdots & \ddots & \vdots & \vdots \\ 0 & 0 & \ldots & A_{r-1,r-1} & A_{r-1,r} \\ 0 & 0 & \ldots & 0 & A_{rr} \end{pmatrix}, \qquad (9.2.12)$$

where $A_{ij} = f_iAf_j$ and all the A_{ii} are prime piecewise domains of the form (8.5.16), for $i, j = 1, 2, \ldots, r$.

Therefore it remains only to prove that all $A_{ji} \neq 0$ for $i > j$. But this statement follows immediately from the form (9.2.12) and proposition 9.2.10. □

Corollary 9.2.13. ([317, Corollary 4.11]). *A serial nonsingular ring is prime if and only if $A_{ij} = e_iAe_j \neq 0$ for all $i, j = 1, 2, ..., n$ where $1 = e_1 + e_2 + \cdots + e_n$ is a decomposition of the identity of A into a sum of pairwise orthogonal primitive idempotents.*

Proof.
Since any serial nonsingular ring is a piecewise domain, the statement follows immediately from lemma 8.5.5. □

Theorem 1.11.5 states that any serial ring A has a classical ring of fractions Q which is a serial ring. In the case when A is a serial nonsingular ring one can obtain full information about the structure of Q.

Lemma 9.2.14. *Any serial nonsingular prime ring A has a classical ring of fractions Q which is a simple Artinian ring.*

Proof.
Since A is a serial ring, A has a classical ring of fractions Q which is also a serial ring, by theorem 1.11.5. Since A is a serial ring, $\text{u.dim}(A_A) < \infty$. So, A is a prime nonsingular ring with $\text{u.dim}(A_A) < \infty$, i.e., A satisfies the condition (2) of Goldie's theorem 5.4.10. Therefore A has a classical quotient ring Q which is semisimple.

Note that the local idempotents of the ring A are local in the ring Q. Therefore the rings A and Q are decomposed simultaneously (i.e., in parallel so to speak) into a direct product of rings. Moreover if the ring Q has nilpotent ideals then so has A. So if A is prime then Q is a simple Artinian ring. □

Theorem 9.2.15. *Any serial nonsingular ring A has a classical quotient ring Q which is Artinian and hereditary. Moreover Q is isomorphic to a finite direct product of rings of the following block-triangular form:*

$$\left(\begin{array}{c|c|c|c} M_{k_1\times k_1}(D) & M_{k_1\times k_2}(D) & \dots & M_{k_1\times k_r}(D) \\ \hline 0 & M_{k_2\times k_2}(D) & \dots & M_{k_2\times k_r}(D) \\ \hline \vdots & \vdots & \ddots & \vdots \\ \hline 0 & 0 & \dots & M_{k_r\times k_r}(D) \end{array}\right), \tag{9.2.16}$$

where D is a division ring.

Proof.

Let A be an indecomposable serial nonsingular ring. By theorem 1.11.5 A has a classical ring of fractions Q which is also a serial ring. From proposition 9.2.11 it follows that the two-sided Peirce decomposition of A has the form (9.2.12) such that $A_{ij} = 0$ and $A_{ji} \neq 0$ for all $i > j$, and all A_i are prime rings. The prime radical N of A is nilpotent, by corollary 8.5.19, and the two-sided Peirce decomposition of N has the following form:

$$N = \begin{pmatrix} 0 & A_{12} & \dots & A_{1,r-1} & A_{1r} \\ 0 & 0 & \dots & A_{2,r-1} & A_{2r} \\ \vdots & \vdots & \ddots & \vdots & \vdots \\ 0 & 0 & \dots & 0 & A_{r-1,r} \\ 0 & 0 & \dots & 0 & 0 \end{pmatrix} \tag{9.2.17}$$

Therefore $A/N \cong A_1 \times A_2 \times \ldots \times A_r$ is a finite direct product of serial nonsingular prime rings. By lemma 9.2.14, $\widetilde{A} = A/N$ is a semiprime Goldie ring. Since A is serial, A contains no infinite sum of non-zero right (left) ideals. Therefore, since A is nonsingular, A is a Rickart ring by theorem 9.2.5, and A satisfies a.c.c. on right and left annihilators by theorem 8.1.12. So A is a nonsingular Goldie ring. Therefore, $\rho(A_A) = \text{u.dim}(A_A) < \infty$ and $\rho({}_AA) = \text{u.dim}({}_AA) < \infty$. Since A is a primely triangular FDI-ring, $C_A(0) = C_A(N)$, by proposition 8.5.28. So A satisfies all conditions of theorem 5.5.28. Therefore the classical quotient ring Q of A is an Artinian serial ring with Jacobson radical R_1 which is equal to the prime radical $N_1 = QN = NQ$.

Now, here is a proof that Q is a nonsingular ring. By theorem 9.2.5 it suffices to show that Q is a right nonsingular ring. Let $Q = \bigoplus_{i,j=1}^{r} Q_{ij}$ be the two-sided Peirce decomposition of Q. Suppose that Q is not right nonsingular. Then by proposition 9.2.7 there exist elements $x \in Q_{ij}$ and $y \in Q_{jk}$ such that $xy = 0$. Since $Q_{ij} = QA_{ij} = AQ_{ij}$, $x = s^{-1}t$ and $y = uv^{-1}$ for some $s, v \in C_Q(0)$ and $0 \neq t \in A_{ij}$, $0 \neq u \in A_{jk}$. Then $0 = xy = s^{-1}tuv^{-1}$, whence $tu = 0$. This contradicts proposition

9.2.7, since A is a nonsingular ring. Thus, Q is a nonsingular ring, and so it is semihereditary, by theorem 9.2.5. Since Q is an Artinian ring, it is hereditary, by proposition 8.2.15.

Thus, Q is an Artinian serial hereditary ring. Then, by proposition 9.1.5, Q is Morita equivalent to a direct product of rings of the form $T_n(D)$, where D is a division ring. Taking into account the two-sided Peirce decomposition (9.2.12) of the ring A, we obtain that the two-sided Peirce decomposition of the ring Q has the form (9.2.16), where $Q_i = M_{k_i \times k_i}(D)$ is the classical quotient ring of A_i, and D is a division ring, $i = 1, 2, \ldots, r$. □

The following theorem yields a full description of serial semihereditary rings.

Theorem 9.2.18. ([122, Theorem 3]). *A serial nonsingular ring A is Morita equivalent to a direct product of rings isomorphic to rings of the following form:*

$$H = \left(\begin{array}{c|c|c|c} A_1 & A_{12} & \ldots & A_{1r} \\ \hline O & A_2 & \ldots & A_{2r} \\ \hline \vdots & \vdots & \ddots & \vdots \\ \hline O & O & \ldots & A_r \end{array}\right), \qquad (9.2.19)$$

where $A_{ij} = M_{k_i \times k_j}(D)$, $i, j = 1, 2, \ldots, r$, *and D is a division ring. Moreover, each* A_i *is a prime serial nonsingular ring with the classical quotient ring of the form* $M_{k_i \times k_i}(D)$ *for* $i = 1, 2, \ldots, r$. *Conversely, all rings of this form are serial and nonsingular.*

Proof.

One can assume that A is an indecomposable ring. Therefore the two-sided Peirce decomposition of A has form (9.2.12), where all A_{ii} are prime serial nonsingular rings. By theorem 9.2.15 the ring A has a classical quotient ring which is Artinian serial and hereditary, and has the block-triangular form (9.2.16), where D is a division ring, k_i is the number of local pairwise orthogonal idempotents in the decomposition $f_i = e_1 + \ldots + e_{k_i}$ for $i = 1, \ldots, r$.

Therefore A_i is an order in $M_{k_i \times k_i}(D)$ and one can assume that $A_i \subseteq M_{k_i \times k_i}(D)$, $A_{ij} \subseteq M_{k_i \times k_j}(D)$, $i, j = 1, \ldots, r$. To show that $A_{ij} = M_{k_i \times k_j}(D)$ for $i \neq j$ it suffices to show that for any two local idempotents $e \in f_i$ and $h \in f_j$ $(i \neq j)$, $eAh = D$. Let $g = e + h$ and $B = gAg$. Write $B_1 = eAe$, $B_2 = hAh$ and $X = eAh$. Then the ring $B = \begin{pmatrix} B_1 & X \\ 0 & B_2 \end{pmatrix}$ is a serial nonsingular ring, and $B \subset T_2(D)$, where $B_i = e_{ii}Be_{ii}$, $i = 1, 2$; $X = e_{11}Be_{22}$, and e_{11}, e_{22} are matrix units. By [146, lemma 13.3.3], X is a uniserial right B_2-module. We show that $\alpha X = X$ for any $0 \neq \alpha \in B_1$. Suppose $\alpha X \neq X$, then for an $y \in X \setminus \alpha X$ there is an inclusion $\alpha X \subset yB_2$, as αX and yB_2 are submodules of a uniserial right B_2-module X. Consider two submodules $(\alpha B_1, \alpha X)$ and $(0, yB_2)$ of the projective uniserial module $(B_1, X) = P_1 = e_{11}B$. Since they do not contain each other, we get a contradiction. This contradiction shows that $\alpha X = X$ for any $\alpha \in A_1$. Since A_1 is an order in D, this means that $dX \subseteq X$ for any $d \in D$. Therefore X is a left vector space over the division ring D. Since $X \subseteq D$, $X = D$.

The theorem is proved. □

Corollary 9.2.20. *Let A be a serial nonsingular ring with prime radical N, and Q the classical quotient ring of A with Jacobson radical J. Then $J = N$.*

9.3 Serial Rings with Noetherian Diagonal

Recall that the **diagonal of a ring** A is the quotient ring $A/Pr(A)$, where $Pr(A)$ is the prime radical of A.

The diagonal of A is always a semiprime ring, by [146, corollary 11.2.5]. By theorem 1.10.7, a serial semiprime right Noetherian ring can be decomposed into a direct product of prime rings. A serial prime right Noetherian ring is also left Noetherian and Morita-equivalent to either a division ring or a ring isomorphic to $H_m(O)$, where O is a discrete valuation ring.

Therefore, it is possible simply to say about the Noetherian diagonal of a serial ring.

Denote $\mathcal{I} = Pr(A)$. Let A be a serial ring with Noetherian diagonal. One can assume that A is a basic ring. Let $\bar{A} = A/\mathcal{I} = \bar{A}_1 \times \ldots \times \bar{A}_r$ be a decomposition of the diagonal of A into a direct product of a finite number of prime rings. Every $\bar{A}_i$ is either a division ring or a ring $H_m(O)$.

Let $\bar{1} = \bar{f}_1 + \ldots + \bar{f}_r$ be the corresponding decomposition of the identity $\bar{1} \in \bar{A}$ into a sum of pairwise orthogonal central idempotents. Since by proposition 1.1.27 $\mathcal{I}$ is a nil-ideal, the idempotents $\bar{f}_1, \ldots, \bar{f}_r$ can be lifted modulo $\mathcal{I}$ preserving their orthogonality, i.e., there is an equality $1 = f_1 + \ldots + f_r$, where $f_i f_j = \delta_{ij}$ and $\bar{f}_i = f_i + \mathcal{I}$ for $i, j = 1, \ldots, r$. The Peirce components are the $A_{ij} = f_i A f_j \subset \mathcal{I}$ for $i \neq j$; $i, j = 1, \ldots, r$. By [146, proposition 11.2.9], $\mathcal{I}_i = e_i \mathcal{I} e_i$ is the prime radical of $A_i = f_i A f_i$, $i = 1, \ldots, r$. Therefore the two-sided Peirce decomposition of $\mathcal{I}$ has the following form:

$$\mathcal{I} = \begin{pmatrix} \mathcal{I}_1 & A_{12} & \ldots & \ldots & A_{1r} \\ A_{21} & \mathcal{I}_2 & \ddots & & A_{2r} \\ \vdots & \ddots & \ddots & \ddots & \vdots \\ A_{r-1,1} & & \ddots & \mathcal{I}_{r-1} & A_{r-1,r} \\ A_{r1} & \ldots & \ldots & A_{r,r-1} & \mathcal{I}_r \end{pmatrix}. \tag{9.3.1}$$

Moreover, $\bar{A} = A/\mathcal{I} = A_1/\mathcal{I}_1 \times \ldots \times A_r/\mathcal{I}_r$, i.e., $\bar{A}_i = A_i/\mathcal{I}_i$ for $i = 1, \ldots, r$.

For a ring A with prime radical $\mathcal{I}$ one can construct the prime quiver, as was done in [146, Section 11.4]. Its which description is given by the following theorem.

Theorem 9.3.2. *Let A be a serial ring with Noetherian diagonal. Then the prime quiver $PQ(A)$ is a disconnected union of cycles and chains.*

Proof.

Denote by $\mathcal{I} = Pr(A)$ the prime radical of A, which has the form (9.3.1). Since $PQ(A) = PQ(A/\mathcal{I}^2)$ and $Pr(A/\mathcal{I}^2) = \mathcal{I}/\mathcal{I}^2$, one can assume that A is a serial ring with $\mathcal{I}^2 = 0$. Consider the diagonal $\overline{A} = A/\mathcal{I} = \overline{A}_1 \times \overline{A}_2 \times \ldots \times \overline{A}_r$, where all the rings $\overline{A}_1, \ldots, \overline{A}_r$ are indecomposable. Let $1 = f_1 + \ldots + f_r$ be the corresponding decomposition of the identity $1 \in A$ into a sum of pairwise orthogonal idempotents which are central modulo $\mathcal{I}$. Note that, since $\mathcal{I}^2 = 0$, $f_i\mathcal{I}f_j$ is a submodule of f_iA, and $f_i\mathcal{I}f_j = f_iAf_j$ for $i \neq j$. Assume that a vertex i is a source of at least two arrows, the ends of which are $s \neq i$ and $t \neq i$. This means that $f_i\mathcal{I}f_s$ and $f_i\mathcal{I}f_t$ are both non-zero. By theorem 1.10.7 one can suppose that each $\overline{A}_i = f_iAf_i/f_i\mathcal{I}f_i$ is either a division ring or a ring of the form $H_m(O)$, where O is a discrete valuation ring. If the idempotent f_i is local, which is always the case if $\overline{A}_i$ is a division ring, then the uniserial module $f_i\mathcal{I}$ contains two submodules $f_i\mathcal{I}f_s$ and $f_i\mathcal{I}f_t$, neither of which is contained in the other, which is impossible. Now assume that $f_iAf_i/f_i\mathcal{I}f_i \simeq H_m(O)$, where O is a discrete valuation ring. Denote by R the Jacobson radical of f_iAf_i and consider a decomposition $f_i = e_1 + \ldots + e_m$ into a sum of pairwise orthogonal local idempotents. Choose the index k such that $e_kAf_s \neq 0$. If $e_k\mathcal{I}f_t \neq 0$, then the uniserial module e_kA contains two submodules, e_kAf_s and $e_k\mathcal{I}f_t$, neither of which is a submodule of the other, which is impossible.

Suppose that $e_k\mathcal{I}f_t = 0$ for all $t \neq s$, in particular, $e_k\mathcal{I}e_k = 0$. If $e_kRe_kAf_s \neq e_kAf_s$, then there are two submodules, $e_kRe_k + e_kRe_kAf_s$ and e_kAf_s of the uniserial module e_kA, neither of which contains the other. Hence, $e_kRe_kAf_s = e_kAf_s$ for every k such that $e_kAf_s \neq 0$. But $e_kRe_k \subset e_kRe_lRe_k + \mathcal{I}$ for every $l \neq k$, hence $e_kRe_k = e_kRe_lRe_k$ and $e_lRe_kAf_s \neq 0$ for every l. Therefore $e_l\mathcal{I}f_t = 0$ for all l, which means that $f_i\mathcal{I}f_t = 0$. A contradiction.

In exactly the same way one can prove that two arrows cannot end at the same vertex. There are two types of finite connected graphs with these properties: a cycle and a chain. The theorem is proved. □

Example 9.3.3.

Let O be a discrete valuation ring with a classical quotient ring which is a division ring D. Consider the ring A with elements of the following form:

$$\begin{pmatrix} \alpha & \beta_1 & \ldots & \beta_{t-1} & \beta_t \\ 0 & \alpha & \ddots & & \beta_{t-1} \\ \vdots & \ddots & \ddots & \ddots & \vdots \\ \vdots & & \ddots & \alpha & \beta_1 \\ 0 & \ldots & \ldots & 0 & \alpha \end{pmatrix},$$

where $\alpha \in O$, $\beta_i \in D$, $i = 1, \ldots, t-1$; and $\beta_t \in D$; or $\beta_t \in D/O$.

Then A is serial with Noetherian diagonal and a nilpotent prime radical. The prime quiver of A is a loop.

From theorem 9.3.2 and proposition 8.5.25 there immediately follows a statement which is analogous to proposition 9.1.3.

Corollary 9.3.4. *Let A be a serial piecewise domain with Noetherian diagonal. Then the prime quiver $PQ(A)$ is a disconnected union of chains.*

From proposition 9.1.5 and corollary 8.5.22 it immediately follows the statement which is analogous to proposition 9.1.3.

Proposition 9.3.5. *If A is a serial piecewise domain with Noetherian diagonal $\overline{A}$. Then $\overline{A}$ is Morita equivalent to a finite direct product of rings of the form D or $H_n(O)$, where D is a division ring and O is a discrete valuation ring.*

Lemma 9.3.6. *Let A be a serial piecewise domain with Noetherian diagonal whose identity decomposes into two local idempotents. Then A is isomorphic to one of the following rings:*

$$(a)\ H_2(O) = \begin{pmatrix} O & O \\ \pi O & O \end{pmatrix}\ ; \qquad (b)\ T_2(D) = \begin{pmatrix} D & D \\ 0 & D \end{pmatrix}$$

$$(c)\ H((O,1),(D,1)) = \begin{pmatrix} O & D \\ 0 & D \end{pmatrix}; \qquad (d)\ H((D,1),(O,1)) = \begin{pmatrix} D & D \\ 0 & O \end{pmatrix};$$

$$(e)\ H((O_1,1),(O_2,1)) = \begin{pmatrix} O_1 & D \\ 0 & O_2 \end{pmatrix},$$

where O, O_1, O_2 are discrete valuation rings with a common division ring of fractions D. All these rings are serial piecewise domains.

Proof.

The ring A can be assumed to be an indecomposable basic ring.

If A is prime, then A is a serial Noetherian prime ring, and, by proposition 9.1.5, A isomorphic to the ring $H_2(O)$, where O is a discrete valuation ring.

If A is not prime, then, by theorem 8.5.17, A is isomorphic to a ring $\begin{pmatrix} A_1 & A_{12} \\ 0 & A_2 \end{pmatrix}$, where A_1, A_2 are prime rings. So, by proposition 9.3.5, A_i is either a division ring or a discrete valuation ring O_i for $i = 1,2$. By theorem 9.2.18, A in an order in an Artinian ring $Q = T_2(D)$ and the Jacobson radical $J = \begin{pmatrix} 0 & D \\ 0 & 0 \end{pmatrix}$ of Q is equal to the prime radical of A. Therefore A_1 and A_2 are orders in D and $A_{12} = D$. Thus we obtain the other cases (b), (c), (d) and (e). □

Theorem 9.3.7. *An indecomposable serial nonsingular ring A with Noetherian diagonal is up to isomorphism Morita equivalent to a ring of the following form:*

$$H((\Delta_1,n_1),\ldots,(\Delta_r,n_r)) = \left(\begin{array}{c|c|c|c} A_1 & A_{12} & \ldots & A_{1r} \\ \hline O & A_2 & \ldots & A_{2r} \\ \hline \vdots & \vdots & \ddots & \vdots \\ \hline O & O & \ldots & A_r \end{array}\right), \tag{9.3.8}$$

where $A_{ij} = M_{n_i \times n_j}(D)$*,* O_i *is a discrete valuation ring with quotient ring* D *which is a division ring, for* $i,j = 1,2,\ldots,r$*. Moreover* $A_i = T_{n_i}(D)$ *if* $\Delta_i = D$ *and* $A_i = H_{n_i}(O_i)$ *if* $\Delta_i = O_i$*. Conversely, all rings of this form are serial nonsingular rings with Noetherian diagonal.*

Proof.
One can assume that A is an indecomposable basic ring. Since A is a serial nonsingular ring, by theorem 9.2.18, there exists a decomposition of $1 \in A$ into a sum of mutually orthogonal idempotents $1 = f_1 + \ldots + f_r$ such that the corresponding two-sided Peirce decomposition of A has form (9.2.19) such that $f_i A f_j = 0$ for $i > j$, all rings $A_{ii} = A_i = f_i A f_i$ are prime and $f_i A f_j = M_{n_i \times n_j}(D)$ for $i < j$, where D is a division ring.

Consequently, by proposition 9.3.5, A_i is either a division ring or a ring of the form $H_m(O)$. So, using lemma 9.3.6, there follows the required form (9.3.8).

Conversely, all rings of this form are serial nonsingular with Noetherian diagonal as follows from theorem 9.2.4. □

9.4 The Krull Intersection Theorem

The Krull intersection theorem is one of the basic results in the theory of commutative rings. Originally it was formulated for ideals in a commutative Noetherian ring by W.Krull in [200] (see also [253]), and then it was extended to Noetherian modules over commutative and noncommutative rings (see [254]).

There are many proofs of the Krull intersection theorem for the commutative case. One of them uses the Artin-Rees lemma, which is very well known as an important tool in commutative algebra.

Theorem 9.4.1. (Artin-Rees Lemma). *Let* A *be a commutative Noetherian ring,* $\mathcal{I}$ *an ideal in* A*,* M *a finitely generated* A*-module and* N *a submodule in* M*. Then there exists a positive integer* m *such that* $\mathcal{I}^n M \cap N = \mathcal{I}^{n-m}(\mathcal{I}^m M \cap N)$ *for all* $n \geq m$*.*

The proof of this theorem uses the Hilbert Basis Theorem and can be found in an almost every textbook on commutative algebra (see, e.g. [8, corollary 10.10]), [24, chapter III, §3]).

Theorem 9.4.2. (Krull Intersection Theorem.) *Let* A *be a commutative Noetherian ring,* $\mathcal{I}$ *an ideal in* A*, and* M *a finitely generated* A*-module. If* $N = \bigcap_{i=1}^{\infty} \mathcal{I}^n M$*, then* $N = \mathcal{I}N$*.*

Proof.

Apply the Artin-Rees lemma to the case $N = \bigcap_{n=1}^{\infty} \mathcal{I}^n M$ and $n = m + 1$. Then $\mathcal{I}^{m+1}M \cap N = \mathcal{I}(\mathcal{I}^m M \cap N)$. Since $\mathcal{I}^m M \cap N \subseteq N$ and $N \subseteq \mathcal{I}^{m+1}M$, we obtain

$$N \subseteq \mathcal{I}^{m+1}M \cap N = \mathcal{I}(\mathcal{I}^m M \cap N) \subseteq \mathcal{I}N \subseteq N$$

which implies that $N = \mathcal{I}N$. □

Corollary 9.4.3. *Let A be a Noetherian commutative ring with Jacobson radical R, $\mathcal{I}$ an ideal in A, and M a finitely generated A-module. Then*

1. *If $\mathcal{I}$ is contained in R then $\bigcap_{i=1}^{\infty} \mathcal{I}^n M = 0$, in particular $\bigcap_{i=1}^{\infty} R^n M = 0$.*
2. *If $\mathcal{I}$ is contained in R then $\bigcap_{i=1}^{\infty} \mathcal{I}^n = 0$, in particular $\bigcap_{i=1}^{\infty} R^n = 0$.*
3. *If A is an integral domain and $\mathcal{I}$ is a proper ideal of A then $\bigcap_{i=1}^{\infty} \mathcal{I}^n = 0$.*
4. *If A is a local ring and $\mathcal{I}$ is a proper ideal of A then $\bigcap_{i=1}^{\infty} \mathcal{I}^n M = 0$.*
5. *If A is a local ring and $\mathcal{I}$ is a proper ideal of A then $\bigcap_{i=1}^{\infty} \mathcal{I}^n = 0$, in particular $\bigcap_{i=1}^{\infty} R^n = 0$.*

Proof.

1. Set $N = \bigcap_{i=1}^{\infty} \mathcal{I}^n M$. Then by theorem 9.4.2 $N = \mathcal{I}N$. If $\mathcal{I} \subseteq R$, then $N = \mathcal{I}N \subseteq RN \subseteq N$. Therefore $N = RN$, which implies $N = 0$ by the Nakayama lemma, since A is a Noetherian ring.
2. This follows as a particular case of 1 with $M = A$.
3. By theorem 9.4.2, if $N = \bigcap_{i=1}^{\infty} \mathcal{I}^n$, then $N = \mathcal{I}N$. Since A is a Noetherian ring, N is a finitely generated A-module. Assume that $\{x_1, x_2, ..., x_s\}$ is a minimal system of generators of N. Since $N = \mathcal{I}N$, there exist elements $a_1, a_2, \ldots, a_s$ such that $x_1 = x_1a_1 + x_2a_2 + \cdots + x_sa_s$, which implies that $x_1(1 - a_1) = x_2a_2 + \cdots + x_sa_s$. Since A is an integral domain, $1 - a_1$ is an invertible element in A. Therefore $x_1 = x_2b_2 + \cdots + x_sb_s$, which contradicts the minimality property of s. Therefore $N = 0$.
4. and 5. These follow from 1 and 2 since any proper ideal of a local ring A is contained in the Jacobson radical A. □

In this section we will prove some version of the Krull intersection theorem for Noetherian modules over a noncommutative ring following A.Caruth [37].

Theorem 9.4.4. *Let M be a Noetherian left A-module, N an A-submodule of M, and $\mathcal{I}$ a central ideal*[4] *in A. Then $x \in \bigcap_{i=1}^{\infty} \mathcal{I}^n M$ if and only if $xa = x$ for some element $a \in \mathcal{I}$.*

Remark 9.4.5. If A is a commutative Noetherian ring and M is a finitely generated A-module, then M is a Noetherian A-module. In this case theorem 9.4.1 follows from theorem 9.4.4.

Lemma 9.4.6. *Let M be a Noetherian left A-module and $\mathcal{I}$ a central ideal in A. Then for any A-submodule of M there is an integer $m \geq 1$ such that $\mathcal{I}^m M \subseteq \mathcal{I}N$.*

Proof.

First it is shown that the ideal $\mathcal{I}$ in A can be assumed to be finitely generated ideal in the center of A.

Let $\mathcal{I}_0 \subseteq \mathcal{I}$, where $\mathcal{I}_0$ is an ideal in A finitely generated by central elements of A. Then $\mathcal{I}_0 M$ is a submodule in M and $\mathcal{I}_0$ can be chosen so that $\mathcal{I}_0 M$ is maximal in the set of all submodules which arise in this way. Let $x \in \mathcal{I}$. Then $xM \subseteq (\mathcal{I}_0 + Ax)M = \mathcal{I}_0 M$ implies $\mathcal{I}M = \mathcal{I}_0 M$ and hence $\mathcal{I}^n M = \mathcal{I}_0^n M$ for all $n \geq 0$. If m is a positive integer such that $\mathcal{I}_0^m M \cap N \subset \mathcal{I}_0 N$, then $\mathcal{I}^m M \cap N = \mathcal{I}_0^m M \cap N \subseteq \mathcal{I}_0 N \subseteq \mathcal{I}N$.

Let $\mathcal{I}$ be a finitely generated central ideal in A with as a generator a set of central elements $x_1, x_2, \ldots, x_t \in A$. Write $C = \mathcal{I}N$. Since M is a Noetherian left A-module, there exist positive integers p_j $(1 \leq j \leq t)$ such that

$$(C + x_1^{p_1} M + \cdots + x_j^{p_j} M) : x_{j+1}^{p_{j+1}} A = (C + x_1^{p_1} M + \cdots + x_j^{p_j} M) : x_{j+1}^{p_{j+1}+1} A \quad (9.4.7)$$

where

$$D : E = \{y \in M \mid ey \in D \text{ for every } e \in E\}$$

for a left A-module D and a central ideal E in A.

Set $b_j = x_j^{p_j}$. We now prove that

$$(b_1 M + \cdots + b_j M) \cap (C : b_1 A + \cdots + b_j A) \subseteq C \quad (9.4.8)$$

when $1 \leq j \leq t$.

The case $j = 1$ is true, since

$$b_1 M \cap (C : b_1 A) = b_1 (C : b_1^2 A) = b_1 (C : b_1 A) \subset C$$

follows from equation (9.4.7) for $j = 1$.

Assume that (9.4.8) is true for $j \leq s - 1$. Then

$$(b_1 M + \cdots + b_s M) \cap (C : b_1 A + \cdots + b_s A) =$$

$$= (b_1 M + \cdots + b_s M) \cap (C : b_s A) \cap (C : b_1 A + \cdots + b_{s-1} A) \subseteq$$

[4]By a **central ideal** of A we mean an ideal of A which can be generated by central elements of A.

$$\subseteq (b_1M + \cdots + b_{s-1}M) \cap (C : b_1A + \cdots + b_{s-1}A) =$$
$$= C + (b_1M + \cdots + b_{s-1}M) \cap (C : b_1A + \cdots + b_{s-1}A) = C$$

using the inductive hypothesis.

Now, set $m = tp$, where $p = \max\{p_1, \dots, p_t\}$. Then

$$\mathcal{I}^m M \cap N \subseteq \mathcal{I}^m M \cap (C : \mathcal{I}) \subseteq C + (b_1M + \cdots + b_tM) \cap (C : b_1A + \cdots + b_tA) \subseteq C = \mathcal{I}N$$

□

Proof of theorem 9.4.4.

Put $N = Ax$, where $x \in \bigcap_{i=1}^{\infty} \mathcal{I}^n M$. Then $Ax = \mathcal{I}^m M \cap Ax \subseteq \mathcal{I}Ax$, which implies that $x = ax$ for some $a \in \mathcal{I}$. □

Another version of the Krull intersection theorem for non-commutative rings connects with weakly Noetherian rings and duo rings.

Definition 9.4.9. A ring A is said to be **weakly Noetherian** if for any arbitrary ideal $\mathcal{I}$ of A and any $\overline{x} = x + \mathcal{I}$ in $A/\mathcal{I}$ there is an integer n such that $\text{r.ann}_A(\overline{x}^n) = \text{r.ann}_A(\overline{x}^m)$, and $\text{l.ann}_A(\overline{x}^n) = \text{l.ann}_A(\overline{x}^m)$ for all $m \geq n$.

It is clear that every Noetherian ring is weakly Noetherian.

A ring A is called a **duo ring** if every one-sided ideal is two-sided.

The following version of the Krull intersection theorem for weakly Noetherian duo rings was proved by O.A.S. Karamzadeh in [183]. We give here this theorem without any proof.

Theorem 9.4.10. *Let A be a weakly Noetherian duo ring, and let $\mathcal{I}$ be a finitely generated ideal in A. Then*

$$\bigcap_{n=1}^{\infty} \mathcal{I}^n = \{a \in A \ : \ a(1 - x) =$$
$$= 0 \ \mathit{for\ some}\ x \in \mathcal{I}\} = \{a \in A \ : \ (1 - y)a = 0 \ \mathit{for\ some}\ y \in \mathcal{I}\}.$$

Corollary 9.4.11. *If A is a weakly Noetherian duo ring and $\mathcal{I}$ is a finitely generated ideal contained in the Jacobson radical R, then* $\bigcap_{n=1}^{\infty} \mathcal{I}^n = 0$.

9.5 Jacobson's Conjecture

This section is devoted to the Jacobson conjecture for various different classes of rings.

Originally the Jacobson conjecture was posed by N. Jacobson in 1956 [160] in the following form:

Let A be a one-sided Noetherian ring, and let R be the Jacobson radical of A. Then the intersection of the finite powers of R is equal to zero, i.e.

$$\bigcap_{n\geq 0} R^n = 0.$$

The Jacobson conjecture in this formulation fails for one-sided Noetherian rings. This is shown by the following counterexample first considered by I.N. Herstein in [148]:

Example 9.5.1. Let

$$A = \begin{pmatrix} S & D \\ 0 & D \end{pmatrix},$$

where S is a commutative Noetherian domain with field of quotients D. Then A is a right Noetherian ring but not left Noetherian ring. The Jacobson radical R of A has the form:

$$R = \begin{pmatrix} J(S) & D \\ 0 & 0 \end{pmatrix},$$

where $J(S)$ is the Jacobson radical of S. Since $J(S)D = D$, we get that

$$\bigcap_{n=1}^{\infty} R^n \supseteq \begin{pmatrix} 0 & D \\ 0 & 0 \end{pmatrix} \neq 0.$$

In particular, if S is the field of rational numbers consisting of all rational numbers with odd denominators and $D = Q$ is the field of all rational numbers, then

$$R = \begin{pmatrix} 2S & Q \\ 0 & 0 \end{pmatrix},$$

and so

$$\bigcap_{n=1}^{\infty} R^n = \begin{pmatrix} 0 & Q \\ 0 & 0 \end{pmatrix} \neq 0.$$

Another class of counterexamples was given by Jategoankar A.V. [164], [165].

Therefore, from that time on the Jacobson conjecture is considered in the following form:

Jacobson's Conjecture. *Let A be a right and left Noetherian ring, and let R be the Jacobson radical of A. Then the intersection of the finite powers of R is equal to zero, i.e.,*

$$\bigcap_{n\geq 0} R^n = 0 \qquad (9.5.2)$$

This conjecture holds for any commutative Noetherian ring, which follows from the Krull intersection theorem considered in the previous section.

There are no general results for all Noetherian rings. But there are some special classes of Noetherian rings for which the Jacobson conjecture holds. These will be considered in this section.

Definition 9.5.3. A ring A with Jacobson radical R is said to satisfy the Jacobson conjecture if the intersection of powers of R is zero.

From corollary 9.4.11 there follows immediately:

Proposition 9.5.4. *A Noetherian duo ring satisfies the Jacobson conjecture.*

First we prove that Noetherian SPSD-rings[5] and Noetherian serial rings satisfy the Jacobson conjecture.

Theorem 9.5.5. *Let A be a Noetherian SPSD-ring with Jacobson radical R. Then* $\bigcap\limits_{i=1}^{\infty} R^i = 0$.

Proof.
Clearly, one can assume that A is a basic ring. Let $1 = f_1 + f_2 + \ldots + f_n$ be a canonical decomposition of the unity of A into a sum of pairwise orthogonal idempotents. Put $A_{ij} = f_i A f_j$ for $i,j = 1,2,\ldots,n$. Then A has the following Peirce decomposition:

$$A = \bigoplus_{i,j=1}^{n} A_{ij} \quad (i,j = 1,2,\ldots,n). \tag{9.5.6}$$

Denote by R_i the Jacobson radical of the ring A_{ii} for $i = 1,2,\ldots,n$. Then the radical R of A has the following Peirce decomposition

$$R = \bigoplus_{i,j=1}^{n} f_i R f_j, \tag{9.5.7}$$

where $f_i R f_i = R_i$ and $f_i R f_j = A_{ij}$ for $i \neq j$; $(i,j = 1,2,\ldots,n)$.
Write

$$\mathcal{I}_k = \begin{pmatrix} R_1^k & A_{12}R_2^k & \cdots & A_{1n}R_n^k \\ A_{21}R_1^k & R_2^k & \cdots & A_{2n}R_n^k \\ \vdots & \vdots & \ddots & \vdots \\ A_{n1}R_1^k & A_{n2}R_2^k & \cdots & R_n^k \end{pmatrix}. \tag{9.5.8}$$

Since $R_i A_{ij} = A_{ij} R_j$ by [146, corollary 14.2.4], $\mathcal{I}_k$ is a two-sided ideal of A and $\mathcal{I}_k \mathcal{I}_r = \mathcal{I}_{k+r}$. Moreover, it is easy to show that $R^n \subseteq \mathcal{I}_1$. Thus $R^{nk} \subseteq (\mathcal{I}_1)^k = \mathcal{I}_k$ and

$$\bigcap_{k=1}^{\infty} R^k \subseteq \bigcap_{k=1}^{\infty} \mathcal{I}_k.$$

By [146, proposition 14.4.10] A_{ii} is a Noetherian uniserial ring and it is either a discrete valuation ring or an Artinian uniserial ring, for $i = 1,2,\ldots,n$. The

[5]Recall that we write **SPSD-ring** for a semiperfect semidistributive ring.

intersection of all natural powers of the Jacobson radical of these rings is zero, by [146, proposition 10.2.1] and the Hopkins theorem 1.1.21. Thus

$$\bigcap_{k=1}^{\infty} A_{ij} R_j^k = 0, \quad i,j = 1,2,\dots,n.$$

Hence the intersection of I_k over all natural k is zero and $\bigcap_{k=1}^{\infty} R^k = 0$. □

Since any serial ring is an SPSD-ring, as an immediate consequence from this theorem there results the next statement.

Theorem 9.5.9 (R.B.Warfield, Jr. [317].) *Let A be a Noetherian serial ring with Jacobson radical R. Then $\bigcap_{i=1}^{\infty} R^i = 0$.*

Remark 9.5.10. Suppose A is an indecomposable semiperfect ring and A/R^2 is a right Artinian ring. If the quiver $Q(A)$ is disconnected then

$$\bigcap_{k=1}^{\infty} R^k \neq 0.$$

Theorem 9.5.11. (R.B.Warfield, Jr. [320, Theorem 5.11], S.Singh [288, Lemma 2.1]). *Let A be a serial ring with Jacobson radical R. Then A is a Noetherian ring if and only if $I = \bigcap_{i=1}^{\infty} R^i = 0$. In particular. A/I is a Noetherian ring for every serial ring A.*

Proof.

The "only if" part is theorem 9.5.8.

As to the inverse statement. Suppose $I = \bigcap_{i=1}^{\infty} R^i = 0$ and let $A_A = P_1^{n_1} \oplus \dots \oplus P_s^{n_s}$ be a decomposition of A_A into a direct sum of principal right A-modules. Observe that for every $m \in \mathbf{N}$ such that $P_i R^m \neq 0$ the inclusion $P_i R^{m+1} \subset P_i R^m$ is proper. Indeed if $P_i R^m = P_i R^{m+1}$ then

$$P_i R^m = P_i R^{m+1} = P_i R^{m+2} = \dots$$

Thus $P_i R^m \subset \bigcap_{n=1}^{\infty} P_i R^n = 0$ and so $P_i R^m = 0$. A contradiction. Obviously, $P_i R^m / P_i R^{m+1}$ is a simple module and if $x \in P_i R^m \setminus P_i R^{m+1}$, then $xA \supset P_i R^{m+1}$ (P_i is a uniserial module) and $xA = P_i R^m$. Therefore P_i is a Noetherian module for $i = 1,\dots,s$. Consequently, A is right Noetherian. Analogously, A is left Noetherian.

In particular, if A is a serial ring, then the ring $B = A/I$ is also serial and the Jacobson radical of B is $J = R/I$. Obviously, the intersection $\bigcap_{i=1}^{\infty} J^i = 0$, so B is a Noetherian ring. □

Remark 9.5.12. Theorem 9.5.4 is false for right Noetherian serial rings as is shown by the following example.

Example 9.5.13.

Let $\mathbf{Q}$ be the field of rational numbers, p a prime integer,

$$\mathbf{Z}_p = \{m/n \in \mathbf{Q} \ : \ (n,p) = 1\}$$

and

$$H(\mathbf{Z}_p, 1, 1) = \begin{pmatrix} \mathbf{Z}_p & \mathbf{Q} \\ 0 & \mathbf{Q} \end{pmatrix}.$$

It is easy to see that the intersection of all natural powers of the Jacobson radical of the ring $H(\mathbf{Z}_p, 1, 1)$ is the ideal

$$\mathcal{I} = \begin{pmatrix} 0 & \mathbf{Q} \\ 0 & 0 \end{pmatrix} \neq 0.$$

The ring $H(\mathbf{Z}_p, 1, 1)$ is serial and right Noetherian and, hence, semidistributive.

The Jacobson conjecture is also true for any Noetherian ring with Krull dimension one. This result was shown by T.H.Lenagan in [216]. We state this theorem for completeness sake without any proof.

Theorem 9.5.14. (T.H. Lenagan [216]). *If A is a Noetherian ring and the right Krull dimension of A does not exceed 1, then*

$$\bigcap_{n \geq 0} R^n = 0,$$

where R is the Jacobson radical of A.

Definition 9.5.15. A right Noetherian ring A is **right bounded** if every essential right ideal of A contains a non-zero two-sided ideal. A **right FBN-ring** A is a right Noetherian ring such that A/P is right bounded for every prime ideal P. A ring which is both a left and right FBN-ring is called an **FBN-ring.**[6]

Obviously, any commutative Noetherian ring is an FBN-ring.

G. Cauchon and A.V. Jategaonkar showed independently that the Jacobson conjecture holds for left Noetherian right FBN-rings.

Theorem 9.5.16. (G. Cauchon [39], A.V. Jategaonkar [167]). *Let A be a left Noetherian right* FBN-*ring. Then*

$$\bigcap_{n \geq 0} R^n = 0.$$

[6]Note that a right FBN-ring is short for a right fully bounded right Noetherian ring. An FBN-ring is short for a fully bounded Noetherian ring.

The following example shows that even two-sided fully bounded rings with Krull dimension one do not necessary satisfy the Jacobson conjecture.

Example 9.5.17. (A.Woodward [327, Example 4.7.1]).

Let S be a commutative discrete valuation ring with the unique maximal ideal M, and U a simple S-module. Let E be the injective hull of U_S. Consider the ring:

$$A = \left\{ \begin{pmatrix} x & y \\ 0 & x \end{pmatrix} \; : \; x \in S, y \in E \right\}$$

Then A is a commutative ring. Let $\mathcal{I}$ be an ideal of A given by

$$\mathcal{I} = \begin{pmatrix} 0 & E \\ 0 & 0 \end{pmatrix}.$$

Then $\text{K.dim}(\mathcal{I}_A) = \text{K.dim}(E_S)$, since $E = E(U_S)$ is Artinian over the commutative Noetherian ring S. Also $A/\mathcal{I} \simeq S$, so $\text{K.dim}(A/\mathcal{I}) = \text{K.dim}(S) = 1$. It follows that $\text{K.dim}(A) = 1$. Thus A is both right and left fully bounded ring with Krull dimension one. However, the Jacobson radical of A has the form:

$$R = \left\{ \begin{pmatrix} m & y \\ 0 & m \end{pmatrix} \; : \; m \in M, y \in E \right\}$$

Since a commutative Noetherian ring satisfies the Jacobson conjecture, $\bigcap\limits_{n=0}^{\infty} M^n = 0$. Since E is injective over the domain S, E is injective and hence $EM = E$. It follows that

$$\bigcap_{n=0}^{\infty} R^n = \begin{pmatrix} 0 & E \\ 0 & 0 \end{pmatrix} \neq 0,$$

so A does not satisfy the Jacobson conjecture.

9.6 Notes and References

The structure of serial right hereditary rings was considered in [146]. The main results on this theory were obtained by V.V.Kirichenko and R.B.Warfield, Jr. (see [187], [188], [189], [317], [321]). Serial piecewise domains were studied by N.Gubareni and M.Khibina in [128].

The structure of nonsingular serial rings and their properties were studied by many authors. The main theorem 9.2.4 was proved by R.B. Warfield in [317]. Proposition 9.2.7 is due to A.Facchini and G. Puninski [83, Lemma 2.2]. M.H.Upham in [310] studied serial rings with Krull dimension one and described the structure of these rings. Also the structure of serial nonsingular rings was considered by O.Gregul' and V.V. Kirichenko in [122] where proposition 9.2.11, lemma 9.2.14, theorem 9.2.15 and theorem 9.2.18 were proved. The proofs of these theorems given in this book are some different from thats in [122].

Serial rings with Noetherian diagonal were studied by N.Gubareni, M.Khibina and V.V. Kirichenko in [129], [128].

The Artin-Rees lemma was proved independently by E. Artin (unpublished) and D.Rees [268] in 1956.

The Jacobson conjecture was posed by N.Jacobson for noncommutative Noetherian rings in 1956 (see [162], p.200). This conjecture holds for any commutative Noetherian ring (see [181], theorem 79). It was proved that this conjecture holds for Noetherian FBN-rings by G.Cauchon in [39] and A.V.Jategaonkar in [166], [167], and for Noetherian rings with Krull dimension one by T.H.Lenagan in [216].

Theorem 9.5.11 was proved by R.B.Warfield, Jr. in [317, theorem 5.11]. The inverse statement of theorem 9.5.11 is due to S.Singh [288, lemma 2.1].

REFERENCES

[1] Albu, T. 1974. Sur la dimension de Gabriel des modules. Algebra Berichte 21, Seminar Kasch-Pareigeis, Munich.

[2] Albu, T. and M.L. Teply. 2001. On the transfinite powers of the Jacobson radical of a DICC ring. J. Korean Math. Soc. 38(6): 1117–1123.

[3] Anderson, D.D. 1975. The Krull intersection theorem. Pacific J. Math. 57: 11–14.

[4] Anderson, D.D. and D. Dobbs (eds.). 1995. Zero-dimensional Commutative Rings. Marcel Dekker, New York.

[5] Anderson, D.D., J. Matijevic and W. Nichols. 1976. The Krull intersection theorem II. Pacific J. Math. 66(1): 15–22.

[6] Anderson, D.D. and K.R. Fuller. 1992. Rings and categories of modules. Second edition, Springer-Verlag, New York.

[7] Assem, I., D. Simson and A. Skowroński. 2006. Elements of the Representation Theory of Associative Algebras. v.I, Techniques of Representation Theory, Cambridge Univ. Press, London Math. Soc. Student Texts 65.

[8] Atiyah, M.F. and I.G. Macdonald. 1969. Introduction to commutative algebra. Addison-Wesley, Massachusetts, London, Amsterdam.

[9] Auslander, M. 1955. On the dimension of modules and algebras, III. Global dimension. Nagoya Math. J. 9: 67–77.

[10] Auslander, M. and I. Reiten. 1996. Syzygy modules for Noetherian rings. Algebra 183(1): 167–185.

[11] Auslander, M., I. Reiten and S. Smalø. 1995. Representation Theory of Artin Algebras. Cambridge University Press, Cambridge.

[12] Azumaya, G. 1950. Corrections and supplementaries to my paper concerning Krull-Remak-Schmidt's theorem. Nagoya Math. J. 1: 117–124.

[13] Baccella, G. 2002. Exchange property and the natural preorder between simple modules over semi-Artinian rings. J. Algebra 253: 133–166.

[14] Baer, R. 1942. Inverses and zero-divisors. Bulletin of the American Math. Soc. 48: 630–638.

[15] Bass, H. 1960. Finitistic dimension and homological generalization of semiprimary rings. Trans. Amer. Math. Soc. 95: 466–488.

[16] Bass, H. 1964. *K*-theory and stable algebra. Publ. Math. IHES. 22: 5–60.

[17] Bautista, R. 1981. On algebras close to hereditary artin algebras. An. Inst. Mat. Univ. Nac. Autónoma México, 21(1): 21–104.

[18] Bautista, R. and D. Simson. 1984. Torsion modules over l-Gorenstein l-hereditary Artinian rings. Comm. Algebra 12(7-8): 899–936.

[19] Bazzoni, S. and S. Glaz. 2006. Prüfer Rings. pp. 55–72. *In*: Multiplicative Ideal Theory in Commutative Algebra, Springer, Berlin.

[20] Bessenrodt, Ch. and G. Törner. 1987. Locally Archimedian right-ordered groups and locally Invariant valuation rings. J. Algebra 105: 328–340.

[21] Birkenmeier, G.F., H.E. Heatherly, J.Y. Kim and J.K. Park. 2000. Triangular Matrix Representations. J. Algebra 230: 558–595.

[22] Birkhoff, G. 1973. Lattice theory. Colloq. Publ. 25. Amer. Math. Soc.

[23] Boisen, M.B. and M.D. Larsen. 1972. Prüfer and valuation rings with zero divisors. Pacific J. Math. 40(1): 7–12.

[24] Bourbaki, N. 1989. Elements of Mathematics. Commutative Algebra. Springer-Verlag, Berlin.

[25] Braun, A. and C.R. Hajarnavis. 2011. The failure of the Artin-Rees property for the Jacobson radical in prime Noetherian ring. J. Algebra 348: 294–301.

[26] Breaz, S., G. Gălugăreanu and P. Schultz. 2011. Modules with Dedekind finite endomorphism rings. Mathematika 76: 15: 28.
[27] Brewer, J.W. 1981. Power Series over Commutative Rings. Marcel Dekker Inc., New York.
[28] Brungs, H.H. and J. Gräter. 1989. Valuation rings in finite-dimensional division algebras. J. Algebra 120: 90–99.
[29] Brungs, H.H. and J. Gräter. 1992. Noncommutative Prüfer and valuation rings. Contemporary Mathematics 131(2): 253–269.
[30] Buchsbaum, D.A. 1955. Exact categories and duality. Trans. Amer. Math. Soc. 80(1): 134.
[31] Butts, H.S. and W.W. Smith. 1967. Prüfer rings. Math. Z. 95: 196–211.
[32] Camillo, V.P. and D. Khurana. 2001. A characterization of unit regular rings. Comm. in Algebra 29(5): 2293–2295.
[33] Camillo, V.P. and H.-P. Yu. 1994. Exchange rings, units and idempotents. Comm. Algebra 22(12): 4737–4749.
[34] Camillo, V.P. and H.-P. Yu. 1995. Stable range one for rings with many idempotents. Trans. Amer. Math. Soc. 347: 3141–3147.
[35] Campbell, J.M. 1973. Torsion theories and coherent rings. Bull. Austr. Math. Soc. 8: 233–239.
[36] Cartan, H. and S. Eilenberg. 1956. Homological algebra. Prinston University Press, Prinston, New York.
[37] Caruth, A. 1993. A short proof of Krull's intersection theorem. Colloquim Math. 64: 153–154.
[38] Catefories, V.C. 1969. Flat regular quotient rings. Trans. Amer. Math. Soc. 138: 241–249.
[39] Cauchon, G. 1974. Sur l'intersection des puissances du radical d'un T-anneau noethérien. C. R. Acad. Sci. Paris, Sér. A. 279: 91–93.
[40] Chandler, B. and W. Magnus. 1982. The History of Combinatorial Group theory: A Case Study in the History of ideas. New York-Heidelberg-Berlin 91–98.
[41] Chase, S.U. 1960. Direct Product of Modules. Trans. Amer. Math. Soc. 97(3): 457–473.
[42] Chase, S.U. 1961. A generalization of the ring of triangular matrix. Nagoya Math. J. 18: 13–25.
[43] Chatters, A.W. 1972. A decomposition theorem for Noetherian hereditary rings. Bull. London Math. Soc. 4: 125–126.
[44] Chatters, A.W. 1979. A note on Noetherian orders in Artinian rings. Glasgow Math. J. 29: 125–128.
[45] Chatters, A.W. 1990. Serial rings with Krull dimension. Glasgow Math. J. 32: 71–78.
[46] Chatters, A.W., A.W. Goldie, C.R. Hajarnavis and T.H. Lenagan. 1979. Reduced rank in Noetherian rings. J. Algebra 61: 582–589.
[47] Chatters, A.W. and C.R. Hajarnavis. 1977. Rings in which every complement right ideal is a direct summand. Quart. J. Math. Oxford Ser. 28(2): 61–80.
[48] Chatters, A.W. and C.R. Hajarnavis. 1980. Rings with Chain conditions. Research Notes in Mathematics 44, Pitman, London.
[49] Chatters, A.W. and P.F. Smith. 1977. A note on hereditary rings. J. Algebra 44: 181–190.
[50] Chen, H. 2000. On stable range condition. Comm. Alg. 28: 3913–3924.
[51] Chen, H. 2001. Exchange rings having stable range one. Int. J. Math. Sci. 25: 763–770.
[52] Chen, H. 2001. Idempotents in exchange rings. New Zealand J. Math. 30: 103–110.
[53] Chen, J. and N. Ding. 1999. Characterization of Coherent rings. Comm. Alg. 27(5): 2491–2501.
[54] Chen, J. and Y. Zhou. 2006. Extensions of Injectivity and coherent rings. Comm. Alg. 34(1): 275–288.
[55] Chiba, K. 2003. Free fields in complete skew fields and their valuations. J. Algebra 263: 75–87.
[56] Cohen, M. and S. Montgomery. 1984. Group-graded rings, smash products and group actions. Trans. Amer. Math. Soc. 282(1): 237–258.
[57] Cohen, M. and L. Rowen. 1983. Group graded rings. Comm. Algebra 11: 1253–1270.
[58] Cohn, P.M. 1956. The complement of a finitely generated direct summand of an abelian group. Proc. Amer. Math. Soc. 7: 520–521.
[59] Cohn, P.M. 1966. Some remarks on the invariant basis property. Topology 5: 215–228.
[60] Cohn, P.M. 1991. Algebra, vol. 3, Wiley.
[61] Crawley, P. and B. Jónsson. 1964. Refinements for infinite direct decompositions of algebraic systems. Pacific J. Math. 14: 797–855.
[62] Curtis, C. and I. Reiner. 1962. Representation theory of finite groups and associative algebras. J. Wiley & Sons, Inc., New York.
[63] Dade, E.C. 1980. Group-Graded Rings and Modules. Math. Z. 174: 241–262.
[64] Dedekind, R. 1887 (1969). Was Sind und Was Sollen die Zahlen? Vieweg, Braunschweig (reprint).

[65] Dinh, H.Q., Asensio P.A. Guil and S.R. Lopez-Permouth. 2005. On the Goldie dimension of hereditary rings and modules. arXiv:math.RA/0512337, 1.

[66] Dokuchaev, M.A., N.M. Gubareni and V.V. Kirichenko. 2011. Rings with finite decomposition of identity. Ukrain. Mat. Zh. 63(3): 319–340.

[67] Drozd, Yu.A. 1975. On serial rings. Math. Zametki. 18(5): 705–710.

[68] Drozd, Yu.A. 1980. The structure of hereditary rings. Math. Sbornik 113(155), N.1(9): 161–172 (in Russian); English translation: 1982. Math. USSR Sbornik 41(1): 139–148.

[69] Drozd, Yu.A. and L.V. Izjumchenko. 1992. Representations of $\mathcal{D}$-species. Ukr. Math. Journal 44: 572–574.

[70] Drozd, Yu.A. and V.V. Kirichenko. 1994. Finite dimensional algebras. Springer-Verlag.

[71] Dubrovin, N.I. 1982. Noncommutative valuation rings. Trudy Moskov. Obshch. 45: 265–280 (in Russian); English translation: 1984. Trans. Moscow Math. Soc. 45: 273–287.

[72] Dubrovin, N.I. 1984. Noncommutative valuation rings in simple finite-dimensional algebras over a field. Mat. Sb. (N.S.) 123(165): 496–509 (in Russian); English translation: 1985. Math. USSR Sb. 51: 493–505.

[73] Dubrovin, N.I. 1991. Noncommutative Prüfer rings. Math. Sbornik 182(9): 1251–1260 (in Russian); English translation: 1993. Math. USSR Sbornik 74(1): 1–8.

[74] Dung, N. and A. Facchini. 1997. Weak Krull-Schmidt for infinite direct sums of uniserial modules. J. Algebra 193: 102–121.

[75] Eilenberg, S., H. Nagao and T. Nakayama. 1956. On the dimension of modules and algebras IV. Nagoya Math. J. 10: 87–96.

[76] Eisenbud, D. 1995. Commutative Algebra. Springer, Berlin.

[77] Eisenbud, D. and P. Griffith. 1971. The structure of serial rings. Pacific J. Math. 36(1): 109–121.

[78] Eisenbud, D. and P. Griffith. 1971. Serial rings. Journal of Algebra 17: 389–400.

[79] Eisenbud, D. and J.C. Robson. 1970. Modules over Dedekind Prime Rings. Journal of Algebra 16: 67–85.

[80] Eisenbud, D. and J.C. Robson. 1970. Hereditary Noetherian Prime Rings. Journal of Algebra 16: 86–104.

[81] Evans, E.G., Jr. 1973. Krull-Schmidt and cancelation over local rings. Pacific J. Math. 46(1): 115–121.

[82] Facchini, A. 1998. Module Theory: endomorphism rings and direct sum decompositions in some classes of modules. Birkhäser, Basel.

[83] Facchini, A. and G. Puninski. 1996. Classical localizations in serial rings. Comm. Algebra 24: 3537–3559.

[84] Faith, C. 2003. Dedekind finite rings and a theorem of Kaplansky. Comm. Algebra 31(9): 4175–4178.

[85] Ferreira, V.O. and É.Z. Fornaroli. Free fields in skew fields. arXiv: 0805.4185v1.

[86] Fossum, R.M., P.A.Griffith and I. Reiten. 1975. Trivial Extensions of Abelian Categories. Lecture Notes in Math. 456.

[87] Fuchs, L. 1963. Partially ordered algebraic systems. Pergamon.

[88] Fuchs, L. 1972. The cancellation property for modules. *In*: Lecture Notes in Math. 246: 191–212.

[89] Fuchs, L. and F. Loonstra. 1971. On the cancellation of modules in direct sums over Dedekind domains. Indag. Math. 33: 163–169.

[90] Fujiwara, T. and K. Murata. 1953. On the Jordan-Hölder-Schreier theorem. Proc. Japan Academy 29(4): 151–153.

[91] Gabriel, P. 1962. Des catégories abéliennes. Bull. Soc. Math. France 90: 323–448.

[92] Garibaldi, S. and D.J. Saltman. 2010. Quaternion algebras with the same subfields. pp. 225–238. *In*: J.-L. Colliot-Thélène et al. (eds.). Quadratic Forms, Linear Algebraic Groups, and Cohomology. Developments in Math. 18, Springer, New York.

[93] Gilmer, R. 1968. Multiplicative ideal theory. Queen's Univ. Press; second ed. 1992.

[94] Gilmer, R. 1969. Two constructions of Prüfer domains. J. Reine Angew. Math. 239-240: 153–162.

[95] Gilmer, R. 1972. On Prüfer rings. Bull. Amer. Math. Soc. 78: 223–224.

[96] Gilmer, R. 1972. Multiplicative ideal theory. Marcel Dekker.

[97] Glaz, S. 2005. Prüfer conditions in rings with zero-divisors. *In*: Lecture Notes in Pure and Appl. Math. 241: 272–281.

[98] Glaz, S. and R. Schwaez. 2011. Prüfer conditions in Commutative Rings. Arab J. Sci. Eng. 36: 967–983.

[99] Goldie, A.W. 1952. The scope of the Jordan-Hölder theorem in abstract algebra. Proc. London Math. Soc. 3(2): 107–113.
[100] Goldie, A.W. 1958. The structure of prime rings with maximum conditions. Proc. Nat. Acad. Sci. USA 44: 584–586.
[101] Goldie, A.W. 1958. The structure of prime rings under ascending chain conditions. Proc. London Math. Soc. 8(3): 589–608.
[102] Goldie, A.W. 1960. Semi-prime rings with maximum condition. Proc. London Math. Soc. 10(3): 201–210.
[103] Goldie, A.W. 1964. Torsion-free modules and rings. J. Algebra 1: 268–287.
[104] Goldie, A.W. 1972. The structure of Noetherian rings. *In*: Lectures on Rings and Modules, Lecture Notes in Mathematics, Springer-Verlag. Berlin 246: 213–321.
[105] Goldie, A.W. 1972. Properties of the idealiser. pp. 161–169. *In*: R. Gordon (ed.). Ring Theory. Academic Press, New York.
[106] Goldie, A.W. 1974. Some recent developments in ring theory. Israel J. Math. 19: 158–168.
[107] Goldie, A.W. and L. Small. 1973. A study in Krull dimension. J. Algebra 25: 152–157.
[108] Goodearl, K.R. 1976. Direct sum properties of quasi-injective modules. Bull. Amer. Math. Soc. 82: 108–110.
[109] Goodearl, K.R. 1976. Rings theory. Nonsingular rings and modules. Marcel Dekker, Inc., New York and Basel.
[110] Goodearl, K.R. 1991. VonNeumann Regular Rings, 2nd edn. Krieger Publ. Co., Malabar, Florida.
[111] Goodearl, K.R. and P. Menal. 1977. Stable range one for rings with many units. J. Pure Appl. Algebra 229: 269–278.
[112] Goodearl, K.R. and L.W. Small. 1984. Krull versus global dimension in Noetherian P.I. rings. Proc. Amer. Math. Soc. 92(2): 175–178.
[113] Goodearl, K.R. and R.B. Warfield, Jr. 1976. Algebras over zero-dimensional rings. Math. Ann. 223: 157–168.
[114] Goodearl, K.R. and R.B. Warfield, Jr. 1982. Krull dimension of differential operator rings. Proc. London Math. Soc. 45: 49–70.
[115] Goodearl, K.R. and R.B. Warfield, Jr. 1989. An introduction to noncommutative Noetherian rings. London Mathematical Society Student Texts 16. Cambridge Univ. Press, Cambridge.
[116] Gordon, R. 1971. Semi-prime right Goldie rings which are direct sums of uniform ideals. Bull. London Math. Soc. 3: 277–282.
[117] Gordon, R. and J.C. Robson. 1973. Krull dimension. Mem. Amer. Math. Soc. 133.
[118] Gordon, R. and L.W. Small. 1972. Piecewise domains. J. Algebra 23: 553–564.
[119] Gräter, J. 1986. Lokalinvariant Bewertungen. Math. Z. 192: 183–194.
[120] Gräter, J. 1986. On commutative Prüfer rings. Arch. Math. (Basel) 46: 402–407.
[121] Gräter, J. 1999. Dubrovin valuation rings and orders in central simple algebras. pp. 151–162. *In*: F.-V. Kuhlmann and M. Marshall (eds.). Valuation Theory and its Applications 1. Amer. Math. Soc.
[122] Gregul', O. and V.V. Kirichenko. 1987. On semihereditary serial rings. Ukr. Math. J. 39(2): 156–161.
[123] Griffin, M. 1969. Prüfer rings with zero divisors. J. Reine Angew. Math. 239/240: 55–67.
[124] Griffin, M. 1974. Valuations and Prüfer rings. Can. J. Math. 26(2): 412–429.
[125] Grothendieck, A. 1957. Sur quelques points d'algèbre homologique. The Tohoku Mathematical Journal. Second Series 9: 119–221.
[126] Gubareni, N. 2011. Valuation and discrete valuation rings. Scientific Research of the Institute of Mathematics and Computer Science 1(10): 61–70.
[127] Gubareni, N. 2014. Semiprime FDI-rings. Journal of Appl. Math. and Comp. Mech. 13(4): 49–59.
[128] Gubareni, N. and M. Khibina. 2007. Serial piecewise domains. Algebra and Discrete Mathematics 4: 59–72.
[129] Gubareni, N. and V.V. Kirichenko. 2006. Serial Rings with T-Nilpotent Prime Radical. Algebra and Representation Theory 9: 147–160.
[130] Gulliksen, T.H. 1974. The Krull ordinal, coprof, and Noetherian localization of large polynomial rings. Amer. J. Math. 96: 324–339.
[131] Guralnick, R. 1991. Cancellation and direct summands in dimension 1. J. Algebra 142: 310–347.
[132] Facchini, A. 1998. Module theory: Endomorphism Rings and direct sum decomposition in some classes of modules. Springer.

[133] Facchini, A. and G. Puninski. 1996. Classical localizations in serial rings. Comm. Algebra 24: 3537–3559.
[134] Fuchs, L. 1949. Über die Ideale arithmetischer Ringe. Comment. Math. Helv. 23: 334–341.
[135] Haghany, A. and K. Varadarayan. 1999. Study of formal triangular matrix rings. Comm. Algebra 27: 5507–5525.
[136] Haghany, A. and K. Varadarayan. 2000. Study of modules over a formal triangular matrix ring. J. Pure and Appl. Algebra 147: 41–58.
[137] Hahn, H. 1907. Über die nicht-archimedischen Größensysteme. Sitzungsberichte der Akademie der Wissenschaften. Math. Naturw. Kl. Band 116, Wien.
[138] Handelman, D. 1977. Perspectivity and cancellation in regular rings. J. Algebra 48: 1–16.
[139] Harada, M. 1966. Hereditary semi-primary rings and triangular matrix rings. Nagoya Math. J. 27: 463–484.
[140] Harada, M. and T. Ishii. 1975. On perfect rings and the exchange property. Osaka J. Math. 121: 483–491.
[141] Harris, N. 1966. Some results on coherent rings. Proc. Amer. Math. Soc. 17(2): 474–479.
[142] Harris, N. 1967. Some results on coherent rings, II. Glasgow Math. Soc. 8: 123–126.
[143] Harris, M.E. 1987. Filtrations, Stable Clifford Theory and Group-Graded Rings and Modules. Math. Z. 196: 497–510.
[144] Hart, R. 1967. Simple rings with uniform right ideals. Journal London Math. Soc. 42: 614–617.
[145] Hart, R. 1971. Krull dimension and global dimension of simple Ore-extensions. Math. Zeitschrift 121: 341–345.
[146] Hazewinkel, M., N. Gubareni and V.V. Kirichenko. 2004. Algebras, rings and modules. Volume 1. Kluwer Acad. Publ.
[147] Hazewinkel, M., N. Gubareni and V.V. Kirichenko. 2007. Algebras, rings and modules. Volume 2. Springer.
[148] Herstein, I.N. 1965. A counterexample in Noetherian rings. Proc. N.A.S. 54(4): 1036–1037.
[149] Hilbert, D. 1903. Grundlagen der Geometrie. Leipzig.
[150] Hill, D. 1992. Left serial rings and their factor skew fields. J. Algebra 146: 30–48.
[151] Hirano, Y., H. Komatsu and L. Mogami. 1986. On Jacobson's conjecture. Math. Journal of Okayama University 28(1): 115–118.
[152] Hodges, T.J. 1984. The Krull dimension of skew Laurent extensions of commutative Noetherian rings. Comm. Algebra 12: 1301–1310.
[153] Hsü, C.S. 1962. Theorems on direct sums of modules. Proc. Amer. Math. Soc. 13: 540–542.
[154] Huang, D. and N. Dung. 1990. A characterization of rings with Krull dimension. J. Algebra 132: 104–112.
[155] Huckaba, J.A. 1988. Commutative rings with zero-divisors. Marcel Dekker.
[156] Hughes, N.J. 1959-1960. The Jordan-Hölder-Schreier theorem for general algebraic systems. Compositio Math. 14(3): 228–236.
[157] Huynh, D. and P. Dan. 1991. On serial Noetherian rings. Archiv der Math. 56(6): 552–558.
[158] Ivanov, G. 1974. Left generalized uniserial rings. J. Algebra 31: 166–181.
[159] Jacobinski, H. 1971. Two remarks about hereditary orders. Proc. Amer. Math. Soc. 28: 1–8.
[160] Jacobson, N. 1945. The radical and semisimplicity for arbitrary rings. Amer. J. Math. 67: 300–320.
[161] Jacobson, N. 1950. Some remarks on one-sided inverses. Proc. Amer. Math. Soc. 1: 352–355.
[162] Jacobson, N. 1956. Structure of rings. Colloquium Publications 37. Providence, Amer. Math. Soc.
[163] Jans, J.P. and T. Nakayama. 1957. On the dimension of modules and algebras VII. Nagoya Math. J. 11: 67–76.
[164] Jategaonkar, A.V. 1968. Left principal ideal domains. J. Algebra 8: 148–155.
[165] Jategaonkar, A.V. 1969. A counter-example in ring theory and homological algebra. J. Algebra 12: 418–440.
[166] Jategaonkar, A.V. 1973. Injective modules and classical localization in Noetherian rings. Bull. Amer. Math. Soc. 79: 152–157.
[167] Jategaonkar, A.V. 1974. Jacobson's conjecture and modules over fully bounded Noetherian rings. J. Algebra 30: 103–121.
[168] Jategaonkar, A.V. 1981. Noetherian bimodules, primary decompositions, and Jacobson conjecture. J. Algebra 71: 379–400.
[169] Jensen, Chr. U. 1964. A remark on arithmetical rings. Proc. Amer. Math. Society 15: 951–954.
[170] Jensen, Chr. U. 1966. Arithmetical rings. Acta Math. Acad. Sci. Hung. 17: 115–123.

[171] Jensen, A. and S. Jøndrup. 1991. Smash products, group actions and group graded rings. Math. Scand. 68: 161–170.
[172] Jespers, E. and P. Wauters. 1988. A general notion of noncommutative Krull rings. J. Algebra 112: 388–415.
[173] Johnson, R.E. 1951. The extended centralizer of a ring over a module. Proc. Amer. Math. Soc. 2: 891–895.
[174] Johnson, R.E. 1957. Structure theory of faithful rings II. Restricted rings. Trans. Amer. Math. Soc. 84: 523–544.
[175] Johnson, R.E. 1960. Structure theory of faithful rings III. Irreducible rings. Proc. Amer. Math. Soc. 11(5): 710–717.
[176] Kanbara, H. 1972. Note on Krull-Remak-Schmidt-Azumaya's theorem. Osaka J. Math. 9: 402–413.
[177] Kaplansky, I. 1952. Modules over Dedekind rings and valuation rings. Trans. Amer. Math. Soc. 72: 327–340.
[178] Kaplansky, I. 1958. On the dimension of modules and algebras, X. A right hereditary ring which is not left hereditary. Nagoya Math. J. 13: 85–88.
[179] Kaplansky, I. 1958. Projective modules. Ann. Math. 68: 372–377.
[180] Kaplansky, I. 1969. Infinite Abelian groups (Second ed.). Univ. of Michigan Press.
[181] Kaplansky, I. 1970. Commutative algebra. Boston, Allyn and Bacon.
[182] Kaplansky, I. 1971. Bass's first stable range condition (mimeographed notes).
[183] Karamzadeh, O.A.S. 1983. On the Krull intersection theorem. Acta Math. Hung. 42: 139–141.
[184] Kelarev, A.V. 2002. Rings constructions and applications. Series in Algebra 9. World Scientific, River Edge, New York.
[185] Kelarev, A.V. and J. Okniński. 1995. Group graded rings satisfying polynomial identities. Glasgow Math. J. 37: 205–210.
[186] Kelarev, A.V. and J. Okniński. 1996. The Jacobson radical of graded PI-rings and related classes of rings. J. Algebra 186(3): 818–830.
[187] Kirichenko, V.V. 1976. Generalized uniserial rings. Mat. sb. 99(141), N.4: 559–581.
[188] Kirichenko, V.V. 1976. Right Noetherian rings over which all finitely generated modules are semichain modules. Dokl. Acad. Nauk Ukrain. SSR, Ser. A 1: 9–12.
[189] Kirichenko, V.V. 1982. Hereditary and semihereditary serial rings (in Russian). Zapiski Nauchnych Sem. LOMI AN SSSR 114: 137–147.
[190] Kirichenko, V.V. 2000. Semiperfect semidistributive rings. Algebras and Representation theory 3: 81–98.
[191] Kirichenko, V.V. and M.A. Khibina. 1993. Semiperfect semidistributive rings. pp. 457–480. *In*: Infinite Groups and Related Algebraic Structures (in Russian). Acad. Nauk Ukrainy, Inst. Mat., Kiev.
[192] Kirkman, E. and J. Kuzmanowich. 1989. Global dimensions of a class of tiled orders. J. Algebra 127: 57–72.
[193] Kokorin, A.I. and V.M. Kopytov. 1974. Fully ordered groups. Halsted Press, John Wiley & Sons.
[194] Kokorin, A.I. and V.P. Kopytov. 1974. Fully ordered groups. Israel Progr. Sci. (Translated from Russian).
[195] Kokorin, A.I. and V.M. Kopytov. 1993. Totally ordered group. *In*: M. Hazewinkel (ed.). Encyclopaedia of Mathematics. KAP 9: 231.
[196] Kopytov, V.M. and N.Ya. Medvedev. 1994. The theory of lattice-ordered groups. KAP.
[197] Krause, G. 1970. On the Krull-dimension of left Noetherian left Matlis-rings. Math. Zeitschrift 118: 207–214.
[198] Krause, G. 1972. On fully left bounded left Noetherian rings. J. Algebra 23: 88–99.
[199] Krause, G. and T.H. Lenagan. 1979. Transfinite powers of the Jacobson radical. Comm. Algebra 7: 1–8.
[200] Krull, W. 1928. Primidealketten in allgemeinen Ringbereichen. S.B. Heidelberg, Akad. Weiss 7. Abh.
[201] Krull, W. 1932. Allgemeine Bewertungstheorie. J. Reine Angew. Math. 167: 160–196.
[202] Krull, W. 1936. Beitrage zur arithmetik kommtativer integritatsbereiche. Math. Zeit. 41: 545–577.
[203] Krull, W. 1951. Jacobsonsche Ringe, Hilbertscher Nullstellensatz, Dimensiontheorie. Math. Zeit. 54: 354–387.
[204] Kurosh, A.G. 1956. The theory of groups 1–2. AMS Chelsea Publ. 1955-1956 (Translated from Russian).

[205] Lady, E. Lee. Finite Rank Torsion Modules Over Dedekind Domains. http://www.math.hawaii.edu/ lee/book/.
[206] Lam, T.Y. 1999. Bass's work in ring theory and projective modules. *In*: T.Y. Lam and A.R. Magid (eds.). Algebra, K-theory, Groups and Education, on the occasion of Hyman Bass's 65th birthday. Contemp. Math. Amer. Math. Soc., Providence, R.I. 243: 83–124.
[207] Lam, T.Y. 1999. Lectures on Modules and Rings. Springer.
[208] Lam, T.Y. 2001. A first course in noncommutative rings. 2nd edn. Graduate Texts in Math. 131. Springer-Verlag, Berlin-Heidelberg-New York.
[209] Lam, T.Y. 2004. A crash course on stable range, cancellation, substitution and exchange. J. Algebra and its Appl. 3(3): 301–342.
[210] Lam, T.Y. 2005. Serre's Problem on projective modules. Springer (Formely, Serre's Conjecture. Lecture Notes in Math. 635).
[211] Larsen, M.D. and A. Mirbagheri. 1971. A note of the intersection of the powers of the Jacobson radical. Rocky Mountain Journal of Mathematics 1(4): 617–622.
[212] Leary, F.C. 1992. Dedekind finite objects in module categories. J. Pure and Appl. Algebra 82: 71–80.
[213] Lemonnier, B. 1972. Sur une classe d'anneaux définie à partir de la déviation. C.R. Acad. Sci. aris Sér. A-B 274: A1688–A1690.
[214] Lemonnier, B. 1978. Dimension de Krull et codeviation. Application au theoreme d'Eakin. Comm. Algebra 6(16): 1647–1665.
[215] Lenagan, T.H. 1973. The nil radical of a ring with Krull dimension. Bull. London Math. Soc. 5: 307–311.
[216] Lenagan, T.H. 1977. Noetherian rings with Krull dimension one. Journal of London Math. Soc. 15(1): 41–47.
[217] Lenagan, T.H. 1978. Reduced rank in rings with Krull dimension. pp. 123–131. *In*: Oystaeyen, F. Van (ed.). Ring Theory, Proceedings of the Conference held in Antwerpen. 1979. Lecture Notes in Pure and Appl. Math. Marcel Dekker.
[218] Lenagan, T.H. 1980. Modules with Krull dimension. Bull. London Math. Soc. 12: 39–40.
[219] Lenzing, H. 1969. Endlich präsentierbar Moduln. Arch. der Math. 20: 262–266.
[220] Levin, J. 1974. Fields of fractions for group algebras of free groups. Trans. Amer. Math. Soc. 192: 339–346.
[221] Levy, L.S. 1983. Krull-Schmidt uniqueness falls dramatically over subrings of $\mathbf{Z} \oplus \cdots \oplus \mathbf{Z}$. Rocky mountain Journal of Math. 13(4): 659–678.
[222] Levy, L.S. 2000. Modules over Hereditary Noetherian prime rings (Survey). Contemporary Mathematics 259: 353–370.
[223] Levy, L.S. and C.J. Odenthal. 1996. Krull-Schmidt theorems in dimension 1. Trans. Amer. Math. Soc. 348(9): 3391–3455.
[224] Levy, L.S. and J.C. Robson. 1999. Hereditary Noetherian prime rings 1: Integrality and simple modules. J. Algebra 218: 307–337.
[225] Levy, L.S. and J.C. Robson. 1999. Hereditary Noetherian prime rings 2: Finitely generated projective modules. J. Algebra 218: 338–372.
[226] Levy, L.S. and J.C. Robson. 2000. Hereditary Noetherian prime rings 3: Infinitely generated projective modules. J. Algebra 225: 275–298.
[227] Lichtman, A.I. 1995. Valuation methods in division rings. Journal of Algebra 177: 870–898.
[228] Lichtman, A.I. 2000. On Universal Fields of Fractions for free algebras. Journal of Algebra 231: 652–676.
[229] Lucas, T.G. 1986. Some results on Prufer rings. Pacific J. Math. 124(2): 333–343.
[230] MacLane, S. 1963. Homology. Springer-Verlag, Berlin-Göttingen-Heidelberg.
[231] Manis, M.E. 1969. Valuations on a commutative ring. Proc. Amer. Math. Soc. 20: 193–198.
[232] Marot, J. 1969. Une gnralisation de la notion d'anneau de valuation. Comptes Rendus Hebdomadaires des Sances de l'Acadmie des Sciences. Sries A et B 268: A1451–A1454.
[233] Marubayashi, H., H. Mijamoto and A. Ulda. 1997. Non-commutative Valuation Rings and Semi-hereditary Orders. Springer-Verlag.
[234] Marubayashi, H. and F. Oystaeyen. 2012. Prime divisors and noncommutative valuation theory. Springer.
[235] Mathiak, K. 1986. Valuations of Skew Fields and projective Hjelmslec spaces. Springer-Verlag.

[236] McCasland Riy, L. and P.F. Smith. 2004. Uniform dimension of modules. Quaterly Journal of Mathematics 55(4): 491–498.
[237] McConnel, J.C. and J.C. Robson. 1987. Noncommutative Noetherian rings. AMS Graduated studies in mathematics 30.
[238] Menini, C. 1986. Jacobson's conjecture. Morita dualisty and related questions. J. Algebra 103: 638–655.
[239] Mewborn, A.C. and C.N. Winton. 1969. Orders in self-injective semiperfect rings. J. Algebra 13: 5–9.
[240] Michler, G. 1969. Structure of semi-perfect hereditary Noetherian rings. J. Algebra 13(3): 327–344.
[241] Michler, G. 1970. Primringe mit Krull-dimension eins. J. Reine Angew. Math. 239/240: 366–381.
[242] Mitchell, B. 1965. Theory of Categories. Boston, MA: Academic Press.
[243] Monk, G.S. 1972. A characterization of exchange rings. Proc. Amer. Math. Soc. 35: 344–353.
[244] Montgomery, S. 1983. Von Neumann finiteness of tensor products of algebras. Comm. Algebra 11: 595–610.
[245] Morandi, P.J. 1991. An approximation theorem for Dubrovin valuation rings. Math. Z. 207: 71–82.
[246] Müller, B. 1991. Uniform modules over serial rings. J. Algebra 144: 94–109.
[247] Năstăsescu, C. and F. van Oystaeyen. Graded Ring Theory. Math. Library 28, North Holland, Amsterdam.
[248] von Neumann, J. 1936. On regular rings. Proc. Nat. Acad. Sci. (USA) 22: 707–713.
[249] Nicholson, W.K. 1977. Lifting idempotents and exchange rings. Trans. Amer. Math. Soc. 229: 269–278.
[250] Nicholson, W.K. 1997. On exchange rings. Comm. Alg. 25(6): 1917–1918.
[251] Nielsen, Pace, P. 2005. Abelian exchange modules. Comm. Algebra 33(4): 1107–1118.
[252] Nielsen, Pace, P. 2006. Countable exchange and full exchange rings. Comm. Algebra 35(1): 3–23.
[253] Northcott, D.G. 1965. Ideal theory. Cambridge University Press, Cambridge.
[254] Northcott, D.G. 1968. Lessons on Rings, Modules and Multiplicities. Cambridge University Press, Cambridge.
[255] Nouazé, Y. and P. Gabriel. 1967. Idéaux premier de l'algèbre enveloppante d'une algèbre de Lie nilpotente. J. Algebra 6: 77–99.
[256] Ore, O. 1933. Theory of non-commutative polynomials. Annals of Math. 34: 480–508.
[257] Palmér, I. 1975. The global dimension of semi-trivial extension of rings. Math. Scand. 37: 223–256.
[258] Palmér, I. and J.-E. Roos. 1971. Formules explicites pour la dimension homologique des anneaux de matrices généralisées. C. R. Acad. Sci. Paris. Sér. A-B 273: A1026–A1029.
[259] Palmér, I. and J.-E. Roos. 1973. Explicit Formulae for the global homological dimensions of Trivial Extensions of Rings. J. Algebra 27: 380–413.
[260] Passman, D.S. 1977. The algebraic structure of group rings. John Wiley & Sons, New York.
[261] Passman, D.S. 1989. Infinite Crossed Products. Academic Press, New York.
[262] Passman, D.S. 1991. A Course in Ring Theory. Wadsworth and Brooks, Cole Math. Series, California.
[263] Pirtle, E.M. 1986. Noncommutative valuation rings. Publications de L'Institut Mathématique 39(53): 83–87.
[264] Popescu, N. 1973. Abelian categories with applications to rings and modules. Boston, MA: Academic Press.
[265] Puninski, G. 2001. Serial rings. Kluwer Acad. Publ.
[266] Puninski, G. 2002. Artinian and Noetherian serial rings. J. of Math. Sciences 110(1): 2330–2347.
[267] Prüfer, H. 1932. Untersuchungen uber teilbarkeitseigenschaften in korpern. J. Reine Angew. Math. 168: 1–36.
[268] Rees, D. 1956. Two classical theorems of ideal theory. Proc. Cambridge Phil. Soc. 52: 155–157.
[269] Reiner, I. 1975. Maximal orders. Academic Press, New York.
[270] Rentschler, R. and P. Gabriel. 1967. Sur la dimension des anneaux et ensembles ordonnés. C. R. Acad. Sci. Paris, Sér. A 265: 712–715.
[271] Ribenboim, P. 1968. Thóerie des Valuations. Presses Univ. Montréal.
[272] Ringel, C.M. 2000. Infinite length modules. Some examples as introduction. *In*: Infinite Length Modules, Bielefeld, 1998, 1–73, Trends in Math., Birkhäuser, Basel, 2000.
[273] Robson, J.C. 1967. Artinian quotient rings. Proc. London Math. Soc. 17(3): 600–616.
[274] Rowen, L.H. 1980. Polynomial Identities in Ring theory. Academic Press, New York.
[275] Rowen, L.H. 1986. Finitely presented modules over semiperfect rings. Proc. Amer. Math. Soc. 97: 1–7.

[276] Rowen, L.H. 1988. Ring Theory, Vol. I. Academic Press.
[277] Ruiz, J.M. 1993. The basic theory of power series. Friedr. Vieweg & Sohn, Braunschweig.
[278] Rump, W. 1996. Discrete posets, cell complexes and the global dimension of tiled orders. Comm. Algebra 24: 55–107.
[279] Sandomerski, P.L. 1968. Nonsingular rings. Proc. Amer. Math. Soc. 19: 225–230.
[280] Sandomerski, P.L. 1969. A note on the global dimension of subrings. Proc. Amer. Math. Soc. 23: 478–480.
[281] Schelter, W. 1975. Essential extensions and intersection theorem. Proc. Amer. Math. Soc. 53: 328–330.
[282] Schiffels, G. 1960. Graduierte Ringe und Modulen. Bonn Math. Schr. 11: 1–122.
[283] Schilling, O.F.G. 1945. Noncommutative valuations. Bull. Amer. Math. Soc. 51: 297–304.
[284] Schilling, O.F.G. 1991. The theory of valuations. Math. Surveys and monographs 4. Amer. Math. Soc.
[285] Schreier, O. 1928. Über den Jordan-Hölderschen Satz. Abhandlungen Math. Semin. Hamburg. Univ. 6(3-4): 300–302.
[286] Shepherdson, J.C. 1951. Inverses and zero divisors in matrix rings. Proc. London Math. Soc. 1: 71–85.
[287] Shul'geifer, E.G. 1972. The Schreier theorem in normal categories. Sibirski Matemat. Zhurnal 13(3): 688–697. English translation: 1972. Sibirian Math. Journal. 13(3): 474–479.
[288] Singh, S. 1982. On a Warfield's theorem on hereditary rings. Arch. Math. 39: 306–311.
[289] Singh, S. 1984. Serial right Noetherian rings. Canad. J. Math. 36(1): 22–37.
[290] Skornyakov, L.A. 1965. On Cohn rings. Algebra i logika (in Russian) 4(3): 5–30.
[291] Small, L.W. 1966. Orders in Artinian rings. J. Algebra 4: 13–41.
[292] Small, L.W. 1966. Hereditary rings. Proc. Nat. Acad. Soc. 55: 25–27.
[293] Small, L.W. 1966. Correction and addendum: Orders in Artinian rings. J. Algebra 4: 505–507.
[294] Small, L.W. 1967. Semihereditary rings. Bull. Amer. Math. Soc. 73: 656–659.
[295] Small, L.W. 1968. Orders in Artinian rings II. J. Algebra 9: 268–273.
[296] Smith, P.F. and A.R. Woodward. 2007. Krull dimension of bimodules. J. Algebra 310: 405–412.
[297] Steinitz, O.F.G. 1912. Rechtickige Systeme und Moduln in algebraischen Zahlkörpern. Math. Ann. 71: 328–354.
[298] Stephenson, W. 1974. Modules whose lattice of submodules is distributive. Proc. London Math. Soc. 28(2): 291–310.
[299] Stock, J. 1986. On rings whose projective modules have exchange property. J. Algebra 103: 437–453.
[300] Talintyre, T.D. 1963. Quotient rings with maximum condition on right ideals. J. London Math. Soc. 38: 439–450.
[301] Talintyre, T.D. 1966. Quotient rings with minimum condition on right ideals. J. London Math. Soc. 41: 141–144.
[302] Tarsy, R.B. 1970. Global dimensions of orders. Trans. Amer. Math. Soc. 151: 335–340.
[303] Teply, M.L. 1997. Right hereditary, right perfect rings are semiprimary. *In*: Advances in Ring theory 313–316.
[304] Tuganbaev, A.A. 1988. Distributive rings and modules. Trudy Moskov. Mat. Obshch. 51: 95–113.
[305] Tuganbaev, A.A. 1998. Semidistributive Modules and Rings. Kluver Acad. Press.
[306] Tuganbaev, A.A. 1999. Distributive Modules and Related Topics. Gordon and Breach Science Publishers, Amsterdam.
[307] Tuganbaev, A.A. 2000. Modules with the exchange properties and exchange rings. pp. 439–459. *In*: M. Hazewinkel (ed.), Handbook of Algebra, Vol. 2. North-Holland, Amsterdam.
[308] Tuganbaev, A.A. 2000. Modules with the exchange property and exchange rings. pp. 439–459. *In*: M. Hazewinkel (ed.). Handbook of Algebra, Vol. 2. North-Holland, Amsterdam.
[309] Tuganbaev, A.A. 2002. Rings and modules with exchange properties. Journal of Math. Sci. 110(1): 2348–2421.
[310] Upham, M.H. 1987. Serial rings with right Krull dimension one. J. Algebra 109: 319–333.
[311] Utumi, Y. 1965. On continuous rings and self-injective rings. Trans. Amer. Math. Soc. 118: 158–173.
[312] Vaserstein, L.N. 1971. Stable rank of rings and dimensionality of topological spaces. Functional Anal. Appl. 5: 102–110.
[313] Vaserstein, L.N. 1984. Bass's first stable range condition. J. Pure Appl. Algebra 34: 319–330.
[314] Vermani, L.R. 2003. Elementary approach to Homological Algebra. Chapman & Hall/CRC.

[315] Wadsworth, A.R. 1999. Valuation theory on finite dimensional division algebras. pp. 385–449. *In*: F.-V. Kuhlmann, S. Kuhlmann and M. Marshall (eds.). Valuation Theory and its Applications Vol. I. Fields Inst. Communications 32, Amer. Math. Soc. Providence, RI, 2002.

[316] Walker, E.A. 1956. Cancellation in direct sums of groups. Proc. Amer. Math. Soc. 7: 898–902.

[317] Warfield, R.B., Jr. 1969. A Krull-Schmidt theorem for infinite sums of modules. Proc. Amer. Math. Soc. 22: 460–465.

[318] Warfield, R.B., Jr. 1969. Decompositions of injective modules. Pacific Journal of Math. 31(1): 263–276.

[319] Warfield, R.B., Jr. 1972. Exchange rings and decompositions of modules. Math. Ann. 199: 31–36.

[320] Warfield, R.B., Jr. 1975. Serial rings and finitely presented modules. J. Algebra 37(2): 187–222.

[321] Warfield, R.B., Jr. 1979. Bezout rings and serial rings. Commun. Algebra 7: 533–545.

[322] Warfield, R.B., Jr. 1980. Cancelation of modules and stable range of endomorphism rings. Pacific J. Math. 91(2): 457–485.

[323] Webber, D.B. 1970. Ideals and modules of simple Noetherian hereditary rings. J. Algebra 16: 239–242.

[324] Wedderburn, J.H.M. 1932. Non-commutative domains of integrity. Journal für die reine und angewandte Mathematik 167.

[325] White, N. (ed.). 1986. Theory of matroids. Cambridge Univ. Press.

[326] Wiegand, R. 1984. Direct sum cancellation over Noetherian rings. pp. 441–466. *In*: Abelian Groups and Modules, Udine. CISM Courses and Lectures 287, Springer, Vienna.

[327] Woodward, A. 2007. Rings and modules with Krull dimension. Thesis Ph.D., University of Glasgow.

[328] Wright, M.H. 1988. Serial rings with right Krull dimension one, II. J. Algebra 117: 99–116.

[329] Wright, M.H. 1989. Krull dimension in serial rings. J. Algebra 124: 317–328.

[330] Wright, M.H. 1989. Certain uniform modules over serial rings are uniserial. Comm. Algebra 17: 441–469.

[331] Wu, T. and Y. Xu. 1997. On the stable range condition of exange rings. Comm. Algebra 25: 2355–2363.

[332] Xue, W. 1990. Exact modules and serial rings. J. Algebra 134: 209–221.

[333] Yamagata, K. 1974. The exchange property and direct sum of indecomposable injective modules. Pacific J. Math. 55: 301–317.

[334] Yamagata, K. 1974. On projective modules with the exchange property. Sci. Rep. Tokyo Kyoiku Daigaku Sect. A 12: 39–48.

[335] Yamagata, K. 1975. On rings of finite representation type and modules with the finite exchange property. Sci. Rep. Tokyo Kyoiku Daigaku Sect. A 13: 1–6.

[336] Yu, H.-P. 1995. Stable range one for exchange rings. J. Pure Appl. Algebra 98: 105–109.

[337] Zassenhaus, H.J. 1934. Zum Satz von Jordan-Hölder-Schreier. Abh. Math. Semin. Hamb. Univ. 10: 106–108.

[338] Zhang, H.B. and W.T. Tong. 2006. The Cancellation Property of Projective Modules. Algebra Colloquium 13(4): 617–622.

[339] Zimmermann-Huisgen, B. and W. Zimmermann. 1984. Classes of modules with the exchange property. J. Algebra 88: 416–434.

INDEX